Proceedings in Life Sciences

Biology of Inorganic Nitrogen and Sulfur

Edited by
H. Bothe and A. Trebst

With 144 Figures

Springer-Verlag
Berlin Heidelberg New York 1981

Professor Dr. Hermann Bothe
Botanisches Institut der Universität Köln
Gyrhofstr. 15,
5000 Köln 41/FRG

Professor Dr. Achim Trebst
Lehrstuhl für Biochemie der Planzen, Ruhr-Universität,
4630 Bochum/FRG

ISBN-13: 978-3-642-67921-6 e-ISBN13-: 978-3-642-67919-3
DOI:10.1007/978-3-642-67919-3

Library of Congress Cataloging in Publication Data. Main entry under title: Biology of Inorganic Nitrogen and Sulfur. (Proceedings in life sciences) Bibliography: p. Includes index. 1. Nitrogen – Metabolism. 2. Sulphur – Metabolism. 3. Nitrogen – Fixation. 4. Microbial metabolism. I. Bothe, Hermann, 1940–. II. Trebst, Achim. [DNLM: 1. Plants – Metabolism. 2. Nitrogen – Metabolism. 3. Sulfur – Metabolism. 4. Nitrogen fixation. QK 881 B615] QR92.N6B56 576'.119214 80–29435

Softcover reprint of the hardcover 1st edition 1981

Typesetting, offsetprinting and bookbinding: Brühlsche Universitätsdruckerei, Giessen.
2131/3130-543210

Preface

Nitrogen and sulfur compounds are continuously synthetized, degraded and converted into other forms in nature. There are many similarities in the principle problems and basic mechanisms of the biology of inorganic nitrogen and sulfur. Many details are not yet understood and hence are the subject of active investigation the world over. In May, 1980, a conference was held in Bochum, Federal Republic of Germany, at which attempts were made to discuss and compare all aspects of both the nitrogen and the sulfur cycle. Lectures were given by internationally recognized experts on the physiology, biochemistry, genetics, and ecology of dinitrogen fixation, of assimilatory and dissimilatory nitrate and sulfate reduction, and of ammonia and sulfide oxidation. In addition, important data were communicated by German scientists of the national program on the Metabolism of Inorganic Nitrogen and Sulfur Compounds, supported by the Deutsche Forschungsgemeinschaft. This book contains all the contributions to the meeting and consequently should be of interest to researchers, teachers, and students in the field.

The members of the German national program on the Metabolism of Inorganic Nitrogen and Sulfur Compounds would like to thank the Deutsche Forschungsgemeinschaft for their generous financial support of the scientific projects during the past four years and for the conference itself. Without this help, the present book would not have been written. The members express their appreciation particularly to Dr. A. Hoffmann of the Deutsche Forschungsgemeinschaft for her invaluable skill and patience in taking care of the projects and scientists. The editors are indebted to all at Springer-Verlag for bringing the book into a form suitable for publication within less than three quarters of a year.

December, 1980 H. BOTHE and A. TREBST

Contents

General Introduction

Nitrogen Metabolism in General

Dinitrogen Fixation

Sulfur Metabolism in General

Contributed Papers: Nitrogen Metabolism

Contributed Papers: Sulfur

Contributors

You will find the addresses at the beginning of the respective contribution

General Introduction

Microorganisms Involved in the Nitrogen and Sulfur Cycles

H.G. SCHLEGEL[1]

The changes of the outer surface of the earth, which have occurred within the last three or four billion years, are by no means only due to plutonic activities, tectonic shifts, weathering and dissolution of minerals and other geochemical processes; rather prokaryotic microorganisms were actively involved. Due to the latter chemical activities the highly reduced atmosphere and lithosphere became oxidized as exemplified by the transformation of ammonia, methane, water, and hydrogen sulfide to dinitrogen, carbon dioxide, oxygen, and sulfate. The precipitation of limestone, deposition of iron oxides, the separation of various minerals and later the deposition of carbon and reduced carbon compounds are at least partially due to microbial activities. The basic metabolism of living organisms developed on the level of prokaryotic life. Prokaryotes occupied ecological niches long before higher forms of life developed and were always present during the evolution of plants and animals. It is, therefore, not surprising that various transformations in the cycle of matter are the domain and the monopoly of prokaryotes still to-day. In fact, the cycles of nitrogen and sulfur in the biosphere mainly deal with the activities of bacteria. Transformations under anaerobic conditions are exclusively brought about by bacteria, and the utilization of inorganic nitrogen and sulfur for biosynthesis is confined to microorganisms and plants.

The metabolism of nitrogen and sulfur compounds has much in common. Both elements can be handled by microorganisms in various states of valency, from their most reduced states (ammonia, hydrogen sulfide) to their most oxidized states (nitrate, sulfate).

The reduced or intermediate forms (NH_3, NO_2^-; H_2S, S) can be used as electron donors for aerobic energy generation and carbon dioxide fixation in the dark (chemolithoautotrophs). The oxidized forms (NO_3^-; NO_2^-; SO_4^{2-}, SO_3^{2-}) can serve as electron acceptors making energy conversion by electron transport phosphorylation under anaerobic conditions ("anaerobic respiration") possible. On the biochemical level analogies exist with respect to the catalysis of 6 electron transfers by a single enzyme such as the reduction of nitrite, dinitrogen, or sulfite. Treating nitrogen and sulfur metabolism together in a symposium is, therefore, well justified.

The organisms involved in the nitrogen and sulfur cycles, their habitats, ecological significance and overall metabolism are adequately presented in recent textbooks (Stanier et al., 1978; Brock, 1979; Gottschalk, 1979; Schlegel, 1980). An up-to-date treatment of the habitats, the isolation and identification of various physiological groups of prokaryotes will appear shortly (Starr et al., 1981).

---

1 Institut für Mikrobiologie, Georg-August-Universität Göttingen, Grisebachstr. 8, 3400 Göttingen, FRG

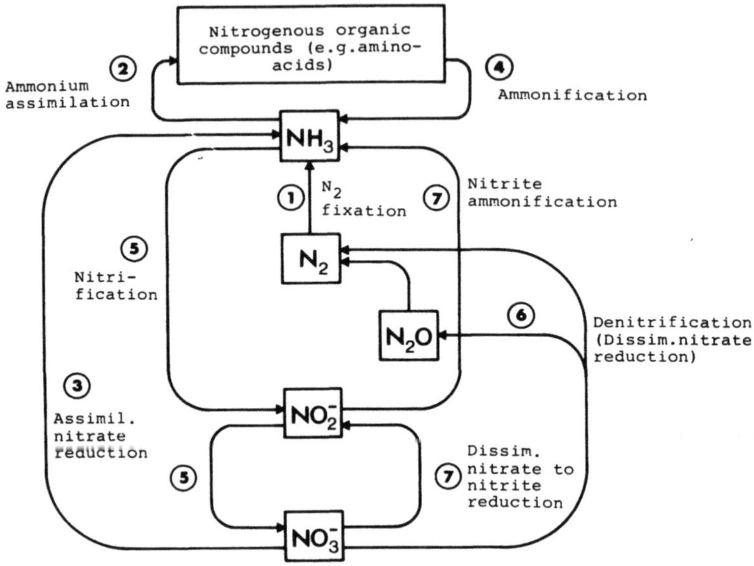

Fig. 1. The nitrogen cycle. The *numbers* indicate the sequence in which the various segments of the cycle are treated in the text

1 Nitrogen Cycle (Fig. 1)

1.1 Nitrogen Fixation

Only prokaryotes are able to fix molecular nitrogen (dinitrogen) and to grow in the absence of a source of combined nitrogen. Nitrogen fixation requires the presence of a special enzyme system, nitrogenase, and particular conditions. The ability to fix nitrogen is distributed among many species of oxygenic and anoxygenic phototrophic bacteria, chemoautotrophic and chemoheterotrophic bacteria, among aerobic and anaerobic bacteria – in short: among representatives of all major physiological groups. There is almost no correlation to the taxonomical units. In culture collections many bacteria are kept which may contain the genetic information for nitrogenase synthesis but have not been tested for the conditions allowing gene expression. The majority of aerobic nitrogen-fixing bacteria are microaerophilic in the sense that when forced to fix nitrogen for growth they tolerate only 0.02 bar oxygen although in the presence of combined nitrogen they grow perfectly well under air. As became known during the last decade, many nonheterocystous cyanobacteria are able to form nitrogenase and fix nitrogen in the light when combined nitrogen is absent and oxygen production is inhibited by DCMU. There may well be many strictly aerobic bacteria which contain the *nif* genes in a kryptic or dormant state, i. e., which are only expressed in the absence of both combined nitrogen and molecular oxygen. These cannot be detected under ordinary cultural conditions; instead incubation at very low relative respiration rates (Schlegel and Vollbrecht, 1980) or without any oxygen in the presence of a fermentable substrate will be required.

The great problem in the field of nitrogen fixation concerns the oxygen sensitivity of nitrogenase (Dixon, 1972, 1978). How do several aerobic bacteria such as *Azotobacter* and the heterocysts in cyanobacteria manage to tolerate the oxygen partial pressure of the air? The discovery of a protecting protein in *Azotobacter* provides a partial answer and encouragement for further work aiming at the transfer of the ability to fix nitrogen to higher plants.

Another fundamental problem concerns the inherent property of nitrogenase to reduce protons in addition to reducing nitrogen. The production of molecular hydrogen, which is apparently obligatorily linked to nitrogen fixation, results in a significant loss of potential energy. The presence of hydrogenases (Schlegel and Schneider, 1978), evidently serving to recycle the hydrogen, in most nitrogen-fixing bacteria becomes understandable (Evans et al., 1978).

Recently close relationships have been detected between the hydrogen-oxidizing bacteria (Aragno and Schlegel, 1981) and the nitrogen-fixing bacteria. Not only several hydrogen bacteria such as *Xanthobacter autotrophicus* (Gogotov and Schlegel, 1974; Wiegel and Schlegel, 1976; Wiegel et al., 1978), *Alcaligenes latus* (Malik et al., in preparation) and other freshly isolated hydrogen bacteria (Malik and Schlegel, 1980) turned out to be able to fix nitrogen, but also several species of well-estblished nitrogen-fixing bacteria such as *Rhizobium japonicum* (Simpson et al., 1979; Lepo et al., 1980) and *Derxia gummosa* (Pedrosa et al., 1980) proved to be able to grow chemolithoautotrophically with hydrogen as electron donor. At least in some strains the autotrophic character becomes only expressed under conditions of nitrogen fixation and not in the presence of combined nitrogen (Malik, personal communication).

How the *nif* genes are regulated, under which conditions they are expressed, how they can be transferred and established in nonnitrogen-fixing organisms and how nitrogenase can be kept working in a foreign environment – these are the questions which should attract skilled enthusiastic scientists.

1.2 Ammonium Assimilation

Ammonium is the sole form of nitrogen which is incorporated into organic compounds. All except a few bacteria and plants are able to utilize ammonia. Many are able to use nitrate, urea or other nitrogenous compounds. Glutamate, aspartate, alanine, serine, and carbamoylphosphate are the products of primary amination. In the synthesis of glutamate either glutamate dehydrogenase or, at limiting ammonium concentrations, the glutamine synthetase-glutamate synthase system are involved. The majority of amino acids are synthesized by transamination.

1.3 Assimilatory Nitrate Reduction

In an aerobic environment either aquatic or terrestric, nitrate is the major form of inorganic nitrogen. Nitrate assimilation requires the reduction of nitrate to ammonia. This is accomplished by a soluble, cytoplasmic enzyme system comprising nitrate reductase B and nitrite reductase. Its formation is induced in the presence of nitrate as the sole nitrogen source. Nitrite reductase catalyzes the reduction of nitrite to ammonium; this requires the transfer of six electrons. The latter are provided via pyridine nucleotides (in fungi and bacteria) or ferredoxin (in plants and bacteria).

1.4 Ammonification

In the course of the mineralization of organic matter the nitrogen present in proteins, purines, pyrimidines and other compounds is released in the form of ammonium. This process is called ammonification. The ability to ammonify is widespread among the bacteria decomposing organic materials either under aerobic or anaerobic conditions. The process does not deserve further comments here.

1.5 Nitrification

Although during mineralization organic nitrogen is released in the form of ammonium the major part of soil and water nitrogen is present as nitrate. The transformation of ammonium to nitrate, nitrification, is the monopoly of bacteria – with almost negligible exceptions. There is no bacterium which is able to convert ammonium directly to nitrate; the oxidation is dependent on the cooperation of two different groups of bacteria, the ammonium oxidizers, recognizable by their genus name Nitroso-(*Nitrosomonas, Nitrosococcus* etc.) and the nitrite oxidizers, recognizable by their genus name Nitro-(*Nitrobacter, Nitrococcus*), both called nitrifiers. The sequential activities of both groups can be described as a short food chain.

These bacteria have their optimum conditions at neutral pH. Only low cell numbers are found in acid soils (Walker and Wickramasinghe, 1979). The question, which organisms achieve ammonium oxidation in acid soils, is still open. A modest heterotrophic nitrification, mainly carried out by *Arthrobacter* and possibly due to a process called co-oxidation, may be responsible for the low rate of ammonium oxidation in acid environments (Focht and Verstraete, 1977).

1.6 Denitrification

Denitrification is a respiratory process with nitrate as electron acceptor; nitrate is reduced to nitrogen via dinitrogen oxide. Many aerobic bacteria are able to denitrify and generate biochemical energy by electrontransport phosphorylation under anaerobic conditions when oxygen is absent and either nitrate, nitrite, or dinitrogen oxide are provided. This process is alternatively called "nitrate respiration." All denitrifiers can as well live with oxygen; there is no obligate denitrifier so far. They contain a complete respiratory system; the enzyme system required for denitrification, nitrate reductase A and nitrite reductase, which is membrane-bound, is, however, only formed in the absence of oxygen. The ability of many denitrifying bacteria to grow with nitrite or dinitrogen oxide indicates that the function of the dissimilatory nitrite reductase is coupled to oxidative phosphorylation.

Denitrification is the only biological process by which combined nitrogen is released as molecular nitrogen to the atmosphere. In the absence of denitrification nitrate would accumulate in the oceans.

1.7 Dissimilatory Nitrate-to-Nitrite Reduction and Nitrite Ammonification

The former term is used here for the process of dissimilatory nitrate reduction (nitrate respiration) which ends with the production of nitrite. Many facultative anaerobic, fermenting bacteria are able to produce nitrate reductase A and to couple nitrite formation with electrontransport phosphorylation. *Escherichia coli*, *Enterobacter*, and several other bacteria are able to benefit from this type of nitrate respiration. But unlike the denitrifiers they are not able to reduce nitrite to dinitrogen. They either excrete nitrite or reduce it to ammonium via the assimilatory nitrite reductase system. This process is called nitrite ammonification. It is not coupled to electrontransport phosphorylation.

The term "nitrate ammonification" was used before the biochemistry of nitrogen metabolism was well understood but then fell out of favor. Only recently the production of significant amounts of ammonium from nitrate has been observed in natural ecosystems and in waste water treatment (Cole and Brown, 1980). From the biochemical point of view it now becomes clear that the enterobacteria and other facultative anaerobes certainly draw benefit from both steps of nitrate ammonification. The first step provides energy via electrontransport phosphorylation under anaerobic conditions (nitrate respiration). In the second step nitrite is involved as an external electron acceptor in the fermentation process relieving the cell from the burden to internally produce electron acceptors from pyruvate to receive the electrons released by the oxidation of glyceraldehyde-3-phosphate. If nitrite is present, it can serve as an electron sink and pyruvate can be used to regenerate extra-ATP via the phosphotransacetylase and acetokinase reactions. After all, the cell draws benefit from both reactions and the old term "nitrite ammonification" comes back into favor.

2 Sulfur Cycle (Fig. 2)

2.1 Sulfide Assimilation and Assimilatory Sulfate Reduction

In living cells sulfur is present mainly in the form of mercaptogroups in sulfur-containing amino acids (1% w/dry wt.). The sulfur source of most prokaryotes and plants is sulfate. Sulfate is reduced by assimilatory sulfate reduction to sulfide, and the latter is incorporated into organic compounds. Sulfate reduction starts with an energy-dependent activation of sulfate to yield phosphoadenosine phosphosulfate; the latter is reduced yielding free sulfite. The reduction of sulfite to sulfide apparently occurs in all organisms as a single step in which six electrons are transferred. In general, iron porphyrin compounds are involved as catalysts. A few organisms not able to reduce sulfate require sulfide or cysteine as a supplement.

2.2 Desulfuration

In the course of the anaerobic decomposition of organic materials the mercaptogroups are released as sulfide; this mineralization process, in which desulfurases are involved, is called desulfuration.

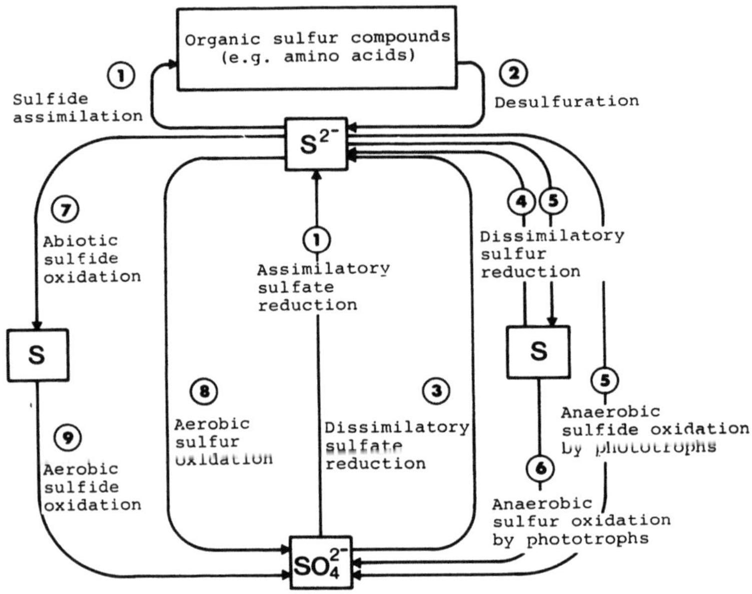

Fig. 2. The sulfur cycle. The *numbers* indicate the sequence in which the various segments are treated in the text

2.3 Dissimilatory Sulfate Reduction

The major amount of hydrogen sulfide produced in nature is due to sulfate reduction occurring under anaerobic conditions. This process, called desulfurication, is brought about by a physiological group of bacteria which use sulfate as an electron acceptor for electrontransport phosphorylation under anaerobic conditions ("sulfate respiration"). Unlike the denitrifying bacteria, the desulfuricants are strictly anaerobic bacteria with a fermentative type of metabolism; in the absence of sulfate several species can generate metabolic energy by fermentation. Those desulfuricants known until 1975 are unable to oxidize their organic substrates (lactate) to CO_2 and H_2O; due to the lack of a complete tricarboxylic acid cycle they excrete acetate. Recently isolated new species can grow on acetate or long chain fatty acids (Widdel and Pfennig, 1977; Pfennig et al., 1981). Most representatives can use molecular hydrogen as electron donor, and at least one species can synthesize the major part of the cell carbon from carbon dioxide (Badziong and Thauer, 1978).

The mechanism of sulfate reduction resembles that of assimilatory sulfate reduction. So far investigated, however, sulfate reduction requires only one phosphorylation step, and adenosinephosphosulfate is directly reduced to sulfite. Sulfite may be reduced to sulfide in a single step by sulfite reductase or via three consecutive steps involving trithionate and thiosulfate. Comparative studies are urgently needed.

2.4 Dissimilatory Sulfur Reduction

While some sulfate-reducing bacteria are not able to use sulfur as an alternate electron acceptor (Biebl and Pfennig, 1977; Pfennig and Biebl, 1981), there is one species, *Desulfuromonas acetoxidans*, which does not use sulfate, but can grow with ethanol or acetate as substrate and flowers of sulfur as electron acceptor (Pfennig and Biebl, 1976; Wolfe and Pfennig, 1977).

2.5 Anaerobic Sulfide Oxidation

Under anaerobic conditions hydrogen sulfide can only be oxidized by phototrophic microorganisms. In anaerobic sulfide oxidation mainly the classical purple and green sulfur bacteria (Chromatiaceae and Chlorobiaceae) are involved. Some cyanobacteria found in the meta- and upper hypolimnion of stratified lakes are able to adapt to anoxygenic photosynthesis, too (Padan, 1979 a,b). Hydrogen sulfide is oxidized in the light either via intermediary accumulation of sulfur droplets inside or outside the cell or directly to sulfate. In eutrophic lakes the sulfate produced may be reduced in situ by the desulfuricants, giving rise to a cycle of secondary biomass production supplied by light energy (Pfennig, 1978).

There are several green sulfur bacteria which oxidize hydrogen sulfide only to sulfur which is deposited outside the cell. The sulfur, which in the form of polysulfide is soluble, can serve as an electron acceptor for the above-mentioned (dissimilatory sulfur reduction) sulfur-reducing bacteria and in this way give rise to another cycle of secondary biomass production supplied by light energy (Wolfe and Pfennig, 1977). Both the hydrogen sulfide-sulfur and the hydrogen sulfide-sulfate cycles make syntrophic associations of bacteria possible.

2.6 Anaerobic Sulfur Oxidation

Under anaerobic conditions sulfur can be oxidized only in the light, and only a small number of purple and green sulfur bacteria are able to use extracellular elemental sulfur either in the form of colloidal sulfur or flowers of sulfur. The bacteria involved are *Ectothiorhodospira, Rhodopseudomonas acidophila, R. palustris, Chlorobium,* and others (Pfennig, 1978).

2.7 Aerobic Abiotic Sulfide Oxidation

Hydrogen sulfide is stable only under anoxic conditions. In the presence of oxygen it does not persist and is oxidized to sulfur. However, the autoxidation of hydrogen sulfide is a slow process, and H_2S and O_2 may coexist for at least a few hours (Beijerinck, 1895; Schlegel and Jannasch, 1981).

2.8 Aerobic Sulfide Oxidation

For long periods hydrogen sulfide and oxygen will be simultaneously available only in areas to which both compounds are continuously supplied, e. g., the mud-water interface of aquatic sediments where hydrogen sulfide providing mud is covered by running water containing oxygen. This kind of constellation provides the habitat of the sulfide-oxidizing nonphotosynthetic bacteria such as *Beggiatoa, Thiothrix, Achromatium, Macromonas, Thiovulum, Thiospira,* and a few others (La Riviére and Schmidt, 1981; Wiessner, 1981). These aerobic sulfide-oxidizing bacteria accumulate sulfur globules intracellularly which disappear when sulfide is lacking. Due to the difficulties of obtaining pure cultures and growing the bacteria in large amounts in the laboratory the biochemistry of aerobic sulfide oxidation by these classical sulfide oxidizers is virtually unknown. With the possibilities of modern technical equipment available nowadays, increasing efforts ought to be made to study and define the conditions for growing these bacteria in the laboratory. After all, it was the study on *Beggiatoa* which initiated the understanding and modern thinking in terms of hydrogen donors, hydrogen transfer, metabolic compartmentation, and comparative analysis of photosynthesis (Winogradsky, 1887).

2.9 Aerobic Sulfur Oxidation

The aerobic oxidation of elemental sulfur is to a large extent the privilege of the sulfur-oxidizing bacteria of the genus *Thiobacillus.* Their habitats are mine drainage waters in which dissolved metal sulfides and, as a product of autoxidation, sulfur are available. Due to sulfur oxidation to sulfate the mine water becomes strongly acid; an eminent acid tolerance is an outstanding character of most thiobacilli (Kuenen and Tuovinen, 1981). A few members of the genus *Thiobacillus* and the genus *Sulfolobus,* in addition to its acidophily characterized by its enormous thermophily (70 °–75 °C) (Brock, 1978, 1981), are faculative chemolithotrophs. Almost everything which is known on the aerobic oxidation of inorganic sulfur compounds to sulfate is due to studies on thiobacilli which can be fairly easily grown as mass culture in the laboratory.

The elucidation of the enzyme reactions involved on the oxidation of hydrogen sulfide and sulfur is difficult due to autoxidation reactions. However, present evidence indicates that sulfide oxidase, sulfur oxidase, sulfite oxidase, adenosinephosphosulfate reductase, and sulfate adenylyltransferase are involved and that part of the liberated electrons enter the respiratory chain on the level of cytochrome c.

The sulfur cycle has enormous ecological impacts. The products of at least two segments of the cycle have a drastic effect on the environment and change the ecosystem completely. (1) In anaerobic sediments or water-bodies, where organic matter is decomposed and the anaerobic food chain is started with fermentations, the presence or absence of sulfate decides the routes of degradation, the organisms involved and the main products (Schlegel and Jannasch, 1981). In presence of sulfate the ubiquitous desulfuricants always win and scavenge the fermentation products as substrates for sulfate reduction (Cappenberg, 1974 a,b; Patel et al., 1978: Khan and Trotter, 1978). The concentration of hydrogen sulfide may become so high that competitive processes such as methanogenesis or acetogenesis are suppressed, and the ecosystem becomes a typical sulfuretum. At low concentrations or in the absence of sulfate the primary fermentation products give rise to secondary fermentations, and finally methane is produced. The reason why freshwater ecosystems and marine ecosystems should be treated separately – in teaching as well as in fundamental and applied research – mainly consists in the unequal distribution of sulfates in both aquatic systems. (2) The second segment of the sulfur cycle, which has a great impact on the environment, concerns aerobic sulfur oxidation and its product sulfuric acid. The acidity of the environment prevents many microbial activities and has a strong selective effect on the possible inhabitants of the ecosystem.

3 Conclusions and Outlook

Both the nitrogen and the sulfur cycles are interconnected with the biological cycles of matter, with the production of biomass by plants and bacteria as well as with the routes of mineralization. For understanding the global cycles, a detailed knowledge of nitrogen and sulfur metabolism and of the organisms catalyzing the respective conversions in the biosphere is necessary. Each segment of the cycles is of fundamental importance and the perpetuation of life on this planet depends on each of them.

In this introductory report several examples have been presented which indicate the incompleteness of our knowledge in the field of the nitrogen and sulfur cycles. Several recent investigations gave rise to rather important modifications of textbook opinion. These investigations did not originate from sophisticated calculations but started with simple fundamental questions, involved careful experimental work applying classical simple techniques, and required only an open unprejudiced mind and the ability to make observations. As long as we know only a few percent of the microorganisms inhabiting the soil and water environments there is no reason to let microbiology be absorbed in molecular biology, biotechnology, and applied chemistry. There are still many microorganisms waiting to be discovered, to tell their life histories, and to reveal their significance in the cycle of matter.

References

Aragno MA, Schlegel HG (1981 in press) In: Starr MP, Stolp H, Trüper HG, Balows A, Schlegel HG (eds) The prokaryotes. A handbook on habitats, isolation and identification of bacteria. Springer, Berlin Heidelberg New York
Badziong W, Thauer RK (1978) Arch Microbiol 117:209–214
Beijerinck MW (1895) Zentralbl Bakteriol Parasitenkd Infektionskr Hyg Abt 2 1:1–9
Biebl H, Pfennig N (1977) Arch Microbiol 112:115–117
Brock TD (1978) Thermophilic microorganisms and life at high temperatures. Springer, Berlin Heidelberg New York
Brock TD (1979) Biology of microorganisms, 4 th edn. Prentice Hall, New York
Brock TD (1981 in press) In: Starr MP, Stolp H, Trüper HG, Balows A, Schlegel HG (eds) The prokaryotes. A handbook on habitats, isolation and identification of bacteria. Springer, Berlin Heidelberg New York
Cappenberg ThE (1974 a) Antonie van Leeuwenhoek J Microbiol Serol 40:285–286

Cappenberg ThE (1974b) Antonie van Leeuwenhoek J Microbiol Serol 40:297–306
Cole JA, Brown CM (1980) FEMS Microbiol Lett 7:65–72
Dixon ROD (1972) Arch Mikrobiol 85:193–201
Dixon ROD (1978) Biochimie 60:233–236
Evans HJ, Emerich DW, Ruiz-Argüeso T, Albrecht SL, Maier RJ, Simpson F, Russel SA (1978) In: Schlegel HG, Schneider K (eds) Hydrogenases, their catalytic activity, structure and function. Goltze, Göttingen
Focht DD, Verstraete W (1977) Adv Microbiol Ecol 1:135–214
Gogotov JN, Schlegel HG (1974) Arch Microbiol 97:359–362
Gottschalk G (1979) Bacterial metabolism. Springer, Berlin Heidelberg New York
Khan AW, Trotter TM (1978) Appl Environ Microbiol 35:1027–1034
Kuenen JG, Tuovinen OH (1981 in press) In: Starr MP, Stolp H, Trüper HG, Balows A, Schlegel HG (eds) The prokaryotes. A handbook on habitats, isolation and identification of bacteria. Springer, Berlin Heidelberg New York
La Rivière JWM, Schmidt K (1981 in press) In: Starr MP, Stolp H, Trüper HG, Balows A, Schlegel HG (eds) The prokaryotes. A handbook on habitats, isolation and identification of bacteria. Springer, Berlin Heidelberg New York
Lepo JE, Hanus FJ, Evans HJ (1980) J Bacteriol 141:664–670
Malik KA, Schlegel HG (1980) FEMS Microbiol Lett 8:101–104
Padan E (1979a) In: Briggs WR, Green PB, Jones RJ (eds) Annual review of plant physiology, vol 30. Annual Reviews Inc Palo Alto, pp 27–40
Padan E (1979b) In: Alexander M (ed) Advances in microbial ecology. Plenum Press, New York
Patel GB, Khan AW, Roth LA (1978) J Appl Bacteriol 45:347–356
Pedrosa FO, Döbereiner J, Yates MG (1980) J Gen Microbiol 119:547–551
Pfennig N (1978) In: Clayton WR, Sistrom WR (eds) The photosynthetic bacteria. Plenum Press, New York
Pfennig N, Biebl H (1976) Arch Microbiol 110:3–12
Pfennig N, Biebl H (1981 in press) In: Starr MP, Stolp H, Trüper HG, Balows A, Schlegel HG (eds) The prokaryotes. A handbook on habitats, isolation and identification of bacteria. Springer, Berlin Heidelberg New York
Pfennig N, Widdel F, Trüper HG (1981 in press) In: Starr MP, Stolp H, Trüper HG, Balows A, Schlegel HG (eds) The prokaryotes. A handbook on habitats, isolation and identification of bacteria. Springer, Berlin Heidelberg New York
Schlegel HG (1980 in press) Allgemeine Mikrobiologie, 5. Aufl. Thieme, Stuttgart
Schlegel HG, Jannasch HW (1981 in press) In: Starr MP, Stolp H, Trüper HG, Balows A, Schlegel HG (eds) The prokaryotes. A handbook on habitats, isolation and identification of bacteria. Springer, Berlin Heidelberg New York
Schlegel HG, Schneider K (1978) In: Schlegel HG, Schneider K (eds) Hydrogenases, their catalytic activity, structure and function. Goltze, Göttingen
Schlegel HG, Vollbrecht D (1980) J Gen Microbiol 117:475–481
Simpson FB, Maier RJ, Evans JH (1979) Arch Microbiol 123:1–8
Stanier RY, Adelberg EA, Ingraham J (1978) General microbiology, 4th edn. Mc Millan, London
Starr MP, Stolp H, Trüper HG, Balows A, Schlegel HG (1981 in press) The prokaryotes. A handbook on habitats, isolation and identification of bacteria. Springer, Berlin Heidelberg New York
Walker N, Wickramasinghe KN (1979) Soil Biol Biochem 11:231–236
Widdel N, Pfennig N (1977) Arch Microbiol 112:119–122
Wiegel J, Schlegel HG (1976) Arch Microbiol 107:139–142
Wiegel J, Wolke D, Baumgarten J, Opitz R, Schlegel HG (1978) Int J Syst Bacteriol 38:573–581
Wiessner W (1981 in press) In: Starr MP, Stolp H, Trüper HG, Balows A, Schlegel HG (eds) The prokaryotes. A handbook on habitats, isolation and identification of bacteria. Springer, Berlin Heidelberg New York
Winogradsky S (1887) Bot Z 45:489–616
Wolfe RS, Pfennig N (1977) Appl Environ Microbiol 33:427–433

Nitrogen Metabolism in General

Nitrogen Metabolism in Plants[1]

L. BEEVERS[2]

1 Introduction

Although plants or plant parts can be cultured on various organic nitrogen forms, particularly amino acids, under natural conditions the majority of plants depend upon inorganic nitrogen sources. Exceptions to this may be various parasitic and semi-parasitic plants which may be provided with organic nitrogenous nutrients of host origin, and insectivorous plants which partially satisfy their nitrogen demands by utilizing organic nitrogen constituents released by lysis of trapped insects.

In the majority of plants the inorganic nitrogen requirements is met by uptake of ammonium and/or nitrate from the soil solution. In addition to these forms of nitrogen certain plants, in symbiotic association with various microorganisms, are able to use atmospheric nitrogen to indirectly fulfill nitrogen needs.

2 Nitrogen Fixation

Symbiotic associations of plants and microorganisms functional in nitrogen fixation occur in a variety of plant types ranging from associations with fungi, pteridophytes, cycads, and flowering plants. The best-characterized association is the legume/Rhizobium system where it is observed that members of the legume subfamilies Mimosoidiae, Cesalpinioidiae, and Papilionatae are able to form functional symbiotic associations. There appear to be no reports of nodulation in the tropical subfamily Swartziodeae that consists only of woody species. Most of the investigations have been performed on the subfamily Papilionatae which contains the economically significant herbaceous forms. Within the species of the subfamily Papilionatae it is generally observed that there is a specificity of invasion of designated species by particular Rhizobium and this has formed the basis of classification of Rhizobium. However a large number of strains and other host relationships such as the Cowpea miscellany and *Lotus* Rhizobium are unclassified (Vincent, 1974). Rhizobium are free-living soil organisms which are usually more numerous in the vicinity of the plant roots and there appears to be a stimulation of their numbers by legumes. The association of a specific Rhizobium and specific host appears to involve interaction between capsular components of the Rhizobium and host-produced glycoproteins (Bal et al., 1978; Dazzo et

1 Supported in part by USDA SEA grant 5901-0410-8-0015-0
2 University of Oklahoma, Norman, OK 73019, USA

al., 1978). Following association of the Rhizobium with the host (the association usually occurs at a root hair, however in *Arachis* infection may occur at the site of emergence of the secondary roots) establishment of successful symbiosis in most instances involves invasion of the host tissue by way of an infection thread formed by invagination of the host cell wall in which the Rhizobium multiplies. Rhizobium are released from the infection thread "into" host cells in vesicles apparently composed of host cell plasma membrane. After this release from the infection thread the Rhizobium changes in shape and becomes enlarged to form the bacteroid. Accompanying this differentiation there is a deposition in the vesicle envelope of leghemoglobin synthesized by the host (Verma et al., 1978). This differentiation of the bacteria into bacteroid and leghemoglobin production is accompanied by the onset of nitrogen fixation capability. In the majority of instances Rhizobial nitrogen fixation only occurs in symbiotic associations. However nitrogen fixation by free-living Rhizobium in the absence of plant host has been reported in Cowpea strains (Pagan et al., 1975).

Root nodule associations occur between organisms other than legumes and Rhizobium. The best-characterized of these is the *Alnus* system, but root nodule symbiosis involving over 150 species from 13 genera (all woody dicotyledons) have been reported (Bond, 1976). It appears that the symbiont is an actinomycete, however the organism has not been isolated and cultured in the free-living state. The infection process has been followed by innoculating plants with suspensions prepared from nodules. Invasion of the host appears to occur by way of root hairs, the endophyte apparently passes through host cell walls by means of localized chemical degradation which facilitates penetration of hyphae. The intracellular hyphae are surrounded by host plasma membrane and a thick pectin-like capsule. Hyphal ends become swollen and form septate club-shaped vesicles within the periphery of the host cells (Newcomb et al., 1978). It appears that these vesicles are the sites of nitrogen fixation (van Straten et al., 1977). It is not established whether deposition of a hemoglobin compound accompanies the onset of nitrogen fixation.

Although nitrogen fixation can be carried out by free-living blue-green algae, these organisms are also able to carry out the process in symbiotic associations. The best-characterized systems are the association of fungi and blue-green algae in lichens. However, similar functional associations occur between the Pteridophyte Azolla and the blue-green alga *Anabaena* (Peters and Mayne, 1974) and associations occur between small cycads and blue-green alga *Nostoc* or *Anabaena* (Halliday and Pate, 1976).

In all of these symbiotic associations nitrogen fixation is apparently mediated by the enzyme nitrogenase. Frequently this conclusion has been based on the measured capacity of the conversion of acetylene to ethylene rather than on demonstration of atmospheric nitrogen fixation; nitrogenase has been isolated from very few symbiotic systems. Studies of nitrogenase from symbiotic associations which have been conducted have been primarily limited to the enzyme from Rhizobium bacteroids. These studies have demonstrated that in similarity to the nitrogenase from free-living organisms, the enzyme from bacteroids is composed of two components, a MoFe protein of molecular weight of about 200,000 and an Fe protein of about 60,000 molecular weight. Although reaction mechanisms have not been extensively investigated in nitrogenases from Rhizobium bacteroids or blue-green algae there is a demonstrated requirement for ATP and reductant for in vitro nitrogen reduction. The reduction apparently pro-

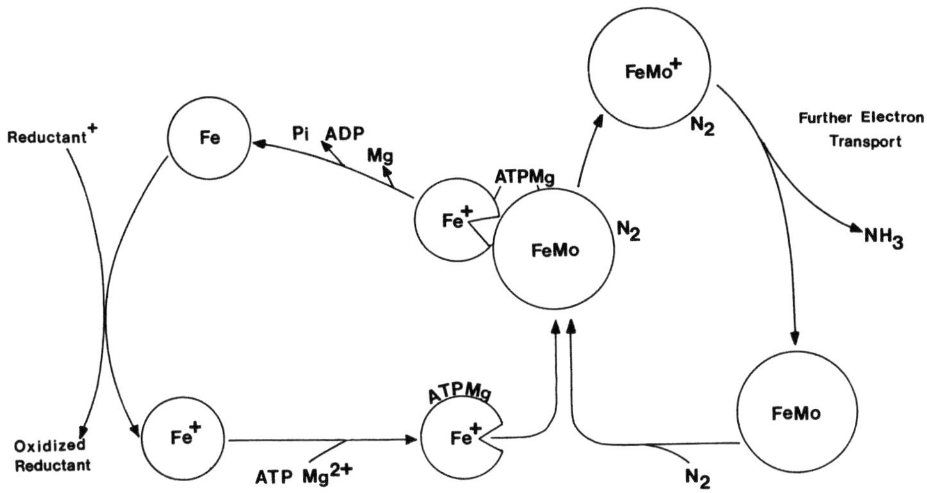

Fig. 1. Simplified scheme for nitrogenase activity. The small Fe-containing protein is reduced by reductant and activated by a ATP-Mg complex. The large FeMo-containing protein complexes with atmospheric nitrogen. The activated Fe joins with the nitrogen-carrying FeMo protein to form the nitrogenase complex in which electrons are transferred to the FeMo protein. After several such cycles NH_3 is released. (Modified from Postgate 1978)

ceeds according to the following scheme (Fig. 1) which has been established from studies of nitrogenase from free-living bacteria and is discussed much more extensively in this volume by Postgate et al. and Zumft.

Noteworthy in this scheme is that the reductant is unidentified and that the product of nitrogen fixation is ammonia. The evidence for ammonia formation as a result of nitrogen fixation has been firmly established on the basis of studies with $^{15}N_2$ (Dilworth, 1974) and more recently $^{13}N_2$ (Wolk et al., 1976).

The source of reductant for nitrogenase activity in bacteroids is not completely resolved. Intact root nodules or nodule homogenates require oxygen for nitrogen fixation, thus indicating a dependence on respiratory processes. Leghemoglobin apparently facilitates diffusion of oxygen to bacteroids while maintaining the level of free oxygen in the nodule at alow enough level to prevent in activation of the nitrogenase system. In addition to oxygen, various Krebs tricarboxylic acid cycle intermediates are effective in stimulating fixation by intact bacteroids, further implicating respiratory intermediates in generating reductant for nitrogen fixation. In vitro nitrogenase activity by cell-free extracts from root nodules can be demonstrated with dithionite as electron donor, however pyruvate, α ketoglutarate, NADH, or NADPH are ineffective. A bacteroid extract fortified with β hydroxybutyrate, NAD, and viologen dye can support nitrogen reduction, leading to the speculation that the viologen substitutes for some natural electron carrier. Some further indication of the occurrence of natural electron carriers is provided with the observation that illuminated chloroplasts can generate reductant for bacteroid nitrogenase. Fractionation of bacteroids demonstrated the presence of a ferredoxin-like component and a flavin component. It is suggested (Yoch, 1979) that the physiological electron donor system for bacteroid nitrogenase involves a flow from

NADPH to bacteroid ferredoxin or flavoprotein (both electron carriers are essential but the sequence of electron flow is not known).

$$\text{NADPH} \xrightarrow{e^-} \text{Bacteroid Ferredoxin, Flavoprotein} \xrightarrow{e^-} \text{Nitrogenase}$$

Production of these electron carriers in the bacteroids is presumably induced by a lowered oxygen content which occurs after leghemoglobin formation. In blue-green algae the electron donor for nitrogen fixation is reduced ferredoxin. Since the heterocysts of blue-green algae are unable to carry out photosystem II of photosynthesis, it is considered that oxidation of respiratory substrates is coupled to the reduction of NADP, ferredoxin, and nitrogenase according to the sequence.

$$\begin{array}{l} \qquad\qquad\qquad\qquad\qquad\qquad\qquad \text{Pyruvate} \\ \qquad\qquad\qquad\qquad\qquad\qquad\qquad\qquad \searrow e^- \\ \text{Pentose Phosphate Cycle Intermediates} \xrightarrow{e^-} \text{NADP} \xrightarrow{e^-} \text{Fd} \xrightarrow{e^-} \text{Nitrogenase.} \end{array}$$

Nothing is known of the source of reductant for nitrogen fixation in non-Rhizobium nodule systems.

3 Nitrate Metabolism

Nitrogen fixation mediated by either free-living organisms or symbiotic associations of nitrogen-fixing organisms and green plants has the capability of furnishing nitrogen to the plant in the form of ammonium. In addition, plants possess the capacity to utilize nitrate as a nitrogen source. There appears to be some variablity in this capacity. Many studies have been performed to determine the relative capabilities of ammonium or nitrate to support plant growth and development. Although it is difficult to generalize, it appears that many calcifuge plants (i. e., those growing in acid conditions) show a perference for ammonium, whereas plants adapted to grow on basic or neutral soils (including most economically important crop plants) seem to prefer nitrate or mixtures of ammonium and nitrate.

3.1 Nitrate Reductase

Nitrate taken up by plants must be reduced to ammonia before the nitrogen can be assimilated into organic form. The process of nitrate reduction may occur predominantly in the roots, as appears to be the case in lupin, or predominantly in the leaves as in *Xanthium* (Pate, 1973). To some extent the ratio of reduction in roots/shoots appears to be a function of nitrate supply, since with increased nitrate availability a progressively greater proportion of nitrate is reduced in the shoot.

Nitrate reduction to ammonium is dependent upon the activities of two enzymes which have been extensively studied in plant tissues and is discussed more extensively in the presentation of Losada et al. of this volume.

The initial metabolism of nitrate involves its reduction to nitrite by the enzyme nitrate reductase. In the majority of higher plants this reduction is mediated by NADH, however an additional enzyme showing activity with both NADPH and NADH has been isolated from soybeans (Jolly et al., 1976), rice (Shen et al., 1976), and corn scutellum (Campbell, 1978). In the tropical legume *Erythrina senegalensis* only the nonspecific NAD(P)H enzyme is present (Stewart and Orebamjo, 1979). The nonspecific enzyme from all sources so far characterized shows a much lower affinity (Km=4.5 m*M* KNO_3) for nitrate than the NADH-dependent enzyme characterized from most other plants (Km=0.1 m*M* KNO_3). The nitrate reductases most extensively purified and characterized from higher plants are NADH-dependent enzymes. The enzyme characteristically has a molecular weight of about 200,000 which, according to recent investigations, is made up of two identical subunits (Kuo et al., 1980). The enzyme is a heme-containing flavo-molybdo-protein which shows diaphorase activity in which NADH can reduce cytochrome c, ferrocyanide or other oxidants. Nitrate reduction can be measured independently of the diaphorase by using reduced viologen dyes or reduced flavins. Nitrate reductase mediated by either reduced flavins or NADH is inhibited by azide or cyanide, however the diaphorase activity is insensitive to these inhibitors. Nitrate reduction mediated by NADH and diaphorase activity is prevented by p-chloromercuribenzoate and these activities are relatively heat-sensitive. In contrast, reduced viologen or $FMNH_2$ stimulated nitrate reduction is insensitive to sulfhydryl inhibitors and is more heat-tolerant. These findings can be interpreted as a two-component enzyme containing a diaphorase function and the nitrate reductase. Although the components have not been physically separated in higher plants, support for the two-component concept is provided by complementation studies of nitrate reductase mutants of tobacco (Mendel and Muller, 1978) and by the observation that the inactive apoenzyme isolated from molybdenum-deficient plants can be activated by addition of a molybdenum-containing factor derived by acid washing nitrate reductase (Rucklidge et al., 1976). The molybdenum-containing factor from higher plants has not been extensively characterized. Preliminary indications from higher plant studies, in agreement with those from fungi, suggest that the factor may be identical to that occurring in other molybdoproteins such as xanthine oxidase and sulfite oxidase. The molybdenum-containing factors from several enzymes has recently been shown to contain pteridine residues which are presumably involved in the transfer of reductant to the molybdenum center of the enzyme (Johnson et al., 1980). The current concept of nitrate reductase can be depicted diagramatically as shown in Fig. 2.

3.1.1 Nitrite Reductase

The nitrite produced by nitrate reductase is further reduced directly to ammonia without the release of free intermediates by the enzyme nitrite reductase. This enzyme has been extensively characterized from extracts of leaf tissue and shown to have a molecular weight of 62,000 and the purified enzyme is not dissociated by sodium dodecyl sulfate. The purified enzyme shows characteristic absorption peaks at 380–390 nm and 570–580 nm and contains three atoms of iron per mol of enzyme. These observations indicate the presence of an iron porphyrin, siroheme, in conjunction with an iron

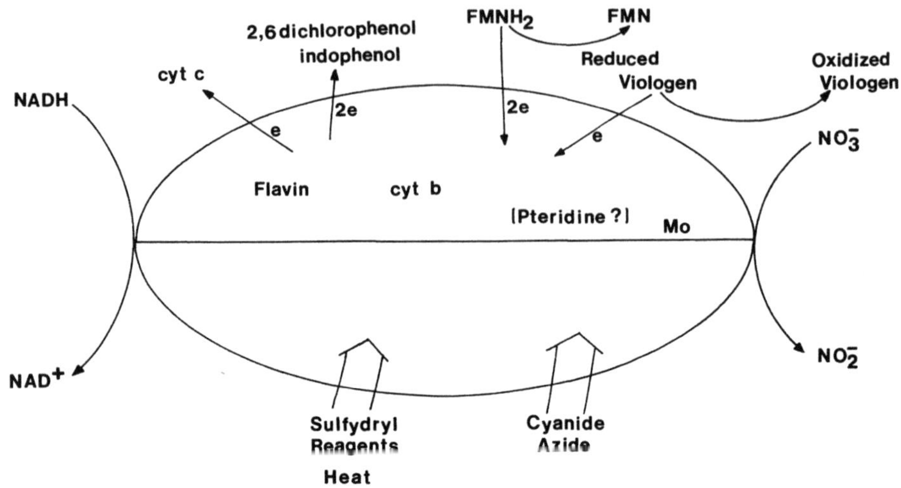

Fig. 2. Diagramatic representation of nitrate reductase activities. The reduction of nitrate with NADH or reduced flavins or reduced viologen dyes as reductant is inhibited by CN^- and azide. Reduction of nitrate with NADH and diaphorase activity with cyt c or 2,6-dichlorophenol indophenol are inhibited by sulfhydryl reagents. In contrast, reduced flavin or reduced viologen dye-mediated reduction of nitrate are unaffected by sulfhydryl reagents. Pteridine is included as a possible intermediate on the basis of the recent demonstration of this component in the molybdenum-containing cofactor from several enzymes including nitrate reductase from *Chlorella*

sulfur center in the enzyme; these components appear to be intimately involved with nitrite reduction (Lancaster et al., 1979). The electron donor for nitrite reduction appears to be reduced ferredoxin which is generated in leaves during noncyclic photophosphorylation. This source of reductant is consistent with the chloroplast location of the enzyme in leaves. Significantly, bundle sheath cells of C_4 plants which have reduced photosystem II capacity and hence a lowered capacity for reduction of ferredoxin have lower levels of nitrate and nitrite reductase than the mesophyll cells (Mellor and Treguna, 1971).

Nitrite reductase activity with reduced viologen dyes or ferredoxin has been isolated from nongreen tissues where the enzyme appears to be located in the protoplast (Dalling et al., 1972). The source of reductant for nitrite reduction in nongreen tissues may be NADPH generated during operation of the pentose phosphate pathway. This contention is based on the finding that nitrite stimulates glucose catabolism by the PPP in seedlings and that a particulate 6-phosphate gluconate dehydrogenase is induced in roots by exposure to nitrate (Emes et al., 1979). While NADPH could be the source of reductant for nitrite reductase the reduced pyridine nucleotide cannot stimulate nitrite reduction in vitro. It appears that in nongreen tissue unidentified electron carriers must be involved to replace the ferredoxin which provides reductant directly to the enzyme in leaves. This lack of definitive information on the nature of the electron donor for nitrite reduction in nongreen cells is reminiscent of the similar deficiency of information on the sequence of electron transport in nitrogen fixation by Rhizobium/legume associations.

Source of Reductant for Nitrate Reduction. Nitrate reduction in leaf tissue is light-dependent. This feature was originally attributed to a requirement of photosynthesis to supply directly reductant for nitrate assimilation. This explanation seems unlikely in view of the pyridine nucleotide specificity and cytoplasmic location of nitrate reductase in leaves. The decline in nitrate utilization in darkness or shade has also been attributed to a decline in extractable nitrate reductase level. While such a decrease in enzyme level has frequently been reported, it appears that the rate of decline is too slow to account for the observed rapid change in nitrate utilization during light-dark transitions (Canvin and Atkins, 1974).

A new understanding of the requirement for light in nitrate assimilation by leaves has developed from in vivo assays of nitrate reductase. In these assays leaf material is incubated with nitrate under anaerobic conditions in the dark and the rate of appearance of nitrite determined. The failure to metabolize nitrite in leaves under the assay conditions is considered to be due to a reduced availability of photosynthetically generated reduced ferredoxin. It is observed that production of nitrite in the in vivo assay is stimulated by inclusion of glycolytic intermediates and Krebs cycle acids; oxygen prevents nitrite production. These findings are interpreted as indicating that reductant generated by glycolysis or Krebs cycle activity is utilized for nitrate reduction. It is proposed (Sawhney et al., 1978; Canvin and Woo, 1979; Hageman et al., 1980) that there is competition between mitochondrial electron transport and nitrate

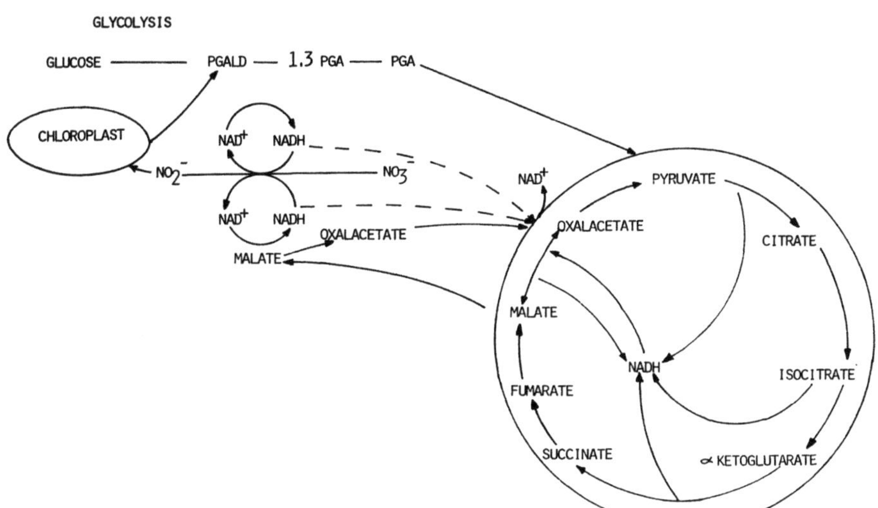

Fig. 3. Proposed source of reductant for nitrate reduction. NADH generated in the cytoplasm during glycolysis can be utilized in nitrate reduction or oxidized by the mitochondrial electron transport chain. Reduced pyridine nucleotide produced during Krebs cycle activities is also oxidized by the mitochondrial electron transport chain. It is suggested that in the light mitochondrial electron transport is inhibited as a result of change in energy charge. In this situation accumulated NADH is oxidized by a reversal of mitochondrial malic dehydrogenase and malate leaves the mitochondria. A cytoplasmic malic dehydroganase converts malate to oxaloacetate and produces NADH to support nitrate reduction. Re-entry of oxalacetate into the mitochondria sustains Krebs cycle activity in the light

reduction for the available NADH. In the dark NADH is oxidized by the mitochondria, thus limiting the availability of reductant for nitrate assimilation. It is argued that NADH oxidation by mitochondria in the light is prevented by a change in energy charge or ATP level. Since NADH is not oxidized in the light it would be available for nitrate reduction. The validity of this scheme outlined in Fig. 3 centers on the light inhibition of mitochondrial electron transport, a concept which is the subject of much controversy. Under conditions of light-inhibited electron transport, an export of malate from the mitochondria and its oxidation during the reduction of nitrate in the cytoplasm provides a mechanism for regenerating NAD and maintaining operation of the Krebs cycle and provision of carbon skeletons necessary for the assimilation of ammonia produced as a consequence of nitrate reduction (Woo et al., 1980).

As attractive as this scheme is, it has some deficiencies. It has been demonstrated that following a transition from anaerobic to aerobic conditions in the dark nitrite is depleted from leaf tissue. However, it has not been convincingly demonstrated that the nitrite nitrogen is assimilated and it is possible that the nitrite is degraded to gaseous oxides of nitrogen. A major reservation concerning the scheme is the assumption that the nitrite which accumulates in the leaves under dark anaerobic conditions does so due to the restricted availability of reduced ferredoxin. However, nitrite also accumulates in roots fed nitrate under anaerobic conditions or treated with uncouplers of electron transport. Thus many of the observations can be interpreted as being due to an impaired nitrite metabolism rather than being attributed to an increased supply of reductant for nitrate assimilation (Lee, 1979).

4 Ammonia Assimilation

Ammonium taken up from the soil solution or produced as a result of nitrate reduction or symbiotic nitrogen fixation is incorporated into organic form. Historically it was considered that this conversion occurred through the activities of the enzyme glutamic acid dehydrogenase (GDH) which in the presence of α ketoglutarate and reduced pyridine nucleotide and ammonium catalyzes the synthesis of glutamate. While this path of incorporation may occur under conditions of high levels of ammonium, it is now generally considered that under most physiological conditions the initial incorporation of ammonium is into glutamine through the activity of glutamine synthetase (GS) and a subsequent conversion of glutamine to glutamate catalyzed by glutamine:2-oxo glutarate amino transferase (GOGAT) (Miflin and Lea, 1976).

$$\text{Glutamate} + NH_3 + \text{ATP} \xrightarrow[\text{Synthetase}]{\text{Glutamine}} \text{Glutamine} + \text{ADP} + \text{Pi}$$

$$\text{Glutamine} + \text{Oxoglutarate} + \underset{\text{or reduced Fd}}{\text{NAD(P)H} + H^+} \xrightarrow{\text{GOGAT}} 2\ \text{Glutamate} + \underset{\text{or oxidized Fd.}}{\text{NAD(P)}^+}$$

This sequence of conversion is consistent with labeling studies in peas and rice seedlings using ^{15}N-labeled nitrogenous sources (Bauer et al., 1977; Yoneyama and Kumazawa, 1975) and the recent sophisticated studies of the Michigan State group using ^{13}N ammonium and ^{13}N nitrate in tobacco cell cultures (Skokut et al., 1978). In the Rhizobium symbiotic association the initial organic products of ^{13}(N) dinitrogen fixation by root nodules have been shown to be glutamine and glutamate (Meeks et al., 1978). It has been established that the appropriate synthetase and GOGAT are localized in the cytosol and that GOGAT is NADH-dependent (Boland et al., 1978). In the majority of nodulated legumes the specific activities of GOGAT and GS exceed the activity of GDH, and GS level is in excess. In nonsymbiotic situations ammonium taken up from the soil solution or produced by nitrate reduction may also be incorporated through the activities of GS and GOGAT. In nongreen tissue GOGAT appears to be NADH-dependent and can be chromatographically separated from the ferredoxin-dependent GOGAT (Matoh et al., 1979) which is localized in the chloroplast (Wallsgrove et al., 1977). Two isozymes of GS have been characterized (Mann et al., 1980). One form appears to be cytoplasmic and functions in ammonium assimilation in the cytoplasm and in nongreen tissue a second form is located in the chloroplast and is probably the main site of ammonia incorporation in leaf tissue. In this location the enzyme has direct access to the products of nitrite reductase activity.

4.1 Asparagine Synthesis

Time course kinetic studies suggest that products of nitrate reduction and symbiotic nitrogen fixation are rapidly incorporated into asparagine. However incorporation of the reduction products into asparagine occurs at a later time than incorporation into glutamine and glutamate (Bauer et al., 1977; Yemm and Willis, 1956) and it appears that asparagine is not synthesized directly by combination of ammonium with aspartic acid. The experimental data appears consistent with the concept that glutamine can serve as the donor group of the asparagine amide group. This proposed synthesis of asparagine by amidation of aspartic acid is supported by the demonstration of Stewart (1979) of the incorporation of radioactivity from aspartate into asparagine in soybean leaves provided with ammonium or glutamine. The ammonium-stimulated conversion of aspartate to asparagine was prevented by inhibitors of glutamine synthetase, further implicating glutamine as the in vivo amide donor for asparagine synthesis. In contrast to the findings of Stewart are other reports (Mitchell and Bidwell, 1970) and research in the author's laboratory that aspartate is not an effective precursor for asparagine synthesis. Enzymatic evidence for the synthesis of asparagine from aspartate with glutamine serving as amide donor has been provided with extracts from cotyledons of germinating lupin and mung bean (Rognes, 1975; Lea and Fowden, 1975; Kern and Chrispeels, 1978). The enzyme involved in this synthesis, asparagine systhetase, has also been demonstrated in extracts of the roots of corn seedlings, but the levels of activity are low and variable (Stulen et al., 1979). Although we can confirm the presence of aspargine synthetase in cotyledons of mung bean, we have been unable to detect the enzyme in extracts from leaves, roots, or cotyledons of pea seedlings. At present there must be some reservations concerning the acceptance of asparagine synthetase as the sole or major mechanism of asparagine synthesis in plants.

4.1.1 Metabolism of Amide Nitrogen

The amides, asparagine and glutamine, serve as transportable forms of nitrogen within the plant and they are frequently present at high levels in the xylem exudate. The amides are capable of providing reduced nitrogen for the synthesis of other nitrogenous components in the plant; particularly important in this regard is the transformation of amide nitrogen into amino nitrogen. In the case of glutamine this conversion can be achieved through the activities of GOGAT. The presence of this enzyme in developing seeds, for instance, provides a mechanism of using amide nitrogen for the de novo formation of amino acids utilized in reserve protein synthesis. (Beevers and Storey, 1976; Sodek and DeSilva, 1977). The amide nitrogen of asparagine is apparently made available for amino acid synthesis in developing seed through the activities of asparaginase which degrades asparagine to aspartate and ammonium (Atkins et al., 1975; Sodek et al., 1980). The released ammonium can be incorporated into glutamine by glutamine synthetase and the amide nitrogen subsequently can be converted into the amino nitrogen of glutamate by the activity of GOGAT. Another mechanism of asparagine catabolism is through the activity of transaminase, this activity seems to occur predominantly in leaf tissue. One of the products of the transaminase reaction is α ketosuccinamic acid which in leaves is rapidly reduced to hydroxy succinamic acid (Lloyd and Joy, 1978). The mechanism by which this compound is metabolized and the means by which the amide nitrogen is made available to subsequent metabolism are not understood.

5 Amino Acid Biosynthesis

Glutamate and aspartate serve as precursors for the synthesis of other amino acids. Many of the enzymes necessary for the synthesis of all of the 20 protein amino acids have been characterized in plants (Miflin and Lea, 1977). However, in only a few instances has the complete sequence of intermediates and appropriate enzymes necessary for the synthesis of a specific amino acid been demonstrated in an individual plant species. To a great extent the amino acids appear to be synthesized in chloroplasts. However, since excised roots and nonphotosynthetic plant cell tissue cultures are able to grow and synthesize amino acids, it is clear that there must be other sites of amino acid biosyntheses.

It is recognized that in leaf tissue glycine and serine may be synthesized in the peroxisome and mitochondria respectively. Glycine is produced by transamination of glyoxylate which is formed in the peroxisome from chloroplast generated glycolate. Glycine is converted to serine in the mitochondria through the activity of glycine decarboxylase and serine hydroxymethyl transferase with the resulting release of photorespiratory CO_2 and ammonia. The ammonia is apparently recovered by the formation of glutamine by a cytosol glutamine synthetase (Keys et al., 1978).

6 Other Nitrogenous Transport and Storage Compounds

In addition to amides, other nitrogenous metabolites may function in nitrogen transport, storage, and interconversion. This is particularly apparent in symbiotic nitrogen fixation where a range of nitrogenous compounds apparently serving transport and storage functions are encountered.

6.1 Ureides

Among these compounds are the ureides allantoin and allantoic acid which have been found to be the most abundant soluble nitrogenous components in xylem sap from the nodulated tropical legumes cowpea (Herridge et al., 1978) and soybean (Matsumoto et al., 1977; Streeter, 1979). The ureides make up a much smaller component of the soluble nitrogen in nodulated temperate legumes, however they may function as soluble nitrogen storage compounds in such woody perennials as sugar maple (Reinbothe and Mothes, 1962). Allantoin is generally considered to be synthesized by the oxidative degradation of purines (Fig. 4). Evidence for operation of this pathway in soybeans is based on the finding of enzymes of purine oxidation (Tajima and Yamomoto, 1975)

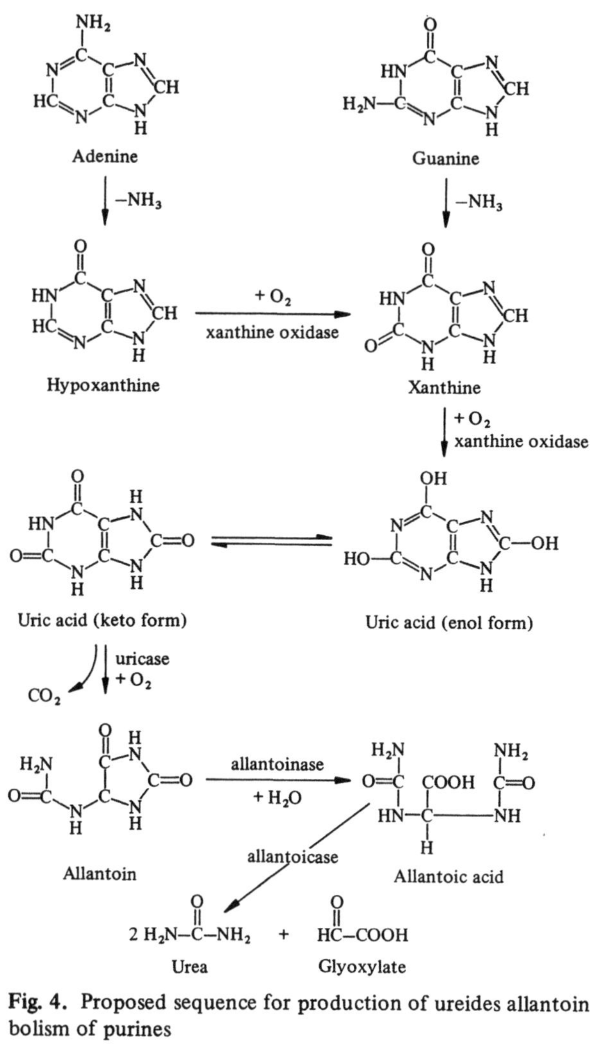

Fig. 4. Proposed sequence for production of ureides allantoin and allantoic acid by aerobic catabolism of purines

and the observation that allopurinol, an inhibitor of xanthine dehydrogenase, prevents ureide production (Fujihari and Yamaguchi, 1978). Xanthine dehydrogenase has been characterized from pea leaves (a plant which does not accumulate ureide) and shown to be a cytoplasmic enzyme (Nguyen and Feierabend, 1978). Uricase appears to be associated with glyoxysomes and peroxisomes (Beevers, 1979). The purines oxidized to give rise to allantoin are presumably synthesized through the complex reaction sequences established in bacterial systems, however to date very few of the necessary enzymes and intermediates have been detected in plants. As a result of the recent interest generated in allantoin synthesis we can anticipate an increased understanding of the pathway of purine biosynthesis in plants. Although in the soybean and presumably other tropical nodulated legumes the biosynthesis of ureides occurs almost exclusively in the nodules it is clear from the observation of ureide storage in woody perennials that the synthesis of these compounds is not confined to symbiotic associations.

Ureide transported from the nodules and released from storage tissues in woody perennials can be degraded to urea and glyoxylate. Allantoin is converted to allantoic acid by the enzyme allantoinase. Although not extensively studied, it is interesting that the occurrence of this enzyme in soybean meal was recorded as early as 1929 (Fosse and Brumel, 1929). Allantoic acid is converted to urea and glyoxylate by the enzyme allantoicase. Urea can be degraded by urease and the ammonium recovered by glutamine synthetase.

There are obviously many missing links in the sequence of ureide biosynthesis and catabolism. We can anticipate consolidation of the proposed schemes in the future following demonstration of the appropriate enzymes and physiological studies demonstrating changes and fluctuations in the levels of suspected intermediates.

6.1.1 Arginine and Citrulline

Arginine frequently functions as a transport and storage compound. In adler citrulline is the principal amino acid accumulated as a result of nitrogen fixation. Synthesis of these basic amino acids proceeds from glutamate via ornithine according to the sequence outlined in Fig. 5. Two mechanisms of acetylation of glutamate are possible, however, according to McKay and Shargool (1977), acetyl ornithine in contrast to acetyl CoA is the preferred acetyl donor. Acetyl glutamate is converted to acetyl glutamic semialdehyde. This acetylated semialdehyde cannot undergo spontaneous cyclization to Δ pyrroline-5 carboxylic acid and in this way glutamate residues are committed to arginine or citrulline biosynthsis instead of being diverted to proline. Acetyl glutamic semialdehyde is converted to acetyl ornithine by the enzyme N-acetyl-L ornithine δ transaminase. Ornithine released during the acetylation of glutamate is converted to citrulline by condensation with carbamyl phosphate in the presence of ornithine transcarbamylase. Apparently this is the extent of the reaction sequence in such species as alder *(Alnus)* which accumulate citrulline; more frequently it appears that this amino acid is converted to arginine through the intermediate argino-succinic acid with the involvement of the enzymes argino-succinate synthetase and argino succinate lyase (Shargool, 1971; Roubelakis and Kliewer, 1978 a).

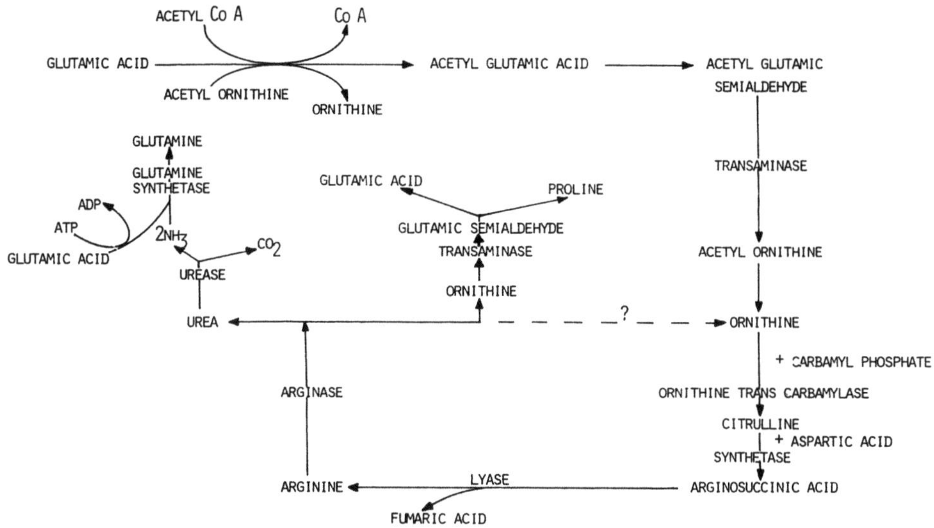

Fig. 5. Proposed pathways for the metabolism of citrulline and arginine

Catabolism of arginine involves its hydrolysis to ornithine and urea by the enzyme arginase (Roubelakis and Kliewer, 1978b). Observations that glutamate pyrroline 5 carboxylic acid and γ amino butyric acid are labeled from exogenously supplied radioactive arginine indicate that the ornithine produced by arginase activity is converted to the intermediate glutamic semialdehyde which can serve as precursor for the other products. The urea produced during arginase activity in higher plants may be degraded to ammonium and carbon dioxide. Ammonia may be utilized in the formation of glutamine, which following conversion to glutamate can be converted to other amino groups by transaminase activity. It has not been established how citrulline is catabolized in those species in which it constitutes a principal transportable form of reduced nitrogen. Exogenously supplied citrulline can be converted to arginine (Hill-Cottingham and Lloyd-Jones, 1973) but it has not been established whether this represents the usual sequence of utilization.

Other Amino Acids. Newly fixed nitrogen may accumulate as γ amino butyric acid in *Trifolium subterraneum* infected with Rhizobium (Gibson, 1976). This metabolite also accumulates under various conditions of stress and in these situations is apparently synthesized by decarboxylation of glutamic acid by the enzyme glutamic acid decarboxylase. Metabolism of γ amino butyric acid has been shown to proceed by a removal of the amino group and the formation of succinic semialdehyde which can be oxidized to succinic acid by an NAD-dependent oxido reductase. The enzyme involved in amino group removal has not been characterized.

In peanuts *(Arachis hypogaea)* γ methylene glutamine makes up the dominant soluble nitrogenous storage and transport component. There is little information available on its mechanism of synthesis or degradation.

7 Conclusions

This brief coverage of nitrogen metabolism in plants has indicated that we now understand the sequence by which inorganic nitrogen derived from nitrogen fixation, or soil nitrate, or ammonium becomes incorporated into organic form. Recent evidence points to the central role of the combined activities of glutamine synthetase and GOGAT in the formation of organic nitrogenous compounds. Plants have developed various strategies for the storage and transport of reduced nitrogen. Our understanding of the synthesis of many of these storage components is fragmentary and there is little information available on the sequence and mechanism of degradation. Ultimately degradation of the storage compounds releases ammonia which is recovered as the amide nitrogen of glutamine, further indicating the central role of the enzyme glutamine synthetase.

Plants show great metabolic diversity in the production of an extensive range of nitrogen-containing secondary metabolites, however time and space preclude discussion of these aspects of nitrogen metabolism which are for the most part unique to higher plants.

References

Atkins CA, Pate JS, Sharkey PJ (1975) Plant Physiol 56:807–812

Bal AK, Shantharam S, Ratnam S (1978) J Bacteriol 133:1393–1400

Bauer A, Joy KW, Urquhart AA (1977) Plant Physiol 59:920–924

Beevers H (1979) Annu Rev Plant Physiol 30:159–193

Beevers L, Storey R (1976) Plant Physiol 57:862–866

Boland MJ, Fordyce AM, Greenwood RM (1978) Aust J Plant Physiol 5:553–559

Bond G (1976) In: Nutman PS (ed) Symbiotic nitrogen fixation in plants. International biological programme. University Press, Cambridge, pp 443–474

Campbell WH (1978) Z Pflanzenphyiol 88:611–616

Canvin DT, Atkins CA (1974) Planta 116:207–224

Canvin DT, Woo KC (1979) Can J Bot 57:1155–1160

Dalling MJ, Tolbert NE, Hageman RH (1972) Biochim Biophys Acta 283:513–519

Dazzo FB, Yanke WE, Brill WJ (1978) Biochim Biophys Acta 539:276–286

Dilworth MJ (1974) Annu Rev Plant Physiol 25:81–114

Emes MJ, Ashihara H, Fowler MW (1979) FEBS Lett 105:370–373

Fosse R, Brunel A (1929) C R Acad Sci 188:426–428

Fujihara S, Yamaguchi Y (1978) Plant Physiol 62:134–138

Gibson AH (1976) In: Newton WE, Nyman CJ (eds) Proc 1st Int Symp Nitrogen Fixation. Pullman, Washington State Univ Press, pp 400–428

Hageman RH, Reed A, Femmer R, Dalling M (1980) Plant Physiol 66: in press

Halliday J, Pate JS (1976) Aust J Plant Physiol 3:349–358

Herridge DF, Atkins JA, Pate JS, Rainbird RM (1978) Plant Physiol 62:495–498

Hill-Cottingham DG, Lloyd-Jones CP (1973) Physiol Plant 29:125–128

Johnson JL, Hainline BE, Rajagopalan KV (1980) J Biol Chem 255:1783–1791

Jolly SO, Campbell WH, Tolbert NE (1976) Arch Biochem Biophys 174:431–439

Kern R, Chrispeels MJ (1978) Plant Physiol 62:815–819

Keys AJ, Bird IF, Cornelius MJ, Lea PJ, Wallsgrove RM, Miflin BJ (1978) Nature (London) 275:741–743

Kuo T, Kleinhofs A, Warner RL (1980) Plant Sci Lett 17:371–381

Lancaster JR, Vega JM, Kamin H, Orme-Johnson NR, Orme-Johnson WH, Krueger RJ, Siegel LM (1979) J Biol Chem 254:1268–1272

Lea PJ, Fowden L (1975) Proc R Soc London Ser B 192:13–26
Lee RB (1979) J Exp Bot 30:119–133
Lloyd NDH, Joy KW (1978) Biochem Biophys Res Commun 81:186–192
Mann AF, Fenten PA, Stewart GR (1980) FEBS Lett 110:265–269
Matoh T, Takahashi E, Ida S (1979) Plant Cell Physiol 20:1455–1459
Matsumoto T, Yatazawa M, Yamamoto Y (1977) Plant Cell Physiol 18:353–359
McKay G, Shargool PD (1977) Plant Sci Lett 9:189–193
Meeks JC, Wolk P, Schilling N, Shaffer PW, Avissar Y, Chien W-S (1978) Plant Physiol 61:980–983
Mellor GE, Tregunna EB (1971) Can J Bot 49:137–142
Mendel RR, Muller AJ (1978) Mol Gen Genet 161:77–80
Miflin BJ, Lea PJ (1976) Phytochemistry 15:873–885
Miflin BJ, Lea PJ (1977) Annu Rev Plant Physiol 28:299–329
Mitchell DJ, Bidwell RGS (1970) Can J Bot 48:2001–2007
Newcomb W, Peterson RL, Callaham D, Torrey JG (1978) Can J Bot 56:502–531
Nguyen J, Feierabend J (1978) Plant Sci Letts 13:125–132
Pagan JD, Child JJ, Snowcroft WR, Gibson AH (1975) Nature (London) 256:406–407
Pate JS (1973) Soil Biol Biochem 5:109–120
Peters GA, Mayne BC (1974) Plant Physiol 53:813–819
Postgate J (1978) Studies in biology, no 92. London, Edward Arnold, p 17
Reinbothe H, Mothes K (1962) Annu Rev Plant Physiol 13:129–150
Rognes SE (1975) Phytochemistry 14:1975–1982
Roubelakis KA, Kliewer WM (1978a) Plant Physiol 62:340–343
Roubelakis KA, Kliewer WM (1978b) Plant Physiol 62:344–347
Rucklidge G, Notton BA, Hewitt EJ (1976) Biochem Soc Trans 4:77–80
Sawhney SK, Naik MS, Nicholas DJD (1978) Nature (London) 272:647–648
Shargool PD (1971) Phytochemistry 10:2029–2032
Shen TC, Funkhouser EA, Guerrero MG (1976) Plant Physiol 58:292–294
Skokut TA, Wolk P, Thomas J, Meeks JC, Shaffer PW, Chien W-S (1978) Plant Physiol 62:299–304
Sodek L, DeSilva WJ (1977) Plant Physiol 60:602–605
Sodek L, Lea PJ, Miflin BJ (1980) Plant Physiol 65:22–26
Stewart CR (1979) Plant Sci Lett 14:269–273
Stewart GR, Orebamjo TO (1979) New Phytol 83:311–320
Streeter JG (1979) Plant Physiol 63:478–480
Stulen I, Israelstam GF, Oaks A (1979) Planta 146:237–241
Tajima S, Yamomoto Y (1975) Plant Cell Physiol 16:271–282
Van Straten J, Akkermans ADL, Roelofsen W (1977) Nature (London) 266:257–258
Verma DPS, Kazazian V, Zogbi U, Bal AK (1978) J Cell Biol 78:919–936
Vincent JM (1974) In: Quispel A (ed) The biology of nitrogen fixation. North-Holland Publishing Company, Oxford, pp 265–341
Wallsgrove RM, Harel E, Lea PJ, Miflin BJ (1977) J Exp Bot 28:588–596
Wolk CP, Thomas J, Shaffer PW, Austin SM, Galonsky A (1976) J Biol Chem 251:5027–5034
Woo KC, Jokinen M, Canvin DT (1980) Plant Physiol 65:433–436
Yemm EW, Willis AJ (1956) New Phytol 55:229–252
Yoch DC (1979) In: Hardy RWF, Bottomley F, Burns RC (eds) A treatise on dinitrogen fixation sections I and II. John Wiley & Sons, New York, pp 605–652
Yoneyama T, Kumazawa TK (1975) Plant Cell Physiol 16:21–26

The Assimilatory Reduction of Nitrate

M. LOSADA, M.G. GUERRERO, and J.M. VEGA[1]

1 Introduction

Nitrate reduction by algae and higher plants is one of the most fundamental biological processes by which inorganic oxidized nitrogen is reduced to ammonia with water as the ultimate reductant at the expense of sun energy (Losada and Guerrero, 1979). With the exception of some blue-green algae, and plants that have a symbiotic association with nitrogen-fixing prokaryotes, photosynthetic organisms derive most of their nitrogen from nitrate. There are besides a variety of nonphotosynthetic organisms, bacteria and fungi, that can also assimilate nitrate, as well as certain free-living anaerobic and aerobic bacteria that can fix molecular nitrogen.

Elementary analysis of plants yields about 2%–10% nitrogen versus 40%–50% carbon on a dry weight basis (Bowen, 1966). Since it has been estimated (Hall, 1979) that 2×10^5 megatons of carbon are fixed annually on a world-wide scale by photosynthesis, it can be roughly calculated that about 2×10^4 megatons of inorganic nitrogen are concomitantly incorporated into organic nitrogen during the process. This figure is about two orders of magnitude higher than present estimates (Burns and Hardy, 1975) for the overall biological and chemical fixation of molecular nitrogen (2×10^2 megatons). It should, moreover, be realized that whereas the reduction of carbon dioxide to the level of carbohydrate requires 4 electrons, the reduction of nitrate to ammonia and its incorporation into amino-nitrogen involve 10 electrons. This means that, on the average, the proportion of electrons going from water to nitrate is only about one fifth of that going to carbon dioxide. In consequence nitrate assimilation is in no way secondary when compared with carbon dioxide assimilation, and certainly much more significant that nitrogen fixation on a quantitative basis.

Mainly on the basis of chemical considerations biochemists have traditionally accepted with insufficient evidence and without much criticism that the assimilatory reduction of nitrate to ammonia should proceed in four consecutive steps, each one involving two electrons, according to the following sequence of reactions:

$$HNO_3 \xrightarrow{2e} HNO_2 \xrightarrow{2e} (HNO) \xrightarrow{2e} NH_2OH \xrightarrow{2e} NH_3 .$$

Although the validity of the first step, i. e., the reduction of nitrate to nitrite, has never been doubted, the further reduction of nitrite to ammonia has always been a controversial subject. Fewson and Nicholas (1961) have proposed, based upon biochem-

1 Departamento de Bioquímica, Facultad de Biología y C.S.I.C., Universidad de Sevilla, Spain

ical studies with microorganisms and plants, that the reduction of nitrite to hyponitrite occurs in two steps, each one involving one electron, via nitric oxide (NO), and other authors have suggested that inorganic nitrogen may be incorporated at the level of hydroxylamine into organic compounds, via oximes (Czygan, 1965; Solomonson and Spehar, 1977).

More critical and intensive physiological, biochemical, and genetic studies with a great variety of organisms at the cellular, subcellular, and molecular levels in different laboratories seem, however, to have firmly established (Beevers and Hageman, 1972; Garrett and Amy, 1978; Hewitt et al., 1976; Kessler, 1964; Losada and Guerrero, 1979; Morris, 1974; Nason, 1962; Syrett and Leftley, 1976; Vennesland and Guerrero, 1979) that the assimilatory nitrate-reducing enzyme system consists only of two metalloproteins, which catalyze the stepwise reduction of nitrate to nitrite and ammonia through an inorganic sequence of reactions unexpectedly much shorter than the ones first postulated:

$$NO_3^- \xrightarrow{2e} NO_2^- \xrightarrow{6e} NH_4^+.$$

Ammonia, the end product of nitrate reduction, controls by itself and as a result of its further incorporation into organic components the key steps involved in nitrate uptake and reduction, behaving as a very efficient antagonist of nitrate in the regulation of its assimilation. Until recently there was only one major pathway to consider for ammonia assimilation, via the reductive amination of 2-ketoglutarate to give glutamate, a reaction catalyzed by glutamate dehydrogenase, but now a new route has been found (Lea and Miflin, 1979). In this second route ammonia is initially incorporated into the amide position of glutamine, via glutamine synthetase, and then transferred to the α position of 2-ketoglutarate to give glutamate, via glutamate synthase.

2 Enzymes of the Nitrate-Reducing System

As mentioned in the preceding section, recent studies carried out in different laboratories have definitely established that, no matter what the type of organism, either bacteria, fungi, algae or higher plants, only two enzymes are involved in the assimilatory reduction of nitrate to ammonia. The molybdoprotein nitrate reductase catalyzes the 2-electron reduction of nitrate to nitrite, and the ironprotein nitrite reductase takes charge of the subsequent 6-electron reduction of nitrite to ammonia, in reactions depending on reduced ferredoxin or pyridine nucleotides as the physiological reductants.

2.1 Assimilatory Nitrate Reductases

According to the specificity for the electron donor two main types of assimilatory nitrate reductase, the enzyme which catalyzes the stoichiometric reduction of nitrate to nitrite, can be distinguished: (1) ferredoxin-nitrate reductase, which is typically present in cyanobacteria (blue-green algae) and also presumably in chemotrophic and photosynthetic bacteria, and (2) NAD(P)H-nitrate reductase, which is characteristically found in eukaryotic organisms. The differential specificity for the physiological reductant is correlated with sensible differences in prosthetic groups and physicochemical properties.

2.1.1 *Ferredoxin-Nitrate Reductase*

According to present evidence (Candau, 1979; Hattori and Myers, 1967, Manzano et al., 1976; Ortega et al., 1976), nitrate reductase from different cyanobacteria cannot accept electrons directly from NAD(P)H but is rather dependent on reduced ferredoxin as the physiological electron donor. The reduction of nitrate to nitrite catalyzed by cyanobacterial ferredoxin-nitrate reductase obeys, thus, the following equation:

$$NO_3^- + 2Fd_{red} + 2H^+ \xrightarrow{2e} NO_2^- + 2Fd_{ox} + H_2O.$$

Also flavodoxin, which physiologically substitutes for ferredoxin under conditions of iron starvation (Bothe, 1977), can act as an electron carrier for the reduction of nitrate to nitrite catalyzed by *Anacystis* nitrate reductase when reduced photosynthetically (Candau, 1979; Manzano, 1977).

By using affinity chromatography on ferredoxin-Sepharose gel as the main step (Manzano et al., 1978), ferredoxin-nitrate reductase of *Anacystis* has been recently purified to homogeneity and partly characterized. The enzyme is a molybdoprotein constituted by only one polypeptide chain with a molecular weight of 75 kdaltons. The absorption spectra of homogeneous preparations of the enzyme exclude the possible presence in its molecule of either flavin or cytochrome, but not of nonheme iron (Candau, 1979).

Though not yet conclusively demonstrated, there are some indications that assimilatory nitrate reductases from both chemotrophic bacteria (Guerrero et al., 1973; Tortolero et al., 1975; Villalobo et al., 1977) and photosynthetic bacteria (Alef and Klemme, 1979; Malofeeva et al., 1975) can also use reduced ferredoxin as the physiological electron donor. These enzymes are active with reduced viologens but cannot use reduced pyridine nucleotide as reductant. In *Azotobacter chroococcum* (Tortolero et al., 1975) and *Ectothiorhodospira shaposhnikovii* (Malofeeva et al., 1975), illuminated higher plant chloroplasts could supply electrons for the reduction of nitrate to nitrite in the presence of nitrate reductase and ferredoxin of the corresponding organism.

Nitrate reductases from the aerobic chemotrophic bacteria *Azotobacter* and *Acinetobacter* have been partly characterized and shown to be molybdoproteins of about 100 kdaltons (Guerrero et al., 1973; Villalobo et al., 1977). The nitrate reductase from the photosynthetic bacterium *Rhodopseudomonas capsulata* has been recently purified and some of its properties studied. A molecular weight of 185 kdaltons has been reported for a protein which seems to be composed of two subunits of 85 kdaltons and to contain one molybdenum atom and two heme molecules of the c-type (Alef and Klemme, 1979).

2.1.2 NAD(P)H-Nitrate Reductase

Nitrate reductase of green algae, higher plants and fungi catalyzes the reduction of nitrate to nitrite by reduced pyridine nucleotides (Losada and Guerrero, 1979; Vennesland and Guerrero, 1979) according to the equation:

$$NO_3^- + NAD(P)H \;+\; H^+ \xrightarrow{2e} NO_2^- + NAD(P)^+ \;+\; H_2O.$$

In general, most pyridine nucleotide-dependent nitrate reductases react with both NADH and NADPH, although to a variable extent with either nucleotide. The enzyme from photosynthetic eukaryotes usually, but not always, shows a certain preference for NADH, whereas NADPH appears to be a better reductant for fungal nitrate reductase. Nevertheless, this cannot be taken as an absolute rule since significant differences regarding specificity are found among enzymes from representatives of the same group of organisms and sometimes even within strains of a common species. Moreover, two types of nitrate reductase exhibiting differential specificity for both nucleotides have been reported to be simultaneously present in the same tissue (Jolly et al., 1976).

In addition to catalyzing the normal physiological reaction, i. e., reduction of nitrate by reduced pyridine nucleotides, NAD(P)H-nitrate reductase exhibits two other activities, which can be assayed separately and selectively inhibited, that involve only partial functions of the overall electron transport chain of the enzyme complex. The diaphorase activity represents the function of the first [NAD(P)H-activating] moiety of the complex, its action resulting in the reduction by NAD(P)H of a variety of 1- and 2-electron acceptors. The so-called terminal nitrate reductase constitutes the second [nitrate-activating] moiety of the complex and is expressed as a pyridine nucleotide-independent activity which results in the reduction of nitrate by reduced flavins or viologens. Both moieties participate jointly and sequentially in the transfer of electrons from NAD(P)H to nitrate (see Losada and Guerrero, 1979).

Purification to electrophoretic homogeneity of a NAD(P)H-nitrate reductase was firstly achieved for the enzyme from the green alga *Chlorella* (Solomonson et al., 1975). After Solomonson (1975) successfully assayed affinity chromatography on Blue Dextran-Sepharose for the purification of *Chlorella* nitrate reductase, this simple and straightforward procedure has been applied to the enzyme from other sources with outstanding results (e. g., De la Rosa et al., 1980; Guerrero and Gutierrez, 1977; Guerrero et al., 1977; Kuo et al., 1980; Notton et al., 1977). Affinity chromatography on FAD-Sepharose has also been employed for the purification of fungal nitrate reductase (Guerrero et al., 1977; Pan and Nason, 1978).

From the available data on highly purified NAD(P)H-nitrate reductase from different sources, the general pattern can be inferred that this enzyme contains FAD, cytochrome b-557 and molybdenum as prosthetic groups, is composed of a variable number of subunits and presents different molecular weight values according to its source. NAD(P)H-nitrate reductase of fungi (Guerrero and Gutierrez, 1977; Pan and Nason, 1978) and higher plants (Kuo et al., 1980; Notton and Hewitt, 1979) are similarly sized (MW around 200 kdaltons) and, exception made of the spinach enzyme which has been proposed to be constituted by two copies of each of two different subunits (Notton and Hewitt, 1979), seem to be composed of only two subunits of about 100 kdaltons. Nitrate reductase of the mold *Neurospora* appears to contain one molybdenum atom and about two cytochrome b-557 molecules per enzyme molecule (Pan and Nason, 1978). Nevertheless, other authors have reported the presence of only one molecule of cytochrome and one molybdenum atom for the *Neurospora* enzyme (Orme-Johnson et al., 1977). Also only one molecule of cytochrome b has been assumed to be present in the enzyme of the yeast *Rhodotorula* (Guerrero and Gutierrez, 1977). Presently available quantitative data about the prosthetic group content of higher plant nitrate reductase are rather scarce (Notton and Hewitt, 1979). Regarding the situation in eukaryotic algae, the NADH-dependent enzyme of *Chlorella* (MW 280 kdaltons) has been shown to contain about three molecules of each FAD and cytochrome b and two molybdenum atoms, and to have three equal subunits of about 90 kdaltons (Giri and Ramadoss, 1979; Solomonson, 1979). In contrast, NAD(P)H-nitrate reductase from *Ankistrodesmus* exhibits a higher molecular weight (460 kdaltons) and appears to be composed of eight similarly sized (58 kdaltons) subunits with an overall content of at least four cytochrome b molecules (De la Rosa et al., 1980).

The participation of FAD in the activity of the diaphorase moiety of the enzyme and of molybdenum in the terminal moiety are now well-established facts. It is also evident that cytochrome b-557 is an ubiquitous component of NAD(P)H-nitrate reductase, as shown by the characteristic absorption spectra with maxima at about 557 nm (α), 527 nm (β) and 423 nm (Soret) for the reduced form, and at 412 nm for the oxidized one. In addition, active participation of this group in the catalytic activity of the enzyme is sustained by the observation that cytochrome b is reduced by NAD(P)H and reoxidized by addition of nitrate. Nevertheless, its location within the enzymatic electron-transport chain is uncertain, although its site of action appears to be placed beyond that of FAD. According to present evidence (Garrett and Amy, 1978; Hewitt et al., 1976; Losada and Guerrero, 1979; Vennesland and Guerrero, 1979), the pathway of electrons from NAD(P)H to nitrate through nitrate reductase from eukaryotes may be depicted as follows:

$$\text{NAD(P)H} \longrightarrow \{\text{FAD} \longrightarrow \text{cyt b-557} \longrightarrow \text{Mo}\} \longrightarrow \text{NO}_3^-.$$

Sulphydryl groups seem to be essential for the diaphorase activity, and evidence has been presented that -SH groups are interposed between NAD(P)H and FAD in the *Neurospora* enzyme (Amy et al., 1977). These authors have also suggested a direct participation of such groups in the electron flow from NAD(P)H to flavin through a disulfide-sulfhydryl transition. It is believed that molybdenum is bound to a small polypeptide, thus forming a low molecular weight "molybdenum cofactor" (Mo-co) which appears to be common to all molybdoenzymes, with the single exception of nitrogenase (Amy and Rajagopalan, 1979; Pienkos et al., 1977). As far as we know, no chemical characterization nor authentication of the Mo-co has been yet carried out in assimilatory nitrate reductases. No general agreement has otherwise being reached regarding the oxidation state change of molybdenum in nitrate reductase during the enzymatic reduction of nitrate to nitrite.

2.2 Assimilatory Nitrite Reductases

Two types of assimilatory nitrite reductase, marked by a well-defined electron donor specificity, have been described: (1) ferredoxin-nitrite reductase, characteristic of photosynthetic organisms, and (2) NAD(P)H-nitrite reductase, typical of nonphotosynthetic organisms. Both enzymes catalyze the stoichiometric reduction of nitrite to ammonia, a reaction which implies the unusual transfer of six electrons, also occurring in the reduction of dinitrogen to ammonia, catalyzed by nitrogenase, and in the reduction of sulfite to sulfide, catalyzed by sulfite reductase. As is the case with nitrate reductase, the differential specificity for the reductant is correlated with significant differences in prosthetic groups and physicochemical properties.

2.2.1 Ferredoxin-Nitrite Reductase

Photosynthetic nitrite reductase presents a marked specificity for ferredoxin as electron donor (Vega et al., 1980a) and catalyzes the following reaction:

$$NO_2^- + 6Fd_{red} + 8H^+ \xrightarrow{6e} NH_4^+ + 6Fd_{ox} + 2H_2O.$$

Flavodoxin can also substitute for ferredoxin as the immediate electron donor for different nitrite reductases (Bothe, 1977; Manzano, 1977; Zumft, 1972).

Ferredoxin-nitrite reductase has been purified to electrophoretic homogeneity from several sources (Cárdenas et al., 1972; Ho et al., 1976; Hucklesby et al., 1976; Vega et al., 1980a; Zumft, 1972), and appears to be constituted by a single polypeptide chain with molecular weight of 60–70 kdaltons. The purified enzymes exhibit similar absorption spectra with peaks in the regions 380–390 nm and 570–580 nm, characteristic of heme-containing proteins. As a typical representative, an enzyme preparation from spinach with a specific activity of 110 units·mg protein^{-1} showed absorption maxima at 276, 386 (Soret), 573 (α), and 690 nm, with an E_{386} of 70 mM^{-1}·cm^{-1}, and A_{276}/A_{386} absorptivity ratio of 1.8. Anaerobic addition of dithionite resulted in the loss of the 690 nm peak and the splitting of the 573 nm absorption band into two broad peaks at 545 and 585 nm (Lancaster et al., 1979; Vega and Kamin, 1977).

Murphy et al. (1974) identified the heme prosthetic group of spinach nitrite reductase as "siroheme," an iron tetrahydroporphyrin of the isobacteriochlorin type with eight carboxylic acid-containing side chains. Electron spin resonance spectroscopy (ESR) has confirmed the presence of siroheme and provided evidence for the additional presence of an iron-sulfur center in the enzyme from higher plants (Aparicio et al., 1975; Cammack et al., 1978; Vega and Kamin, 1977). Lancaster et al. (1979) established that spinach nitrite reductase contains a tetranuclear iron-sulfur center (4Fe-4S) and suggested that the enzyme contains one siroheme and one (4Fe-4S) cluster per molecule. These prosthetic groups had been previously found as components of *Escherichia coli* sulfite reductase (Rueger and Siegel, 1976).

The midpoint redox potential of the siroheme prosthetic group of higher plant nitrite reductase is, for an n=1 titration, -50 mV (pH 7.8) for the spinach enzyme (Stoller et al., 1977) and -120 mV (pH 8.5) for the marrow enzyme (Cammack et al.,

1978). The iron-sulfur center shows a very negative midpoint potential: assuming an n=1 titration, values of -550 mV (pH 9.0) for the spinach nitrite reductase (Stoller et al., 1977) and -615 mV (pH 8.7) for the marrow enzyme (Cammack et al., 1978) have been reported.

Spectrophotometric and ESR studies in the absence and presence of inhibitors, such as cyanide or carbon monoxide, support that siroheme is the enzyme site where nitrite is bound in order to start the catalytic cycle. Nitrite reacts relatively slowly with the oxidize spinach enzyme forming a 1:1 complex (Vega and Kamin, 1977). Nitrite reductase from spinach also forms a 1:1 complex with ferredoxin (Knaff et al., 1978). Both the iron-sulfur center and the siroheme, prereduced with dithionite, are rapidly reoxidized upon addition of nitrite, strongly supporting a role for these groups in the mechanism of nitrite reduction to ammonia (Lancaster et al., 1979).

Kinetic and steady-state turnover experiments suggest that nitrite is bound to siroheme and reduced to ammonia without the accumulation of free nitrogen compounds of intermedia redox state. The pathway of electron flow from reduced ferredoxin to nitrite, via nitrite reductase, may be represented as follows (Vega et al., 1980a):

$$Fd_{red} \longrightarrow \{(4Fe\text{-}4S) \longrightarrow \text{siroheme}\} \longrightarrow NO_2^- .$$

2.2.2 *NAD(P)H-Nitrite Reductase*

The fungal nitrite reductase (Garrett and Amy, 1978) presents a marked specificity for reduced pyridine nucleotides as electron donors and requires FAD for maximal activity. It catalyzes the following reaction:

$$NO_2^- + 3NAD(P)H + 5H^+ \xrightarrow{6e} NH_4^+ + 3NAD(P)^+ + 2H_2O.$$

The purified NAD(P)H-nitrite reductase from *Neurospora* catalyzes in vitro other electron-transfer reactions, which involve only part of the overall electron transport chain of the enzyme complex and can be separately assayed. The first moiety is known as NAD(P)H-diaphorase, and the second one as dithionite-nitrite reductase. Both activities participate jointly in the transfer of electrons from NAD(P)H to nitrite (Garrett and Amy, 1978; Vega, 1976). This enzyme complex has a molecular weight of 290 kdaltons and is composed of two similar-sized subunits (Garrett and Amy, 1978). Its absorption spectrum with wavelength maxima at 280, 390 (Soret), and 580 (α) nm is indicative of a heme-protein (Greenbaum et al., 1978).

FAD specifically protects against enzyme inactivation and is required for the NAD(P)H-diaphorase activity (Vega, 1976). In addition, complete reduction of the enzyme-bound heme by NAD(P)H only occurs in the presence of FAD (Lafferty and Garrett, 1974; Vega et al., 1975). The heme chromophore of NAD(P)H-nitrite reductase was also identified as siroheme (Vega et al., 1975). The nitrite reductase-bound siroheme can be reduced by the physiological electron donor NAD(P)H plus FAD. Nitrite can form a complex with the reduced enzyme which shows absorption bands at 560 and 585 nm, thus indicating that siroheme may be involved in binding the substrate. Inhibition studies with cyanide, sulfite, arsenite and especially carbon monoxide support this point of view (Vega et al., 1975). No experimental evidence has yet been reported indicating the presence of an iron-sulfur center as a prosthetic group of fungal nitrite reductase.

As was the case with ferredoxin-nitrite reductase, present data also suggest that nitrite is reduced to ammonia by NAD(P)H-nitrite reductase while held as a ligand to the siroheme, no free nitrogen compounds of intermedia redox state having been detected (Garrett and Amy, 1978). Although the precise nature of the 6-electron transfer by which this transformation is achieved remains still unknown, the following minimum linear scheme of electron flow in assimilatory *Neurospora* nitrite reductase may be proposed:

$$\mathrm{NAD(P)H} \longrightarrow \{\mathrm{FAD} \longrightarrow \mathrm{siroheme}\} \longrightarrow \mathrm{NO_2^-}.$$

In summary, both ferredoxin-nitrite reductase and NAD(P)H-nitrite reductase share in common siroheme as the functional prosthetic group where nitrite is bound and reduced to ammonia. There are, however, basic differences between these two enzymes that should be emphasized: (1) specificity for electron donor; (2) development in nonphotosynthetic organisms of an additional FAD-dependent NAD(P)H-diaphorase moiety, which is required for the electron transfer from reduced pyridine nucleotide to the nitrite reducing moiety, and (3) molecular size.

Information regarding the assimilatory nitrite reductase from bacteria is still very scanty. The enzyme from *E. coli* (Coleman et al., 1978; Kemp and Atkinson, 1966) and *Azotobacter* (Vega et al., 1973) exhibits a high specificity for NADH as electron donor. In spite of its marked in vitro instability Coleman et al. (1978) have achieved purification until homogeneity of the *E. coli* enzyme, which presents a molecular weight of 190 kdaltons and is composed of two similar subunits. Apparently the bacterial enzyme differs in composition from the fungal enzyme, since no absorbance maximum was found in the visible region and only flavin was detected. However, iron has been reported to have a functional role in the formation of active nitrite reductase in *Azotobacter* (Guerrero and Vega, 1975).

3 Nitrate Reduction in Light and Darkness

Nitrate is the main source of nitrogen for higher plants and can also be utilized by algae and a variety of fungi and bacteria as an alternate source of nitrogen. It is nowadays widely accepted that, although these types of living cells can use, in general, a great diversity of reducing compounds, either inorganic, like molecular hydrogen, or organic, like carbohydrates and organic acids, for the dark reduction of nitrate to ammonia, only green cells, in particular, can use water as reductant in the light. More precisely, it can be asserted at the present time that all the reducing power utilized by every green plant, including its nongreen tissues, and by the living world on the whole, comes ultimately from water. Although secondary to a certain degree, the question remains, however, debatable in some cases how directly the reduced coenzymes generated photosynthetically can couple in vivo with the nitrate-reducing system.

Scientists, predisposed by historical reasons, have habitually accepted that the reduction of nitrate in green cells is a process far removed from the light reactions of photosynthesis and dependent on organic substrates as the immediate source of reducing power. Now, it is well established (Losada, 1979; Losada and Guerrero, 1979) that the photosynthetic reduction of nitrate is, at least in certain aspects, even more directly coupled to the light reactions of photosynthesis than that of carbon dioxide. It should, moreover, be realized that the range of plant physiological events depending on light energy is in no way limited to carbon dioxide assimilation or even to those processes which take place directly inside the chloroplasts or the green cells. Nitrogen and carbon metabolism are nevertheless so much intertwined from their initial inorganic stages that they cannot easily be independently analyzed. We certainly owe to the work of Warburg (1964), in spite of some of his pertinacious misinterpretations, much of our knowledge on the not yet completely clarified role played by CO_2 in the photosynthetic reactions implied in nitrate uptake, reduction and assimilation

3.1 Nitrate Reduction in Vivo

For their versatility in being able to use different sources of energy, carbon and nitrogen, and their susceptibility to be cultured as cell suspensions and to be examined under precise experimental conditions, the contribution of microorganisms to our knowledge of the mechanism and control of nitrate assimilation can hardly be exaggerated, thus deserving an especial and detailed prospect.

In 1920 Warburg and Negelein made, under rather unphysiological conditions, the fundamental discovery that living *Chlorella* cells suspended in a CO_2-free highly acid "nitrate mixture," containing 0.1 *M* KNO_3 and 0.01 *M* HNO_3, evolved oxygen in the light with the concomitant reduction of nitrate to ammonia, according to the equation:

$$HNO_3 + H_2O \xrightarrow{\text{light}} NH_3 + 2O_2 .$$

Since, however, the same authors found that the alga cells could also reduce nitrate in darkness with the simultaneous oxidation of some of their organic components to CO_2, they interpreted the photochemical reduction of nitrate to ammonia on the basis of a dark reaction which yielded CO_2 and a photolysis of CO_2 which yielded O_2, according to the following equations:

$$HNO_3 + 2C + H_2O \xrightarrow{\text{dark}} NH_3 + 2CO_2$$

$$2CO_2 \xrightarrow{\text{light}} 2C + 2O_2$$

$$\overline{HNO_3 + H_2O \xrightarrow{\text{light}} NH_3 + 2O_2 .}$$

In the dark, the rate of O_2 uptake increased about 40% when the alga cells were transferred to the nitrate mixture, but the rate of CO_2 output increased still more to give a CO_2/O_2 quotient of about 1.5. The carbon dioxide production in excess of the oxygen uptake was labeled "extra CO_2," and, after a period of several hours, was stoichiometric with the amount of ammonia excreted to the medium, in accordance with the equation written above. Upon illumination with saturating light, the algal suspension in the nitrate mixture showed an extra-O_2 evolution and excretion of ammonia at rates two or three times faster than the extra-CO_2 and ammonia produced in the dark, even though no CO_2 was provided. The acceleration of nitrate reduction in the light was explained in terms of increased permeability to HNO_3, which allowed acceleration of the intracellular reduction of nitrate by the organic cell constituents.

Cramer and Myers (1949) extended the work of Warburg and Negelein (1920) to conditions of normal photosynthetic growth, in which the product of nitrate reduction is subsequently assimilated in cell material instead of being excreted. *Chlorella* grew rapidly in the dark with glucose as the only carbon source, and nitrate reduction was noted as an increase in the CO_2/O_2 quotient to about 1.6. Uptake of nitrate was characteristically accompanied by an increase in pH, whereas uptake of ammonium decreased the pH, as previously reported by Pratt and Fong (1940). Under conditions of light limitation, the quotient CO_2/O_2 was much smaller in the presence (0.6–0.8) than in the absence of nitrate (0.9–1.0). This change was caused by a decreased rate of CO_2 uptake, the rate of oxygen evolution being identical in the two cases. Since elementary analysis of *Chlorella* yielded about 50% C and 10% N, on an ash-free dry basis, Cramer and Myers (1949) concluded that, in the overall photosynthetic metabolism, nitrate reduction and carbon dioxide reduction were competing processes, which shared the reducing power supplied by the photochemical reactions in comparable proportions, i. e., 1 versus 2.2. Cramer and Myers (1949) confirmed also the important observation of Pratt and Fong (1940) that, although *Chlorella* will assimilate nitrogen from either nitrate or ammonium, when both sources of inorganic nitrogen are supplied together only ammonium is utilized and no nitrate reduction occurs.

The formulation by Van Niel (1941) that photosynthesis consists essentially in the photolysis of H_2O, thus relegating the assimilation of CO_2 to nonphotochemical reactions, led him to postulate that the photochemical reduction of nitrate, discovered by Warburg and Negelein (1920), could be interpreted as a process in which nitrate acts directly as an alternate hydrogen acceptor, in a reaction essentially analogous to the photochemical reduction of CO_2, with nitrate replacing CO_2. In 1953 Van Niel et al. demonstrated in fact that, at high light intensity, suspensions of *Chlorella* cells, supplied with nonlimiting concentrations of CO_2, produced oxygen at a greater rate when nitrate was simultaneously present, the quotient CO_2/O_2 being considerably lower than in the absence of nitrate, even though the rate of CO_2 assimilation was not reduced.

The CO_2 requirement claimed by Warburg (1964) for the in vivo reduction of nitrate was proved to be, however, unquestionable, at least at low pH. In 1965, Warburg repeated his classical experiments under more favorable conditions and found that illuminated cell suspensions of *Chlorella*, at pH 4.5, in argon free of CO_2 and O_2 only produced oxygen in the presence of nitrate when CO_2 was present (Warburg et al., 1965). In other words, if the CO_2 formed by the respiratory activity of the cells was removed with caustic potash, no O_2 was evolved. Leaving aside Warburg's interpretation, now known to be wrong, and since the same results were also obtained when ferricyanide substituted for nitrate, the phenomenon might perhaps be better ascribed, according to present knowledge, either to a CO_2 requirement for allowing the entrace of the substrates into the cell or corresponding cells site(s), or to its compulsory reduction by the light-reduced coenzymes to serve, in its turn, as an indispensable redox intermediate in the overall reducing process (see below). In this connection, the experiment of Davis (1953) was also relevant. Illuminated *Chlorella* suspensions in Warburg vessels at pH 5, without atmospheric CO_2 (KOH in the side arm), did not evolve O_2 in the presence of nitrate, and glucose, in combination with inorganic phosphate, was required for evolution of O_2 coupled to the reduction of nitrate. Using whole *Ankistrodesmus* cells suspended in a nitrate medium, at pH 6.5, with KOH in the center well, Kessler (1957a, 1964) corroborated the CO_2 requirement for nitrate photoreduction. This CO_2 effect might today be considered, in principle, different from two other CO_2 effects on the HILL reaction that are well established now (Govindjee and Van Rensen, 1978; Metzner, 1978): the CO_2 requirement for the noncyclic photosynthetic electron flow, precisely discovered by Warburg and Krippahl in 1958, and the stimulation of photophosphorylation by added bicarbonate (Punnet, 1965).

Kessler (1953) and Syrett (1955) demonstrated that, in the dark, not only the reduction of nitrate was accompanied by the production of extra-CO_2 by oxidation of carbohydrates but also the reduction of nitrite, according to the equations:

$$HNO_3 + 2(CH_2O) \xrightarrow{dark} NH_3 + 2CO_2 + H_2O$$

$$HNO_2 + 3/2(CH_2O) \xrightarrow{dark} NH_3 + 3/2CO_2.$$

The photochemical reduction of nitrite in green algae was discovered by Kessler (1957a) with *Ankistrodesmus*. In the light nitrite was reduced to ammonia with the concomitant evolution of stoichiometric amounts of O_2, according to the equation:

$$HNO_2 + H_2O \xrightarrow{light} NH_3 + 3/2O_2.$$

The reaction (Kessler, 1957a, 1964) was stimulated in the presence of CO_2, but in a lesser degree (two-fold) than the photoreduction of nitrate (ten-fold). In eukaryotic algae, the in vivo reduction of nitrite, which depends on ferredoxin, should certainly be more closely coupled to the light reactions of photosynthesis than the reduction of nitrate, which depends on pyridine nucleotide. Kessler and Zumft (1973) and Kulandaivelu et al. (1976) showed in fact that the addition of nitrite to intact cells of different algae species under various experimental conditions decreased chlorophyll fluorescence more markedly than did the addition of nitrate.

The photosynthetic reduction of nitrate and nitrite to ammonia by normal cells of the green alga *Scenedesmus* was achieved by Bongers (1958) in the absence of CO_2 at pH values between 7.5 and 9.0 (no alkali in the center well of the Warburg flask). Oxygen was produced at a fairly uniform rate for hours, and stoichiometric amounts of ammonia were concomitantly excreted to the medium. In a control suspension without oxidized nitrogen or carbon, the evolution of oxygen was negligible. In the light-limiting range (below 100 $W \cdot m^{-2}$) oxygen production was independent of the nature of the oxidant, and a linear relationship existed between oxygen output and light intensity. At saturation, the rates of oxygen evolution corresponding to nitrate, nitrite, and CO_2 were in the approximate proportion of 1:1.3:2.3.

Bongers' cells presumably lacked carbon reserves with which to make carbon skeletons for ammonia assimilation. As shown by Syrett and Morris (1963) and by Thacker and Syrett (1972a) carbon-depleted cells of *Chlorella* and *Chlamydomonas* reduced in the light nitrate to ammonia, which accumulated in the absence of CO_2 (pH 6.1–7.0; KOH in the center well of the Warburg vessel). When either CO_2 or glucose was present to provide a carbon source no ammonia accumulated since it was assimilated. This therefore confirmed as a very interesting phenomenon the initial observation made in fungi by Morton (1956) that the addition of ammonium has no effect on nitrate reduction with carbon-starved cells, which do not assimilate ammonium, but immediately inhibits nitrate assimilation when either CO_2 or glucose are present to provide a carbon source (Morris, 1974; Syrett and Leftley, 1976; Syrett and Morris, 1963; Thacker and Syrett, 1972a). Similar rapid effects of ammonium on nitrate uptake have been observed with other prokaryotic and eukaryotic algae, fungi, and higher plants. As will be discussed in Section 4, nitrate uptake and nitrate reduction are separate steps in nitrate assimilation, and the inhibition of nitrate utilization is not due to ammonia per se but to a product of its assimilation.

With regard to the CO_2 or glucose requirement for nitrate reduction at low pH observed by Warburg et al. (1965), Davis (1953), and Kessler (1957a, 1964) but not by Bongers (1958), Syrett and Morris (1963), and Thacker and Syrett (1972a) at higher pH, the observation of Eisele and Ullrich (1977) may be relevant. These authors studied the effect of pH and CO_2 or glucose on nitrate uptake and reduction by *Ankistrodesmus* cells in the light and showed that these processes were strongly inhibited when the pH decreased, but that this pH dependence was abolished in the presence of CO_2 or glucose. Grant and Turner (1969) have reported, nevertheless, that whereas *Chlorella* and *Tetraselmis suecica* assimilated some nitrate in light, at pH 7.8, in the absence of external CO_2, *Dunaliella tertiolecta* and *Phaeodactylum tricornutum* did not. All these species required CO_2 to assimilate nitrate at maximum rates in light (Grant and Turner, 1969) as also did normal cells of *Chlamydomonas* at pH 7 (Thacker and Syrett, 1972a).

The long-standing problem of the connection between photosynthesis and nitrate reduction has also been studied, using intact cells, in blue-green algae. In the light *Anabaena* reduced nitrate and nitrite to ammonia with the production of stoichiometric amounts of oxygen. CO_2 accelerated the reduction of nitrate and nitrite at high light intensities, and photoreduction of nitrite was markedly lowered by inhibitors of the noncyclic photosynthetic electron flow from water (Hattori, 1962). Nitrate reduction to nitrite by a mutant of the marine blue-green alga *Agmenellum,* which was impaired in its ability to reduce nitrite, was strictly light-dependent and only slightly stimulated by carbon dioxide (Stevens and Van Baalen, 1973). The photosynthetic reduction of nitrate by living cells of the blue-green alga *Anacystis* allows a continuous and prolonged production of ammonia in the presence of methionine sulfoximine (Ramos et al., 1980). As will be shown in Section 4, ammonium does not interfere with the uptake and reduction of nitrate when its further assimilation is blocked. Using methionine sulfoximine-treated *Anacystis* cells, Guerrero and Flores (1980) have determined that two mol of oxygen are evolved per mol of nitrate reduced (1.5 mol when nitrite is the substrate) in a process which is light-dependent and that does not take place in the absence of CO_2, even at high pH values.

Certain bacteria containing the enzyme hydrogenase are able to use molecular hydrogen for the reduction of nitrate and nitrite to ammonia (Stephenson and Stickland, 1931). The stoichiometry of these reactions was established by Woods in 1938. During the overall process the bacterium *Clostridium welchii* and two strains of *E. coli* consumed 4 molecules of hydrogen per molecule of nitrate reduced, whereas 3 molecules of hydrogen were consumed per molecule of nitrite reduced, in accord with the following partial and total equations:

$$HNO_3 + H_2 \longrightarrow HNO_2 + H_2O$$

$$HNO_2 + 3H_2 \longrightarrow NH_3 + 2H_2O$$

$$HNO_3 + 4H_2 \longrightarrow NH_3 + 3H_2O.$$

McNall and Atkinson (1957) obtained a strain of *E. coli* which could utilize either nitrate or nitrite as the source of nitrogen for growth and could reduce nitrate and nitrite to ammonia with the uptake of stoichiometric amounts of hydrogen.

Using the green algae *Ankistrodesmus* and *Scenedesmus,* which are able to activate hydrogen after an adaptation period of several hours, Kessler (1957b, 1964) showed the reduction of nitrate and nitrite to ammonia by molecular hydrogen in the dark and under anaerobic conditions. In such algae and certain strains of *Chlorella* (Damaschke and Lübke, 1958) the reduction of nitrite proceeded much faster than that of nitrate. At pH 6.3 CO_2 was not required for the nitrite reaction to proceed but, in its presence, the initial rate kept its constant value, which otherwise declined with time. The CO_2 effect was clearly catalytic since in the dark and under anaerobic conditions no CO_2 assimilation took place (Kessler, 1957b, 1964). By contrast, according to Stiller (1966) the reduction of nitrite with molecular hydrogen by *Chlorella* cells did go to completion at pH 6.4 without added CO_2 but at pH 4.5 was strictly dependent on CO_2 addition. Thus, a similar phenomenon to that described above for nitrate reduction is apparently responsible for this pH-dependent CO_2 effect on nitrite reduction in living cells.

In contrast to the green algae just mentioned, the blue-green algae *Anabaena* and *Anacystis* were unable to use molecular hydrogen for the reduction of nitrite in the dark. The reaction occurred, however, in the light (Hattori, 1963; Peschek, 1979).

With respect to the green tissues of higher plants, Burström (1943) claimed that leaves of wheat seedlings had a unique photosynthetic nitrate-reducing system, but Mendel and Visser (1951) reported that two types of systems were operative in tomato-leaf discs: one in the light, and another in the dark, depending on some organic substrates as electron donors. On the other hand, Delwiche (1951) found that the rate of nitrate assimilation in detached tomato leaves was not affected by the presence or absence of light, thus suggesting that no process specifically associated with photosynthesis was involved. However, more recently Canvin and Atkins (1974) studied the assimilation of nitrate and nitrite in leaves of barley, wheat, corn, and bean, and concluded that the assimilation of nitrate into amino-nitrogen was strictly dependent on light and ceased abruptly when the light was extinguished. Incorporation of nitrite was also markedly dependent on light. The photochemical reduction of nitrite coupled to stoichiometric evolution of oxygen in higher plants was discovered by Vanecko and Varner (1955) with wheat leaves and confirmed by Marré et al. (1953) using *Elodea* leaves and plants of *Wolfia.*

C_4 plants leaves are characterized by possessing two cell types, mesophyll and bundle-sheath cells, with distinct photochemical and biochemical functions. Apparently nitrate reduction occurs predominantly (Mellor and Tregunna, 1971; Neyra and Hageman, 1978; Rathnam and Das, 1974; Rathnam and Edwards, 1976) or exclusively (Harel et al., 1977; Moore and Black, 1979) in the mesophyll cells. Moore and Black (1979) reported that isolated crabgrass mesophyll cells were capable of light-dependent O_2 evolution in the presence of nitrite but not with nitrate.

In higher plants, nitrate reduction occurs not only in leaves but also in roots, embryos, cotyledons, scutella, aleurone cells, pollen grains, etc. (Beevers and Hageman, 1972; Hewitt et al., 1976). Willis and Yemm (1955) have described, in roots, the changes in O_2 absorption and CO_2 production which followed the addition of nitrate. With nitrogen-starved barley roots the process was characterized by an increase in CO_2 evolution which was greater than the increase in O_2 consumption. The capacity of roots to retain and to reduce nitrate within their tissue has been shown by Minotti and Jackson (1970) to be related to their carbohydrate content in wheat seedlings roots.

3.2 Nitrate Reduction at the Subcellular Level

Nitrate reduction to ammonia is by itself a dark exergonic reaction involving eight electrons and catalyzed by only two enzymes operating in series which depend on ferredoxin and/or pyridine nucleotides as electron donors (see Sect. 2). In blue-green algae both enzymes are ferredoxin-dependent and tightly bound to the chlorophyll-containing lamellae, whereas in fungi they depend on pyridine nucleotide and appear free in the cytoplasm. In eukaryotic algae and higher plants, only nitrite reductase is ferredoxin-dependent, whereas nitrate reductase is a pyridine nucleotide-enzyme. There seems to be general agreement about the intracellular localization of nitrite reductase as a soluble plastidial enzyme, but there is contradictory evidence regarding nitrate reductase (Beevers and Hageman, 1972; Garrett and Amy, 1978; Hewitt et al., 1976; Losada and Guerrero, 1979; Morris, 1974; Syrett and Leftley, 1976; Vennesland and Guerrero, 1979).

We shall now discuss the most relevant and pertinent facts regarding the light and dark reduction of nitrate and nitrite at the subcellular level. Perhaps the first and simplest one to be considered is that the electron carriers required for the nitrate-reducing enzymes, namely ferredoxin and pyridine nucleotide, can be reduced either in the light with water by *direct* coupling with the electron transport chain of noncyclic photophosphorylation or in the dark with molecular hydrogen or with reduced carbon compounds, during their dissimilatory dehydrogenation via glycolysis and the Krebs cycle. The second fact is that the reducing power generated either photosynthetically or by dark reactions has to be transferred, either by redox coenzymes, redox metabolites or reduced substrates, to the cell site(s) or tissues where the nitrate reduction process takes place.

Since, in 1953, Nason and coworkers isolated and characterized the nitrate reductase of *Neurospora* (Nason and Evans, 1953) and soybean leaves (Evans and Nason, 1953) as a NAD(P)H-dependent flavin-containing enzyme, the coupling between light and dark systems which can supply both reduced pyridine and flavin nucleotides and the nitrate reductase system has been reported by numerous authors.

Evans and Nason (1953) and Jagendorf (1956) first succeeded in coupling the photochemical reduction of $NADP^+$ by soybean and spinach grana with the reduction of nitrate by a purified nitrate reductase obtained from soybean leaves and *Neurospora*, respectively. It was further shown (Losada and Paneque, 1966; Paneque et al., 1969) that illuminated spinach grana, supplemented with spinach ferredoxin and $NADP^{\oplus}$ reductase could couple, in the presence of NAD^+ and nitrate reductase purified from spinach leaves, the reduction of nitrate to nitrite with the stoichiometric oxidation of water to oxygen. Losada's group showed moreover that flavin nucleotides could also mediate as electron carriers in the reduction of nitrate by nitrate reductase either in the dark, with an H_2-hydrogenase system (Del Campo et al., 1965) or an NADPH-NADP reductase system (Losada et al., 1965), or in the light, with an illuminated grana system (Losada et al., 1965; Ramirez et al., 1964). A possible more direct role of flavins in the photochemical reduction of nitrate was suggested by Stoy (1955) in order to explain the specially favorable influence of blue light upon nitrate reduction in wheat leaves, a phenomenon that he attributed to light absorption by a yellow pigment of the flavin type. Stoy (1956) himself observed that riboflavin could act in the absence of any other photochemically active pigment as a light-absorbing catalyst and supply electrons for the reduction of nitrate with the aid of purified nitrate reductase from wheat leaves. Stoy's demonstration that flavins can act by themselves as electron donor photosystems for the reduction of nitrate has been more recently corroborated by coupling with different nitrate-reducing enzymic systems (Candau et al., 1980; Llobell et al., 1980; Zumft et al., 1980).

Whether or not photochemically reduced pyridine and flavin nucleotides can directly couple with the nitrate reductase system in vivo is still a matter of controversy, as is the intracellular localization of the enzyme. Although nitrate reductase has been found to be present, at least in part, in the chloroplast (Coupé et al., 1967; Del Campo et al., 1963; Grant et al., 1970; Plaut and Littan, 1974; Rathnam and Das, 1974; Rathnam and Edwards, 1976), it is frequently referred to as a cytoplasmic enzyme (Dalling et al., 1972a; Ritenour et al., 1967; Wallsgrove et al., 1979). Alternatively, peroxisomes have been considered as the intracellular site of nitrate reductase (Lips and Avissar, 1972). Rathnam and Das (1974) claimed that nitrate reductase was associated with the chloroplast envelope membranes, but Wallsgrove et al. (1979) found no evidence for this kind of association. Lips and Avissar (1972), on their side, have suggested that nitrate reductase seems to be located on the membrane of the microbodies.

It deserves moreover to be mentioned that isolated intact chloroplasts have been shown to reduce nitrate in the light with a concomitant evolution of oxygen (Grant and Canvin, 1970; Heber and French, 1968; Swader and Stocking, 1971). According to these latter authors, however, the ability to reduce nitrate was lost after further purification unless a soluble component was added back to the resulting chloroplast preparation.

Assuming an extraplastidial localization of nitrate reductase, and in order to explain how the photosynthetically generated reducing power can move out of the chloroplasts, several chloroplasts-cytoplasmic shuttle systems, e. g., triose phosphate/phosphoglycerate, malate/oxaloacetate, have been proposed (Neyra and Hageman, 1978; Stocking and Larson, 1969).

Mortenson et al. (1962) and Valentine et al. (1963) found that ferredoxin participates in the reduction of nitrite to ammonia with H_2 by *Clostridium pasteurianum* preparations. Huzisige et al. (1963) and Losada et al. (1963) obtained from spinach leaves and spinach chloroplasts, respectively, a soluble ferredoxin-dependent enzyme which catalyzed the reduction of nitrite by illuminated spinach grana. The reduction of nitrite to ammonia in the light was shown to be a new type of noncyclic, ferredoxin-dependent photophosphorylation coupled with the formation of stoichiometric amounts of oxygen and ATP. Ferredoxin-nitrite reductase also catalyzed the reduction of nitrite in the dark with either an H_2-hydrogenase system or a NADPH-NADP reductase system (Losada et al., 1963; Paneque et al., 1964; Ramirez et al., 1966). Although the enzyme was localized in the chloroplasts, it leaked out when these organelles were repeatedly washed with an isotonic solution (Ramirez et al., 1966). Similar results were reported not only for spinach but also for vegetable marrow and maize by Betts and Hewitt (1966) and Joy and Hageman (1966).

The intracellular localization of ferredoxin-nitrite reductase in chloroplasts from different C_3 and C_4 plants seems now to be firmly established (Dalling et al., 1972a; Grant et al., 1970; Harel et al., 1977; Rathnam and Das, 1974; Rathnam and Edwards, 1976; Ritenour et al., 1967; Wallsgrove et al., 1979) although Lips and Avissar (1972) have claimed that it is located within the microbodies. Recent work from different laboratories has moreover asserted that intact chloroplasts isolated from a variety of plants are not only able to reduce nitrite in the light to ammonia but also to incorporate it into amino-nitrogen (Anderson and Done, 1978; Grant and Canvin, 1970; Magalhaes et al., 1974; Miflin, 1974; Neyra and Hageman, 1974; Plaut et al., 1977; Swader and Stocking, 1971).

With regard to nongreen tissues, the characterization and intracellular localization of nitrate and nitrite reductase is still a matter worth further investigation. It seems to be well established that in nonchlorophyllous cells both NAD(P)H-nitrate reductase and nitrite reductase are the typical enzymes also found in green cells, but the physiological reductant for nongreen tissue nitrite reductase remains unknown, since all attempts to extract ferredoxin from them have so far failed (Beevers and Hageman, 1972; Dalling et al., 1973; Hewitt et al., 1976; Ida et al., 1974; Sanderson and Cocking, 1964 a, b). The enzyme level in roots may be comparable with that found in leaves (Sanderson and Cocking, 1964 a, b) or even higher (Miflin, 1967). Whereas Dalling et al. (1972b) reported that in wheat roots nitrite reductase was localized in proplastids and nitrate reductase was a normal soluble enzyme, Miflin (1970) and Bourne and Miflin (1970) concluded that in barley roots both enzymes resided in a particle, nitrate reductase utilizing succinate as electron donor, and nitrite reduction apparently depending on ATP. However, Blevins et al. (1976) have objected to these results and suggested that they might be due to bacterial contamination.

Guerrero et al. (1971) have shown that the nitrate-reducing system operating in a colorless *Chlorella* mutant, aerobically grown in darkness with glucose, seems to be identical to that operating in normal cells autotrophically grown in the light. Ferredoxin, isolated from the dark-grown cells, was active in mediating the electron transfer from the NADPH-NADP reductase system to the nitrite-nitrite reductase system isolated from the same cells. On the other hand, Kessler et al. (1970) and Ahmed and Morris (1968) have claimed that the reduction of nitrite by green algae does require ATP. The mechanism of this reaction remains unknown both at the cellular and the subcellular levels (Kessler, 1964; Morris, 1974).

The stoichiometric photochemical reduction of nitrate and nitrite to ammonia with water as the electron donor by chlorophyll-containing particles of blue-green algae is of outstanding significance in the present situation because it shows in the simplest way and beyond any objection (Losada, 1979; Losada and Guerrero, 1979) that nitrate reduction can be directly coupled with the noncyclic electron transport chain of photosynthesis, i. e., without the concourse of carbohydrates as intermediates (Candau et al., 1976; Ortega et al., 1976). In these prokaryotic organisms both enzymes of the nitrate-reducing system are tightly bound to the photosynthetic lamellae and depend on ferredoxin as the electron donor (Hattori and Myers, 1966; Hattori and Uesugi, 1968; Guerrero et al., 1974; Manzano et al., 1976).

4 Regulation of Nitrate Utilization

The capacity for nitrate utilization in different organisms has been shown to vary in response to changes in environmental conditions such as light, temperature, pH, CO_2, and oxygen levels, nitrogen source, and other factors (Beevers and Hageman, 1972; Garrett and Amy, 1978; Hewitt et al., 1976; Losada and Guerrero, 1979; Morris, 1974; Syrett and Leftley, 1976; Vennesland and Guerrero, 1979). Many of the reports regarding these effects do not allow, however, a direct conclusion about the nature of the control mechanisms operating in each case. Several control mechanisms acting at different levels seem to contribute to the overall regulation of nitrate utilization in order to ensure both the proper operation of this relevant process, upon which the provision of reduced nitrogen rests, and its poise with other related metabolic pathways.

Regulation of nitrate reduction has an important relative weight in the overall regulation of nitrate utilization. Mechanisms for controlling in vivo nitrate reduction include regulation of enzyme synthesis and degradation, and of enzyme activation and inactivation. Besides, the availability of the specific reductants and the competence with other metabolic processes that simultaneously make use of them will doubtless limit the rate of in vivo nitrate (and nitrite) reduction. Feedback inhibition of the nitrate-reducing pathway by the reaction products of both enzymes and by metabolites implied in the assimilation of ammonium might equally have a regulatory role in nitrate reduction, as could also be the case with intermediate metabolites or final products of different pathways, especially of those related with carbon metabolism.

Total activity of nitrite reductase in different cells and tissues appears to be several-fold higher than that of nitrate reductase and, accordingly, accumulation of nitrite is seldom observed. The reduction of nitrate, rather than that of nitrite, seems thus more likely to be the rate-controlling step in nitrate reduction. These and other considerations have led Beevers and Hageman (1972) to conclude that nitrate reductase is the logical point to effect regulation of the input of reduced nitrogen for the organism. In fact, nitrate reductase appears to be a target for many of the control mechanisms of nitrate utilization, and most of the studies on regulation of nitrate utilization have been focused on this enzyme. Nevertheless, as more evidence is being gained about the participation of specific transport systems in the entrace of nitrate into cells and tissues, the significance of the nitrate uptake step in the regulation of nitrate utilization is presently starting to be widely recognized. Once nitrate enters the organism across the barrier of the plasma membrane, and before it is reduced, it can be confined into storage organelles or be transferred to the site where reduction takes place. The occurence of these intermediate steps between uptake and reduction, which are more significant as the complexity of the organism increases, needs also to be considered, as both compartmentation and translocation can also play a role in controlling the availability of the substrate for the enzymes of the nitrate-reducing pathway.

The mechanisms by which different factors affect nitrate utilization mainly through their effects on nitrate uptake, on nitrate reduction, or on both processes, will be considered below. Especial attention will be dedicated to the effects of ammonium, the end product of nitrate reduction, which, while exhibiting no inhibitory effects in vitro on the activity of nitrate reductase and nitrite reductase, behaves in vivo as a very active antagonist of nitrate metabolism in practically all types of organisms. Although much information is presently available on this so-called "ammonium effect," the molecular basis of the mechanism(s) of ammonium action is far from being fully understood (Beevers and Hageman, 1972; Garrett and Amy, 1978; Hewitt et al., 1976; Losada and Guerrero, 1979; Morris, 1974; Syrett and Leftley, 1976; Vennesland and Guerrero, 1979). Nevertheless, it is becoming more and more evident that ammonium assimilation rather than ammonium itself seems to be responsible for many facets of ammonium inhibition on nitrate utilization. This fact emphasizes once more the tight connection existing between nitrate reduction and the further assimilation of the resulting ammonium.

4.1 Control of Nitrate Uptake

Estimates of the high affinity of whole algal cells for nitrate (Eppley and Rogers, 1970; Syrett and Leftley, 1976) and observations that some cells and tissues can accumulate nitrate at concentrations far higher than those in the outer medium have been taken as evidence supporting the participation of effective uptake mechanisms in nitrate utilization. In many cases, especially in unvacuolated unicellular organisms, such as many algae, nitrate reduction almost immediately follows nitrate uptake; therefore, the separate investigation of the nitrate uptake step usually represents a difficult task. This close relationship between nitrate uptake and nitrate reduction has led to the suggestion that both processes are mechanistically linked, being functions of the same molecule, i. e., nitrate reductase (Butz and Jackson, 1977). Current evidence strongly argues against this idea, since a close correlation between nitrate uptake and nitrate reduction does not always exist. Moreover, nitrate uptake has been shown to take place in cells which do not contain a functional nitrate reductase (Doddema et al., 1978; Heimer and Filner, 1970, 1971; Schloemer and Garrett, 1974).

The approach of using tungstate-treated cells or nitrate reductase-less mutants has allowed the study of the properties of nitrate uptake systems in tobacco XD cells (Heimer and Filner, 1970, 1971) and in the fungus *Neurospora* (Schloemer and Garrett, 1974). Taking advantage of the differential behavior of uptake and reduction, the characteristics of nitrate uptake systems have also been investigated in different cells containing active nitrate reductase, such as algae (Calero et al., 1980; Eisele and Ullrich, 1977; Flores and Guerrero, 1980; Serra et al., 1978), fungi (Goldsmith et al., 1973), and in higher plants seedlings (Chantarotwong et al., 1976; Jackson et al., 1973; Minotti and Jackson, 1970; Rao and Rains, 1976), and roots (Neyra and Hageman, 1975), among others. In all instances, the process of nitrate uptake appears to have an active component which requires metabolic energy.

For fungi (Goldsmith et al., 1973; Schloemer and Garrett, 1974) and a variety of higher plant species (Ashley et al., 1975; Heimer and Filner, 1971; Jackson et al., 1973; Neyra and Hageman, 1975) nitrate has been reported to behave as an inducer of the nitrate uptake system. Nevertheless, high linear rates of nitrate uptake have been observed to develop very rapidly upon nitrate addition to nitrogen-starved algae (Syrett and Hipkin, 1973), tobacco cells (Behrend and Mateles, 1975), maize roots (Ivanko and Ingversen, 1971), and seedlings of wheat (Ashley et al., 1975; Minotti and Jackson, 1970), and *Arabidopsis* (Doddema et al., 1978), indicating that the system might be either constitutive or subjected to repression by ammonium or products of its assimilation. Nitrate uptake activity in ammonium-treated *Penicillium* mycelium developed rapidly in the presence of nitrate, but the reappearance also took place in medium without added nitrogen, although more slowly and to a lower level than with nitrate (Goldsmith et al., 1973). Different amino acids, but not urea or ammonium, repressed the formation of the nitrate transport system in tobacco cells (Heimer and Filner, 1971), and *Neurospora* (Schloemer and Garrett, 1974), whereas no nitrate uptake activity developed in *Penicillium* (Goldsmith et al., 1973) in the presence of ammonium even when nitrate was simultaneously present.

In addition to this inhibitory effect on the development of the nitrate uptake system, the addition of ammonium or certain amino acids to *Penicillium* mycelium in nitrate medium resulted in rapid inactivation of the nitrate uptake system. The decay was protected by cycloheximide, suggesting the involvement of proteins in the inactivating process and/or ammonium-mediated repression of the synthesis of a protein component of the nitrate uptake system with a high turnover rate (Goldsmith et al., 1973).

The prevention by ammonium of nitrate utilization is a widely recognized fact, whose occurrence is far from being restricted to fungal systems (Morton, 1956), information about this antagonic effect of ammonium on nitrate uptake being available for a large variety of organisms, such as the duckweed *Spirodela* (Ferguson and Bollard, 1969), seedlings of wheat (Minotti et al., 1969), rice (Sasakawa and Yamamoto, 1978; Shen, 1969), and barley (Rao and Rains, 1976), cell cultures of rose (Mohanty and Fletcher, 1976), tobacco (Heimer and Filner, 1971), soybean and wheat (Bayley et al., 1972), diatoms (Cresswell and Syrett, 1979; Eppley and Rogers, 1970; Serra et al., 1978), green algae (Cramer and Myers, 1949; Pistorius et al., 1978; Pratt and Fong,

1940; Syrett and Leftley, 1976; Syrett and Morris, 1963; Thacker and Syrett, 1972a; Tischner and Lorenzen, 1979), and blue-green algae (Flores, Guerrero and Losada, in preparation; Ohmori et al., 1977). Exceptions to the general rule have been noted, however, for the systems of *Neurospora* (Subramanian and Sorger, 1972), barley roots (Smith and Thompson, 1971), and plants of the duckweed *Lemna minor* (Orebamjo and Stewart, 1975a). Notwithstanding, Schloemer and Garrett (1974) have reported that ammonium behaved as a noncompetitive inhibitor with respect to nitrate of the *Neurospora* nitrate transport system.

Leaving aside these latter cases, the general pattern is repetitively observed that uptake of nitrate from solutions containing both ammonium and nitrate only starts once ammonium has been exhausted or its concentration greatly reduced (Bailey et al., 1972; Ferguson and Bollard, 1969; Mohanty and Fletcher, 1976; Morton, 1956; Shen, 1969; Syrett and Morris, 1963). Several workers have suggested that such an inhibition might be exerted at the level of nitrate reductase activity (e. g., Ferguson, 1969; Morton, 1956; Syrett and Morris, 1963), but the consensus has been recently reached that, independently of causing inactivation or repression of nitrate reductase in a latter stage (see Sect. 4.2), the presence of ammonium leads to an immediate inhibition of nitrate uptake. This has been particularly well studied in prokaryotic and eukaryotic algae where the addition of low concentrations of ammonium to cells actively consuming nitrate results in a prompt cessation of nitrate uptake (Cramer and Myers, 1949; Flores, Guerrero and Losada, in prep.; Ohmori et al., 1977; Pistorius et al., 1978; Syrett and Morris, 1963; Syrett and Leftley, 1976; Thacker and Syrett, 1972a; Tischner and Lorenzen, 1979), without a significant change in the cellular level of nitrate reductase activity for the time that nitrate utilization remains inhibited (Flores, Guerrero and Losada, in prep.; Ohmori et al., 1977; Pistorius et al., 1978; Syrett and Leftley, 1976; Tischner and Lorenzen, 1979). The inhibition is typically reversible, and the ability of the cells to utilize nitrate is restored immediately after exhaustion of ammonium in the outer medium, both in green (Cramer and Myers, 1949; Syrett and Morris, 1963; Syrett and Leftley, 1976; Thacker and Syrett, 1972a) and blue-green algae (Flores, Guerrero and Losada, in prep.; Ohmori et al., 1977). Again, similar effects are observed with fungi (Morton, 1956). Confirming previous suggestions, Cresswell and Syrett (1979) have clearly demonstrated the nitrate transport step to be the target of the short-term ammonium inhibition of nitrate utilization. Using cells of the diatom *Phaeodactylum tricornutum,* these authors have shown that the presence of ammonium in the medium produces an immediate reversible inhibition of nitrate accumulation within the cells.

The short-term inhibitory effect of ammonium on nitrate uptake by algae appears to obey ammonium assimilation rather than ammonium itself. The proposal of an organic product of ammonia assimilation as mediator of ammonium in causing inhibition of nitrate assimilation was raised by Morton (1956) and Syrett and Morris (1963) on the ground of a lack of effect of ammonium on nitrate uptake by carbohydrate-starved cells unless a carbon source was added. Recent experiments carried out with blue-green algae (Flores, Guerrero and Losada, in prep.) have shown that in the presence of specific inhibitors of ammonium assimilation (methionine sulfoximine and azaserine) ammonium does not exhibit any inhibitory effect on nitrate uptake. These results bespeak an involvement of ammonium incorporation into amino acids in the control of nitrate utilization. The regulation of nitrate uptake by ammonium appears thus to rest on a highly responsive feed-back system for the modulation of the rate of nitrate transport in response to changes in pools of, as yet unidentified, metabolites involved in ammonium assimilation.

The CO_2-requirement for nitrate utilization by unicellular algae, already discussed in Section 3.1, might also be located in the transport of nitrate into the cells. In order to explain the low nitrate uptake activity at low pH values of *Ankistrodesmus* cells in the absence of CO_2 but not in its presence, Eisele and Ullrich (1977) have proposed the participation of a carrier-mediated transport of nitrate, the carrier being labile at low pH values but becoming stabilized by CO_2 (or glucose). Preliminary experiments carried out with cells of the blue-green alga *Anacystis,* in which nitrate utilization is strictly dependent upon the presence of CO_2 even at high pH values, also suggest the nitrate transport system to be the locus responsible for the CO_2-requirement. Methionine sulfoximine-treated *Anacystis* cells, which take up and reduce nitrate but excrete most of the resulting ammonium to the medium as their capacity for assimilating it is severely hampered, still exhibit a CO_2-requirement for nitrate uptake, whereas the photosynthetic reduction of nitrate carried out by algal membrane preparations freely proceeds at normal rates in the absence of CO_2 (Flores and Guerrero, 1980).

From the above considerations it can be concluded that nitrate uptake seems to be a very important factor in the regulation of nitrate utilization, determining the rate of the overall process even under conditions in which the cellular level of nitrate reductase activity remains unaffected. In addition to the above-mentioned data for the situation of presence of ammonium, some authors have also concluded that, under normal conditions, the rate of in vivo nitrate reduction is more a function of the rate of nitrate uptake than of the amount of nitrate reductase present in the cell or tissue (e. g., Chantarotwong et al., 1976; Eisele and Ullrich, 1977) and not the opposite, as was generally accepted previously. Nevertheless, for those higher plants in which nitrate reduction is totally or predominantly confined to the shoot, it appears that nitrate reduction coupled to malate synthesis in the shoot controls the uptake of nitrate by the roots, malate moving down to the root system in which it is oxidized, and yields bicarbonate which exchanges for nitrate (Benzioni et al., 1971; Kirkby and Armstrong, 1980). In other words, even though control of nitrate reduction through availability of substrate by nitrate uptake appears to be a common situation, nitrate reduction can also have a regulatory role in nitrate uptake in certain cases.

4.2 Control of Nitrate Reduction

It seems obvious that the activity level of the enzymes of the nitrate-reducing pathway, and particularly that of nitrate reductase in the cells or tissues has to play an important role in determining the actual rate of in vivo nitrate reduction. Nevertheless, as discussed above, the enzyme activity level is not the sole factor involved in determining the rate of the process, since situations can be envisaged, and in fact do occur, in which, despite the presence of a high activity level, the availability of the specific reductant or substrate can limit the potential for reducing nitrate. In the preceding item evidence has been especially provided showing how nitrate uptake can effectively limit the supply of substrate for nitrate reduction. In addition, translocation from the uptake site to the main reduction site, e. g., from root to shoot, can also regulate nitrate availability to nitrate reductase and hence the rate of nitrate reduction. Moreover, the existence of compartmentation within cells or tissues may also determine a limited availability of substrate. In this context, the existence of two different nitrate pools, namely the storage or nonmetabolic pool (perhaps in the vacuole) and the active or metabolic pool, is a widely accepted concept for the intracellular distribution of nitrate in higher plants (Ferrari et al., 1973; Hageman, 1979; Heimer and Filner, 1971; Oaks, 1979). It should be also noted that, in addition to its role as substrate, nitrate can also have a positive effect on stabilization of nitrate reductase against inactivation, and on stimulation of nitrate reductase synthesis and activity, at least in certain cases (see below).

A big range of reported stimulating effects of light on nitrate reduction (see e. g., Beevers and Hageman, 1972, and Sect. 3) appears to affect most, if not all, of the different factors considered above. Light plays a decisive role not only in the generation of reductants but also in determining the availability of nitrate to the site of reduction through positive effects on uptake, translocation, redistribution of pools and another energy-requiring processes, such as nucleic acid and protein synthesis. Many of these light effects, including those on enzyme synthesis and activation, can in fact be mimicked by added carbohydrates or other carbon compounds (see e. g., Oaks, 1979; Roth-Bejerano and Lips, 1973; Thacker and Syrett, 1972b). It has also been proposed that many light-induced effects connected with nitrate metabolism in higher plants, which include rhythmic changes in enzyme activity level, involve the participation of phytochrome and/or blue light-absorbing pigments (Jones and Sheard, 1979).

An indirect but rather significant estimation of the actual capacity for nitrate reduction in higher plant tissues appears to be that based in considering nitrate reduction as the difference between nitrate taken up and nitrate accumulated (Chantarotwong et al., 1976). The so-called in vivo assay procedure, but which should more properly be referred to as in situ assay, involves the measurement of nitrite excreted by intact tissues under dark-anaerobic conditions (Ferrari and Varner, 1970; Ferrari et al., 1973) and is widely employed at present, especially for higher plant tissues. Although this type of assay does not seem to provide a good assessment for nitrate reduction, it might, however, be adequate for the estimation of nitrate pools and nitrate reductase activity level. This procedure is, of course, much more accurate than in vitro activity assays in crude extracts, where many factors can affect the significance of the results. In situ assays involving freezing and thawing (Syrett, 1973) or treatments with organic solvents (Flores, Guerrero and Losada, in prep.) are at present successfully used for determinations of nitrate reductase activity in algal cells.

In any event, the detection of a decrease in enzyme activity in response to changes in environmental or physiological conditions does not necessarily imply the occurrence of a parallel change of the in vivo nitrate reduction, since the decreased activity level might still suffice to sustain the rate of nitrate reduction present under the initial conditions. This is not difficult to understand since in many circumstances the nitrate reductase activity level is in considerable excess of that actually required to account for the observed in vivo rates, especially when limiting factors others than nitrate reductase are determining the rate of nitrate reduction.

Changes in nitrate reductase activity level have been observed in response to the influence of a variety of factors (Beevers and Hageman, 1972; Garrett and Amy, 1978; Hewitt et al., 1976; Losada and Guerrero, 1979; Morris, 1974; Syrett and Leftley, 1976; Vennesland and Guerrero, 1979) but it has not always been possible to determine whether they were brought about by changes in the net amount of the enzyme protein or by changes in enzyme activity. In the following sections we will discuss available information about such observed changes that could give an insight into the mechanisms involved.

4.2.1 Enzyme Synthesis

In general, the level of nitrate reductase is high in organisms grown on nitrate and low if grown on ammonium, nitrite reductase usually following a similar pattern. Both reductases thus behave as adaptive enzymes, and, although an obligate requirement for nitrate as the specific inducer of both enzymes has been repeatedly claimed, it is now firmly established that the synthesis of nitrate and nitrite reductase can certainly take place in the absence of nitrate in different types of organisms.

De novo synthesis of the enzymes of the nitrate-reducing system has been clearly shown to occur in the absence of nitrate, e. g., in nitrogen-free media or in the presence of certain amino acids or other nitrogen compounds, in unicellular algae, both prokaryotic (Herrero et al., 1980) and eukaryotic (Diez et al., 1977; Herrera et al., 1972; Morris and Syrett, 1963; Rigano et al., 1974; Syrett and Hipkin, 1973; Vega et al., 1971). The assimilatory nitrate reductase of the aerobic bacterium *Acinetobacter* is also synthetized de novo upon transfer from ammonium to nitrogen-free medium (Villalobo et al., 1977). Likewise, in *Penicillium,* Morton (1956) has shown that nitrate reductase activity develops to a significant level upon transfer of ammonium-grown mycelium to nitrogen-free medium. In higher plants, nitrate reductase is usually regarded as a nitrate-inducible enzyme (Beevers and Hageman, 1972; Hewitt et al., 1976). Nevertheless, considerable enzyme levels are sometimes found in media from which nitrate is absent, as is the case in seedlings of maize (Champigny, 1963) and cotton (Radin, 1974), and in excised barley roots (Smith and Thompson, 1971). In tobacco XD cells, Heimer and Riklis (1979 a) have demonstrated an RNA synthesis-dependent increase in nitrate reductase activity following the removal of the nitrogen source from the medium, and have stated that the initial development of nitrate reductase activity in these cells is independent of nitrate. Müller and Grafe (1978) have also reported the presence of significant levels of nitrate reductase in tobacco cells grown in the absence of nitrate. Moreover the stimulating effect of nitrate on the nitrate reductase level of higher plants seems not to be specific of this compound, since de novo synthesis of the enzyme could also be elicited, in the absence of nitrate, by a variety of unrelated substances (Knypl, 1979) including cytokinins (e. g., Hirschberg et al., 1972; Kende and Shen, 1972) and organic nitro-compounds (Shen, 1972). Therefore, at least in these organisms, the proposed role of nitrate as an inducer is difficult to be maintained.

Although it has been proposed that light is required for the synthesis of nitrate reductase in photosynthetic tissues (Beevers and Hageman, 1972; Hewitt et al., 1976), it has been conclusively demonstrated that the enzyme is present at a considerable level in *Chlorella* and tobacco cells grown in the dark (Guerrero et al., 1971; Morris and Syrett, 1963; Müller and Grafe, 1978) and in etiolated barley leaves (Roth-Bejerano and Lips, 1973) in the presence of an adequate carbon source.

The mechanism by which nitrate (or other compounds) enhances the level of nitrate reductase protein is not well defined. Although for mold fungi the term "nitrate induction" of nitrate reductase is frequently used (Garrett and Amy, 1978), it has been proposed that in *Neurospora* the role played by nitrate is not that of a true inducer, since it appears not to be essential for the transcription of nitrate reductase genes, but rather that its presence results in both stimulation of transcription and protection of the enzyme against inactivation (Premakumar et al., 1979). Alternatively, it has been proposed that in *Aspergillus* nitrate reductase acts as a repressor of its own synthesis unless nitrate is present, in an example of autoregulation (Cove and Pateman, 1969). Using tobacco XD cells, Zielke and Filner (1971) provided an elegant demonstration that the increase in nitrate reductase level in the presence of nitrate implied de novo synthesis of the enzyme. Nevertheless, since nitrate reductase synthesis was also shown to occur under "noninducing" conditions, in which a decay in activity took place, these authors were unable to decide whether the effect of nitrate was due to an increased rate of synthesis or to a reduced decay of enzyme activity. Ingle (1968), working with radish cotyledons, could not detect any difference between presence or absence of nitrate with regard to incorporation of radioactive precursors into nitrate reductase protein, even though the activity was higher in the presence of nitrate.

The effects of inhibitors of transcription and translation on the nitrate-promoted increase in nitrate reductase level are sometimes contradictory. Whereas the nitrate-promoted development of nitrate reductase in some cases (e. g., Ingle, 1968; Jones et al., 1978; Shen, 1969) is sensitive to the action of inhibitors of both RNA and protein synthesis, the same is not true for other systems. For example, inhibition of RNA synthesis by actinomycin D did not result in prevention of the development of nitrate reductase in response to nitrate either in embryos of *Agrostemma* (Kende and Shen, 1972) or in germinating cotton seedlings (Radin, 1974). Cycloheximide usually inhibits increases in nitrate and nitrite reductase activity involving enzyme synthesis both in higher plants (Ingle, 1968; Jones et al., 1978; Sluiters-Scholten, 1973) and in eukaryotic algae (Hipkin and Syrett, 1977; Losada et al., 1970), indicating that both enzymes of the nitrate-reducing pathway are synthetized on 80 S cytoplasmic ribosomes.

The antagonic effect of ammonium on nitrate metabolism with regard to the synthesis of the enzymes of the nitrate-reducing pathway is evident in many organisms. Although the role of ammonium as a repressor of nitrate and nitrite reductase has been questioned, especially for higher plants (Beevers and Hageman, 1972; Hewitt et al., 1976), available data speak in favor of this compound, or products of its assimilation, as being active in the regulation of the level of the corresponding proteins through a repressive action on their synthesis. The situation is particularly clear for lower plants where ammonium usually overrides the stimulating effect of nitrate, enzyme activity being totally absent or present only at basal levels both in ammonium- or ammonium nitrate-containing medium. Such a situation holds in the case of some chemotrophic bacteria (Villalobo et al., 1977), unicellular blue-green algae (Herrero et al., 1980) and eukaryotic algae (Diez et al., 1977; Herrera et al., 1972; Losada et al., 1970; Rigano et al., 1974; Syrett and Hipkin, 1973; Vega et al., 1971), indicating repression of nitrate (and nitrite) reductase caused by the presence of ammonium. The stage of enzyme synthesis affected by ammonium or ammonium derivatives appears to be transcription in unicellular blue-green algae (Herrero et al., 1980; Stevens and Van Baalen, 1974), whereas a post-transcriptional control has been proposed to operate in green algae (Hipkin and Syrett, 1977), where it appears that ammonium-grown cells contain preformed mRNA for nitrate reductase synthesis. For fungi, it is accepted that the presence of ammonium leads to repression of nitrate reductase acting at the transcriptional level (Garrett and Amy, 1978; Premakumar et al., 1979). An organic nitrogenous compound whose generation from ammonium requires the participation of glutamine synthetase, rather than ammonium itself, seems to be the true repressor of nitrate reductase synthesis. "Ammonium repression" is no longer evident in cells without an active glutamine synthetase, such as certain *Neurospora* mutants (Dunn-Coleman et al., 1979; Premakumar et al., 1979) or methionine sulfoximine-treated *Anacystis* cells (Herrero et al., 1980). Premakumar et al. (1979) have suggested glutamine to be the putative nitrate reductase repressor. In addition, these workers have also shown that glutamine prevents the synthesis of the short half-life mRNA for nitrate reductase (Premakumar et al., 1978).

Also in certain higher plant systems, ammonium and/or amino acids have been shown to act as repressors of nitrate reductase synthesis. The presence of ammonium or certain amino acids usually leads to decreased nitrate reductase level, and these compounds have been shown to prevent the nitrate-promoted increase of nitrate reductase, not only in roots (Frith, 1972; Oaks, 1979; Oaks et al., 1977; Radin, 1975; Smith and Thompson, 1971), but also in cell cultures of tobacco, tomato and soybean (Bayley et al., 1972; Behrend and Mateles, 1975; Heimer and Filner, 1970), and in duckweed plants (Ferguson, 1969; Orebamjo and Stewart, 1975 a). Nevertheless, exceptions to this general rule have been observed in germinating rice seedlings (Shen, 1969) and cell cultures of wheat (Bayley et al., 1972) and rose (Mohanty and Fletcher, 1976). Ammonium repression of nitrate reductase has been conclusively shown to occur in plants of *Lemna* (Orebamjo and Stewart, 1975 a), and it has been proposed that the effect of ammonium is indirect through a product of its assimilation acting at the level of transcription. Post-transcriptional control by amino acids has otherwise been suggested to take place in cultured tobacco cells (Heimer and Riklis, 1979 b) where nitrate reductase repression by amino acids had been earlier recognized (Heimer and Filner, 1970).

From the information discussed above the contention can be held that ammonium-mediated repression is a widely distributed mechanism for the control of the synthesis of the enzymes of the nitrate-reducing system, although a stimulating effect of nitrate on enzyme synthesis, which could mean induction for certain groups of organisms, has also to be taken into account. In this context it is worth commenting on a differential way of regulating nitrate reductase synthesis which is observed in prokaryotes able to use dinitrogen as an alternate nitrogen source to nitrate, a behavior that could be of interest from the standpoint of Comparative Biochemistry and Physiology. From the available data on nitrate reductase activity level in relation to the nitrogen source, it can be concluded that nitrate reductase in diazotrophs, both aerobic chemotrophic bacteria and blue-green algae, appears to behave as a nitrate-inducible enzyme (Guerrero et al., 1973; Herrero, Flores and Guerrero, in prep.; Ohmori and Hattori, 1970), whereas ammonium-mediated repression seems to be the control mechanism operative in those species that are not able to fix molecular nitrogen (Herrero et al., 1980; Villalobo et al., 1977) but rather depend on nitrate as the nitrogen source in their natural environment.

4.2.2 Enzyme Activity

In addition to the long-term changes in the amount of enzyme that might be ascribed to controlled synthesis, the occurrence of more or less rapid variations of the nitrate reductase activity level in response to the presence of ammonium or to changes in particular environmental factors has likewise to be taken into account, since it also contributes decisively to determining the actual enzyme level. In some cases, proteolytic degradation seems to be involved, whereas reversible enzyme inactivation by specific protein factors or metabolites has been shown to be implied in others (Beevers and Hageman, 1972; Hewitt et al., 1976; Losada and Guerrero, 1979; Vennesland and Guerrero, 1979).

Nitrate reductase decay has been observed after transferring different fungi to ammonium or nitrogen-free medium. The ammonium-induced decay seems to depend upon protein synthesis (Goldsmith et al., 1973; Lewis and Fincham, 1970) and is protected by nitrate (Subramanian and Sorger, 1972). Specific nitrate reductase inactivators that exhibit protease activity have been detected in *Neurospora* (Sorger et al., 1978). The occurrence of proteases that preferentially inactivate nitrate reductase has also been shown in higher plant tissues, especially in roots (Wallace, 1978; Yamaya et al., 1980). The lower nitrate reductase levels found in older parts of the root appear to be correlated with higher activity level of inactivating factors (Oaks, 1979; Wallace, 1978). The participation of such factors is also thought to be involved in the rapid activity decay of nitrate reductase in leaf tissue in response to transferring leaves to darkness or CO_2-free atmosphere (Kannangara and Woolhouse, 1967; Travis et al., 1969).

Another type of proteinaceous nitrate reductase-inactivating factors that inhibit enzyme activity but do not seem to cause a degradation of the nitrate reductase protein has been found to be present in leaf tissue (Jolly and Tolbert, 1978; Sherrard et al., 1979), rice seedlings (Kadam et al., 1974), and cell cultures of different plant species (Yamaya and Ohira, 1976). Such factors seem to cause nitrate reductase inactivation through binding to the enzyme protein (Yamaya et al., 1980), and its action has been proposed to be involved in reversible activity changes in response to light-dark transitions (Jolly and Tolbert, 1978; Sherrard et al., 1979). In vivo reversible inactivation by ammonium of nitrate reductase in *Lemna* plants has also been explained by a reversible protein-protein interaction (Orebamjo and Stewart, 1975b).

Relevant information about reversible inactivation of nitrate reductase, especially in response to the presence of ammonium, has been obtained from work with algal systems (Losada and Guerrero, 1979; Vennesland and Guerrero, 1979) where the so-called "metabolic interconversion," implying the existence of the enzyme in two metabolically interconvertible forms, either active or inactive, has been clearly shown to occur. The relative proportion of each form appears to be dependent upon the intracellular concentration of specific metabolites which in turn are affected by changes in environmental conditions. Particularly clear is the effect of ammonium, whose presence in the medium leads to a complete conversion of the active to the inactive form in a middle-term effect (1–2 h), reactivation being achieved upon ammonium removal in a process independent of de novo protein synthesis. Inactivation implies activity loss of the terminal moiety of the NAD(P)H-nitrate reductase complex, and seems to involve reduction of a functional group (presumably molybdenum) since in vitro incubation of the active enzyme under reducing conditions also results in inactivation. The inactive nitrate reductase appears to be identical to that obtained in vivo, and in vitro reactivation of the inactive form, obtained by either in vivo or in vitro treatments, can be achieved upon treatment with oxidizing compounds, nitrate among them. This general behavior has been observed for the enzyme of different species of green algae (Diez et al., 1977; Gómez-Moreno et al., 1972; Herrera et al., 1972; Lorimer et al., 1974; Losada, 1976; Losada et al., 1970, 1973; Maldonado et al., 1973), the blue-green alga *Nostoc* (Ortega et al., 1977) and the thermophilic red alga *Cyanidium* (Rigano et al., 1974; Rigano and Aliotta, 1975). In the case of *Cyanidium*, treatments which caused loss of the diaphorase activity or affected the integrity of the enzyme protein have also been reported to reactivate the ammonium-inactivated nitrate reductase (Rigano et al., 1974; Rigano and Aliotta, 1975).

Regarding the mechanism of this ammonium-promoted inactivation of nitrate reductase, different interpretations have been raised. For *Cyanidium* it has been suggested that in the enzyme complex there is a labile moiety bearing an inhibition site for ammonium (or a product of its assimilation) which on destruction results in the reactivation of the enzyme (Rigano et al., 1974; Rigano and Aliotta, 1975). Recent experiments carried out with methionine sulfoximine-treated cells indicate that ammonium itself is not the actual effector of nitrate reductase inactivation in *Cyanidium* (Rigano et al., 1979). For the metabolic interconversion of nitrate reductase, Losada and coworkers have proposed that the ammonium effect results from its uncoupling action on photosynthetic phosphorylation, through the increase in the cellular levels of reducing power and ADP (Losada, 1976). Three main lines of evidence seem to support this proposal: (1) other uncouplers, such as methylamine and arsenate parallel the action of ammonium, but simultaneous inhibition of the noncyclic electron flow prevents the inactivation caused by either of these compounds, (2) enzyme inactivation following addition of methylamine is concomitant with an increase in the cellular levels of the reduced pyridine nucleotides and the uncharged form of adenine nucleotides, and (3) active nitrate reductase can be converted in vitro into the inactive form by NAD(P)H, with ADP showing a cooperative effect for this inactivation and nitrate protecting against it. The active form is regarded as an oxidized state of the enzyme, which is converted into the reduced inactive form under appropriate conditions (Losada, 1976; Losada and Guerrero, 1979). On the other hand, Vennesland and coworkers have proposed that the inactive form of nitrate reductase represents a cyanide complex of the reduced enzyme, as the in vivo inactivated enzyme shares common properties with the inactive form of the enzyme resulting from in vitro treatment with NADH and cyanide, and it releases stoichiometric amounts of cyanide upon reactivation in the presence of nitrate (Lorimer et al., 1974; Vennesland and Guerrero, 1979). In this context, Solomonson and Spehar (1977, 1979) have recently proposed a model for the ammonium-promoted inactivation of nitrate reductase in *Chlorella*, through the generation of cyanide.

Inactivation of *Chlamydomonas* nitrate reductase has been shown to take place in response to transfer of the cells to darkness (Thacker and Syrett, 1972b; Vega et al., 1980b). Activation-inactivation of nitrate reductase following light-dark transitions has also been observed in synchronous cultures of *Chlorella* (Hodler et al., 1972; Tischner and Hüttermann, 1978). The dark-inactivated enzyme of *Chlamydomonas* appears to have properties similar to those of the reduced form which results from ammonium treatment in vivo (Vega et al., 1980b).

Evidence of the possible occurrence of nitrate reductase reversible inactivation in higher plants without an involvement of protein factors has as yet received only scanty experimental support. Nevertheless, in vitro inactivation by incubation with NADH and reactivation by ferricyanide, properties which might be indicative of the occurrence of metabolic interconversion, has been reported for higher plant nitrate reductase (Leong and Shen, 1979; Palacián et al., 1974; Trinity and Filner, 1979; Wallace, 1975). Trinity and Filner (1979) have also shown that nitrate reductase extracted from nitrate-grown tobacco cells is partially inactive, activity increasing upon incubation under adequate conditions, and Benzioni and Heimer (1977) have reported that temperature affects reversibly in vivo nitrate reductase activity in barley leaf strips.

Aparicio et al. (1976) have reported that in vitro inactivated nitrate reductase from *Chlorella* and spinach leaves can be rapidly activated by blue light, but not by red light, and have suggested a physiological role for the quality of light in modulating nitrate reductase activity, the flavin prosthetic group being the light-absorbing pigment.

To date, interconversion between active and inactive forms of the second enzyme of the nitrate-reducing system, i. e., nitrite reductase, has only been reported to occur in vitro, using enzyme preparations of the bacteria *Azotobacter* and *E. coli*. The process is apparently related to a redox change of the enzyme protein (Vega et al., 1973).

Acknowledgments. The experimental work from the authors' laboratory reported herein was supported in part by grants from the Centro de Estudios de la Energia (Spain) and the Philips Research Laboratories (The Netherlands). The authors wish to thank Mrs. Antonia Friend for helpful secretarial assistance.

References

Ahmed J, Morris I (1968) Biochim Biophys Acta 162:32–38
Alef K, Klemme JH (1979) Z Naturforsch 34c:33–37
Amy NK, Rajagopalan KV (1979) J Bacteriol 140:114–124
Amy NK, Garrett RH, Anderson BM (1977) Biochim Biophys Acta 480:83–95
Anderson JW, Done J (1978) Plant Physiol 61:692–697
Aparicio PJ, Knaff DB, Malkin R (1975) Arch Biochem Biophys 169:102–107
Aparicio PJ, Roldán JM, Calero F (1976) Biochem Biophys Res Commun 70:1071–1077
Ashley DA, Jackson WA, Volk RJ (1975) Plant Physiol 55:1102–1106
Bayley JM, King J, Gamborg OL (1972) Planta 105:15–24
Beevers L, Hageman RH (1972) In: Giese AC (ed) Photophysiology, vol 7. Academic Press, New York London, p 85
Behrend J, Mateles RI (1975) Plant Physiol 56:584–589
Benzioni A, Heimer YM (1977) Plant Sci Lett 9:225–231

Benzioni A, Vaadia Y, Lips H (1971) Physiol Plant 24:288–290
Betts GF, Hewitt EJ (1966) Nature (London) 210:1327–1329
Blevins DG, Lowe RH, Staples L (1976) Plant Physiol 57:458–459
Bongers LHJ (1958) Neth J Agric Sci 6:79–88
Bothe H (1977) In: Trebst A, Avron M (eds) Encyclopedia of plant physiology, New series, vol 5, Photosynthesis I. Springer, Berlin Heidelberg New York, p 217
Bourne WF, Miflin BJ (1970) Biochem Biophys Res Commun 40:1305–1310
Bowen HJM (1966) In: Trace elements in biochemistry, Academic Press, London New York
Burns RC, Hardy RWF (1975) In: Nitrogen fixation in bacteria and higher plants. Springer, Berlin Heidelberg New York
Burström H (1943) Ann Agric Coll Sweden 11:1–50
Butz RG, Jackson WA (1977) Phytochemistry 16:409–417
Calero F, Ullrich WR, Aparicio PJ (1980) In: Senger H (ed) The blue light syndrome. Springer, Berlin Heidelberg New York, p 411
Cammack R, Hucklesby DP, Hewitt EJ (1978) Biochem J 171:519–526
Candau P (1979) PhD thesis. Univ Sevilla, Sevilla
Candau P, Manzano C, Losada M (1976) Nature (London) 262:715–717
Candau P, Manzano C, Guerrero MG, Losada M (1980) Photobiochem Photobiophys 1:167–174
Canvin DT, Atkins CA (1974) Planta 116:207–224
Cárdenas J, Barea JL, Rivas J, Gómez-Moreno C (1972) FEBS Lett 23:131–135
Champigny ML (1963) Physiol Veg 1:139–169
Chantarotwong W, Huffaker RC, Miller BL, Granstedt RC (1976) Plant Physiol 57:519–522
Coleman KJ, Cornish-Bowden A, Cole JA (1978) Biochem J 175:483–493
Coupé M, Champigny ML, Moyse A (1967) Physiol Veg 5:271–291
Cove DJ, Pateman JA (1969) J Bacteriol 97:1374–1378
Cramer M, Myers J (1949) J Gen Physiol 32:93–102
Cresswell RC, Syrett PJ (1979) Plant Sci Lett 14:321–325
Czygan FC (1965) Planta 64:301–311
Dalling MJ, Tolbert NE, Hageman RH (1972a) Biochim Biophys Acta 283:505–512
Dalling MJ, Tolbert NE, Hageman RH (1972b) Biochim Biophys Acta 283:513–519
Dalling MJ, Hucklesby DP, Hageman RH (1973) Plant Physiol 51:481–484
Damaschke K, Lübke M (1958) Z Naturforsch 13b:134–135
Davis EA (1953) Plant Physiol 28:539–544
De la Rosa MA, Diez J, Vega JM, Losada M (1980) Eur J Biochem 106:249–256
Del Campo FF, Paneque A, Ramirez JM, Losada M (1963) Biochim Biophys Acta 66:450–452
Del Campo FF, Paneque A, Ramirez JM, Losada M (1965) Nature (London) 205:387–388
Delwiche CC (1951) J Biol Chem 189:167–175
Diez J, Chaparro A, Vega JM, Relimpio AM (1977) Planta 137:231–234
Doddema H, Hofstra JJ, Feenstra WJ (1978) Physiol Plant 43:343–350
Dunn-Coleman NS, Tomset AB, Garrett RH (1979) J Bacteriol 139:697–700
Eisele R, Ullrich WR (1977) Plant Physiol 59:18–21
Eppley RW, Rogers JN (1970) J Phycol 6:344–351
Evans HJ, Nason A (1953) Plant Physiol 28:233–254
Ferguson AR (1969) Planta 88:353–363
Ferguson AR, Bollard EG (1969) Planta 88:344–352
Ferrari TE, Varner JE (1970) Proc Natl Acad Sci USA 65:729–736
Ferrari TE, Yoder OC, Filner P (1973) Plant Physiol 51:423–431
Fewson CA, Nicholas DJD (1961) Nature (London) 190:2–7
Flores E, Guerrero MG (1980) Abstr II Congr Fed Eur Soc Plant Physiol, Santiago (Spain) 1980
Frith GJT (1972) Plant Cell Physiol 13:1085–1090
Garrett RH, Amy NK (1978) Adv Microb Physiol 18:1–50
Giri L, Ramadoss CS (1979) J Biol Chem 254:11703–11712
Goldsmith J, Livoni JP, Norberg CL, Segel IH (1973) Plant Physiol 52:362–367
Gómez-Moreno C, Aparicio PJ, Palacián E, Losada M (1972) FEBS Lett 26:11–14
Govindjee, Van Rensen JJS (1978) Biochim Biophys Acta 505:183–213

Grant BR, Canvin DT (1970) Planta 97:227–246
Grant BR, Turner IM (1969) Comp Biochem Physiol 29:995–1004
Grant BR, Atkins CA, Canvin DT (1970) Planta 94:60–72
Greenbaum P, Prodouz KN, Garrett RH (1978) Biochim Biophys Acta 526:52–64
Guerrero MG, Flores E (1980) Abstr V Congr Photosyn, Kallitea (Greece) 1980
Guerrero MG, Gutierrez M (1977) Biochim Biophys Acta 482:272–285
Guerrero MG, Vega JM (1975) Arch Microbiol 102:91–94
Guerrero MG, Rivas J, Paneque A, Losada M (1971) Biochem Biophys Res Commun 45:82:89
Guerrero MG, Vega JM, Leadbetter E, Losada M (1973) Arch Mikrobiol 91:287–304
Guerrero MG, Manzano C, Losada M (1974) Plant Sci Lett 3:273–278
Guerrero MG, Jetschmann K, Völker W (1977) Biochim Biophys Acta 482:19–26
Hageman RH (1979) In: Hewitt EJ, Cutting CV (eds) Nitrogen nutrition of plants. Academic Press, London New York, p 591
Hall DO (1979) Sol Energy 22:307–328
Harel E, Lea PJ, Miflin BJ (1977) Planta 134:195–200
Hattori A (1962) Plant Cell Physiol 3:355–369
Hattori A (1963) In: Microalgae and photosynthetic bacteria. University Press, Tokyo, p 485
Hattori A, Myers J (1966) Plant Physiol 41:1031–1036
Hattori A, Myers J (1967) Plant Cell Physiol 8:327–337
Hattori A, Uesugi I (1968) In: Shibata K, Tamamiya A, Jagendorf AT, Fuller RC (eds) Comparative biochemistry and biophysics of photosynthesis. University Press, Tokyo, p 201
Heber U, French CS (1968) Planta 79:99–112
Heimer YM, Filner P (1970) Biochim Biophys Acta 215:152–165
Heimer YM, Filner P (1971) Biochim Biophys Acta 230:362–372
Heimer YM, Riklis E (1979a) Plant Sci Lett 16:135–138
Heimer YM, Riklis E (1979b) Plant Physiol 64:663–664
Herrera J, Paneque A, Maldonado JM, Barea JL, Losada M (1972) Biochem Biophys Res Commun 48:996–1003
Herrero T, Flores E, Guerrero MG (1980) Abstr II Congr Fed Eur Soc Plant Physiol, Santiago (Spain) 1980
Hewitt EJ, Hucklesby DP, Notton BA (1976) In: Bonner J, Varner JE (eds) Plant biochemistry. Academic Press, London New York, p 633
Hipkin CR, Syrett PJ (1977) J Exp Bot 28:1270–1277
Hirschberg K, Hübner G, Borriss H (1972) Planta 108:333–337
Ho CH, Ikawa T, Nisizawa K (1976) Plant Cell Physiol 17:417–430
Hodler M, Morgenthaler JJ, Eichenberger W, Grob EC (1972) FEBS Lett 28:19–21
Hucklesby DP, James DM, Banwell MJ, Hewitt EJ (1976) Phytochemistry 15:599–603
Huzisige H, Satoh K, Tanaka K, Hayasida T (1963) Plant Cell Physiol 4:307–322
Ida S, Mori E, Morita Y (1974) Planta 121:213–224
Ingle J (1968) Biochem J 108:715–723
Ivanko S, Ingversen J (1971) Physiol Plant 24:59–65
Jackson WA, Flesher D, Hageman RH (1973) Plant Physiol 51:120–127
Jagendorf AT (1956) Arch Biochem Biophys 62:141–150
Jolly SO, Tolbert NE (1978) Plant Physiol 62:197–203
Jolly SO, Campbell W, Tolbert NE (1976) Arch Biochem Biophys 174:431–439
Jones RW, Sheard RW (1979) In: Hewitt EJ, Cutting CV (eds) Nitrogen nutrition of plants. Academic Press, London New York, p 521
Jones RW, Abbott AJ, Hewitt EJ, Best GR, Watson EF (1978) Planta 141:183–189
Joy KW, Hageman RH (1966) Biochem J 100:263–273
Kadam SS, Gandhi AP, Sawhney SK, Naik MS (1974) Biochim Biophys Acta 350:162–170
Kannangara CG, Woolhouse HW (1967) New Phytol 66:553–561
Kemp JD, Atkinson DE (1966) J Bacteriol 92:628–634
Kende H, Shen TC (1972) Biochim Biophys Acta 286:118–125
Kessler E (1953) Flora 140:1–38
Kessler E (1957a) Planta 49:505–523

Kessler E (1957b) Arch Mikrobiol 27:166–181
Kessler E (1964) Annu Rev Plant Physiol 15:57–72
Kessler E, Zumft WG (1973) Planta 111:41–46
Kessler E, Hofmann A, Zumft WG (1970) Arch Mikrobiol 72:23–26
Kirkby EA, Armstrong MJ (1980) Plant Physiol 65:286–290
Knaff DB, Smith JM, Malkin R (1978) FEBS Lett 90:195–197
Knypl JS (1979) In: Hewitt EJ, Cutting CV (eds) Nitrogen nutrition of plants. Academic Press, London New York, p 141
Kulandaivelu G, Spiller H, Böger P (1976) Plant Sci Lett 7:225–231
Kuo T, Kleinhofs A, Warner RL (1980) Plant Sci Lett 17:371–381
Lafferty MA, Garrett RH (1974) J Biol Chem 249:7555–7567
Lancaster JR, Vega JM, Kamin H, Orme-Johnson NR, Orme-Johnson WH, Krueger RJ, Siegel LM (1979) J Biol Chem 254:1268–1272
Lea PJ, Miflin BJ (1979) In: Gibbs M, Latzko E (eds) Encyclopedia of plant physiology, New series, vol 6 Photosynthesis II. Springer, Berlin Heidelberg New York, p 445
Leong CC, Shen TC (1979) Experientia 35:584–585
Lewis CM, Fincham JRS (1970) J Bacteriol 103:55–61
Lips H, Avissar Y (1972) Eur J Biochem 29:20–24
Llobell A, Bárcena JA, Vila R, Paneque A (1980) Z Pflanzenphysiol 98:271–276
Lorimer GH, Gewitz HS, Völker W, Solomonson LP, Vennesland B (1974) J Biol Chem 249:6074–6079
Losada M (1976) J Mol Catal 1:245–264
Losada M (1979) Bioelectrochem Bioenerg 6:205–225
Losada M, Guerrero MG (1979) In: Barber J (ed) Photosynthesis in relation to model systems. Elsevier/North-Holland Biomedical Press, Amsterdam New York, p 365
Losada M, Paneque A (1966) Biochim Biophys Acta 126:578–580
Losada M, Paneque A, Ramirez JM, Del Campo FF (1963) Biochem Biophys Res Commun 10:298–303
Losada M, Ramirez JM, Paneque A, Del Campo FF (1965) Biochim Biophys Acta 109:86–96
Losada M, Paneque A, Aparicio PJ, Vega JM, Cárdenas J, Herrera J (1970) Biochem Biophys Res Commun 38:1009–1015
Losada M, Herrera J, Maldonado JM, Paneque A (1973) Plant Sci Lett 1:31–37
Magalhaes AC, Neyra CA, Hageman RH (1974) Plant Physiol 53:411–415
Maldonado JM, Herrera J, Paneque A, Losada M (1973) Biochem Biophys Res Commun 51:27–33
Malofeeva IV, Kondratieva EN, Rubin AB (1975) FEBS Lett 53:188–189
Manzano C (1977) PhD thesis. Univ Sevilla, Sevilla
Manzano C, Candau P, Gómez-Moreno C, Relimpio AM, Losada M (1976) Mol Cell Biochem 10: 161–169
Manzano C, Candau P, Guerrero MG (1978) Anal Biochem 90:408–412
Marré E, Forti G, Bianchetti R, Parisi B (1963) In: Colloques Internationaux CNRS no. 119, La photosynthèse, CNRS, Paris, p 557
McNall EG, Atkinson DE (1957) J Bacteriol 74:60–66
Mellor GE, Tregunna EB (1971) Can J Bot 49:137–142
Mendel JS, Visser DV (1951) Arch Biochem Biophys 32:158–169
Metzner H (1978) In: Metzner H (ed) Photosynthetic oxygen evolution. Academic Press, London New York, p 59
Miflin BJ (1967) Nature (London) 214:1133–1134
Miflin BJ (1970) Planta 93:160–170
Miflin BJ (1974) Planta 116:187–196
Minotti PL, Jackson WA (1970) Planta 95:36–44
Minotti PL, Williams DC, Jackson WA (1969) Planta 86:267–271
Mohanty B, Fletcher S (1976) Plant Physiol 58:152–155
Moore R, Black CC (1979) Plant Physiol 64:309–313
Morris I (1974) In: Stewart WDP (ed) Algal physiology and biochemistry. Blackwell, Oxford London Edinburgh Melbourne, p 583

Morris I, Syrett PJ (1963) Arch Mikrobiol 47:32–41
Mortenson LE, Valentine RC, Carnahan JE (1962) Biochem Biophys Res Commun 7:448–452
Morton AG (1956) J Exp Bot 7:97–112
Müller AJ, Grafe R (1978) Mol Gen Genet 161:67–76
Murphy MJ, Siegel LM, Tove SR, Kamin H (1974) Proc Natl Acad Sci USA 71:612–616
Nason A (1962) Bacteriol Rev 26:16–41
Nason A, Evans HJ (1953) J Biol Chem 202:655–673
Neyra CA, Hageman RH (1974) Plant Physiol 54:480–483
Neyra CA, Hageman RH (1975) Plant Physiol 56:692–695
Neyra CA, Hageman RH (1978) Plant Physiol 62:618–621
Notton BA, Hewitt EJ (1979) In: Hewitt EJ, Cutting CV (eds) Nitrogen assimilation of plants. Academic Press, London New York, p 227
Notton BA, Fido RJ, Hewitt EJ (1977) Plant Sci Lett 8:165–170
Oaks A (1979) In: Hewitt EJ, Cutting CV (eds) Nitrogen nutrition of plants. Academic Press, London New York, p 217
Oaks A, Aslam M, Boesel I (1977) Plant Physiol 59:391–394
Ohmori K, Hattori A (1970) Plant Cell Physiol 11:873–878
Ohmori M, Ohmori Z, Strotmann H (1977) Arch Microbiol 114:225–229
Orebamjo TO, Stewart GR (1975a) Planta 122:27–36
Orebamjo TO, Stewart GR (1975b) Planta 122:37–44
Orme-Johnson WH, Jacob GS, Henzi MT, Averill BA (1977) In: Raymon KN (ed) Bioinorganic chemistry II, Advances in chemistry series, vol 162. American Chemical Society, Washington, p 389
Ortega T, Castillo F, Cárdenas J (1976) Biochem Biophys Res Commun 71:885–891
Ortega T, Castillo F, Cárdenas J, Losada M (1977) Biochem Biophys Res Commun 75:823:831
Palacián E, De la Rosa F, Castillo F, Gómez-Moreno C (1974) Arch Biochem Biophys 161:441–447
Pan SS, Nason A (1978) Biochim Biophys Acta 523:297–313
Paneque A, Ramirez JM, Del Campo FF, Losada M (1964) J Biol Chem 239:1737–1741
Paneque A, Aparicio PJ, Cárdenas J, Vega JM, Losada M (1969) FEBS Lett 3:57–59
Peschek GA (1979) Biochim Biophys Acta 548:187–202
Pienkos PT, Shah VK, Brill WJ (1977) Proc Natl Acad Sci USA 74:5468–5471
Pistorius EK, Funkhouser EA, Voss H (1978) Planta 141:279–282
Plaut Z, Littan A (1974) In: Avron M (ed) Proceedings of the third international congress on photosynthesis. Sept 1974. Elsevier, Amsterdam, p 1507
Plaut Z, Lendzian K, Bassham JA (1977) Plant Physiol 59:184–188
Pratt R, Fong J (1940) Am J Bot 27:735–743
Premakumar R, Sorger GJ, Gooden D (1978) Biochim Biophys Acta 519:275–278
Premakumar R, Sorger GJ, Gooden D (1979) J Bacteriol 137:1119–1126
Punnet T (1965) Plant Physiol 40:1283:1284
Radin JW (1974) Plant Physiol 53:458–463
Radin JW (1975) Plant Physiol 55:178–182
Ramirez JM, Del Campo FF, Paneque A, Losada M (1964) Biochem Biophys Res Commun 15: 297–302
Ramirez JM, Del Campo FF, Paneque A, Losada M (1966) Biochim Biophys Acta 118:58–71
Ramos JL. Guerrero MG, Losada M (1980) Abstr II Congr Fed Eur Soc Plant Physiol, Santiago (Spain) 1980
Rathnam CKM, Das VSR (1974) Can J Bot 52:2599–2605
Rathnam CKM, Edwards GE (1976) Plant Physiol 57:881–885
Rao KP, Rains DW (1976) Plant Physiol 57:55–58
Rigano C, Aliotta G (1975) Biochim Biophys Acta 384:37–45
Rigano C, Aliotta G, Violante U (1974) Plant Sci Lett 2:277–281
Rigano C, Rigano VDM, Vona V, Fuggi A (1979) Arch Microbiol 121:117–120
Ritenour GL, Joy KW, Bunning J, Hageman RH (1967) Plant Physiol 42:233–237

Roth-Bejerano N, Lips H (1973) New Phytol 72:253–257
Rueger DC, Siegel LM (1976) In: Singer TP (ed) Flavins and flavoproteins. Elsevier Scientific Publishing Company, Amsterdam New York, p 610
Sanderson GW, Cocking EC (1964a) Plant Physiol 39:416–422
Sanderson GW, Cocking EC (1964b) Plant Physiol 39:423–431
Sasakawa H, Yamamoto Y (1978) Plant Physiol 62:665–669
Schloemer RH, Garrett RH (1974) J Bacteriol 118:259–269
Serra JL, Llama MJ, Cadenas E (1978) Plant Physiol 62:991–994
Shen TC (1969) Plant Physiol 44:1650–1655
Shen TC (1972) Plant Physiol 49:546–549
Sherrard JH, Kennedy JA, Dalling MJ (1979) Plant Physiol 64:640–645
Sluiters-Scholten CMTH (1973) Planta 113:229–223
Smith FW, Thompson JF (1971) Plant Physiol 48:219–240
Solomonson LP (1975) Plant Physiol 56:853–855
Solomonson LP (1979) In: Hewitt EJ, Cutting CV (eds) Nitrogen nutrition of plants. Academic Press, London New York, p 199
Solomonson LP, Spehar AM (1977) Nature (London) 265:373–375
Solomonson LP, Spehar AM (1979) J Biol Chem 254:2176–2179
Solomonson LP, Lorimer GH, Hall RL, Borchers R, Bailey JL (1975) J Biol Chem 250:4120–4127
Sorger GJ, Premakumar R, Gooden D (1978) Biochim Biophys Acta 540:33–47
Stephenson M, Stickland LH (1931) Biochem J 25:205–214
Stevens SE, Van Baalen C (1973) Plant Physiol 51:350–356
Stevens SE, Van Baalen C (1974) Arch Biochem Biophys 161:146–152
Stiller M (1966) Plant Physiol 41:348–352
Stocking CR, Larson S (1969) Biochem Biophys Res Commun 37:278–282
Stoller ML, Malkin R, Knaff DB (1977) FEBS Lett 81:271–274
Stoy V (1955) Physiol Plant 8:963–986
Stoy V (1956) Biochim Biophys Acta 21:395:396
Subramanian KN, Sorger GJ (1972) J Bacteriol 110:538–546
Swader JA, Stocking CR (1971) Plant Physiol 47:189–191
Syrett PJ (1955) Physiol Plant 8:924–929
Syrett PJ (1973) New Phytol 72:37–46
Syrett PJ, Hipkin CR (1973) Planta 111:57–64
Syrett PJ, Leftley JW (1976) In: Sunderland N (ed) Perspectives in experimental biology, vol 2. Pergamon Press, Oxford New York London Paris, p 221
Syrett PJ, Morris I (1963) Biochim Biophys Acta 67:566–575
Thacker A, Syrett PJ (1972a) New Phytol 71:423–433
Thacker A, Syrett PJ (1972b) New Phytol 71:435–441
Tischner R, Hütermann A (1978) Plant Physiol 62:284–286
Tischner R, Lorenzen H (1979) Planta 146:287–292
Tortolero M, Vila R, Paneque A (1975) Plant Sci Lett 5:141–145
Travis RL, Jordan WR, Huffaker RC (1969) Plant Physiol 44:1150–1156
Trinity PM, Filner P (1979) Plant Physiol 63:S 133
Valentine RC, Brill WJ, Wolfe RS, San Pietro A (1963) Biochem Biophys Res Commun 10:73–78
Vanecko S, Varner JE (1955) Plant Physiol 30:388–390
Van Niel CB (1941) Adv Enzymol 1:263–328
Van Niel CB, Allen MB, Wright BE (1953) Biochim Biophys Acta 12:67–74
Vega JM (1976) Arch Microbiol 109:237–242
Vega JM, Kamin H (1977) J Biol Chem 252:896–909
Vega JM, Herrera J, Aparicio PJ, Paneque A, Losada M (1971) Plant Physiol 48:294–299
Vega JM, Guerrero MG, Leadbetter E, Losada M (1973) Biochem J 133:701–708
Vega JM, Garrett RH, Siegel LM (1975) J Biol Chem 250:7980–7989
Vega JM, Cárdenas J, Losada M (1980a) Methods Enzymol 69:255–270
Vega JM, Florencio FJ, De la Rosa MA (1980b) Abstr V Congr Photosyn, Kallitea (Greece) 1980
Vennesland B, Guerrero MG (1979) In: Gibbs M, Latzko E (eds) Encyclopedia of plant physiology, New series, vol 6, Photosynthesis II. Springer, Berlin Heidelberg New York, p 425

Villalobo A, Roldán JM, Rivas J, Cárdenas J (1977) Arch Microbiol 112:127–132
Wallace W (1975) Biochim Biophys Acta 377:239–250
Wallace W (1978) Biochim Biophys Acta 524:418–427
Wallsgrove RM, Lea PJ, Miflin BJ (1979) Plant Physiol 63:232–236
Warburg O (1964) Annu Rev Biochem 33:1–14
Warburg O, Krippahl G (1958) Z Naturforsch 13b:509–514
Warburg O, Negelein E (1920) Biochem Z 110:66–115
Warburg O, Krippahl G, Jetschmann C (1965) Z Naturforsch 20:993–996
Willis AJ, Yemm EW (1955) New Phytol 54:163–181
Woods DD (1938) Biochem J 32:2000–2012
Yamaya T, Ohira K (1976) Plant Cell Physiol 17:633–641
Yamaya T, Oaks A, Boesel IL (1980) Plant Physiol 65:141–145
Zielke HR, Filner P (1971) J Biol Chem 246:1772–1779
Zumft WG (1972) Biochim Biophys Acta 276:363–375
Zumft WG, Castillo F, Hartman KM (1980) In: Senger H (ed) The blue light syndrome. Springer, Berlin Heidelberg New York

Dissimilatory Nitrate Reduction

F.R. WHATLEY[1]

1 Nitrate Respiration

A number of bacteria employ nitrate and other oxides of nitrogen not simply as a source of nitrogen for synthetic purposes leading to the production of proteins, nucleic acids and other nitrogen-containing compounds (assimilatory nitrate reduction) but rather as a terminal electron acceptor alternative to oxygen. Such organisms reduce nitrate to nitrite and further to dinitrogen gas (N_2) in dissimilatory nitrate reduction. They may be said to carry out a nitrate respiration (Haddock and Jones, 1977). The amount of energy available in going from the level of an "average" electron originating from an average substrate (e^-) to nitrate (NO_3^-) is approximately +18 kcal. when the final product is N_2. The energy available between e^- and sulfate (SO_4^{2-}) is +3.4 kcal. (H_2S as final product) and from e^- to CO_2 it is +2.4 kcal. (CH_4 as final product). Since the amount of energy available from e^- to nitrate is quite large it is perhaps not surprising to discover that there are a number of denitrifying bacteria capable of exploiting the use of nitrate as a terminal oxidant, and many more than sulfate reducers or methane formers. The pathway from nitrate to N_2 gas in nitrate respiration is summarised in Fig. 1 (Payne, 1973).

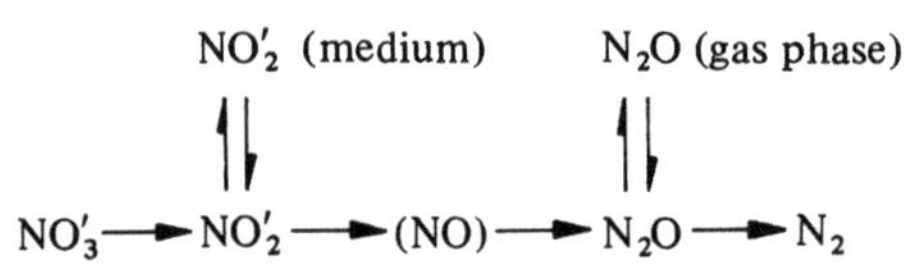

Fig. 1. The pathway of nitrate to dinitrogen gas in nitrate respiration

When *Paracoccus denitrificans* is allowed to grow on a nitrate medium in a closed environment, nitrite (NO_2^-) accumulates temporarily in the medium but can be reutilised as nitrate becomes exhausted. After the initial accumulation of nitrite, nitrous oxide can be detected in the gas space and may accumulate to quite a large extent. Subsequently N_2 gas is released as the final reduction product of nitrate in dissimilatory nitrate reduction. Figure 1 also suggests, and this can be experimentally realized, that *Paracoccus* and other denitrifiers can be grown with NO_3^-, NO_2^- or N_2O as terminal electron acceptor in place of oxygen. The intermediate (NO) has been detected as an in vitro product of the reduction of nitrite by isolated nitrite reductase (cytochrome

---

1 School of Botany, University of Oxford, South Parks Road, Oxford OX1 3RA, Great Britain

c.d.) but there is no evidence for its accumulation, temporary or otherwise, in the medium during a normal growth experiment. It is possible that it is always maintained as a bound intermediate on the nitrite reductase enzyme.

2 Induction of Enzymes of Nitrate Respiration

The enzymes of nitrate respiration are only formed where conditions for dissimilatory nitrate reduction are present, i. e., they are inducible. Anaerobic conditions alone are sufficient to induce all the enzymes of the dissimilatory nitrate reduction sequence but the presence of nitrate or nitrite intensifies the response and greater amounts of both nitrate and nitrite reductases are then formed. The response is also greatly intensified if sodium azide is added. This compound is not metabolized but is an inhibitor of nitrate reductase (Payne, 1973). There is no good evidence to show that the formation of nitrite is necessary for the induction of nitrite reductase even though nitrite accumulation is a common feature of growth of denitrifiers and many other bacteria in a growth medium. It is likely that the temporary nitrite accumulation in cultures is a control phenomenon and is not directly related to the total amount of enzyme protein synthesized. Incubation under anaerobic conditions in the presence of N_2O also leads to an enhanced production of both nitrate and nitrite reductases, indicating an operon for the whole nitrate respiration complex (Stouthamer, 1976a). In addition to the denitrifying bacteria many organisms, e. g., *Escherichia coli* K_{12}, are able to couple the single step reduction of nitrate to nitrite to their normal energy metabolism and appear to make use of this first step of nitrate reduction for the conservation of energy as ATP (Garland et al., 1975). Subsequently *E. coli,* which cannot reduce nitrite to nitrogen, reduces nitrite through the equivalent of an assimilatory pathway yielding ammonia (NH_3), and the latter is discarded into the medium. This is a nitrate ammonification reaction. (Schlegel, personal communication.)

3 Ecology of Denitrification

When a decomposing organism, e. g., a leaf, falls on the soil its nitrogen is initially released as ammonia (NH_3 or NH_4^+). The ammonia passes into the upper layers of the soil and because the clay particles are negatively charged NH_4^+ readily becomes adsorbed. In this form it is available for nitrogen uptake by plants (Fig. 2). However, this rich source of potential energy does not go unexploited and many bacteria, e.g., *Nitrosomonas* and *Nitrobacter,* co-operatively oxidise NH_4^+ to NO_3^- within the upper aerobic region of the soil. The nitrate is of course a negatively charged ion, NO_3^-, which is no longer adsorbed by the soil particles. This is also true of any nitrate supplied as an agricultural fertiliser. In the upper layers of the soil which can be exploited by the roots of crop plants (the "useful" zone) both ammonia and nitrate can be taken up by the plant and incorporated into proteins and other nitrogen-containing substances. In the plant, nitrate is first reduced to ammonia by assimilatory nitrate reductase and ammonia is used directly for incorporation into amino acids, proteins and other compounds. There are strong indications that nitrate is not always selected by the plant

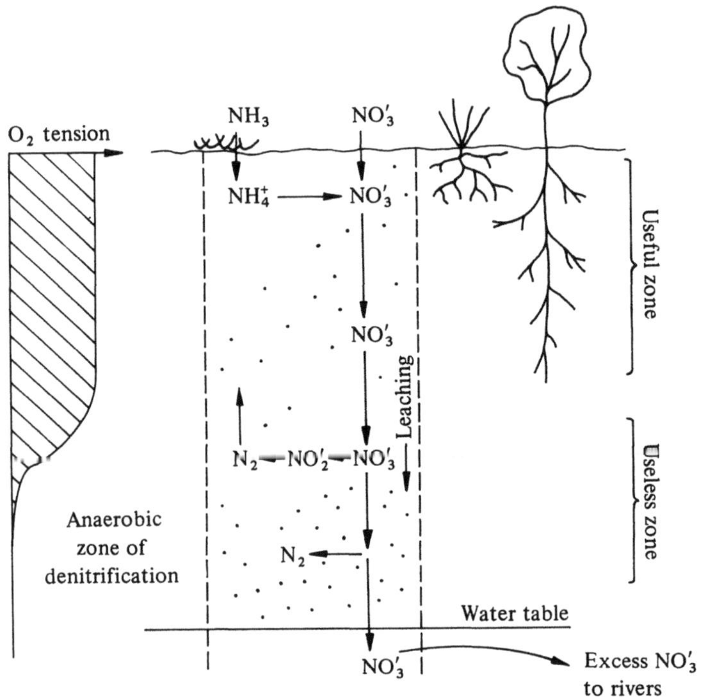

Fig. 2. The ecology of denitrifying bacteria. The *dots* in the soil profile indicate the frequency with which denitrifiers can be isolated from soil samples at different depths

and that ammonia may be preferentially taken up. Since nitrate is not adsorbed on the soil particles it is readily leached to a lower level and eventually finds itself below the area of exploitation by the roots of plants. Such nitrate has now effectively left the "useful" zone and has entered a "useless" zone of the soil. This nitrate may apparently either accumulate in the lower layers of the soil, or enter the normal water drainage system, carrying nitrate into rivers or lakes. However, should it reach the lower layers of the soil the nitrate moves into a zone in which oxygen is deficient and in this anaerobic zone a large denitrification takes place. Denitrifying bacteria attack the nitrate, using it as a terminal electron acceptor in place of oxygen, and dinitrogen gas is produced as the end product of the reduction. The N_2 gas readily diffuses back into the atmosphere and of course some of it is made available again as fixed nitrogen by free-living or symbiotic bacteria using the nitrogenase reaction. Where excess nitrate which has escaped denitrification is lost into the water table and appears in a water supply from which drinking water is taken, a serious potential problem exists. Nitrate is not permitted in drinking water above certain minimal quantities and these levels may easily be exceeded under particular circumstances e. g., the water supply in East Anglia is generally in excess of the permitted quantity as a consequence of the high levels of nitrogen fertilisers used in cereal production in this area. If nitrate is present in excessive amounts it becomes reduced to nitrite, and nitrosamines may be formed in the gut.

The nitrosamines are potentially hazardous compounds and their presence has been associated with the occurrence of cancer, particularly in children.

3.1 Water Purification

One of the methods available for treatment of water contaminated with nitrate is to make use of denitrifying bacteria (Payne, 1973). Anaerobic cultivation of samples of activated sewage sludge to which methanol has been added, leads to the selection of fairly pure cultures of the stalked bacterium *Hyphomicrobium.* This organism, which is a budding bacteria, readily attaches itself to a suitable substrate by a fine stalk. If offered methanol and oxygen it is capable of oxidising methanol to CO_2 and water, and readily removes any oxygen present in the water. Under the anaerobic conditions thus produced this same bacterium can use nitrate instead of oxygen to oxidise methanol to CO_2 and water, reducing the nitrate to N_2 gas. A schematic diagram of a theoretical water purification plant involving *Hyphomicrobium* is shown in Fig. 3.

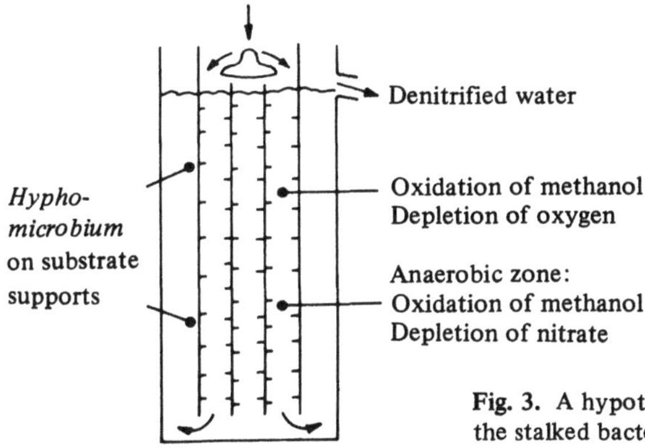

Fig. 3. A hypothetical water purification scheme using the stalked bacterium, *Hyphomicrobium*

In practice the contaminated water will almost always contain individuals of *E. coli* (and other organisms) which will become associated with the *Hyphomicrobium* on the substrate. Under anaerobic conditions the *E. coli* will carry out a nitrate ammonification (a form of nitrate respiration) in parallel with the normal denitrification to N_2 gas, and the removal of nitrate in this mixed system may yield a mixture of ammonia and N_2 gas. Under certain circumstances as much as 30% of the total nitrogen may appear in the denitrified water as ammonia, although some of this will be removed for the growth of new cell material by the nitrate-respiring bacteria (Schlegel, personal communication). It is necessary to calculate accurately the amount of methanol required just to reduce the oxygen and nitrate. If too much were added the purified water would now contain an amount of methanol which, although readily oxidised by bacteria in later stages of the water purification process, would contribute to an excessive and unwanted biological oxygen demand.

4 Nitrate Utilisation and Taxonomy

The accumulation of nitrite by organisms offered nitrate under anaerobic conditions has been used as a convenient test for assimilatory nitrate utilisation of many bacteria. Bergey's Manual lists many species which are able to use nitrate in a dissimilatory way but which cannot take it further than nitrite. Some species of genera like *Escherichia, Spirillum, Propionobacterium, Corynebacterium,* and *Streptococcus* are able to accumulate nitrite readily in this way. Others are able to reduce nitrate dissimilatively all the way to nitrogen gas (denitrification) and they are easily recognised not only by their ability to accumulate NO_2^- temporarily but also by their later production of N_2 gas. Examples of denitrifying bacteria include species of *Paracoccus, Thiobacillus, Pseudomonas, Hyphomicrobium, Alcaligenes,* and *Bacillus.* Although the ability to accumulate nitrite might appear to be of potential use in taxonomy the distribution of the ability to do so is so random as to make it unhelpful. Its principal value is in the positive recognition of strains of one species which may or may not denitrify or form nitrite and for which positive criteria are not otherwise available. This is most likely to be of importance in medical studies, e. g., in the separation and identification of *Mycobacterium tuberculosis* strains from closely related species.

5 Denitrifying Bacteria

Although there are many denitrifiers, much of the biochemical work has been done on *Escherichia coli, Proteus mirabilis* (nitrite formers), *Pseudomonas denitrificans, Paracoccus (Micrococcus) denitrificans, Alcaligenes faecalis,* and *Bacillus licheniformis* (denitrifiers).

5.1 Escherichia coli

This typical facultative aerobe can adapt to the use of nitrate rather than oxygen as terminal electron acceptor (Haddock and Jones, 1977). It lacks the typical denitrifying nitrite reductase which has cytochrome c.d as a principal constituent. If neither oxygen nor nitrate is available then *E. coli* can obtain ATP by fermentation. Figure 4 shows

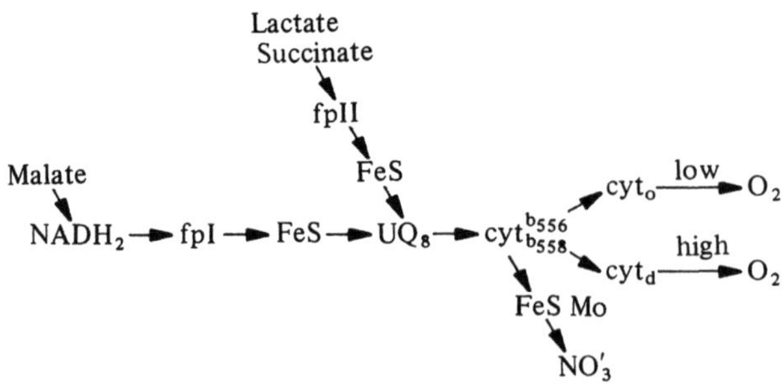

Fig. 4. Intermediates of electron transport in *E. coli* arranged in a linear sequence

the electron transport pathway which operates in *E. coli.* The sequence of electron transfer from reduced NAD through flavoprotein, FeS compounds, and menaquinone or ubiquinone to cytochrome b has been clearly distinguished by many workers. Malate contributes electrons to pyridine nucleotide, while lactate and succinate contribute electrons at the level of the quinone. Electrons are passed from cytochrome b to oxygen either by cytochrome d (a system with a high affinity for oxygen) or through cytochrome o (a system with a low affinity for oxygen). In the absence of oxygen, nitrate reductase is formed and its production is accompanied by the accumulation of a new cytochrome b component, cytochrome b-555. Normally the induction of nitrate reductase is accompanied in *E. coli* by the formation of formate dehydrogenase, which donates electrons to the electron transport chain at the FeS level. When grown on nitrate, the preferred substrate for *E. coli* is formate, although this organism can use other substrates more slowly. The transfer of electrons is accompanied by the formation of ATP (Garland et al., 1975). The nitrate respiration, like normal aerobic respiration in bacteria, takes place in the plasmamembrane, where the intermediates of electron transport are localized. The membrane shows good evidence of sidedness and it is possible to arrange the intermediates in terms of a loop system, in which the various steps of the reduction involve the transfer of hydrogen atoms from inside to outside and the transfer of electrons from the outside to the inside. A consequence of such an arrangement is that protons accumulate at the outer surface of the membrane and may be released into the medium (Mitchell, 1968). The return of the electrons through the proton translocating ATPase, which spans the membrane from outside to inside, causes the conversion of the energy of the proton gradient into the chemical energy of ATP. A typical electron transport system in *E. coli* arranged in the form of proton generating loops is indicated in Fig. 5.

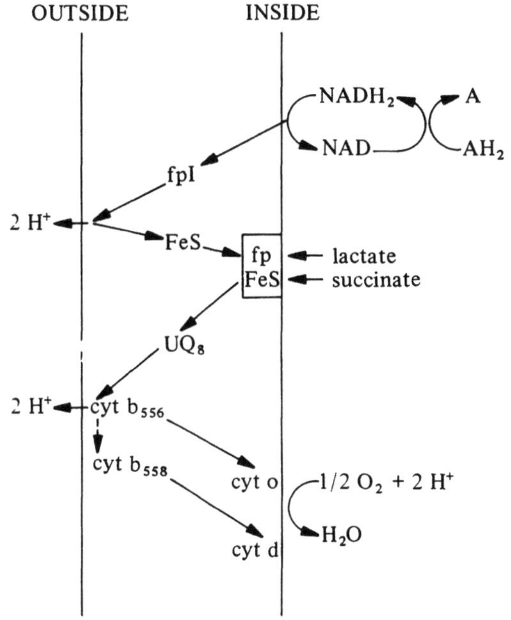

Fig. 5. Intermediates of electron transport in *E. coli* arranged in loops in a bacterial membrane to show proton-transfer capability

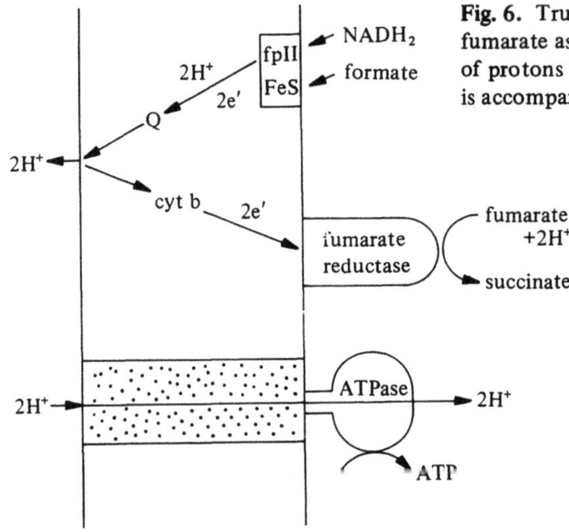

Fig. 6. Truncated electron transport in *E. coli* using fumarate as a terminal electron acceptor. The return of protons through the proton-translocating ATPase is accompanied by ATP formation

An additional electron transport system formed by *E. coli* is one in which formate or pyridine nucleotide donates electrons through a menaquinone loop via cytochrome b to fumarate, using the fumarate reductase which is located on the inner side of the membrane. This is a special case of a very truncated electron transport pathway in which fumarate acts as a terminal electron acceptor like oxygen, but unlike oxygen (or nitrate) it is an organic rather than an inorganic electron acceptor; the latter is typical of conventional oxidative metabolism. If the correct substrates are present, *E. coli* can therefore maintain an active proton translocating respiratory chain anaerobically using either nitrate or fumarate as the terminal acceptor (Fig. 6).

5.2 Paracoccus denitrificans

The typical free-living denitrifiers like *Pseudomonas denitrificans* and *Paracoccus denitrificans* are obligate aerobes, in that they cannot normally ferment and they maintain an essentially aerobic mode of life. They are typical nitrate respirers. Like *E. coli, Paracoccus denitrificans* only forms nitrate reductase (and nitrite reductase) when oxygen is absent. The presence of oxygen suppresses the induction of both the reductases. (Stouthamer, 1976a).

The electron transport scheme involved in the oxidative metabolism of *Paracoccus denitrificans* is shown in Fig. 7 (John and Whatley, 1977). The electron transport pathway of *Paracoccus*, shown in heavy type, is extraordinarily similar to the electron transport system in mitochondria: this seems to be typical of several of the denitrifying bacteria. Hydrogens from malate enter at the level of $NADH_2$ or $NADPH_2$; those from glucose-6-phosphate as $NADPH_2$, and those from succinate, formate and lactate at the level of UQ_{10}. Hydrogen may enter at the level of $NADH_2$ or possibly at the level of cytochrome c, as does the artificial donor, reduced dichlorophenolindophenol. A suggested topology within the membrane of the electron transport pathway in *Paracoccus* again indicates the sidedness of the membrane (Fig. 8). The occurrence of

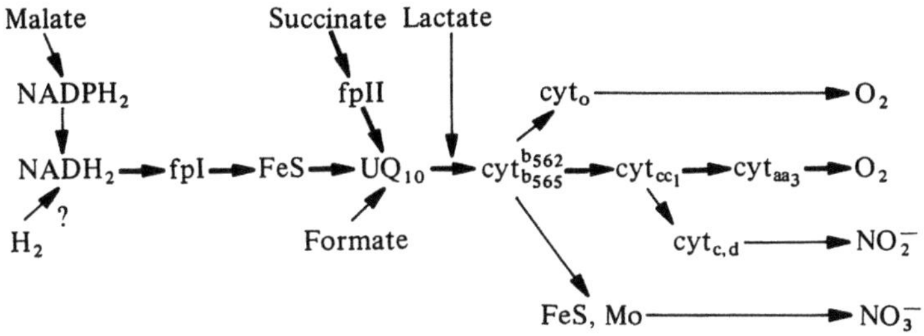

Fig. 7. Electron transport in aerobic and anaerobic respiration in *Paracoccus denitrificans.* Intermediates connected with *heavy arrows* are constitutive, those with *light arrows* are inducible only under specified conditions

loops of electron transport intermediates makes proton pumping possible, as is indicated in the diagram, and ATP synthesis results from the activity of the proton translocating ATPase. The components shown connected by light arrows in Fig. 7 are only induced by appropriate conditions (e. g., anaerobiosis in the case of nitrate reductase), whereas the components connected by heavy arrows are constitutive. The inducible nitrate reductase is an iron sulphur protein which contains molybdenum. Its properties indicate that it is in the class of nitrate reductase A described by Pichinoty and Piechaud (1968) and is induced only when nitrate respiration is required. On returning from nitrate respiration to ordinary aerobic conditions no more nitrate reductase is formed,

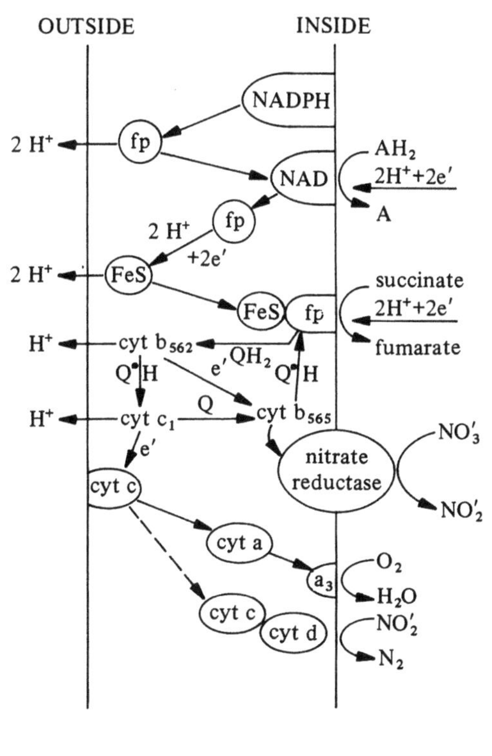

Fig. 8. Intermediates of electron transport arranged in loops in a *Paracoccus* membrane to show proton-transfer capability. The Q cycle is according to Mitchell (1975). An additional proton flux between cytochrome c and cytochrome aa_3 (not shown) may be due to a direct proton pump mechanism not involving a loop

although the nitrate reductase which has been synthesised remains active for some time. In addition to nitrate reductase A, *Paracoccus* also contains nitrate reductase B. This enzyme is a constitutive enzyme and its formation does not depend on anaerobic conditions. It appears that this enzyme is also a FeS:Mo enzyme, but its function is concerned with the reduction of nitrate to nitrite, which is then further reduced in the assimilatory nitrate reductase system to yield ammonia. *Paracoccus* is able to grow with nitrate both as terminal electron acceptor and as nitrogen source, although there is usually an extended lag in the onset of growth when it is offered only nitrate. This lag is significantly reduced if nitrate is added together with glutamate as reduced nitrogen source, and nitrate reductase A is readily synthesised. When glutamate is added the operation of nitrate reductase B is no longer the factor limiting the introduction of amino groups into protein synthesis. The regulation of the two streams of nitrate metabolism when nitrate is both sole nitrogen source and electron acceptor is potentially an interesting one, but virtually nothing is known about the control systems involved. The nitrite reductase, which is also induced when the organism is growing on either nitrate or nitrite, is a cytochrome c.d enzyme. It obtains electrons for nitrite reduction from cytochrome c.

The components of the electron transport chain in *Paracoccus denitrificans* can be arranged in the form of loops whereby proton accumulation outside can be accomplished (outwards in the intact cell); the accumulated protons are returned via ATPase to give ATP synthesis. Vesicles derived from the plasmamembrane of *Paracoccus denitrificans* formed by osmotic shock following lysozyme treatment of whole cells take up a configuration which is either right side out or inside out with respect to the normal configuration. Growth on glucose plus oxygen for some unexplained reason gives mostly right side out particles, whereas growth on succinate plus nitrate yields principally inside out particles (Burnell et al., 1975). If necessary a complete separation of right side out from inside out particles can be attempted experimentally by the use of ADP-sepharose columns. Right side out particles cannot react with the bound ADP and pass straight through the column without hindrance because their ATPase (ADP-binding site) is inside the vesicles. By contrast inside out particles can react with the bound ADP since their ATPase is external and so they become adsorbed onto the column. After washing with buffer the inside out particles can be eluted with ADP solutions (Ferguson et al., 1978). A preparation of inside out particles can carry out electron transport coupled to ATP but right side out particles cannot do this, since ADP cannot pass through the membrane and is therefore unavailable. Phosphorylating particles show good respiratory control and react to inhibitors in exactly the same way as mitochondria (John and Whatley, 1977).

Those obligate aerobes which are also denitrifiers appear not to be able to carry out fermentation simply because they normally do not synthesise the NAD-dependent lactate dehydrogenase. The situation is indicated in Fig. 9.

Glucose yields glucose-6-phosphate, which in the case of *Paracoccus denitrificans* is converted to glyceraldehyde-3-P. This becomes oxidised to phosphoglyceric acid (PGA) at the expense of NAD and this oxidation is accompanied by the substrate phosphorylation of one molecule of ADP per PGA formed. Further transformation of PGA to phosphoenolpyruvate makes available a second substrate phosphorylation, when PEP is converted to pyruvate in a reaction coupled to ATP synthesis. However further glucose cannot be oxidised in the absence of an acceptor of hydrogens from the $NADH_2$, which accumulates in this system. These can be provided experimentally by oxygen in the presence of a membrane vesicle preparation, or by nitrate plus membranes, or by lactate dehydrogenase together with pyruvate. The reduced products are respectively water, nitrite, and lactate. In this in vitro system there is a clear demonstration of the lack of lactate dehydrogenase. The ATP accumulated is produced not by oxidative phosphorylation through an electron transport system but by substrate phosphorylation and its formation is quite insensitive to respiratory uncouplers, such as dinitrophenol (Fig. 9).

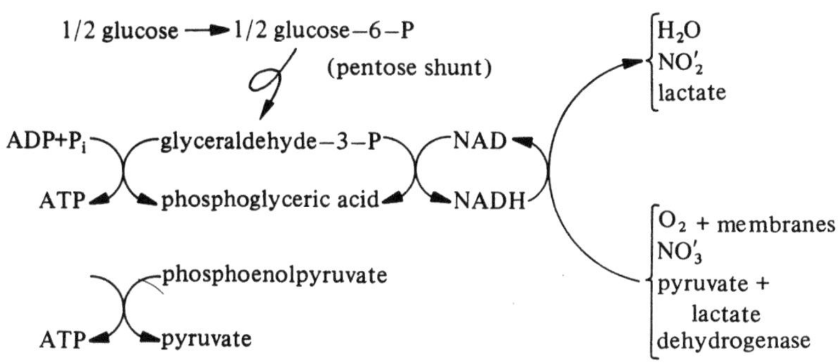

Fig. 9. Substrate phosphorylation in crude extracts of *Paracoccus* depends in vitro on regeneration of $NADH_2$ by O_2, NO_3, or by added pyruvate and lactate dehydrogenase

6 Switching Between Oxygen and Nitrate

After adaptation to nitrate, *Paracoccus denitrificans* and other nitrate-adapted organisms maintain their nitrate reductase for some time following their return to oxygen, although the production of new enzyme protein is prevented. If offered the choice between oxygen or nitrate it is clear that both *Paracoccus* and *E. coli* prefer to use oxygen (John, 1977). Experiments with a nitrate/oxygen electrode, in which a continuous trace of the oxygen and nitrate status can be simultaneously monitored, a clear demonstration of a switching phenomenon may be seen. When cells are added to the system oxygen is initially used up. When oxygen reaches zero concentration nitrate utilisation commences and continues, but only until a further supply of oxygen is made available by the addition of H_2O_2 (in the presence of catalase). As soon as oxygen is available nitrate respiration ceases and oxygen uptake ensues. This switching can be made many times. A similar switching occurs with nitrite. A possible explanation of the switching phenomenon, proposed by P. John, is based on the kinetics of oxidation and reduction of cytochrome b or cytochrome c in the electron transport chain (Fig. 7). When oxygen is available, electrons are actively removed from cytochrome c via cytochrome a and the oxidised cytochrome c continues to abstract electrons from cytochrome b. Assuming a constant supply of electrons from substrate via $NADH_2$ etc. one can easily see that the ratio of oxidised cytochrome b to reduced cytochrome b in the presence of oxygen will be predominantly in favour of the oxidised form. The amount of reduced cytochrome b available for oxidation through nitrate reductase is apparently insufficient to produce any measurable rate of reaction. When oxygen is all used up the cytochrome c and hence cytochrome b will become more fully reduced. Under these circumstances electrons can now pass into the FeS:Mo protein and on to nitrate at a rate which is essentially still controlled by the rate of electrons coming from the substrate. When oxygen is made available the system returns to the initial equilibrium and nitrate reduction can no longer take place. A similar explanation is proposed for the nitrite/oxygen switching reaction. In the case of *E. coli* the use of succinate gives a low rate of reaction on the addition of cells (succinate is not a preferred substrate). When all the oxygen is used up the use of nitrate commences but it is instantly turned off on the further addition of H_2O_2 to supply oxygen. However in the case of formate (formate dehydrogenase is induced by the same anaerobic conditions that induce nitrate reduction) the electron flux from the donor is an order of magnitude greater than with succinate. Under conditions where formate is the donor, *E. coli* uses nitrate at a measurable rate even in the presence of oxygen, but when all the oxygen has been used up the rate of electron transport to nitrate does increase. The addition of H_2O_2 lowers the rate of nitrate utilisation to its initial rate.

If chlorate, an analogue of nitrate, is added to inside out particles, chlorate is reduced to chlorite, just as nitrate is reduced to nitrite. However if whole cells of *Paracoccus* are offered chlorate they are only able to use chlorate very slowly whereas oxygen is an effective electron acceptor, as in normal respiration. The addition of H_2O_2 to supply oxygen makes no difference to the rate of utilisation of chlorate by intact *Paracoccus* cells. If the cell membranes are broken at the end of such a cycle by the addition of the detergent Triton X-100, the chlorate can now come in contact with the nitrate reductase and a rapid utilisation of chlorate is measured. The uptake is stopped by the addition of sodium azide, which is a good inhibitor of nitrate reductase. This experiment is taken to indicate clearly that nitrate reductase is located on the inner face of the *Paracoccus* cell membrane. Garland et al. (1975) have suggested that in *E. coli*, nitrate reductase spans the membrane rather than being located only on the inside. They have suggested that it is therefore accessible from the outside and that no nitrate carrier is involved, the nitrate reductase in this case acting directly as a proton pump. There would thus be no requirement for a loop of electron transport intermediates in the membrane for this step.

7 Proton Fluxes Associated with Nitrate Metabolism

Recently Kristjansson et al. (1978) have determined the yield of protons accumulated by denitrifiers in response to small oxidant pulses provided by oxygen, nitrate, nitrite or nitrous oxide. These measurements are an extension of the important observations of Scholes and Mitchell (1970), Meijer et al. (1977), and Lawford (1978) on the proton translocating yield of electron transport with O_2 or NO_3^-. In the presence of valinomycin, which permits a full realisation of the total proton yield by exchanging potentially "bound" protons with potassium, Kristjansson et al. show that the yields of protons per two electrons oxidised is: for oxygen = 7.5, for nitrate = 4.3, for nitrite = 3.7, and for N_2O = 4.5. They also show that a series of fifteen consecutive injections give reproducible results, indicating that there was no lack of substrate in *Paracoccus* or *Pseudomonas denitrificans* under their conditions and that this type of experiment is valid. The ratio of approximately 2:1 in terms of proton yield agrees in general with the greater efficiency of growth observed on oxygen than on nitrate (Stouthamer, 1976b). Van Verseveld et al. (1977) observed a greater growth yield from *Paracoccus* with oxygen than nitrate but could not readily distinguish nitrate from other nitrogen oxides (although nitrite in higher concentrations is an uncoupler of oxidative phosphorylation). The observation is also in line with the $P:2e^-$ values reported by John and Whatley (1970) of 1.5 with O_2 and 1.0 with nitrate.

In the absence of valinomycin new sets of observations are reported by Kristjansson et al. (1978). By repeated applications of nitrate or nitrite pulses the initial acidification of the medium, observed reproducibly in the presence of valinomycin, was seen to be changed over a sequence of three or four pulses in the absence of valinomycin into an alkalinisation of the medium. This occurs when normal nitrate respiration is partly inhibited by developing a membrane potential (which K^+ would have discharged in the presence of valinomycin but not of course in its absence). When the membrane is fully charged and respiration fully inhibited there appears to be one proton (H^+) taken up for each nitrate molecule (NO_3^-). This proton uptake serves eventually to prevent alkalinisation of the internal medium when nitrate is used up in ni-

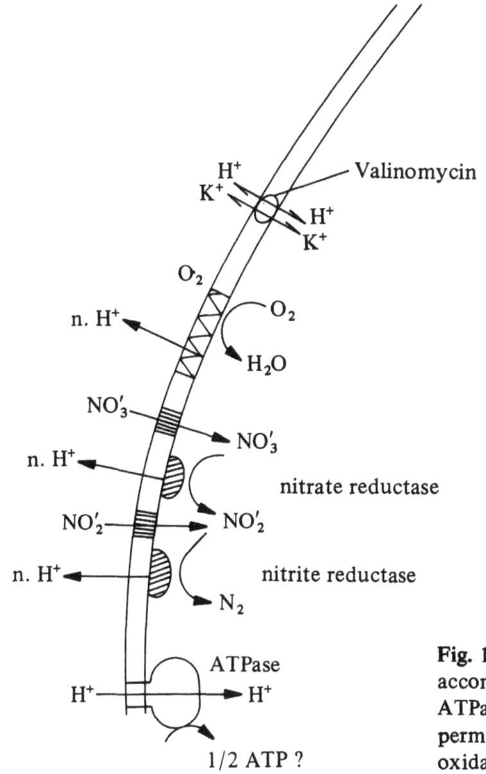

Fig. 10. Facilitated nitrate uptake in *Paracoccus* is accompanied by stoichiometric uptake of protons via ATPase. Proton excretion in presence of valinomycin permits observation of $H^+:e^-$ ratios with pulses of oxidant. See text for details

trate respiration, since one NO_3^- is replaced by one HO^- and this must be neutralised. The proton apparently enters through the normal proton translocating ATPase, since the alkalinisation reaction is stopped by the addition of dicyclohexylcarbodiimide (DCCD). This inhibitor is supposed to prevent ATP formation by plugging up the proton translocating subunit, F_0, of the membrane bound ATPase.

Figure 10 summarises these effects. The nitrate uptake is shown to occur through a specific carrier, the oxidation of nitrate to take place on nitrate reductase on the inner plasmamembrane and the nitrite to be further metabolised on the inner membrane to N_2 gas. The uptake of protons is associated directly with nitrate uptake, the energy for the proton uptake being derived by the generation of HO^- when nitrate is fully reduced to N_2. The proton uptake may of course be coupled to the formation of a fraction of an ATP molecule. Valinomycin permits the escape of protons which might otherwise have been used in charging the membrane. Oxygen is shown as readily entering the system by diffusion and the electron transport chain is indicated as a loop system capable of translocating eight protons per oxygen atom used in respiration. These experiments are consistent with an internal utilisation of nitrate, confirming previous conclusions based on chlorate experiments and those of Kristjansson and Hollocher (1979) that the nitrate reductase is internally sited on the plasmamembrane of *Paracoccus*. The nitrate-transporting system is poorly characterised. Nitrate uptake appears to operate by a facilitated diffusion and not as a proton symport of the type used for sulfate and phosphate uptake, and perhaps for other metabolites.

8 Concluding Remarks

These observations, made on *Paracoccus* and *Pseudomonas denitrificans* as representative of the class of denitrifying bacteria (nitrate respirers) emphasise the importance of the precise location of specific electron carriers within the bacterial membrane system (Figs. 5 and 8) and are consistent with the occurrence of loops of intermediates arranged for proton translocation on the Mitchell (1968) model. Oxygen and nitrate are shown as largely interchangeable, but since the oxides of nitrogen abstract electrons from more reduced intermediates than oxygen (cytochrome b and c, rather than cytochrome aa_3) the yield of protons and hence total energy-conserving potential with the nitrogen oxide is significantly lower than with oxygen. The system to oxygen is constitutive, though when nitrate is being used there is usually less cytochrome oxidase produced. However nitrate respiration is an inducible system, only called up (at some expense to the general energy economy of the cell) when oxygen is not available. The distribution of cytochrome components also alters when nitrate respiration is operating, cytochrome b (for nitrate reductase) and cytochrome c.d (for nitrite reductase) both increasing greatly and cytochrome aa_3 (cytochrome oxidase) being somewhat diminished. The ability to switch back to oxygen in preference to nitrate or nitrite even in fully adapted cells implies that the bacteria find it important to conserve the maximum amount of energy available, and the suppression of synthesis of all the enzymes of nitrate respiration under aerobic conditions has implications for energy saving.

The ecological consequences of the operation of the denitrifying organisms in the soil are, of course, essentially "accidental," since the bacteria are merely exploiting nitrate as a less satisfactory substitute for the oxygen which they prefer. But their importance in the nitrogen cycle and their role in preventing a disastrous accumulation of unwanted nitrate where it could not be exploited by plants cannot be under-rated.

References

Burnell JN, John P, Whatley FR (1975) Biochem J 150:527–536
Ferguson SJ, Kell DB, John P (1978) FEMS Microbiol Lett 3:151–154
Garland PB, Clegg RA, Boxer D, Downie JA, Haddock BA (1975) In: Quagliariello E, Papa S, Palmieri F, Slater EC, Siliprandi N (eds) Electron transfer chains and oxidative phosphorylation. North Holland-American Elsevier, Amsterdam
Haddock BA, Jones CW (1977) Bacterial respiration. Bacteriol Rev 41:47–49
John P (1977) J Gen Microbiol 98:231–238
John P, Whatley FR (1970) Biochim Biophys Acta 216:342–352
John P, Whatley FR (1977) Biochim Biophys Acta 463:129–153
Kristjansson JK, Hollocher TC (1979) J Bacteriol 137:1227–1233
Kristjansson JK, Walter B, Hollocher TC (1978) Biochemistry 17:5014–5019
Lawford HG (1978) Can J Biochem 56:13–22
Meijer EM, Van Verseveld HW, Van der Beek EG, Stouthamer AH (1977) Arch Microbiol 112: 25–34
Mitchell P (1968) Chemiosmotic coupling and energy transduction. Glynn Research Ltd., Bodmin, UK
Mitchell P (1975) FEBS Lett 59:137–139
Payne WJ (1973) Bacteriol Rev 37:409–452
Pichinoty F, Piechaud M (1968) Ann Inst Pasteur, Paris 114:77–98
Scholes P, Mitchell P (1970) J Bioenerg 1:309–323
Stouthamer AH (1976a) Adv Microb Physiol 14:315–375
Stouthamer AH (1976b) Yield studies in micro-organisms. Merrow Publ Co, Watford, England
Van Verseveld HW, Meijer EM, Stouthamer AH (1977) Arch Microbiol 112:17–23

Dinitrogen Fixation

Genetics of Dinitrogen Fixation

W.J. BRILL[1]

1 Introduction

The only organism that has been used for the detailed analysis of the genes involved with dinitrogen fixation (*nif* genes) is *Klebsiella pneumoniae*. This organism is closely related to the well-studied *Escherichia coli* and *Salmonella typhimurium*. Unlike these latter species, certain *K. pneumoniae* strains are able to grow in media free of fixed nitrogen compounds (Mahl et al., 1965). Such strains fix dinitrogen from the air. Dinitrogen fixation by *K. pneumoniae* only occurs anaerobically because oxygen is a potent inactivator of nitrogenase. Genetic techniques that have been developed for *E. coli*, such as generalized transduction by phage P1 or gene inactivation by random insertion of phage Mu have been used to develop the genetics of dinitrogen fixation in *K. pneumoniae* (Streicher et al., 1971; Bachhuber et al., 1976; Rao, 1976; Elmerich et al., 1978). This chapter will focus on recent work that has been published from this laboratory. Earlier work and work from other laboratories have been reviewed (Brill, 1975; Elmerich 1979; Beringer, 1980).

2 Arrangement of nif Genes

Mutations that specifically prevent *K. pneumoniae* from growing on dinitrogen are clustered in one region of the chromosome. The clustering of *nif* genes was directly indicated by the fact that this region, incorporated into a transmissible plasmid, induces growth on dinitrogen by *E. coli* when that plasmid is introduced in *E. coli* (Dixon and Postgate, 1972). Many mutations have been located with respect to each other by deletion mapping experiments (MacNeil et al., 1978 a,b). Deletions were isolated from strains containing a Mu phage integrated within the *nif* region. The particular Mu used has a thermolabile repressor. Deletions obtained in this manner were used to locate point mutations and to order them with respect to each other. A similar study, confirming the map, has been accomplished by use of transposons (Merrick et al., 1980).

The number of complementation groups was determined by cis-trans complementation between point mutations in a Rec$^-$ (recombination defective) host. It was sur-

1 Department of Bacteriology and Center for Studies of Nitrogen Fixation, University of Wisconsin, Madison, WI 53706/USA

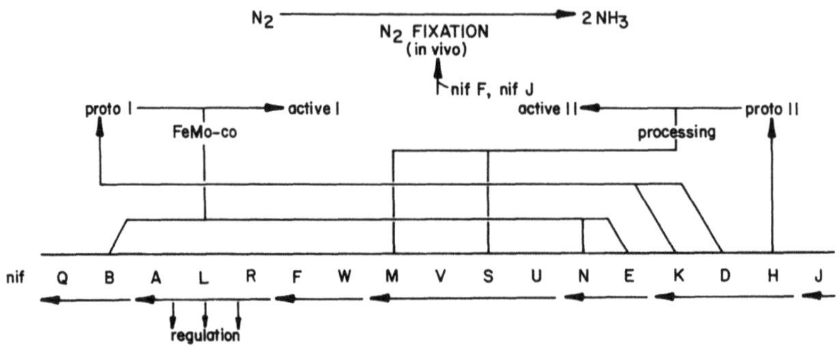

Fig. 1. Order, transcriptional organization, and function of the *Klebsiella pneumoniae nif* genes

prising to find 13 complementation groups (MacNeil et al., 1978b), since that meant that at least 13 *nif* genes are required for dinitrogen fixation. This seems to be a large number of genes required to code for nitrogenase, a protein with only three different polypeptides. These complementation groups were further analyzed with regard to the number and polarity of operons (MacNeil et al., 1978b). Complementation tests with polar mutations (e. g., Mu insertions) indicates that the genes are organized into seven different operons. All of the operons are transcribed in the same direction (see Fig. 1). This was confirmed by analyzing strains with *lacZ* (coding for β-galactosidase) fused onto *nif* genes (D. MacNeil, unpubl. results).

Further analysis of the *nif* region indicated that several more *nif* genes are present. For instance, the *nif*A complementation group contains Mu insertions randomly dispersed within that complementation group; however, point mutations are clustered only in the part of *nif*A that is distal to the operator portion of that operon (MacNeil et al., 1978b). This is explained by a gene (*nif*L) that is operator proximal to *nif*A, but is not essential for dinitrogen fixation under the laboratory growth conditions. In other words, nonpolar mutations in *nif*L are not picked up in Nif^- mutants but polar mutations inactivate the function of the *nif*A product and thus prevent expression of *nif*A, which is essential for dinitrogen fixation. In fact, some strains containing Mu insertions in *nif*L are capable of reverting to Nif^+. Mu insertions in *nif*A do not revert to yield the Nif^+ phenotype.

3 Products of nif Genes

The products of 15 of the *nif* genes have been determined by analysis of two-dimensional polyacrylamide gels of extracts from Nif^- mutants (Roberts et al., 1978; Roberts, unpubl. results). If some mutations in a gene alter the charge or molecular weight of a polypeptide, then that gene must code for the synthesis of that polypeptide. One unusual observation is that one of the nitrogenase polypeptides (component II) has two species of slightly different molecular weights (Roberts et al., 1978). Those two species are coded by a single gene (*nif*H) because a mutation causing a molecular weight change in one species simultaneously changes the molecular weight of the other.

Several *nif* genes seem to be involved in component II modification and it should be interesting to determine if any of these genes yields a component II comprising only one of these species.

The products of *nif*D and *nif*K comprise a complex (component I of nitrogenase) because many mutations in one gene cause the other gene product to become susceptible to intracellular proteolysis. This was not surprising because component I is isolated as a stable complex. The genes *nif*E and *nif*N also code for proteins that complex with each other since stability of one product is dependent on the structure of the other.

In the case of the *nif*F-coded product, the protein may have two different molecular weights; the smaller polypeptide has the catalytic activity (Nieva-Gómez et al., in press) but the larger one may be a read-through product and may play a regulatory role (Roberts, unpubl. results).

4 Functions of nif Gene Products

The functions of the *nif*-encoded proteins were determined by preparing cell-free extracts of mutants and adding pure component I, pure component II, or pure iron-molybdenum cofactor (FeMo-co) to reconstitute activity in vitro (Roberts et al., 1978). FeMo-co can be extracted from component I with solvents and it has been shown to consist of 1Mo:8Fe:6S (Shah and Brill, 1977). FeMo-co is the active site of nitrogenase (Rawlings et al., 1978; Shah et al., 1978). For genes (*nif*B, *nif*Q, *nif*N, and *nif*E) are required for FeMo-co synthesis (Roberts et al., 1978). Mutants with lesions in these genes can be reconstituted to fix dinitrogen in vitro when FeMo-co is added to the extract. At least one gene, *nif*M, codes for a protein that must play a role in processing the component II structural protein. The structural proteins of nitrogenase are coded by *nif*K, *nif*D, and *nif*H. These genes comprise one operon and this allows the cell to make the proteins in ratios that are optimal for nitrogenase activity. A situation of excess component I, for instance, results in inhibition of dinitrogen fixation (Shah et al., 1975).

5 Regulation of nif Expression

5.1 Fixed Nitrogen

Nitrogenase synthesis is regulated by fixed nitrogen (Parejko and Wilson, 1970), possibly through the action of glutamine synthetase (Streicher et al., 1974; Tubb, 1974). When a good nitrogen source, such as ammonium, is available no nitrogenase activity is detected (Parejko and Wilson, 1970). In fact, none of the *nif*-encoded proteins is detected in ammonium-grown *K. pneumoniae* (Roberts et al., 1978). Certain mutations in the genes that code for or regulate glutamine synthetase (*gln* genes) prevent nitrogenase synthesis even in the absence of fixed nitrogen (Streicher et al., 1974; Tubb, 1974). Other *gln* mutations allow derepression of nitrogenase synthesis (Shanmugam et al., 1975). Also, the glutamine synthetase inhibitor, methionine sulfoximine, added to ammonium-grown cells, derepressed nitrogenase synthesis (Gordon and Brill, 1974). The actual mechanism by which *gln* products control nitrogenase synthesis, as well as the synthesis of enzymes involved in the degradation of nitrogen-containing metabolites (e. g., histidine, proline), is not yet understood.

Most mutations in *nif*A prevent synthesis of all of the *nif*-encoded proteins with the exception of the *nif*L-encoded proteins (Roberts, unpubl. results). Certain *nif*A mutations have the same pleiotropic-negative effect on the other operons but allow synthesis of an inactive *nif*A product. Therefore, *nif*A codes for a positive regulatory effector that is required to derepress all of the other *nif* operons with the exception of *nif*RLA.

The adenine analog, 6-cyanopurine, turns colonies purple when an active *nif*A product is synthesized in the cells (MacNeil and Brill, 1978). It is not known whether the *nif*A product, itself, induces the purpling of 6-cyanopurine or whether the *nif*A product causes another gene product to catalyze this reaction. If another gene product is specifically involved, it is not coded by a gene in the *nif* region of the chromosome. This reaction is prevented under conditions of ammonium repression. The color reaction has been used to determine if an induced deletion eliminates a part of the *nif*RLA operon. Mutants derepressed for nitrogenase synthesis in the presence of ammonium have been isolated as purple colonies on ammonium-containing medium. In many cases, such mutants have lesions in the *gln* genes (unliked to *nif*). However, some derepressed mutants have mutants within *nif*RLA (D. MacNeil, upubl. results). These derepressed mutants have been useful for determining the target of action by the *gln* product(s). It seems that *gln* factor(s) only act to turn off transcription of the *nif*RLA operon. Once the production of *nif*A protein is stopped, then the remaining operons cannot be turned on because they require action of the *nif*A product. A derepressed mutant, caused by a lesion in *nif*RLA retains its capacity to fix dinitrogen in ammonium, even when the strain contains a gln^- mutation. That *gln* mutation, by itself, prevents derepression of nitrogenase even when the cells are nitrogen-starved.

5.2 Oxygen

Dinitrogen-fixing *K. pneumoniae* cells exposed to air rapidly lose their ability to fix dinitrogen. Nitrogenase is rapidly inactivated by oxygen (St John et al., 1974). To prevent futile derepression of nitrogenase in air when cells are starved of nitrogen, the bacteria have a mechanism to repress nitrogenase synthesis in the presence of oxygen. This regulation is mediated via the *nif*L protein (Roberts, unpubl. results). In fact, certain *nif*L mutations cause the *nif* proteins to be synthesized in air. Of course, the proteins are inactivated and then degraded via intracellular proteases. A model for this regulation involves the *nif*L protein interacting with the *nif*A protein to prevent the *nif* A protein from turning on synthesis of the other *nif*-encoded proteins.

6 nif DNA

The entire *nif* region has been cloned onto an amplifiable plasmid (Pühler et al., 1979). Segments of *nif* also have been individually cloned (Cannon et al., 1977, 1979), and lambda *nif* specialized transducing phage are available (MacNeil et al., in press). The size of the *nif* region is approximately 23 kb.

The *lacZ* gene, coding for β-galactosidase, has been fused onto each of the seven operons (MacNeil, unpubl. results). A large number of operator and promoter mutations have been isolated (Zhu, unpubl. results) and their nucleotide sequences are being compared. This work should yield interesting information with regard to regulatory gene structure, binding of RNA polymerase, binding of the *nif*A product, and the regulatory phenotype. Cloned fragments of *nif* also are being used to examine, directly, the regulation of transcription by fixed nitrogen and oxygen.

It is interesting that the *nif*HDK operon from *K. pneumoniae* hybridizes to DNA on a large plasmid in *Rhizobium* (Nuti et al., 1979). All other dinitrogen-fixing bacteria that have been examined also contain DNA that hybridizes to *nif*HDK (Mazur et al., 1980; Ruvkan and Ausubel, 1980), but no other *K. pneumoniae nif* fragments hybridize. Therefore, the genes specifying the structure of nitrogenase are the only *nif* genes that remain conserved among the great variety of dinitrogen-fixing bacteria.

7 Conclusions

There is obviously a great deal more work that needs to be completed in order to understand the genetics, regulation, and biochemistry of dinitrogen fixation in *K. pneumoniae.* The only other organism that has been used to study the genetics of *nif* is *Azotobacter vinelandii* (Page and Sadoff, 1976; Bishop and Brill, 1977; Page, 1978). In that case, *nif* genes do not seem to be clustered as they are in *K. pneumoniae* (Bishop and Brill, 1977). Even though mapping techniques are available in *Rhizobium,* no data is available with regard to the organization and regulation of *nif* (reviewed in Beringer, 1980). The next few years are expected to result in a burst of activity in this research area and it will be very interesting to determine how different organisms synthesize nitrogenase and regulate that synthesis.

References

Bachhuber M, Brill WJ, Howe MM (1976) J Bacteriol 128:749–753
Beringer JE (1980) J Gen Microbiol 116:1–7
Bishop PE, Brill WJ (1977) J Bacteriol 130:954–956
Brill WJ (1975) Annu Rev Microbiol 29:109–129
Cannon FC, Riedel GE, Ausubel FM (1977) Proc Natl Acad Sci USA 74:2963–2967
Cannon FC, Riedel GE, Ausubel FM (1979) Mol Gen Genet 174:59–66
Dixon RA, Postgate JR (1972) Nature (London) 237:102–103
Elmerich C (1979) Physiol Veg 17:883–906
Elmerich C, Houmard J, Sibold L, Manheimer I, Charpin N (1978) Mol Gen Genet 165:181–189
Gordon JK, Brill WJ (1974) Biochem Biophys Res Commun 59:967–971
MacNeil D, Brill WJ (1978) J Bacteriol 136:247–252
MacNeil D, Howe MM, Brill WJ, J Bacteriol (in press)
MacNeil T, Brill WJ, Howe MM (1978a) J Bacteriol 134:821–829
MacNeil T, MacNeil D, Roberts GP, Supiano MA, Brill WJ (1978b) J Bacteriol 136:253–266
Mahl MC, Wilson PW, Fife MA, Ewing WH (1965) J Bacteriol 89:1482–1487
Mazur BJ, Rice D, Haselkorn R (1980) Proc Natl Acad Sci USA 77:186–190

Merrick M, Filser M, Dixon R, Elmerich C, Sibold L, Houmard J (1980) J Gen Microbiol 117: 509–520
Nieva-Gómez D, Roberts GP, Klevicks S, Brill WJ, Proc Natl Acad Sci USA (in press)
Nuti MP, Lepidi AA, Prakash RK, Schilperoort RA, Cannon FC (1979) Nature (London) 282: 533–535
Page WJ (1978) Can J Microbiol 24:209–214
Page WJ, Sadoff HL (1976) J Bacteriol 125:1080–1087
Parejko RA, Wilson PW (1970) Can J Microbiol 16:681–685
Pühler A, Burkardt HJ, Klipp W (1979) Mol Gen Genet 176:17–24
Rao RN (1976) J Bacteriol 128:356–362
Rawlings J, Shah VK, Chisnell JR, Brill WJ, Zimmermann R, Münck E, Orme-Johnson WH (1978) J Biol Chem 253:1001–1004
Roberts GP, MacNeil T, MacNeil D, Brill WJ (1978) J Bacteriol 136:267–279
Ruvkun GB, Ausubel FM (1980) Proc Natl Acad Sci USA 77:191–195
Shah VK, Brill WJ (1977) Proc Natl Acad Sci USA 74:3249–3253
Shah VK, Chisnell JR, Brill WJ (1978) Biochem Biophys Res Commun 81:232–236
Shah VK, Davis LC, Brill WJ (1975) Biochim Biophys Acta 384:353–359
Shanmugam KT, Chan I, Morandi C (1975) Biochim Biophys Acta 408:101–111
St John RT, Shah VK, Brill WJ (1974) J Bacteriol 119:266–269
Streicher S, Gurney E, Valentine RC (1971) Proc Natl Acad Sci USA 68:1174–1177
Streicher SL, Shanmugam KT, Ausubel F, Morandi C, Goldberg RB (1974) J Bacteriol 120:815–821
Tubb RS (1974) Nature (London) 251:481–485

Rhizobium Genetics

W. HEUMANN[1]

1 Introduction

The bacterial family of Rhizobiaceae, consisting of the two genera *Rhizobium* and *Agrobacterium,* creates a challenge to geneticists and biochemists because of its genetic and regulatory complexity. Both genera live as obligate aerobes saprophytically in soil and both are further characterized by their intimate association and cohabitation with plants. *Rhizobia* invade by a complex process of recognition and infection the roots of legumes inducing the cortex cells to remeristematisize and to build a new anatomically well-defined structure, the root nodule. The mechanism of this redifferentiation is not understood, but it seems plausible that the microsymbiont interferes directly or indirectly with the plant genome, inducing the growth and differentiation of new plant cells and regulating their structural organization with the result of nodule formation. This process of genetic regulation of the plant genome by the bacteria continues, the new meristem being located on the tip of the growing nodule. In the inner tissue of the differentiated nodule, the domicile of the bacteria is established and now, by a process which is again not understood, the plant regulates the bacterial genome, turning off the genomic regions responsible for cell propagation and turning on the genes responsible for nitrogen fixation and for delivery of the NH_4^+ formed to the plant cytosol. This process is accompanied by shaping the bacterial rods into enlarged branched bacteroids and by the formation of leghemoglobin which is the genetical product of both partners, the plant producing the globin (Sidloi Lumbroso et al., 1978) and the bacteria the heme cofactor (Cutting and Schulman, 1969). The leghemoglobin, being located in the plant cytosol and outside on the membrane surrounding the bacteroids, (Desh et al., 1976) transports oxygen to the ATP generating machinery of the bacteroids necessary for the N_2 reduction. As a result of the *Rhizobium* invasion the plant utilizes the bacteria to cover its own nitrogen supply, whereas the differentiated bacteroids finally die: the plant genetically parasitizes the bacteria.

The interaction of bacteria and host plant in the *Agrobacterium* infection is also regulated on the genetical level, but it is differently organized and leads to plant tumor formation (crown gall). *Agrobacteria* invade plants through occasional wounds. Probably all dicotyledons can act as hosts. The bacteria induce the propagation of plant cells on the infection site, but in contrast to the *Rhizobium* system the resulting tissue is unorganized, consisting of a chaotic mixture of parenchyme and tube cells: the tu-

1 Institut für Mikrobiologie und Biochemie, Lehrstuhl Mikrobiologie, Universität Erlangen-Nürnberg, Egerlandstr. 7, 8520 Erlangen/FRG

mor colonized by bacteria. Occasionally during propagation on the host or in tissue culture tumor cells can partly redifferentiate into roots or shoots called teratoma (Drummond, 1979). The genetic regulation of plant cells by bacteria is in this case much better understood than in the *Rhizobium* system, probably because it is much simpler. Virulent *Agrobacterium* strains always inherit a large extrachromosomal element called Ti plasmid, part of which, the T DNA, is transferred to plant cells and maintained there, indefinitely inducing oncogenic transformation. The T DNA is replicated and probably transcribed in the tumor tissue, it codes beside for the malignant properties for novel amino acids called opines. These can be used as nitrogen and energy source by the colonizing bacteria but not by the producing plant. This type of parasitism by *Agrobacteria,* induced by the possession of the Ti plasmid is called "opine farming" by Tempé et al. (1979). The second step of the *Rhizobium* system, the regulatory interference into the bacterial genome by the plant, does not occur during *Agrobacterium* tumor formation, except that the opines induce the conjugative transfer of Ti-plasmids between *Agrobacterium* cells. In this case the bacteria induce the metabolic machinery of the plant to produce substances for their own growth: the bacteria genetically parasitize the plant.

A comparison between the *Rhizobium* and the *Agrobacterium* system reveals the much higher evolved complexity of the *Rhizobium* symbiosis. The latter consists of at least a two-step genetical interference between bacterial and plant genome: first the induction of nodule formation, a genetical regulation of the plant genome by the bacteria and second the induction of nitrogen fixation, a genetical regulation of the bacterial genome by the plant. The *Agrobacterium* tumor induction on the other hand is only a one-step genetical interference into the plant genome by a bacterial genetic trait leading to the proliferation of an unorganized mixture of plant cells producing a special bacterial food, the opines.

Regarding the very close relationship between *Rhizobium* and *Agrobacterium* and the similarities and differences in the plant interaction of these two genera it seems reasonable to investigate their genetics simultaneously and to compare the results with each other.

2 Rhizobium Conjugation

2.1 Chromosomal Transfer Mobilized by P1 Plasmids

During recent years increasing efforts have been undertaken to elucidate the genetics of *Rhizobium,* mainly because of the agricultural importance of its symbiotic nitrogen fixation. Originally linkage mapping experiments by chromosomal transfer failed because no indigenous fertility systems could be found in *Rhizobium* wild-type strains eliciting a satisfactory transfer rate.

However, experiments to mobilize chromosomal transfer by plasmids of foreign origin turned out to be successful. Datta et al. (1971) discovered the almost unrestricted transferability of the R plasmid RP4 to all gram-negative bacteria. This broad host range plasmid has been originally isolated from *Pseudomonas aeruginosa;* it inherits beside genes responsible for its own replication and transfer, resistance markers for ampicillin, tetracycline, and kanamycin (Burkardt et al., 1979). It is classified in the incompatibility group P1. Meade and Signer (1977) performed mapping experiments with RP4 mobilized chromosomal transfer in *Rhizobium meliloti* and reported a recombination rate of $10^{-5}-10^{-6}$ per donor. At almost the same time Haas and Holloway (1976) found a fertility factor in *P. aeruginosa*, R68.45, with enhanced chromosome donor ability (Cda). Riess et al. (1980) reported that R68.45 differs from RP4 only by a short amplified region. R68.45 has been used successfully by Kondorosi et al. (1977) in *R. meliloti*, by Beringer and Hopwood (1976) and by Beringer et al. (1978b) in *R. leguminosarum* and by Casadesus and Olivares (1979) in *R. meliloti* to construct chromosomal maps. These efforts, recently summarized by Beringer (1980), always resulted in a circular linkage map. A comparison of these maps is difficult because partly different markers were used. But nevertheless some linkage groups show similarities.

A similar approach in mobilizing chromosomal transfer in *Rhizobium* has been undertaken by constructing R-primes inheriting a fragment of *Rhizobium* DNA. It was expected that a more efficient transfer could be achieved by the region of homology. Jacob et al. (1976) constructed RP4-prime plasmids by in vitro insertion of DNA fragments of *R. leguminosarum*. Such recombinant plasmids were no more efficient at chromosome mobilization than the unmodified plasmid. However recent experiments with in vitro constructed RP4-prime plasmids conducted by Julliot and Boistard (1979) were successful. Creating DNA homology between plasmid and *R. meliloti* chromosome DNA they were able to achieve single-origin polarized transfer at high frequency. Johnston et al. (1978b) and Kiss et al. (1980) observed the spontaneous in vivo formation of R68.45-primes inheriting *R. meliloti* chromosome sections. The broad host range of the R-prime plasmids was still retained and used to transfer *Rhizobium* genes or integrated *Rhizobium* phage DNA into other gram-negative bacteria and to study their expression.

2.2 Transfer of Symbiotic Genes with Indigenous Rhizobium Plasmids

Mainly two recent results were crucial for the further development of *Rhizobium* genetics: the detection of naturally occurring transmissible plasmids in *R. leguminosarum* coding for bacteriocin production (Hirsch, 1979) and the development of transposon mutagenesis (Beringer et al., 1978a). This method is based on the construction of so-called suicide plasmids RP4:: Mu which has originally been used by Van Vliet et al. (1978) to carry resistance transposons into *Agrobacterium tumefaciens* causing deletion mutants. Using this procedure Beringer et al. (1978a) integrated transposon Tn5 coding for kanamycin resistance into different sites of the chromosome of *R. leguminosarum, R. trifolii*, and *R. phaseoli* and also into indigenous plasmids of *R. leguminosarum* (Johnston et al., 1978a). Insertion of Tn5 into one of the bacteriocinogenic plasmids mentioned above allowed the selection of plasmid transconjugants simply by selection for kanamycin resistance. The transfer of one of these bacteriocinogenic Tn5 plasmids into a noninfective *R. leguminosarum* strain restored its infectivity, and the transfer into *R. trifolii* and *R. phaseoli* induced these strains to nodulate peas, but not in *R. meliloti* or in *Agrobacterium tumefaciens*. Brewin et al. (1980) investigated the symbiotic characters of *R. leguminosarum* strains which could be transferred by these plasmids in more detail, using different *R. leguminosarum* mutants impaired in infectivity or effectivity as recipients. It appears obvious from these investigations that continuation of research along these lines will finally lead to the understanding of the genetics of symbiosis of *Rhizobia*.

3 Plasmids of Rhizobiaceae

The two genera of the Rhizobiaceae, *Agrobacterium* and *Rhizobium*, inherit large plasmids. Zaenen et al. (1974) isolated these extrachromosomal elements of virulent *A. tumefaciens* strains. They have a molecular weight (MW) of $100-160 \times 10^6$ and carry the genetic information for oncogenicity, they are called Ti-plasmids (tumor-inducing). Nuti et al. (1977) have shown that different *Rhizobium* species also carry large extrachromosomal elements. These plasmids are generally twice as large as the Ti plasmids, their MW ranging between 90 and 350×10^6. In addition several smaller plasmids were often found in different *Agrobacteria* and *Rhizobia*, for review see Prakash et al. (1978). In the meantime maps of Ti-plasmids have been constructed, indicating the position of several genetic regions, for review see Schilperoort et al. (1979). In contrast very little is known so far about the genetic information coded in the large *Rhizobium* plasmids, but in analogy to the *Agrobacterium* system it is suspected that much, if not all, of the genetic information responsible for the plant interaction is plasmid coded also in *Rhizobium*. The first genetic evidence for extrachromosomal inheritance of symbiotic genes provided by the Beringer group has been described in the foregoing chapter. Nuti et al. (1979) reported recently biophysical results indicating that *Rhizobium leguminosarum* nitrogenase genes are also plasmid encoded. They could show that cloned DNA fragments of the *Klebsiella pneumoniae* nitrogenase structural genes hybridized with restriction fragments of *Rhizobium* plasmid encoded. They could show further combined efforts by genetic and biophysical methods will elucidate in much more detail the structure and function of the *Rhizobium* plasmids.

4 The Complex Genome of the Rhizobiaceae

4.1 Formation of Convertants and Segregants

Occasionally we observed variations of pure, bacteriologically controlled cultures of wild-type *Rhizobia* and wild-type *Agrobacteria* as a result of unphysiological treatment and after nitrite mutagenesis (stress conditions). Since at the first glance these observations seem to be in disagreement with the basic concept of pure cultures, we studied the different types of variants and the conditions of their appearance in more detail under strong bacteriological control. For this purpose we used as *Rhizobium* wild type mainly *R. meliloti*, a strain kindly provided by Dr. Dénarié, numbered 20-1 in our laboratory. *R. meliloti* is genetically the most homogeneous of all *Rhizobia*, shown, for instance, by *Xho*I DNA restriction analysis of several strains of different origin (Merkt, 1979). As *Agrobacterium* wild type we analyzed mostly a strain with the laboratory number 18-1, a derivative of the octopine-strain B6 used in many laboratories. We tested its oncogenicity under greenhouse conditions with *Bryophyllum* as host plant and under continuous microbiological control with potato discs as host.

The conditions and the efficiency of inducing the strain variations mentioned are different for different strains. The following procedure is relatively successful: growth of the wild-type bacteria overnight in nutrient broth, storage of the washed moist pellet at 30 °C for 8–10 days, culture of a sample in Jensen plant nutrient solution (1942) free of combined nitrogen except for 1 mmol KNO_3 in a rotary shaker for again 8–10 days, repeated plating of culture samples on nutrient agar plates as a test for colony variation. As a result of this treatment *Rhizobium* as well as *Agrobacterium* wild-type strains produce a characteristic pattern of new strains, the appearance of which can only be explained by reorganization of their wild-type genomes, though we do not understand so far the mechanism of this process.

The pattern of new strains formed is characteristic, and these new strains are stable under normal growth conditions, so that their properties can be analyzed and defined, however, under the stress conditions described above the process of reorganization can continue or reverse. Reversible reorganization is called conversion, irreversible reorganization is defined as segregation respectively (Table 1).

4.2 Description of the Conversion and Segregation Process

As Table 1 shows, stress treatment of *R. meliloti* 20-1 *wild*-type leads to the reversible conversion to *nif*-type and *car*-type (bact). The frequency of this process is hardly to define. Cell growth during treatment in Jensen-KNO_3 solution is almost zero. If, for instance, five parallel samples are tested, usually the appearance of new strains starts simultaneously in all five samples, beginning with *nif*-type and continuing to *car*-type (bact), so that finally three different strains grow in the test platings. The *nif* strains can easily be recognized by their colony shape, since *nif*-type cells are able to swarm on the surface of 1.5% agar plates. *car*-Type strains are discernible by their bright yellow carotenoid pigmentation. *car*-Type (bact) is further characterized by the cell shape. The cells resemble nodule bacteroids except that *car*-type bacteroids grow by budding (Fig. 1). If the experiment is started with *nif*-type strains, *car*-type (bact) colonies are usually the first ones to appear and later on also revertants to the *wild*-type are formed. The reversion of *car*-type (bact) to *wild*-type has never been observed under stress conditions. However in nodulation experiments after prolonged growth of lucerne plants inoculated with *car*-type (bact) strains, under bacteriologically controlled conditions, occasionally some samples nodulate and the reisolation of bacteria from nodules shows the wild-type revertants. *car*-Type (bact) strains are relatively unstable on nutrient agar plates and segregate spontaneously into *car*-type (st), rod-shaped. This process is not reversible. Since the other strains listed in the line of minimal strains are not as easily discernible from the respective strains in the *Rhizobium* line as bacteroid and rod growth in the carotenoid strains, *car*-type (st) serves as starting material for stress treatment to induce conversion in the minimal line. As a result similar convertants appear as in the *Rhizobium wild*-type line. The *pro wild*-type is structurally homologous to the *wild*-

Table 1. The conversion and segregation pattern of the Rhizobiaceae is shown in the table in three horizontal lines, the *Rhizobium* line, the *Agrobacterium* line, and the line of minimal strains in between. The interchange of strains along the horizontal arrows is generally reversible, though the reversion rate in the direction of the solid arrows is more frequent than that in the opposite direction. Reversible interchange is defined as conversion, whereas the formation of minimal strains from both the *Rhizobium* and *Agrobacterium* line is irreversible; it is called segregation. This implies that conversion means genome-reorganization without gain or loss of DNA, whereas the irreversibility of segregation is probably caused by loss of genetic information. The table further shows that the row of minimal strains is pertinent to both the *Rhizobia* and *Agrobacteria*. Correlation of strains on the basis of conjugational fertility, DNA density assayed by analytical ultracentrifugation, DNA similarity compared by *Xho*I restriction enzyme analysis, phage typing and other characters, is indicated by surroundings of the respective strains with dashed lines.

The table shows results of *Rhizobium meliloti* and *Agrobacterium tumefaciens*. However, since some of the conversion and segregation steps have already been reproduced with strains of *R. leguminosarum* and *R. lupini*, there are indications of the general validity of the pattern.

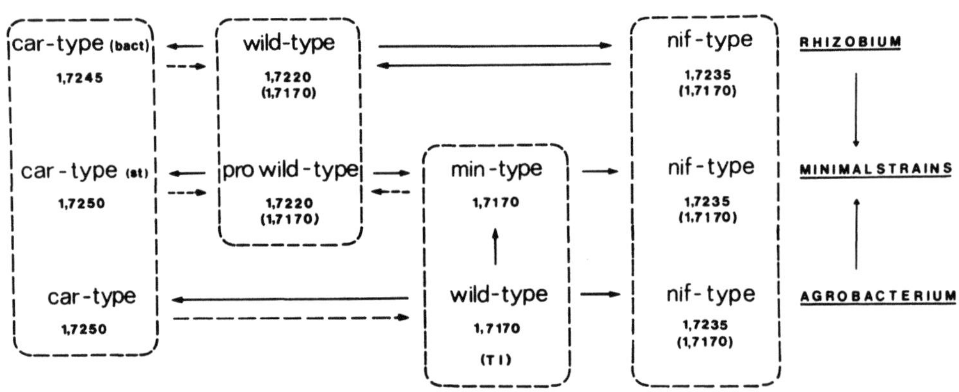

CONVERSION AND SEGREGATION OF RHIZOBIUM AND AGROBACTERIUM

Symbols:

car-type	=	yellow colonies pigmented by carotenoid formation
car-type (bact)	=	cells growing in form of budding bacteroids instead of rods, Fig. 1
car-type (st)	=	rod shaped cells (Stäbchen)
wild-type *(Rhizobium)*	=	effectively nodulating strain
wild-type *(Agrobacterium)*	=	oncogenic in *Bryophyllum* and potato discs
pro wild-type	=	growth characteristics identical with *wild*-types, except nodulation or tumor indication
nif-type	=	strains highly effective in free-living nitrogen fixation activity on nitrogen-free agar plates under atmospheric conditions
min-type	=	minimal-type strain. DNA *Xho*I restriction pattern almost identical with the *Agrobacterium* wild-type pattern, but not nodulating and not oncogenic

The figures given under the type symbols are the DNA density values ς in g/cm^3, assayed by analytical ultracentrifugation, the figures in parenthesis designate plasmid DNA densities. TI = *Agrobacterium*, TI-plasmid

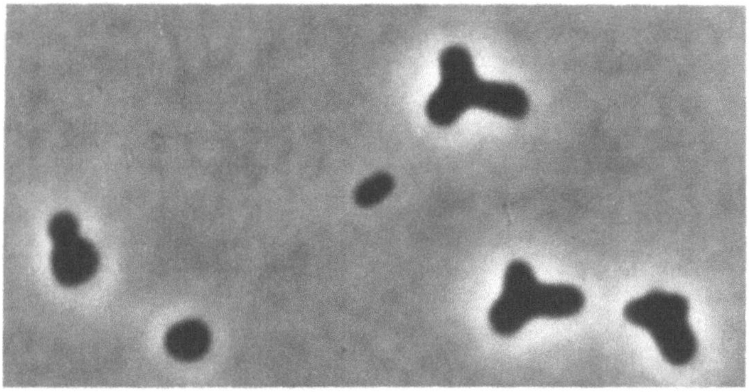

Fig. 1. Bacteroid strain *car*-type (bact) 33-1. The cells grow by budding. (Phase contrast)

type and the *nif*-type in the minimal line to the *nif*-type in the *Rhizobium* line. Conversion of *car*-type (st) further produces a new strain called *min*-type (minimal), which has its structural counterpart in the *Agrobacterium* line, though it can be produced from *Rhizobium wild*-type and *Agrobacterium wild*-type as well. A reversion of the *nif*-type in the minimal line to the *min*-type has never been observed. A similar conversion pattern as in *wild*-type *Rhizobium* can be observed in *wild*-type *Agrobacterium*. It can converse into a *car*-type and *nif*-type segregate into a *min*-type. *min*-Type strains made from *Agrobacterium* and from *Rhizobium* cannot be differentiated by any criteria.

4.3 Characterization of the Convertants and Segregants

(For sake of simplicity in the following the strain symbols will be preceded with an *R*, *M* or *A* depending to what line in Table 1 the respective strain belongs, to *Rhizobium*, Minimal or *Agrobacterium* respectively.)

4.3.1 Nodulation

All strains listed in Table 1 were tested for effective nodulation with lucerne seedlings. The experiments were carried out in Erlenmeyer flasks covered with glass cups preventing contamination. The flasks were bacteriologically controlled during the experiment. *R wild*-type strain 20-1 nodulates lucerne, effective nodules are visible usually after 10–14 days, the seedlings never show signs of nitrogen starvation under these conditions. *R nif*-type as such does not nodulate. However, after a prolonged period of growth – more than two months – especially when the seedlings are kept in agar medium, *R nif*-type reverts to *wild*-type and nodulates the plants so that these, already yellowish by nitrogen starvation, start growing again, developing numerous effective nodules. A similar process can be observed in nodulation experiments carried out with *R car*-type (bact) though the reversion rate is less frequent with this convertant.

Fig. 2. *Agrobacterium tumefaciens* tumors on potato disc, three months after inoculation

4.3.2 Tumor Induction

A wild-type is the sole strain of the scheme which induces tumors in *Bryophyllum* leaves and stems and on the surface of potato discs as well (Fig. 2). However, *R wild*-type 20-1 likewise induces growth on potato disc similar to *Agrobacterium* tumors, but not with the same efficiency (Fig. 3). Repeated passages of *R wild*-type on potato discs did not increase the growth-inducing efficiency. *M pro* wild type finally very often induces the growth of stems and roots on potato discs (Fig. 4).

4.3.3 Conjugational Fertility

All *M car*-type (st) strains show high conjugational fertility. Transfer, recombination and replication mechanisms were investigated in detail (Fig. 5). The results are reviewed in Heumann (1979). *M car*-types (st) received by conversion of *R wild*-types of different *Rhizobium* species seem to be identical with regard to the conjugation system and all other characteristics tested; obviously the host specificity is lost during segregation. *M car*-type exhibits also conjugational chromosome transfer to *R car*-type (bact) strains and to *A car*-type strains though the latter can act as recipients only. No indigenous fertility system with a reasonable transfer frequency has been observed in the other strains listed in Table 1.

Fig. 3. Tumorous growth on potato disc induced by *Rhizobium meliloti* 20-1, three months after inoculation

Fig. 4. Stems and roots on potato disc induced by inoculation with *pro wild*-type strain 32-1, three months after inoculation

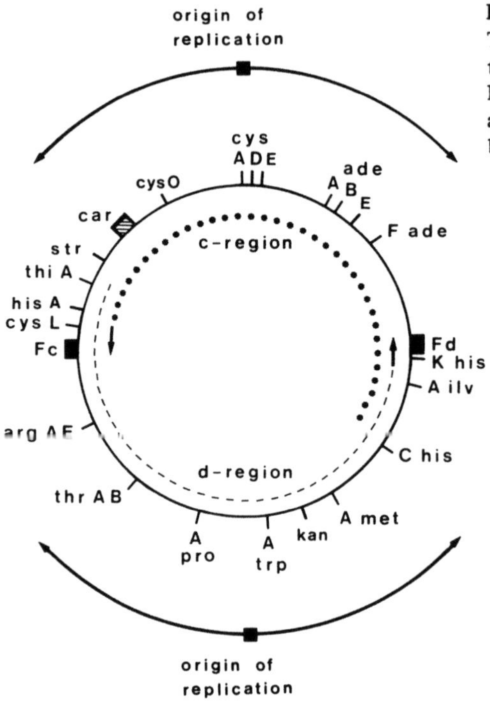

Fig. 5. Chromosome map of *car*-type (st) strains. The conjugation system is characterized by two transfer regions, the c and d regions, originating at Fc and Fd respectively. The replication starts always at one of two starting points and proceeds bidirectionally (Heumann, 1979).

4.3.4 Phage Typing

Virulent phages were isolated for *M car*-type (st) and *A wild*-type. The search for virulent phages lysing *R wild*-type 20-1 and *R nif*-type has not been successful so far. Phage strain 1-56-1 forming clear plaques on *M car*-type (st) lyses similarly *R car*-type (bact) and *A car*-type, but not the other strains. Phage 18-1-1, forming clear plaques on *A wild*-type lyses *M min*-type with identical efficiency, but does not react with the other strains.

4.3.5 DNA Restriction Pattern

Merkt (1979) tested in our laboratory different restriction endonucleases for the production of a clear-cut fragment pattern of the total DNA of all strains listed in Table 1. We finally chose *Xho*I, an enzyme described by Gingeras et al. (1978). We isolated the enzyme in our laboratory from *Xanthomonas holcicola* (Rösch, 1979). *Xho*I cleaves the total DNA of all strains completely and produces relatively little smear in the background in the electropherogram. *Xho*I restriction analysis turned out to be a sensitive tool for the comparison of the DNA of the different strains. The *Xho*I restriction pattern of each strain type isolated in different independent experiments with partly different methods as a control were completely identical (not shown). A comparison of the restriction pattern of different strains, shown in Fig. 6, gave the following results: *R wild*-type 20-1 and *M pro wild*-type 32-3 have very similar DNA according to the *Xho*I fragment pattern. The same high similarity is revealed between the *Xho*I

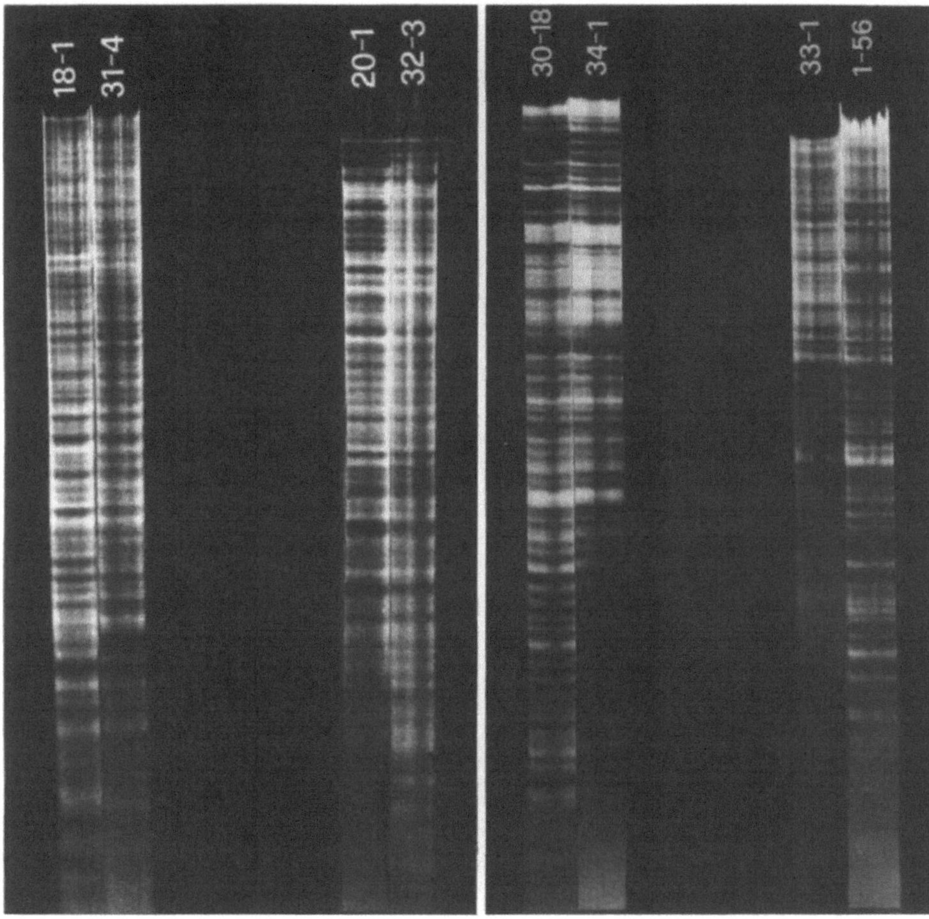

a b

Fig. 6 a, b. *Xho*I restriction pattern of total cell DNA of different strains listed in Table 1. The electrophoretic fragment patterns are ordered in pairs of high similarity.

20-1	*R. meliloti wild*-type;	*30-18*	*Rhizobium nif*-type;
32-3	*pro wild*-type, minimal strain;	*34- 1*	*nif*-type, minimal-strain;
18-1	*Agrobacterium wild*-type;	*33- 1*	*Rhizobium car*-type (bact);
31-4	*min*-type, minimal strain;	*1-56*	*car*-type strains, minimal strain

fragmentation of *R car*-type (bact) 33-1 and *M car*-type strains 1-56 as well as between *R nif*-type 30-18 and *M nif*-type 34-1. The fragment pattern of *A wild*-type 18-1 and *M min*-type 31-4 are also very similar though few bands of *A wild*-type 18-1 seem to be missing in the pattern of *M min*-type 31-4. The results of *Xho*I DNA restriction analysis show that the DNA of the different strains is very little changed during segregation, they confirm the phage typing results. Conversion, however, must be a process which changes the *Xho*I restriction pattern of the strains fundamentally so that no similarities between strains in the horizontal lines in Table 1 can be observed.

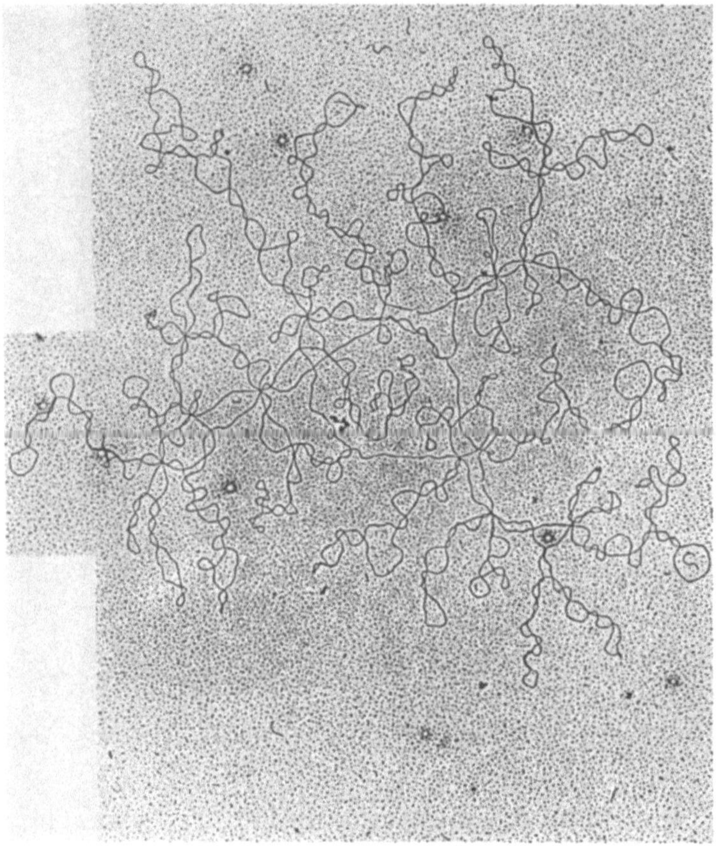

Fig. 7a. *Rhizobium meliloti* large plasmid, contour length 90 μm. Preparation and electron-micrograph by Merkt (1979)

4.3.6 DNA Analysis by Analytical Ultracentrifugation

DNA density measurements in CsCl density gradient centrifugation indicate that some of the DNA of the strains listed in Table 1 have a small peak at $\varsigma = 1{,}717\ g/cm^3$. Preparative separation of the low density DNA in Cesiumchloride-Ethidiumbromide density centrifugation and electron microscopical preparations of this DNA revealed large open circular DNA molecules beside plasmid DNA of smaller size. The size of the large plasmids assayed by contour length measurement was between MW 160 and $220 \cdot 10^6$, (Fig. 7), that of the smaller plasmids MW $24–30 \cdot 10^6$, and MW $10 \cdot 10^6$ as well. Small size plasmids were found especially in *R nif*-type preparations (Merkt, 1979). *M pro wild*-type and *M nif*-type DNA give the same density profile as the respective strains in the *R* line. We were neither able to isolate plasmid DNA in any of the *car*-types nor in the *min*-type DNA preparation. Since these strains converse reversibly into strains inheriting plasmid DNA, reversible integration of plasmid DNA into the chromosome must occur. The density of the chromosomal DNA of the strains listed in the *Rhizobium* line in Table 1 increases from *wild*-type $\varsigma = 1.7220\ g/cm^3$ over *nif*-type $\varsigma = 1.7235\ g/cm^3$ to *car*-type (bact) $\varsigma = 1.7245\ g/cm^3$. We do not understand so far the DNA reorganization mechanism which causes this phenomenon. The similarity of

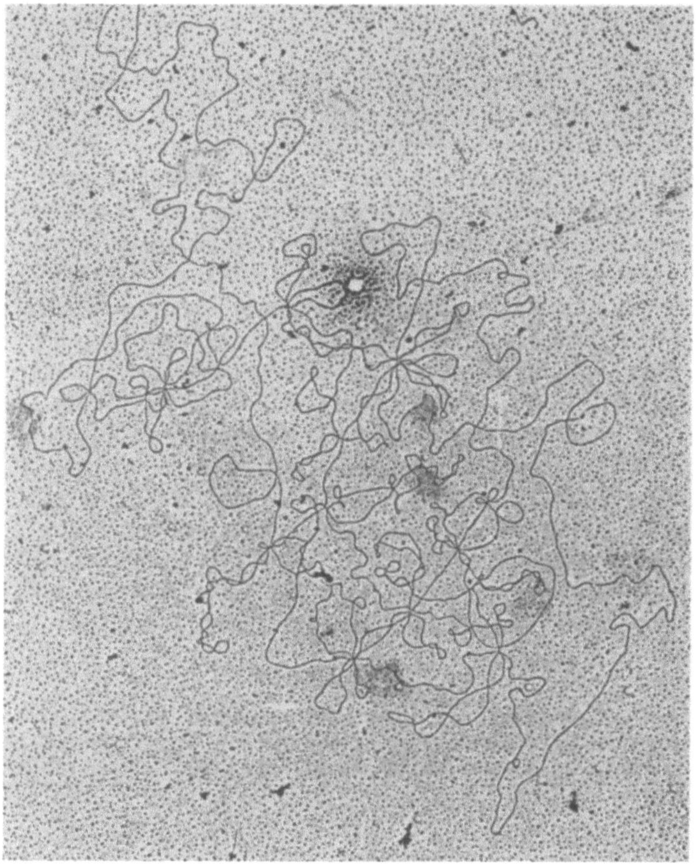

Fig. 7 b. Large plasmid isolated from *Rhizobium nif*-type, contour length 100 μm. Preparation and electronmicrograph by Merkt (1979)

strains surrounded by dashed lines is again confirmed by the DNA density profiles of these strains.

4.3.7 Nitrogen Fixation by the nif-Types

The high nitrogen-fixing activity of the two *nif*-type strains on solid agar plates free of combined nitrogen is especially remarkable. The nitrogenase activity continues on plates over more than eight weeks, it results in 15–20 nmol ethylene/mg protein · min. The nitrogenase of these strains seems to be protected against atmospheric oxygen by a thick and tenacious layer of exopolysaccharides around the cells. No nitrogen fixation can be observed in liquid medium, probably because the polysaccharide layer dissolves into the medium. The genetic *nif* region of these strains hybridizes with cloned *Klebsiella nif* DNA (Jäckel, 1980). These *nif*-type strains seem to be a useful model for further study of the *Rhizobium nif* genes and their nitrogenase.

5 Conclusion

The most plausible explanation of the results of this study seems the conception that the wild-type genome of the Rhizobiaceae developed during evolution by the combination of genomes of at least two differently evolved bacteria, a diazotrophic and a carotenoid-pigmented strain. The conjugation system of *car*-type (st) strains might serve as a model for such genome combinations. After chromosomal transfer the exogenote is inserted into the recipient chromosome, forming a tandem merodiploid structure, which segregates by intrachromosomal recombination. The segregants listed in the middle line in Table 1 and defined as "minimal strains" probably represent the gene pool for the Rhizobiaceae, *Rhizobium* and *Agrobacterium* originally formed by genome combinations. That the wild-type lines of these two genera are connected to these minimal strains by several genetic and molecular characteristics is one of the results of this study. The next question is: what causes these minimal strains to become *Rhizobium* with a specific host range or tumor-inducing *Agrobacterium* respectively? We have no reliable experimental results so far to answer this question. But one could speculate that during a primary infection of legumes or other dicotyledons by minimal strains these were transformed by the respective plant DNA into host-specific *Rhizobium* or oncogenic *Agrobacterium: pro wild*-type would be transformed into *Rhizobium* and *min*-type into *Agrobacterium.* A detailed analysis of the genomes involved will help to solve this question.

Acknowledgments. This work is the result of the scientific engagement of many coworkers. I have to thank Mrs. A. Hofmann for patiently writing and rewriting the manuscript, Mrs. E. Wagner for skillful technical assistance, and Miss G. Hauke for excellent photographic work.

References

Beringer JE (1980) The development of *Rhizobium* genetics. The fourth Fleming lecture. J Gen Microbiol 116:1–7

Beringer JE, Hopwood DA (1976) Chromosomal recombination and mapping in *Rhizobium leguminosarum.* Nature (London) 264:291–293

Beringer JE, Beynon JL, Buchanan-Wollaston AV, Johnston AWB (1978a) Transfer of drug-resistance transposon Tn5 to *Rhizobium.* Nature (London) 276:633–634

Beringer JE, Hoggan SA, Johnston AWB (1978b) Linkage mapping in *Rhizobium leguminosarum* by means of R plasmid-mediated recombination. J Gen Microbiol 104:201–207

Brewin NJ, Beringer JE, Buchanan-Wollaston AV, Johnston AWB, Hirsch PR (1980) Transfer of symbiotic genes with bacteriocinogenic plasmids in *Rhizobium leguminosarum.* J Gen Microbiol 116:261–270

Burkardt HJ, Riess G, Pühler A (1979) Relationship of group P1 plasmids revealed by heteroduplex experiments: RP1, RP4, R68, and RK2 are identical. J Gen Microbiol 114:341–348

Casadesus J, Olivares J (1979) Rough and fine linkage mapping of the *Rhizobium meliloti* chromosome. Mol Gen Genet 174:203–209

Cutting JA, Schulman HM (1969) The site of heme synthesis in soybean root nodules. Biochem Biophys Acta 192:486–493

Datta N, Hedges RW, Shaw EJ, Sykes RB, Richmond MH (1971) Properties of an R factor from *Pseudomonas aeruginosa.* J Bacteriol 108:1244–1249

Desh P, Verma S, Bal Araga K (1976) Intracellular site of synthesis and localization of leghemoglobin in root nodules. Proc Natl Acad Sci USA 73:3843–3847

Drummond M (1979) Crown gall disease. Nature (London) 281:343–347
Gingeras TR, Myers PA, Olson JA, Handberg FA, Roberts RJ (1978) A new specific endonuclease present in *Xanthomonas holcicola* and *Brevibacterium luteum*. J Mol Biol 118:113–122
Haas D, Holloway BW (1976) R factor variants with enhanced sex factor activity in *Pseudomonas aeruginosa*. Mol Gen Genet 147:234–251
Heumann W (1979) *Rhizobium lupini* genetics. In: Arber W et al. (eds) Current topics in microbiology and immunology, vol 88. Springer, Berlin Heidelberg New York, pp 1–23
Heumann W, Springer R (1977) Formation of merodiploid clones by conjugation in *Rhizobium lupini*. Mol Gen Genet 150:73–79
Hirsch PR (1979) Plasmid determined bacteriocin production by *Rhizobium leguminosarum*. J Gen Microbiol 113:219–228
Jacob AE, Cresswell JM, Hedges RW, Coctzee JN, Beringer JE (1976) Properties of plasmids constructed by the in vitro insertion of DNA from *Rhizobium leguminosarum* or *Proteus mirabilis* into RP4. Mol Gen Genet 147:315–323
Jäckel B (1980) unpublished results
Jensen HL (1942) Nitrogen fixation in leguminous plants. I General characteristics of root nodule bacteria isolated from species of *Medicago* and *Trifolium* in Australia. Proc Linn Soc NSW 66: 98–108
Johnston AWB, Beynon JL, Buchanan-Wollaston AV, Setchell SM, Hirsch PR, Beringer JE (1978a) High frequency transfer of nodulating ability between strains and species of *Rhizobium*. Nature (London) 276:635–636
Johnston AWB, Setchell SM, Beringer JE (1978b) Interspecies crosses between *Rhizobium leguminosarum* and *R. meliloti:* formation of haploid recombinants and of R-primes. J Gen Microbiol 104:209–218
Julliot JS, Boistard P (1979) Use of RP4-prime plasmids constructed in vitro to promote a polarized transfer of the chromosome in *Escherichia coli* and *Rhizobium meliloti*. Mol Gen Genet 173: 289–298
Kiss GB, Dobo K, Dusha I, Breznovits A, Orosz L, Vincze E, Kondorosi A (1980) Isolation and characterization of an R-prime plasmid from *Rhizobium meliloti*. J Bacteriol 141:121–128
Kondorosi A, Kiss GB, Forrai T, Vincze E, Banfalvi Z (1977) Circular linkage map of the *Rhizobium meliloti* chromosome. Nature (London) 268:525–527
Meade HM, Signer ER (1977) Genetic mapping of *Rhizobium meliloti*. Proc Natl Acad Sci USA 74:2076–2078
Merkt B (1979) Biophysikalische und biochemische Charakterisierung des *Rhizobium* Genoms. Diss Erlangen, p 62
Nuti MP, Ledeboer AM, Lepidi AA, Schilperoort RA (1977) Large plasmids in different *Rhizobium* species. J Gen Microbiol 100:241–248
Nuti MP, Lepidi AA, Prakash RK, Schilperoort RA, Cannon F (1979) Evidence for nitrogen fixation (nif) genes on indigenous *Rhizobium* plasmids. Nature (London) 282:533–535
Prakash RK, Hooykaas PJJ, Ledeboer AM, Kijne J, Schilperoort RA (1978) Detection, isolation and characterization of large plasmids in *Rhizobium*. In: Orne-Johnston WH, Newton WE (eds) Proc IIIrd Int Symp Nitrogen Fixation. University Park Press, Madison
Riess G, Holloway BW, Pühler A (1980) R.68.45, a plasmid with chromosome mobilizing ability (Cma) carries a tandem duplication of 2120 bp. Genet Res (in press)
Rösch A (1979) unpublished results
Schilperoort RA, Hooykaas PJJ, Klapwijk PM, Koekman BP, Nuti MP, Ooms G, Prakash RK (1979) Characters of large plasmids in Rhizobiaceae involved in the interaction with plant cells. In: Timmis KN, Pühler A (eds) Plasmids of medical, environmental and commercial importance. Elsevier/North Holland Biomedical Press, Amsterdam New York, p 339
Sidloi-Lumbroso R, Kleiman L, Schulman HM (1978) Biochemical evidence that leghemoglobin genes are present in the soybean but not in the *Rhizobium* genome. Nature (London) 273: 558–560
Springer A, Heumann W (1978) Genetical analysis of merodiploidy in *Rhizobium lupini*. Mol Gen Genet 165:57–63

Tempé J, Guyon P, Tepfer D, Petit A (1979) The role of opines in the ecology of the Ti-plasmids of *Agrobacterium.* In: Timmis KN, Pühler A (eds) Plasmids of medical, environmental and commercial importance. Elsevier/North Holland Biomedical Press, Amsterdam New York, p 353

Van Vliet F, Silva B, van Montagu M, Schell J (1978) Transfer of RP4: :Mu plasmids to *Agrobacterium tumefaciens.* Plasmid 1:446–455

Zaenen J, van Larebeke N, Teuchy H, van Montagu M, Schell J (1974) Supercoiled circular DNA in crown gall inducing *Agrobacterium* strains. J Mol Biol 86:109–127

Some Aspects of the Physiology of Dinitrogen Fixation

J.R. POSTGATE, R.R. EADY, R.A. DIXON, S. HILL, D. KAHN, C. KENNEDY, P. PARTRIDGE, R. ROBSON, and M.G. YATES[1]

1 Introduction

The physiology of an organism, or of a biological process, results from the integration of its biochemical and genetic features in the living organism. It can be approached descriptively, by a study of the responses of the organism or process to environmental changes, but more sophisticated understanding of physiology requires the integration of genetical and biochemical knowledge.

In practice, descriptive physiology usually precedes analytical physiology. Biological nitrogen fixation provides a good example in which physiological constraints observed empirically during the first half of the century can now be rationalized, at least in part, in terms of the peculiar biochemistry of nitrogenase and the complexity of the genetic apparatus necessary to cope with this enzyme system. The prevalence of biocoenoses involving mutualistic or symbiotic associations of diazotrophs with higher organisms adds a further order of complexity to the physiology of biological nitrogen fixation in the natural environment; the physiology of such associations will not be discussed further since it is additional to the physiology of diazotrophy per se.

The basic biochemical features of nitrogenase which determine the physiology of diazotrophy are the following.

1. Need for a specific electron transport system, usually involving a flavodoxin and/or a ferredoxin, to donate electrons to the nitrogenase proteins.

2. Need to expend ATP during nitrogen fixation at the surprising rate of some 2 to 2.5 molecules ATP hydrolysed/electron transferred by the enzyme to its substrate.

3. Need to incorporate Fe and S into several of the proteins of the nitrogenase system and, in addition, to incorporate Mo into the terminal protein.

4. Obligation to evolve H_2 during nitrogen fixation.

5. Need to exclude O_2 from the nitrogenase proteins.

6. Need to restrict nitrogenase synthesis to situations in which it is economic (e. g., in the absence of fixed N) and practicable (low or zero oxygen tension).

Though the reasons for these biochemical features are far from well understood, the many ways in which they relate to the special physiological properties of diazotrophs have been discussed and reviewed in several publication. Therefore a few examples only will be recapitulated here, as follows.

The need for low-potential ferredoxins of flavodoxins, which are autoxidizible, may well be exploited in *Azotobacter* to dispose of electrons when nitrogen fixation

1 ARC Unit of Nitrogen Fixation, University of Sussex, Brighton, BN1 9RQ/UK

is "switched off" in response to oxygen stress (Scherings et al., 1977; Robson and Postgate, 1980). In *Klebsiella,* it seems likely that electron transport proteins are specified by genes within the *nif* cluster (St. John et al., 1975; Roberts et al., 1978; Hill and Kavanagh, 1980).

The need for ATP accounts in part for low substrate yields of nitrogen-fixing as against ammonia-assimilating populations of the same microbe (see, for example, Hill et al., 1972).

The need for Mo, known for many decades (Bortels, 1930), involves a gene product also needed for nitrate reductase synthesis (Kennedy and Postgate, 1977; Skotnicki and Rolfe, 1977); some form of Mo may regulate nitrogenase synthesis (below).

The obligation to evolve H_2, seemingly an intrinsic property of nitrogenase (see Robson and Postgate, 1980), leads to physiologically inefficient nitrogen fixation; some organisms use an uptake hydrogenase to re-utilise the H_2 from further nitrogen fixation, thus augmenting their physiological efficiency (see Evans et al., 1980; Walker and Yates, 1978; Robson and Postgate, 1980).

The need to exclude oxygen arises principally from the destructive effect of O_2 on the nitrogenase proteins and also from the autoxidizibility of at least some of the specific electron transport proteins. It has led to the evolution of physiological stratagems for excluding O_2 ranging in complexity from simple respiration to elaborate compartmentation (see Postgate, 1975; Stewart, 1975; Robson and Postgate, 1980).

Finally, the energy cost and oxygen sensitivity of nitrogenase are reflected in a complex regulatory system in which both fixed N (as NH_4^+) or O_2 regulate synthesis and sometimes function of the enzyme (see Merrick et al., 1980; Robson and Postgate, 1980).

It must be clear that analysis of the physiology of diazotrophy in biochemical-genetical terms has now become too large a subject to be reviewed in the space available here. Therefore this contribution will concentrate on three areas where recent developments have increased our understanding of the physiology of the diazotrophic cell.

2 Oxygen as a Regulator of Nitrogenase

2.1 Conformational Protection and Oxygen Regulation in Azotobacter

The oxygen relations of *Azotobacter* were originally rationalized (Dalton and Postgate, 1969 a, b) in terms of respiratory protection, which enabled nitrogenase to function protected from O_2 by aerobic respiration, backed up by conformational protection, whereby the oxygen-sensitive sites of nitrogenase became physically protected from O_2 damage when the O_2 stress was strong enough to overcome respiratory protection. In the conformationally protected state, nitrogenase was presumed to be inactive but undamaged. A phenomenon whereby nitrogenase activity became reversibly "switched" on or off in response to decreased or increased O_2 stress (Drozd and Postgate, 1970 a, b) was interpreted as an expression of conformational protection (e. g., Postgate, 1975). However, it may equally have been an expression of the autoxidizibility of electron carriers (Yates and Jones, 1974; Scherings et al., 1977). The question is still open.

The oxygen-tolerant "particulate" nitrogenase which could be extracted from *Azotobacter* (Bulen et al., 1965; Kelly, 1969 a; Reed et al., 1974) was regarded as conformationally protected, and examination of such oxygen-tolerant extracts has considerably clarified the nature of conformational protection. Though various protective agents had been observed (Yates, 1970), the discovery (Haaker and Veeger, 1977; Scherings et al., 1977) that "Shethna's protein" or "*A. vinelandii* FeS protein II" formed a 1:1 complex with purified *A. vinelandii* nitrogenase and passively protected the enzyme proteins from O_2 damage seems to have physiological relevance. The protective protein is an Fe_2S_2 protein of M.W. 24,000 and E_0' of -0.225 mV and it binds to nitrogenase in its oxidized form. A comparable protein of M.W. 14,000 has been obtained from *A. chroococcum* (Robson, 1979 a) and is also protective. Protection requires equimolar Mg^{2+} and is associated with aggregation of the nitrogenase-protective protein complex to give the "particulate" nitrogenase of earlier workers (Haaker and Veeger, 1977; Scherings et al., 1977). The plausible view has been proposed that the oxidized form of the protective protein, in the presence of Mg^{2+}, binds to the nitrogenase proteins, forming an inactive complex in which they are protected from O_2 damage; disappearance of O_2 allows formation of the reduced species of the protective protein and its consequent dissociation from the nitrogenase proteins, followed by renewed nitrogenase activity (Scherings et al., 1977).

NH_4^+ is the classical regulator of nitrogenase biosynthesis in all nitrogen-fixing bacteria. The reality of conformational protection raises questions regarding its possible role as a regulator of nitrogenase fuction and/or synthesis. Dalton and Postgate (1969 a, b) proposed the idea of conformational protection partly because chemostat experiments indicated that nitrogenase activity was regulated in response to the *duration* of the exposure of populations of *A. chroococcum* to oxygen. Thus, chemostat populations fixed similar amounts of N_2 per unit culture volume over a 3-fold range of growth rates (see Hill et al., 1972). Dr. J.W. Drozd (unpubl. results) obtained a hint that biosynthesis rather than function of nitrogenase was regulated by oxygen because the maximum acetylene reduction rate of whole *A. chroococcum* was proportional to dilution rate and this point was later established clearly by Robson (1979 b) who showed regulation of pulse-labelled nitrogenase peptide synthesis by O_2. Some evidence exists that O_2 regulates nitrogenase synthesis in *Rhizobium japonicum* (see Robson and Postgate, 1980).

Robson's (1979 b) demonstration of O_2 regulation of nitrogenase biosynthesis in *A. chroococcum* has been further studied. O_2 repressed synthesis more rapidly than NH_4^+: cell suspensions were repressed by O_2 within 2 min compared with some 15 min for NH_4^+ repression; derepression from O_2 was equally rapid. The minimum oxygen tension required to repress nitrogenase synthesis was that which had "switched off" nitrogenase activity by 20-fold (i. e., to 5% of maximum activity), so function of nitrogenase is more sensitive to O_2 than synthesis in this organism (Fig. 1). Rifampicin at 100 μg/ml inhibited mRNA synthesis (according to ^3H-uracil incorporation) and also general ^{35}S-labelled protein synthesis, but nitrogenase synthesis was detectable immediately after the relief of O_2 repression even in the presence of rifampicin. Thus O_2 regulates by a different mechanism from NH_4^+, one which probably operates at the translational level. *Nif* mRNA apparently has an unusually long half-life in rifampicin-treated *A. chroococcum.*

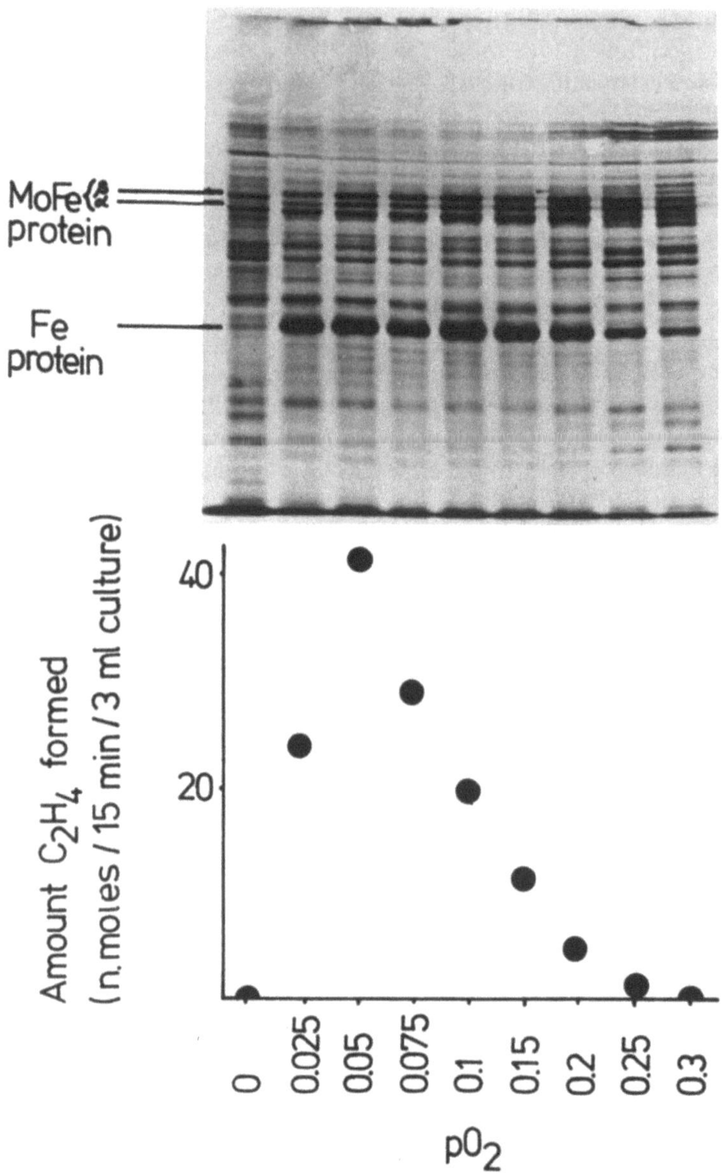

Fig. 1. Effects of oxygen tension on synthesis and activity of *Azotobacter chroococcum* nitrogenase. Cultures of *A. chroococcum* NCIB 8003 fixing N_2 at various oxygen tensions were pulse-labelled and examined by SDS-gel electrophoresis for nitrogenase peptides (see Robson, 1979a). The direction of migration is downwards; the faster-moving of the O_2-repressed proteins is the Fe-protein of nitrogenase (Ac2), the slower-moving pair is the molybdoprotein (Ac1). The *lower curve* shows the nitrogenase activities, assessed by acetylene reduction, of the populations shaken in flasks at the same oxygen tensions (cf. Drozd and Postgate, 1970a).

Regulation of nitrogenase biosynthesis by O_2 in *A. chroococcum* thus seems established. Conformational protection provides in principle a means of regulation of function of nitrogenase, if the proposition that the conformationally protected enzyme is inactive is valid. Plots of nitrogenase activity against oxygen tension give the familar bell-shaped curve shown in Fig. 1 (see also Drozd and Postgate, 1970 a) and it is tempting to regard the descending part of such a curve as an expression of nitrogenase activity being partially regulated by conformational protection. However the existence of the O_2-dependent switch-on-off phenomenon, as well as uncertainty about its mechanism (above), complicates interpretation of such experiments. Activity might equally be regulated by partial diversion of electrons through an autoxidizible carrier (above), or the population may be an inhomogeneous mixture of fully active and fully "switched-off" cells. Scherings et al. (1977) showed that oxidized FeS protein II caused a lag in the oxidation of flavodoxin hydroquinone by crude *A. vinelandii* nitrogenase in vitro, which they interpreted as implying direct regulation of nitrogenase function by the protective protein in the intact cell.

2.2 Conformational Protection and Oxygen Regulation in Klebsiella

K. pneumoniae is an anaerobic nitrogen fixer. One can in fact handle derepressed *K. pneumoniae* in air without damaging its nitrogenase and the crude nitrogenase in cell-free extracts is only slowly inactivated in air (Kelly, 1969 b); these facts indicate that some equivalent of conformational protection may exist in this organism. Hill (1976 a) obtained evidence for some degree of O_2-dependent reversible switch-on-off of nitrogenase activity in *K. pneumoniae.* No definitive evidence for a protective protein has been published and the question of regulation of function by O_2 has not otherwise been studied. Regulation of nitrogenase synthesis by O_2 (St. John et al., 1974; Eady et al., 1978) is, however, well established in this organism. Eady et al. (1978) used pulse labelling to study its kinetics and showed that a mutant constitutive for NH_4^+-regulation did not escape O_2 repression, thus implying differences in the regulatory mechanisms. Study of O_2 as a regulator in any depth requires the use of an oxystat to control the dissolved O_2 tension and is complicated because the acetylene test may reflect either damage to nitrogenase or its regulation. A new lead arose from the genetic analysis of the *nif* gene cluster of *K. pneumoniae* (see Merrick et al., 1980; W. J. Brill, this vol., p. 81). The *nif* cluster is now known to consist of 15-17 linked genes in 7-8 transcriptional units (Fig. 2) and many of the gene products have been identified on polyacrylamide gels. The products of the structural genes for nitrogenase, *nif*H, D, and K, were those shown by Eady et al. (1978) to be regulated by O_2; we have observed more recently that the *nif*J product is also regulated by O_2. Two adjacent *his*-proximal genes in Fig. 2 have regulatory functions (*nif*A and *nif*L). An important advance has been the availability of methods to fuse the *lac*Z gene into various *nif* operons, by exploiting a derivative of bacteriophage Mu called Mud (Ap*lac*) which carries this gene as well as ampicillin resistance (see Merrick et al., 1980). *Lac*Z specifies the enzyme β-galactosidase, which can be detected by a simple colour test, so in principle the regulation of any operon into which it is fused can be studied independently of the regulation of the others simply by estimating β-galactosidase activity. Recent findings in our laboratories have involved a combination of genetic and physiological studies; some data which have a direct bearing on O_2 as a regulator of *nif* expression are as follows.

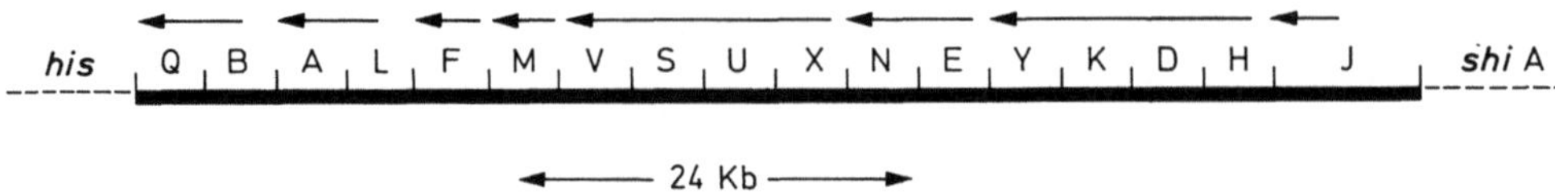

Fig. 2. The *nif* gene cluster
The *nif* gene cluster of the *Klebsiella pneumoniae* chromosome (early 1980). The letters signify the component genes (cistrons); the adjacent genes are *his* (for biosynthesis of histidine) and *shi*A (for uptake of shikimic acid). The *arrows* define the seven operons and the directions of transcription. The products of many have been identified on gel electrophoretograms and functions can be assigned to a few as below

Gene	MW of peptide (KD)	Function or other comment
Q	?	?
B	25	Involved in FeMoco synthesis or processing
A	57	Regulatory
L	45	Regulatory
F	17	Involved in electron input into nitrogenase
M	28	Involved in processing the Fe protein
V	42	Influences specificity of nitrogenase
S	45	?
U	32	?
X	18	*
N	50	Like B
E	46	Like B
Y	19	*
K	60	Specifies β subunit of MoFe-protein
D	56	Specifies α subunit of MoFe-protein
H	35	Specifies subunit of Fe-protein
J	120	Involved in electron input?

* X and Y are new genes detected by "protein mapping" of cloned *nif* fragments (A. Pühler and W. Klipp, unpublished)

When the *lac*Z gene was fused into *nif*H or *nif*L, expression was repressed normally by NH_4^+. The effects of various dissolved oxygen tensions (dO_2) are illustrated in Table 1. O_2 repressed expression of the *lac*Z: :*nif*H fusion at a dO_2 of 0.5% saturation by 3-5-fold (lines 1, 3), as would be expected from its effect on nitrogenase synthesis. However, O_2 at that concentration did not repress expression of *lac*Z fused into *nif*L, though some repression was observed at higher dissolved oxygen tensions (Table 1, line 1). These observations support earlier evidence that the mechanism of O_2 regulation differs from that of NH_4^+, though the process may change according to the dO_2. Since the *lac*Z fusion in *nif*H is polar (Fig. 2), the products of *nif*K, D, and Y are absent, as well as *nif*H product, so we conclude that neither the nitrogenase peptides nor the *nif*Y product are involved in O_2 regulation of *nif.*

We observed that some *nif*$^+$ revertants of *nif*L mutants (e. g., *nif*2400 in Table 1, line 4) escaped from oxygen repression and continued to synthesize nitrogenase at a

Table 1. Influence of oxygen on β-galactosidase activity when *lac*Z is fused into *K. pneumoniae* *nif*L or *nif*H

Fusion allele	Additional *nif* chromosomal mutations		β-galactosidase activity[b] under O_2 as a percentage of that under N_2 in cultures maintained at a dO_2 (% saturation) of		
			0.5	3.0	5.0
1 *nif*L 2,782(p)[a]	Δ *his-nif*		123	68	54
2 *nif*H 2,783(p)	Δ *his-nif*		32	–	–
3 *nif*H 2,783(c)[a]	–		10	–	–
4 *nif*H 2,783(c)[a]	*nif*L 2,265, *nif* 2,400		107	–	–
5 *nif*H 2,783(c)[a]	–	plus Nif$^+$	3	–	–
6 *nif*H 2,783(c)[a]	*nif*L 2,265, *nif* 2,400	plasmid	2	–	–

[a] p denotes fusion on a plasmid, c denotes fusion on the chromosome

[b] Cultures were grown in glucose minimal medium plus excess NH_4^+ overnight at 30 °C under N_2, harvested by centrifugation and re-suspended in nitrogen-free medium at 30 °C. β-galactosidase activity was measured after 4.5 h exposure either to N_2 or to the indicated dO_2 in an oxystat supplied with air plus N_2

– not tested

dO_2 of 0.5% saturation. This phenotype was confirmed using strains in which a chromosomal *nif*H: :*lac* fusion was joined to the *nif*L reversion mutation; in these strains β-galactosidase synthesis occurred in the presence of O_2 (line 4). Most importantly, O_2 repression of *lac* was restored in these strains by introduction of a wild-type *nif* plasmid (Table 1, lines 5 and 6). These results suggest a model in which the *nif*L gene product acts as a negative control protein that prevents expression of other *nif* operons in the presence of O_2. This model is consistent with the expression of *lac*Z in *lac*Z: :*nif*L strains at low dO_2 mentioned above.

3 Hydrogenase and the Efficiency of Azotobacter

The physiological efficiency of *Azotobacter* as compared with other nitrogen-fixing bacteria is well known. Intact cells apparently consume some 5 mol ATP/mol N_2 fixed compared with 20–30 mol ATP consumed by anaerobic organisms (Hill et al., 1972; Anderson and Shanmugam, 1977; Hill, 1976b). These data were obtained from yield studies of chemostat cultures or, in one case, examination of an NH_4^+-constitutive strain of *K. pneumoniae.* In the case of *Azotobacter,* the number is subject to considerable error because growth yields are influenced by oxygenation, but the conclusion still stands that the apparent ATP efficiency of nitrogen fixation in this organism is, in the absence of O_2 stress, exceptionally high both compared with other organisms and with its own cell-free enzyme (which consumes about 15 ATP/N_2 in vitro). This high

Table 2. Amounts of N fixed per gram carbon source consumed by some nitrogen-fixing systems

System and conditions	mg N/g C-source	Ref.
K. pneumoniae P-limited, anaerobic in chemostat	3.8–4.7	Bergersen (1974)
K. pneumoniae O_2-limited in chemostat	9–14.5	Hill (1976a)
K. pneumoniae nif constitutive, anaerobic, resting	19.5	Anderson and Shanmugam (1977)
C. pasteurianum batch cultures	1–16	Mishustin and Shilnikova (1971) p 261
C. pasteurianum chemostat, sucrose	11.8	Daesch and Mortenson (1968)
Bacillus polymyxa anaerobic batch cultures	1.5–3	Mishustin and Shilnikova (1971) p 336
Desulfovibrio desulfuricans in anaerobic chemostat	6.2–6.6	Hill et al. (1972)
Xanthobacter (Mycobacterium) flavus	14–18	Mishustin and Shilnikova (1971) p 353
A. chroococcum, various batch cultures in air	13–20	Jensen (1954)
A. chroococcum, batch cultures in air	7.4	Parker (1954)
A. chroococcum, at 0.04 atm O_2	21.6	Parker (1954)
A. chroococcum chemostat		
cultures at 0.3 atm O_2	24	Dalton and Postgate (1969a)
cultures at 0.1 atm O_2	38	Dalton and Postgate (1969a)
cultures at 0.01 atm O_2	43	Dalton and Postgate (1969a)
R. leguminosarum in pea nodules	67.8	Minchin and Pate (1973)

relative efficiency is illustrated at a coarser level in Table 2, where the yields of fixed N/mg carbohydrate consumed are presented; only the functioning bacteroid in the legume symbiosis exceeds oxygen-limited *Azotobacter* in gross nitrogen-fixing efficiency.

An important factor affecting the efficiency of aerobic nitrogen fixation is oxygen; aspects of this truism have been discussed above. A second factor influencing the efficiency of aerobic nitrogen fixation is the possession of hydrogenase, which re-cycles the hydrogen produced by nitrogenase. The importance of possession of hydrogenase in the legume symbiosis, proposed by Dixon (1972), has been much clarified by the work of H.J. Evans and his colleagues: ability on the part of the symbiotic bacteria to form hydrogenase in the nodule lessens and can eliminate loss of H_2 from the nodule and consequently increase the physiological and, indeed, agronomic efficiency of the symbiosis. This topic has been adequately reviewed elsewhere (e. g., Evans et al.,

1980). A comparable situation was revealed in *Azotobacter* by Smith et al. (1976) and occurs in other free-living nitrogen-fixing bacteria including phototrophs (see Vignais, 1978). In principle, hydrogenase scavenges hydrogen produced as a by-product of nitrogenase action and re-utilizes it. This process may augment efficiency by one or more of four processes: prevention of inhibition of nitrogenase by H_2; enhancement of respiratory protection of nitrogenase through the H_2/O_2 reaction; use of the H_2 as an additional reductant for nitrogenase; regeneration of ATP through the H_2/O_2 reaction [see Vignais (1978) and Robson and Postgate (1980) for references].

A role of hydrogenase in the economy of the nitrogen fixation process may now be accepted, though its details in individual instances may remain to be elucidated. It provides a rational basis for the hydrogenase-nitrogenase correlation which has been recognised for several decades (e. g., Wilson, 1958). Yates and Walker (1980) reviewed their extensive chemostat study of hydrogenase activity in *A. chroococcum,* which revealed that the physiological status of the population influenced hydrogenase content and that populations with inadequate hydrogenase were readily obtained in chemostats which evolved H_2 during growth. Wilson's group (e. g., Lee and Wilson, 1943; Green and Wilson, 1953) presented evidence some decades ago that hydrogenase activity was regulated along with nitrogenase activity in *Azotobacter,* though NH_4^+-repressed populations still possessed substantial hydrogenase activity. A re-investigation of this question in our laboratories has shown the following information.

1. Oxygen- or sulphate-limited populations of *A. chroococcum* show low hydrogenase activities which are independent of whether the N-source is N_2 or NH_4^+. Such populations evolve H_2 during growth. Nitrogenase activity is, of course, strongly repressed by NH_4^+.

2. Carbon-limited populations show high hydrogenase activities with N_2 as the N source but low activities with NH_4^+. Nitrogenase activities show the familiar repression by NH_4^+.

3. Nitrogenase activity precedes hydrogenase activity during the transition from NH_4^+-supported to N_2-supported growth of C-limited populations, suggesting that the product of nitrogenase action, H_2, induces hydrogenase synthesis.

4. Batch cultures growing with N_2 show low activity during growth which increases in the early stationary phase, presumably because the population then becomes carbon-limited.

Table 3 illustrates the first two of these findings. The essential conclusion is that hydrogenase is largely repressed in conditions of carbon-sufficiency but is regulated according to the N-source in conditions of carbon limitation. Exogenous H_2 augments the hydrogenase activity of NH_4^+-repressed populations, supporting the view that endogenous H_2 is a hydrogenase inducer. Carbon metabolite repression of hydrogenase synthesis has also been proposed in *Rhizobium* since cultures in vitro show higher hydrogen-oxidizing activities on media of low carbon content (Maier et al., 1978; Lim, 1978; Lim and Shanmugam, 1979). Teleologically, it is reasonable for organisms to be metabolically extravagant in conditions where supplies of carbon metabolite are sufficient. The question remains open why hydrogenase is never completely repressed, either by metabolite or ammonium.

Table 3. Hydrogenase activities of *Azotobacter* chroococcum in continuous culture in various conditions

Chemostats run for at least 5 doublings at D=0.1 h^{-1} to reach steady states. Hydrogenase activities of all suspensions measured manometrically under H_2 with methylene blue (28 m*M*) as acceptor and EDTA (2.4 m*M* + NaF (90 m*M*) to ensure permeability thereto.

Physiological status of cells	n-mol H_2 absorbed/min/mg dry wt. cells
Sucrose limited, N_2-fixing	230
Sucrose-limited, NH_4^+-assimilating	50
SO_4^{2-}-limited, N_2-fixing	43
SO_4^{2-}-limited, NH_4^+-assimilating	44

4 Molybdenum in the Regulation of Nitrogenase Synthesis

The well-established need for molybdenum shown by nitrogen-fixing systems (Bortels, 1930) was rationalized in the 1960's with the discovery that the larger nitrogenase protein contains molybdenum. It is located in a fragment containing also Fe and sulphide-S which has been given the trivial name "FeMoco" (Shah and Brill, 1977) and may well be involved in enzyme function (Smith, 1977). Mobilization of Mo in *Clostridium pasteurianum* requires an uptake system which is specific to nitrogenase and associated with a storage protein (see Elliott and Mortenson, 1977). Nitrate reductase is another molybdenum enzyme, though its Mo-containing fragment ("Moco") is chemically different from the FeMoco of nitrogenase. Nevertheless, its Mo-processing system may sometimes be shared for nitrogenase synthesis, because *Klebsiella nif* genes in *E. coli* require the *nar*D gene product for expression (Kennedy and Postgate, 1977; Skotnicki and Rolfe, 1977). The *nar*D product is involved in processing Mo for nitrate reductase synthesis.

Several suggestions that molybdenum might have a regulatory function on *nif* have been published. In *Azotobacter vinelandii*, Nagatani and Brill (1974) proposed that Mo regulated synthesis of Av1 but not Av2 on the basis of measurements of both activity and immunologically cross-reacting material. Cardenas and Mortenson (1975) used activity and moybdate incorporation studies to conclude that Mo is essential for nitrogenase biosynthesis as well as activity in *C. pasteurianum* (see also Elliott and Mortenson, 1977). In *K. pneumoniae*, Brill et al. (1974) reached a similar conclusion because Mo-deficient organisms showed neither nitrogenase activity nor cross-reacting material to anti-sera to nitrogenase proteins. However, Kennedy and Postgate (1977) observed that nitrogenase peptides were present in normal amounts, though enzymically inactive, in the absence of exogenous Mo in the *nar*D mutant of *E. coli* carrying *Klebsiella nif* genes. Nagatani and Haselkorn (1978) observed synthesis of inactive nitrogenase proteins in Mo-starved *Plectonema boryanum*.

We have investigated putative regulation of *nif* by Mo in *K. pneumoniae* further. Physiological studies with the wild-type *K. pneumoniae* indicate the following points.

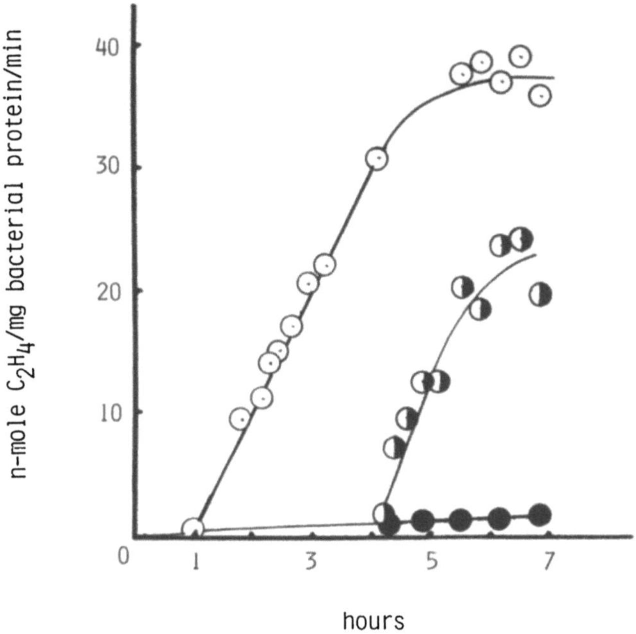

Fig. 3. Involvement of molybdenum in the derepression of nitrogenase. *Klebsiella pneumoniae* strain M5a1 was grown at 30 °C in NFDM medium (Cannon et al., 1974) prepared with components depleted of traces of molybdenum by extraction with 8-hydroxyquinoline (Kelly and Lang, 1970) but containing 1.5 mg $(NH_4)_2SO_4$/ml. Acid-washed glassware was used. Cells were harvested at room temperature, resuspended at 8×10^8 cells/ml in degassed Mo-deficient NFDM containing 100 μg aspartate/ml. At time zero two 10 ml samples of suspension were withdrawn and syringed into flasks containing argon +0.1 atm acetylene, one with 100 m*M* Na_2MoO_4; all flasks were shaken at 30 °C. A third flask containing molybdate and acetylene was set up after 4 h 10 min. Gas samples were removed periodically for gas chromatographic assay of ethylene formed (cf. Fig. 1). ○ With molybdate ◑ Molybdate added late ● Without molybdate

1. Organisms can grow in specially purified media without added molybdate provided NH_4^+ is available.

2. Such populations rapidly form active nitrogenase when transferred to NH_4^+-free media suitable for derepression of *nif*, provided molybdate is supplied (Fig. 3).

3. Such populations do not form active nitrogenase when transferred to Mo-free derepression medium (Fig. 3), though nitrogenase peptides are detectable immunologically and by pulse labelling. Kp1 is found at about 3% and Kp2 at 10% of the concentrations observed in Mo-sufficient conditions.

4. Addition of molybdate during incubation in N-free medium leads to rapid formation of active, acetylene-reducing nitrogenase (Fig. 3). Protein synthesis de novo may be involved, since Mo-activation is prevented completely by pre-treatment with chloramphenicol.

These findings could be subject to various interpretations: that molybdate is a regulator, that a molybdenum transport or storage protein regulates, that Kp1 itself regulates. To clarify the situation we have examined the effect of molybdate on strains carrying *nif::lacZ* fusion. Haploid strains were constructed of genotype *Δnif/nif::lac* (i. e., with chromosomal *nif* deleted but carrying plasmids with *lac* fusions in *nif*B, A, L, F, M, S, U, N, E, H or J). They all had similar β-galactosidase activities when derepressed, whether molybdate was supplied or not. Therefore molybdate itself is not a regulator of any of the known *nif* operons. Merodiploid strains of genotype nif^+*/nif::lac* (i. e., similar but without the *nif* deletion) were also constructed, the majority expressed *lac*Z to the same extent as haploid strains, again independently of added molybdate. The exception was a nif^+*/nif*H::*lac* strain which had the regular β-galactosidase activity in the absence of moybdate but which showed a 7-fold enhancement of activity in its presence. This observation implies that the *nif*HDKY operon is regulated autogenously by at least one of its own gene products.

As explained earlier, the *lac*Z fusion in *nif*H is polar: it prevents synthesis of Kp1 and the *nif*Y product as well as of Kp2 (cf. Fig. 2). Since one or more products of the HKDY operon, as well as molybdate, are required for maximal transcription from the *nif*H promotor it is likely that a molybdoprotein, most probably Kp1, is involved in the autogenous regulation of this operon.

5 Conclusion

Three broad aspects of the physiology of nitrogen fixation by certain free-living heterotrophs have been briefly reviewed and up-dated here. They form part of ongoing studies of nitrogen fixation in our and others' laboratories which will probably modify and extend the results reported here even while this contribution is in press. They nevertheless illustrate the value of exploiting combined genetical and biochemical knowledge to study physiological problems in depth.

Acknowledgements. We acknowledge the assistance of Dawn Allitt, Eugene Kavanagh, Marie Hawkins, Linda Witts, Martin Wingfield, and Frances Campbell.

References

Anderson K, Shanmugam KT (1977) J Gen Microbiol 103:107–122
Bergersen FJ (1974) J Gen Microbiol 84:412–414
Bortels H (1930) Arch Mikrobiol 1:333–342
Brill WJ, Steiner AL, Shah VK (1974) J Bacteriol 118:986–989
Bullen WA, Burns RC, LeComte JR (1965) Proc Natl Acad Sci USA 53:532–539
Cannon FC, Dixon RA, Postgate JR, Primrose SB (1974) J Gen Microbiol 80:227–239
Cardenas J, Mortenson LE (1975) J Bacteriol 123:978–984
Daesch G, Mortenson LE (1968) J Bacteriol 110:103–109
Dalton H, Postgate JR (1969a) J Gen Microbiol 54:463–473
Dalton H, Postgate JR (1969b) J Gen Microbiol 56:307–319
Dixon ROD (1972) Arch Mikrobiol 85:193–201
Drozd JW, Postgate JR (1970a) J Gen Microbiol 60:427–429
Drozd JW, Postgate JR (1970b) J Gen Microbiol 63:63–73
Eady RR, Issack R, Kennedy C, Postgate JR, Ratcliffe HD (1978) J Gen Microbiol 104:277–285
Elliott BB, Mortenson LE (1977) In: Newton WE, Postgate JR, Rodriguez-Barrueco C (eds) Recent developments in nitrogen fixation. Academic Press, London New York, pp 205–217

Evans HJ, Emerich DW, Ruiz-Argueso T, Maier RJ, Albrecht SL (1980) In: Newton WE, Orme-Johnson WH (eds) Nitrogen fixation, vol II. Symbiotic associations and blue-green algae. Univ Park Press, Baltimore, pp 69–86
Green M, Wilson PW (1953) J Bacteriol 65:511–517
Haaker H, Veeger C (1977) Eur J Biochem 77:1–10
Hill S (1976a) J Gen Microbiol 93:335–345
Hill S (1976b) J Gen Microbiol 95:297–312
Hill S, Kavanagh EP (1980) J Bacteriol 141:470–475
Hill S, Drozd JW, Postgate JR (1972) J Appl Chem Biotechnol 22:541–558
Jensen HL (1954) Bacteriol Rev 18:195–214
Kelly M (1969a) Biochim Biophys Acta 171:9–22
Kelly M (1969b) Biochim Biophys Acta 191:527–540
Kelly M, Lang G (1970) Biochim Biophys Acta 223:86–104
Kennedy CK, Postgate JR (1977) J Gen Microbiol 98:551–557
Lee SB, Wilson PW (1943) J Biol Chem 151:377–385
Lim ST (1978) Plant Physiol 62:609–611
Lim ST, Shanmugam KT (1979) Biochim Biophys Acta 584:479–492
Maier RJ, Campbell NER, Hanus J, Simpson FB, Russell SA, Evans HJ (1978) Proc Natl Acad Sci USA 75:3258–3262
Merrick M, Dixon R, Eady R, Hill S, Kennedy C, Iaccarino M (1980) In: Nitrogen fixation. Annu Proc Phytochem Soc Eur 18th. Academic Press, London New York, pp 377–394
Minchin FR, Pate JS (1973) J Exp Bot 24:259–271
Mishustin EN, Shilnikova VK (1971) Biological fixation of atmospheric nitrogen (English translation). MacMillan, London, pp 420
Nagatani HH, Brill WJ (1974) Biochim Biophys Acta 360:160–166
Nagatani HH, Haselkorn R (1978) J Bacteriol 134:597–605
Parker CA (1954) Nature (London) 173:780–781
Postgate JR (1975) In: Quispel A (ed) North Holland, Amsterdam, pp 663–686
Reed DW, Toia RE, Roveed D (1974) Biochem Biophys Res Commun 58:20–26
Roberts GP, McNiel T, MacNiel D, Brill WJ (1978) J Bacteriol 136:267–279
Robson RL (1979a) Biochem J 181:569–575
Robson RL (1979b) FEMS Microbiol Lett 5:259–262
Robson RL, Postgate JR (1980) Annu Rev Microbiol 34:183–208
Scherings G, Haaker H, Veeger C (1977) Eur J Biochem 77:621–630
Shah VK, Brill WJ (1977) Proc Natl Acad Sci USA 74:3249–3253
Skotnicki M, Rolfe B (1977) Biochem Biophys Res Commun 78:726–733
Smith BE (1977) J Less Common Met 54:465–475
Smith LA, Hill S, Yates MG (1976) Nature (London) 262:209–210
Stewart WDP (1975) In: Quispel A (ed) North Holland, Amsterdam, pp 696–718
St. John RT, Shah VK, Brill WJ (1974) J Bacteriol 119:266–269
St. John RT, Johnston HM, Seidman C, Garfinkel D, Gordon J, Shah VK, Brill WJ (1975) J Bacteriol 121:759–765
Vignais PM (ed) (1978) Biochimie 60:217–341
Walker CC, Yates MG (1978) Biochimie 60:225–231
Wilson PW (1958) In: Rutland WC (ed) Encyclopedia of plant physiology, vol 8. Springer, Berlin Heidelberg New York, pp 9–42
Yates MG (1970) FEBS Lett 8:281–285
Yates MG, Jones CW (1974) Adv Microb Physiol 11:97–136
Yates MG, Walker CC (1980) In: Newton WE, Orme-Johnson WH (eds) Nitrogen fixation, vol I. Free-living systems and chemical models. Univ Park Press, Baltimore, pp 95–110

The Biochemistry of Dinitrogen Fixation

W.G. ZUMFT[1]

1 Introduction

Biochemical exploration of the metabolic pathway of dinitrogen reduction has produced a wealth of information, from which, to cope with the available space, only recent data pertinent to the enzyme itself will be presented. Developments until about 1975/1976 were discussed in detail at two occasions (Zumft and Mortenson, 1975; Zumft 1976) akin to many other authoritative reviews of the subject and will not be repeated here. A valuable source of information have been, and will continue to be, the proceedings of the dinitrogen fixation biennials (Newton and Nyman, 1976; Newton et al., 1977; Newton and Orme-Johnson, 1980). Three monographs provide a broader coverage beyond the enzyme into areas of chemistry and physical chemistry (Postgate, 1971; Burns and Hardy, 1975; Hardy et al., 1979). Problems of the chemistry related to molybdenum as essential trace metal for dinitrogen reduction were recently compiled in a single volume (Newton and Otsuka, 1980) and also covered in a more personal account (Pratt, 1979).

2 The Enzyme and its Components

Nitrogenase, belatedly acknowledged by the enzyme commission as reduced ferredoxin: dinitrogen oxidoreductase (ATP-hydrolyzing) [EC 1.18.2.1], has already split its personality, and the two components of nitrogenase are now also called (di)nitrogenase reductase (usually known as iron protein or Fe protein) and (di)nitrogenase sensu stricto (the molybdenum-iron protein or MoFe protein) (Hageman and Burris, 1978a). The reasoning for this nomenclature is outlined in the section on catalysis (see below).

Newly added characterizations of the nitrogenase components, i.e., those of *Azotobacter chroococcum* (Yates and Planqué, 1975), *Bacillus polymyxa* (Emerich and Burris, 1978a), *Xanthobacter autotrophicus* (formerly *Corynebacterium autotrophicum)* (Berndt et al., 1978) and even the first characterized representative of the cyanobacteria, *Anabaena cylindrica* (Hallenbeck et al., 1979), have only corroborated the previous picture of the highly conservative nature of the enzyme. A modification of nitrogenase has so far been found only in three members of photosynthetic bacteria

---

1 Institut für Botanik, Universität Erlangen-Nürnberg, Schloßgarten 4, 8520 Erlangen/FRG

and in the nonphotosynthetic *Azospirillum lipoferum* (see below). A break-through was the achievement of a cell-free nitrogenase activity from the ecologically important symbiosis in alder nodules (Benson et al., 1979), which is caused by a *Frankia*-type actinomycete. Other major groups with putative biochemical interest are the methane-oxidizing bacteria and the single diazotrophic species of the thiobacilli, *Thiobacillus ferrooxidans;* they have not yet been studied in a cell-free system. Within the pseudomonads, previously considered to be incapable of nitrogen fixation, one species was recently claimed to do so (Zolg and Ottow, 1975), as well as a green alga from a hot spring (Yamada and Sakaguchi, 1980).

The isolation procedures for practically all nitrogenases have followed similar patterns of well-established methods, however, under the constant threat of oxygen damage to the highly labile proteins. Recent attempts to alleviate this situation were the development of purification of nitrogenase at sub-zero temperatures in the presence of ethylene glycol (Tsopanakis et al., 1979), and the use of immunoadsorbent chromatography (Ferrante and Nicholas, 1976). These methods still await acceptance by other laboratories. The nitrogenase complex of the obligate aerobically nitrogen-fixing *Azotobacter* has the fortunate property of being "oxygen stable" ($t_{1/2}$ in air >4 days) in association with a low molecular weight iron-sulfur protein (Haaker and Veeger, 1977; Robson, 1979).

Heterologous crosses result when nitrogenase is reconstituted in vitro from components that originate from different organisms. The most extensive study of enzyme hybridization was done by Emerich and Burris (1978b). From eight different organisms, with their nitrogenase components purified to or near to homogeneity, 55 heterologous crosses (of possible 56) were made and assayed for enzymatic activity. In 49 cases a complex formation between the components was found, 45 of which generated a fully or partially active nitrogenase. The crosses were made from obligate aerobic and anaerobic bacteria, from photosynthetic and symbiotic bacteria, and from microaerophilic and facultative anaerobic bacteria, underlining the high evolutionary stability of the enzyme. The inactive crosses involved always one component from *Clostridium pasteurianum,* suggesting that this nitrogenase must have intrinsically different binding characteristics. There is, however, a heterologous, catalytically active cross-reaction between *C. pasteurianum* and *A. cylindrica* (Tsai and Mortenson, 1978), and *C. pasteurianum* and *Klebsiella pneumoniae* (Smith et al., 1976). The activity of heterologous complexes depended on pH, component ration, and protein concentration, factors which are operative also in homologous enzyme reconstitutions (Shah et al., 1975; Davis et al., 1975). Because heterologous crosses, in part even when they are inactive, can give information on the stoichiometry of nitrogenase, the nature of electron transfer, rate-limiting steps, or the role of ATP, the crosses between *K. pneumoniae* and *A. chroococcum* (Thorneley et al., 1975), *K. pneumoniae* and *C. pasteurianum* (Smith et al., 1976), and the catalytically inactive complex between *A. vinelandii* and *C. pasteurianum* (Emerich and Burris, 1976; Emerich et al., 1978) were studied under these mechanistic aspects.

The Fe protein has a molecular weight around 60,000 [57, 674 for the clostridial Fe protein from amino acid composition including also the prosthetic group (Tanaka et al., 1977)], i. e., approximately 1/4 that of the MoFe protein. Two tightly binding, identical subunits constitute the protein monomer as found in solution. The primary structure of the subunit with 273 amino acids has been established and has not revealed a relationship to the bacterial ferredoxins (the term azoferredoxin should be dropped therefore from usage). The molecule has six cysteinyl residues which are randomly distributed over the amino acid chain; the positions of those that bind the iron-sulfur cluster remain to be identified. The protein has also a large number of repetitive sequences of amino acid doublets and triplets (Tanaka et al., 1977).

The MoFe proteins have molecular weights slightly over 200,000 and consist of an assembly of four subunits. For the component from *K. pneumoniae* (Kennedy et al., 1976), *A. vinelandii* (Lundell and Howard, 1978), and *C. pasteurianum* (Zumft et al., 1980) it was established by peptide mapping or terminal sequences that the molecular structure is $\alpha_2\beta_2$. Conclusions from detergent electrophoresis alone are insufficient to establish the quarternary structure of this protein (Kennedy et al., 1976). The subunits can be separated from each other in large scale by electrophoretic procedures or by protein derivatization prior to ion exchange chromatography. The clostridial α subunit, although it contains only three tryptophan residues, can be easily distinguished from the smaller β-subunit (which contains none) by derivative spectrophotometry (Zumft et al., 1980).

3 The Iron Centers

Chemical analysis and spectroscopic investigation of the Fe protein have provided compelling evidence for the iron-sulfur nature of the prosthetic group (Zumft and Mortenson, 1975; Zumft, 1976; Newton et al., 1977; Newton and Orme-Johnson, 1980).

The Fe protein contains four atoms of iron and an equivalent amount of inorganic sulfur per protein dimer. An elegant approach to identify the type of iron-sulfur structure was the development of the cluster displacement or extrusion technique. Displacement of the iron-sulfur core with thiols and spectroscopic comparison of the extruded material with known clusters, together with a near quantitative removal of the iron from the protein, was taken as evidence for the existence of a single Fe_4S_4 cluster in the Fe protein (Orme-Johnson et al., 1977; Gillum et al., 1977). Since both polypeptides of the Fe protein are identical, the cluster should be bound equally by both peptides to preserve molecular symmetry, unless the biosynthetic insertion of the cluster is asymmetric to begin with. The clostridial Fe protein is mechanistically (Zumft et al., 1974) and stoichiometrically (Ljones and Burris, 1978b) a one-electron carrier, contrasting, for example, with clostridial ferredoxin which is mechanistically a one-electron carrier but transfers a total of two electrons.

The low integrations of EPR spectra of Fe proteins from different species (0.2–0.45 spins/mol) were discussed in detail by Orme-Johnson et al. (1977) and particularly focused on the question of a second paramagnetic center. Computer simulation of the EPR spectrum of *K. pneumoniae* has indeed led to the prediction of such an additional center (Lowe, 1978) whose nature, if corroborated, remains to be identified.

Thorneley et al. (1976) found addition of a second electron in a kinetically slow phase to the Fe protein of *A. chroococcum* interpreted this, however, as a nonphysiological event. Affinity methods for purification (e. g., nucleotide, dye-ligand, or ferredoxin matrices) should be applied to this protein to obtain a population with complete metal centers. From the removal of Fe with bathophenanthrolinedisulfonate from clostridial Fe protein, with a catalytical activity in the upper bracket ($1.7\ \mu mol \cdot min^{-1} \cdot mg^{-1}$), it was estimated that little more than half of the protein was in an active form (Ljones and Burris, 1978a).

Despite some criticism concerning the analytical data for this most labile iron-sulfur protein ($t_{1/2} \approx 20$ s in air for the clostridial Fe protein), the low integration values, and the total number of transferable electrons (which ultimately all relates to iron), the present picture of the Fe protein as a one-electron carrier, with one Fe_4S_4 cluster, oscillating probably between the $(Fe_4S_4)^{2-}$ and $(Fe_4S_4)^{3-}$ state, with two binding sites for $MgATP^{2-}$, and a stoichiometry of 2 mol ATP hydrolyzed per mol electrons transfered, fits well together and blends into the current overall understanding of nitrogenase catalysis.

The nature of the iron-sulfur centers in the MoFe protein is far more complex; in particular, this component contains a mixed-type iron-molybdenum center which will be discussed in the next chapter. The MoFe protein gives a fine example of a biological structure where quantitative Mössbauer spectroscopy provided far-reaching structural insights (Münck et al., 1975; Zimmermann et al., 1978). Quantitation of the iron environments with this technique was better than 2%. Three out of four spectroscopically distinguishable components could be attributed to physically separable iron-sulfur clusters which were classified in four species of Fe_4S_4 clusters (called P-type), two species belonging to the mixed-type cluster (called M-type) and a minor species called S-type (Zimmermann et al., 1978). This work has provided convincing evidence that the number of iron atoms in the MoFe protein is 30 ±2. The iron atoms in the Fe_4S_4 clusters are spectroscopically of two types present in a ratio of 1:3 and they are also electronically somewhat distinct from the Fe_4S_4 clusters of bacterial ferredoxins. This is also evident in their dichroitic properties (Stephens et al., 1979). Despite the modification by the protein environment, the P-clusters could be extruded with thiols and related to a Fe_4S_4 structure (Orme-Johnson et al., 1977). A less ambiguous way was cluster extrusion by o-xylyl-α,α'-dithiol, subsequent ligand exchange with p-trifluoro-methylbenzenethiol and identification of the cluster as Fe_4S_4 with the aid of synthetic standards by the chemical shift in ^{19}F NMR spectroscopy (Kurtz et al., 1979). It is interesting to note that the MoFeS complex (M-cluster) could not be extruded from the MoFe protein by thiols. In addition to evidence for iron-sulfur clusters from Mössbauer spectroscopy. EPR signals were observed in the CO-inhibited nitrogenase (Davis et al., 1979) or as transient species (Yates and Lowe, 1976) which resembled those of low molecular weight iron-sulfur proteins and most likely were not signals which originated from the M-cluster.

4 The Iron-Molybdenum Cofactor

Molybdenum in form of a low-molecular-weight compound could be removed from acid-treated MoFe protein by solvent extraction (Shah and Brill, 1977). Because of the historic development of this field, which was outlined in an early paper on the isolation of molybdenum compounds from clostridial MoFe protein (Zumft, 1974), the molybdenum part of nitrogenase was called a cofactor, although its real status might turn out to be that of a prosthetic group. The hypothesis of the interchangeability of molybdenum cofactors among all Mo-containing enzymes has not resisted experimental proof, and certainly sets nitrogenase apart from the other Mo-enzymes (Pienkos et al., 1977). The isolation of a mutant devoid of its Mo-compound which synthesizes a MoFe protein without the molybdenum cofactor, and can therefore be used as a test system for the solvent extracted low molecular weight Mo-compounds, cannot be overestimated in its importance for the vigorous development of this area of bioinorganic chemistry (Shah and Brill, 1977).

The characteristic EPR signal of dithionite reduced MoFe protein (nitrogenase can be identified by it spectroscopically even inside intact cells) is a S=3/2 system that originates from the FeMo-cofactor (Rawlings et al., 1978; Huynh et al., 1979). The MoFe protein contains two such cofactors with a presumed composition of its central core of $MoFe_6S_6$ (spectroscopic and analytical data do not agree on this point). The FeMo-cofactor ties up at least 12 Fe atoms out of the total of 30 and represents a novel type of conjugated molybdenum-iron-sulfur cluster.

From acid/base treated clostridial MoFe protein spectroscopically highly characteristic Mo-compounds were isolated (Zumft, 1974), which were later identified as tetra- and tri-thiolato complexes of molybdenum (Zumft, 1978). The conditions under which these compounds were isolated would not allow a de novo synthesis from groups of the native protein, leading to the conclusion that the MoFe protein contains a tetrasulfido complex of molybdenum coupled to iron-sulfur clusters (Zumft, 1978). The drastic isolation procedure maintains only the core structure around the molybdenum but not an intact iron-sulfur moiety, which results in the loss of reconstitutive capability in the mutant test and in the loss of acetylene-reducing activity. The FeMo-cofactor in situ does not reduce acetylene with borohydride (no single activity has ever been demonstrated for the MoFe protein alone), however, once separated from the protein it yields about 8% of activity on a basis of molybdenum versus complete nitrogenase (Shah et al., 1978). Cyclopropene is also reduced by the FeMo-cofactor with BH_4^-, but distinct from nitrogenase, only cyclopropane is formed and no propene (McKenna et al., 1979).

With the advent of X-ray absorption spectroscopy through experimental and theoretical advances, the MoFe protein was among the very first metalloproteins to be examined. The absorption edge energy (for Mo around 2,015 eV), and in particular analysis of the spectral fine structure beyond the absorption edge (EXAFS) provides information about the absorbing atoms, their number, and the atomic distances within the first coordination sphere of the absorber. The method cannot, however, provide information about spatial arangements and little beyond the first coordination sphere. This is a genuine draw-back of the method, which has to be kept in mind when attempting to build models or even when comparing model compounds with the protein, for which the recent identification of the molybdenum-cofactor of xanthine oxidase and other Mo-enzymes as pteridine derivative provides a striking illustration (Johnson et al., 1980).

X-ray absorption spectroscopy was carried out with the clostridial MoFe protein (Cramer et al., 1978a), the MoFe protein and the isolated FeMo-cofactor from *A. vinelandii* (Cramer et al., 1978b). Spectra of the three sources were highly similar and led to the conclusion that molybdenum is co-ordinated by three to four sulfur atoms at 2.35 Å distance, one or two sulfur atoms at a slightly larger distance, and two or three iron atoms at 2.72 Å. The quintessence of these measurements is presented in the following two models (Cramer et al., 1978a):

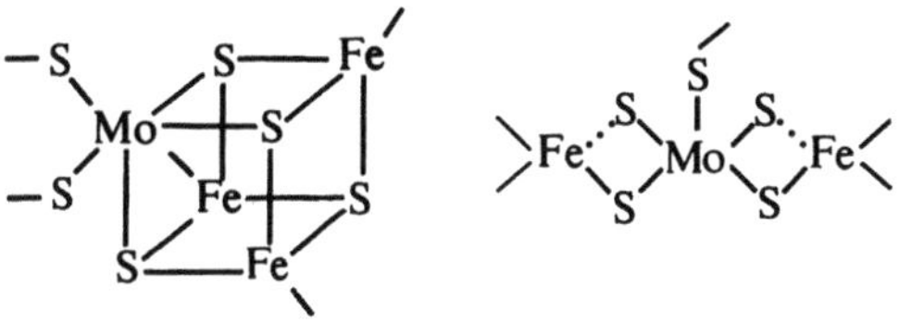

The isolated FeMo-cofactor is redox active; it could be dyeoxidized and re-reduced by dithionite (Burgess et al., 1980a). The presumptive achievement of the fully reduced state, which is EPR-silent (also referred to as super-reduced) by photochemical reduction (Rawlings et al., 1978), could not be repeated so far, rather it was shown that EDTA which is used as reductant in the photochemical system, by itself produces an EPR-silent state (Burgess et al., 1980a). Addition of thiophenol to the isolated FeMo-cofactor sharpened its EPR signal and made it more similar to the spectrum of native MoFe protein.

Synthetic approaches to the structure of the FeMo-cofactor have led to cubane iron-sulfur clusters of the sum formula $[Mo_2Fe_6S_9(SC_2H_5)_8]^{3-}$ (Wolff et al., 1979) or $[Mo_2Fe_6S_8(SPh)_9]^{3-}$ (Christou et al., 1978). In these complexes one Mo atom replaces one Fe atom in a corner of the cluster as shown in the model above, and is bridged via three additional sulfur ligands to a second identical cluster. EXAFS data of these compounds closely resemble that of the native MoFe protein. The open model (above right) was also synthesized in form of the $(Ph_4P)_2$ $[Cl_2FeS_2MoS_2FeCl_2]$ complex (Coucouvanis et al., 1980a). A smaller unit containing only the MoS_4 core, bridged to one FeS_2 group, $[(S_5)FeS_2MoS_2]^{2-}$, resembled spectrophotometrically closely the acid/base treated clostridial MoFe protein (Coucouvanis et al., 1980b), from which the thiolato complexes were obtained (Zumft, 1978). In still another attempt to fuse the presently available data into a feasible model, a structure was proposed where to Fe_4S_4 clusters were joined via a tetra-thiolato unit of molybdenum (Teo and Averill, 1979).

5 Catalysis

The versatility of nitrogenase towards substrates has greatly facilitated the study of practically all aspects of this enzyme. Alternative substrates were mainly discovered until the mid-sixties, among which acetylene has proved particularly useful, due to its ease of reduction by nitrogenase and facile analysis. Apart from the parent compound, practically all substrates in the nitrile, isonitrile, and alkyne series are being reduced most sluggishly, with relative rates versus dinitrogen of several orders of magnitude lower. Whereas CH_3CCH is still being reduced, CH_3CCCH_3 cannot anymore be accommodated at the active site for reduction (Burns and Hardy, 1975). Cyclopropene

(c-C_3H_4) is the newest addition to substrates for nitrogenase. It is reduced readily by the purified enzyme, as well as by intact cells and the isolated FeMo cofactor (McKenna et al., 1979, 1980). Because of its interesting stereochemistry of reduction, it might serve as an active site probe for nitrogenase and the FeMo cofactor. In ^2H_2O, ⩾90% [*cis*-1,2^2H_2] cyclopropane was formed; if any, the *trans* isomer accounted for less than 5%–10%. The simultaneously formed propene did not show this high stereoselectivity, [*trans*-1,3-$^2H_2$]-*propene,* [*cis*-1,3-$^2H_2$]propene, and [2,3-$^2H_2$]propene were found in a ratio of 2:1:1 (McKenna et al., 1980).

Diazene and hydrazine were repeatedly suggested as enzyme-bound intermediates in the reduction of dinitrogen. On acid or alkali addition to a complete nitrogenase system under dinitrogen (but not under argon or with cyanids as substrate) the formation of hydrazine was detected by reaction with p-dimethylaminobenzaldehyde (Thornely et al., 1978). The originally enzyme-bound intermediate was not necessarily hydrazine, the latter rather being formed during hydrolysis, but nevertheless this finding provided suggestive evidence for the presence of a dinitrogen-hydride intermediate. As further evidence for the diazene-hydrazine pathway of N_2 reduction, the discovery by Bulen of hydrazine as nitrogenase substrate could be confirmed, and indirect evidence for the existence of an intermediate at the diazene level was provided by a painstaking reinvestigation of HD formation by nitrogenase (Burgess et al., 1980b). The key to the understanding of the nitrogenase-catalyzed "exchange" reaction is hidden in the fact that HD formation *does* require electrons. Total electron flow under nitrogen, (N_2-fixing conditions), argon (hydrogen-evolving conditions), and nitrogen plus hydrogen (N_2-fixing but H_2-inhibited conditions) remains constant. In the presence of D_2, when accounting total electron flow by the products being formed, this constancy of electron flow can only be shown when one electron is assigned to each HD molecule formed. Mechanistically, electron allocation was proposed to be derived from a diazene intermediate according to the following equations (Burgess et al., 1980b):

$$E + N_2 + 2H^+ + 2e^- \rightarrow E\text{-}N_2H_2$$
$$E\text{-}N_2H_2 + D_2 \rightarrow E + N_2 + 2HD$$
$$\text{Sum: } 2H^+ + 2e^- + D_2 \rightarrow 2HD.$$

The main feature of this mechanism is the existence of a futile cycle that consumes energy and reductant. This futile cycle, if it exists, should have dramatic consequences among the hydrogen bacteria when growing autotrophically on dihydrogen and dinitrogen. It is unlikely that an organism provides molecular means to suppress this reaction; the ubiquitous linkage of dinitrogen reduction with dihydrogen evolution speaks against this. An indirect protection by creating a hydrogen sink at the active site of nitrogenase, however, could be provided by hydrogenase in addition to its role in energy metabolism and hydrogen recycling. Since inhibition of nitrogenase by hydrogen requires electron transport and ATP hydrolysis, one can attempt a few predictions how the hydrogen bacteria could respond to changes in the partial pressure of hydrogen (pH_2). A higher pH_2 could lead to an increase in the scavenging reaction, i. e. activity or level of hydrogenase. At the same time the ATP/e_2 ratio for the nitrogenase reaction should decrease, as more energy is needed to supply the cell with the same amount of ammonium. To compensate for the increased energy need, the oxygen consumption of diazotrophically growing cells should increase in response to higher pH_2.

As the most simple expression of these effects, an increase in pH_2, once it is beyond the regulatory capacity of these bacteria, should lead to growth inhibition of a diazotrophic culture but not of a culture growing on ammonium as N-source. Although for the behavior of a single parameter alternative explanations are possible, the concerted response should provide support on the cellular level for the above outlined mechanistic interpretation of hydrogen inhibition. A few scattered data in the literature can indeed be taken as preliminary evidence to support this idea (De Bont and Leijten, 1976; Berndt and Wölfle, 1978).

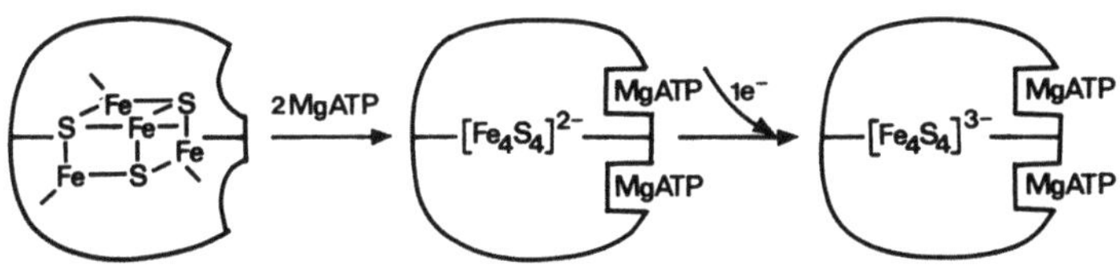

Fig. 1. Simplified scheme of the initial reaction events involving the Fe protein

The initial events for nitrogenase catalysis occurring on the Fe protein are graphically shown in Fig. 1. Two binding sites for $MgATP^{2-}$ on the Fe protein were previously identified by gel filtration and by EPR, and a conformational change of the protein was proposed as a result of nucleotide binding (Zumft, 1976). Studying nucleotide-dependent chelation rates of iron with bathophenanthroline-disulfonate, Ljones and Burris concluded that both ATP-binding sites have to be occupied before the protein undergoes the conformational change, and that the two sites have different affinities for $MgADP^-$(Ljones and Burris, 1978a). Four equivalent binding sites for Mg^{2+} (K_d=1.7 ±0.3 m*M*) and the formation of ternary protein-metal nucleotide complexes were shown by proton NMR (Bishop et al., 1977). Since these results conflicted with the demonstration of two binding sites by other methods, it was suggested that the binding of two $MgATP^{2-}$ molecules (one per each subunit) would hinder sterically binding of the nucleotide-metal complex at the two other sites.

Kinetic studies provided evidence that $MgATP^{2-}$ binds prior to the reduction of the protein (Watt and Burns, 1977). The rate constant for reduction of the oxidized Fe protein by SO_2^- in the absence of $MgATP^{2-}$ was considerably greater ($k > 10^8$ $M^{-1} s^{-1}$) (Thorneley et al., 1976) than that of SO_2^- utilization during nitrogenase catalysis ($k = 3 \cdot 10^5$ $M^{-1} s^{-1}$) (Watt and Burns, 1977). $MgADP^-$ in turn inhibited reduction of oxidized Fe protein by dithionite (Yates et al., 1975). These differing rates suggest that rate-limiting conformational changes occur in the Fe protein, which limit its rapid reduction particularly during the steady state where it is present largely in oxidized form (Mortenson and Thorneley, 1979).

Electron transfer from the Fe protein to the MoFe protein requires complex formation and there is wide agreement on the notion that electron transfer within this complex is intimately coupled to ATP hydrolysis. Titrations of the MoFe protein with the Fe protein and vice versa were interpreted in terms of models for stable and transient complexes with a 1:1 and 2:1 stoichiometry (molecules Fe protein per molecule MoFe protein). The experimental data were fitted by assuming a transient complex of the two nitrogenase components, with both 1:1 and 2:1 complexes being active (Hageman and Burris, 1978b). A dissociation of this complex after each electron transfer was concluded from the fact that hydrogen evolution showed a lag phase in the order of the turnover time of nitrogenase, whereas ATP hydrolysis was linear from the start of the reaction (Hageman and Burris, 1978a). The lag phase was found to be dependent on the turnover time of the MoFe protein but not on that of the Fe protein. During this lag period also the EPR signal of the MoFe protein decreased (Hageman and Burris, 1979). This in turn indicated that electrons were transferred at random from the Fe protein to the pool of MoFe protein molecules during the transient complex formation. According to this notion the Fe protein then merely serves as a reductase for the MoFe protein, whereas the role of the latter is confined to substrate reduction. This argumentation has substantiated the new nomenclature of the nitrogenase components mentioned in section 2 (Hageman and Burris, 1978a). Throughout this article, however, the conventional nomenclature is still being used.

The direct coupling of electron transfer with ATP hydrolysis is also evident from the identical rates of phosphate release and oxidation of the Fe protein (followed spectrophotometrically at 420 nm) during the pre-steady state (Eady et al., 1978). The minimal number of ATP molecules hydrolyzed for two electrons transferred by nitrogenase from *A. vinelandii* was four (Watt et al., 1975). This number was confirmed with the enzyme from *K. pneumoniae* (Imam and Eady, 1980). ATP/e_2 ratios were shown to respond dramatically to temperature and pH, so that at low pH and low temperature electron transfer and ATP hydrolysis were considerably uncoupled (Watt et al., 1975; Imam and Eady, 1980). The obligatory requirement of ATP for nitrogenase catalysis has now intrigued many researchers for nearly 20 years. Numerous and diverse mechanisms were envisaged in the past (cf. Zumft, 1976), the majority of which favors the generation of a highly reduced species through conservation of the negative free energy change of ATP hydrolysis. This hypothesis was recently further elaborated (Stiefel, 1977; Ljones, 1979).

Part of nitrogenase catalysis certainly can be understood from a potentiometric point of view, despite the lasting scarcity of reliable data (Zumft et al., 1974; O'Donnell and Smith, 1978; Watt, 1980). If one adopts a rigorous potentiometric view of nitrogenase, substrate reduction will occur whenever the required low-potential redox state of the enzyme has been reached. Within this framework other requisites of the nitrogenase reaction, including ATP, serve only the purpose of achieving this highly negative redox state. Contrary to the general belief of ATP lowering the redox potential, one can also argue that its role consists in the opposite, viz., raising the redox potential to a less negative value to allow electron transfer between the two components. Both, reductive and oxidative hypothesis can be based on a unifying concept of energy-transducing redox systems (Losada, 1978).

The rationale for a reducing species with extremely low redox potential was provided by Stiefel (1977) who estimated a potential in the vicinity of -1.0 V if diazene is an intermediate in the reduction of dinitrogen. The Fe protein with a redox potential around -0.3 V (or even in form of its ATP-complex of -0.4 V – this change being most likely a solvent effect on the redox group and perhaps not a property of all Fe proteins will not be able to reduce a species with a redox potential significantly lower than -04 V. It is likely that a highly reducing species is indeed part of nitrogenase catalysis

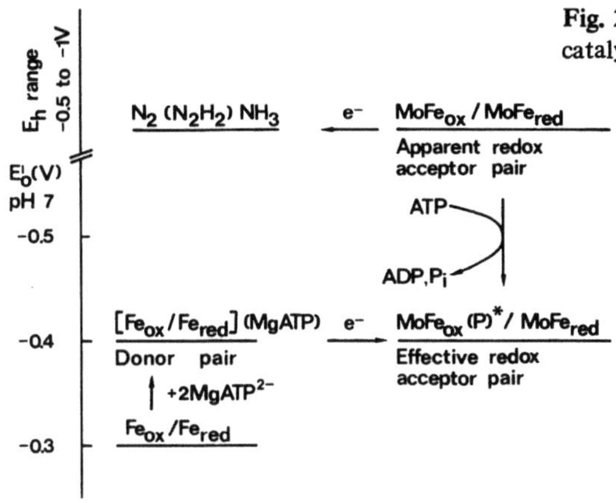

Fig. 2. Model for the role of ATP in nitrogenase catalysis. For explanation of the figure see text

since threshold potentials around -0.4 V were measured, above which no catalytic activity of nitrogenase was observed (Evans and Albrechts, 1975; Watt and Bulen, 1976). Moreover, the FeMo cofactor exists in a so-called "super-reduced" form, although its redox potential could not yet be measured. The electron donation for substrate reduction is being assigned to the oxidized and reduced (=super-reduced) form of the MoFe protein which would have the required low redox potential (Fig. 2). A more definite assignment of the electron donor pair to one of the different metal groups of the MoFe protein is avoided intentionally since we do not know which types of redox equilibria exist among the metal centers. This low potential redox pair is not accessible to electron donation from the Fe protein and is therefore called the "apparent electron acceptor pair." The main postulate of the proposed hypothesis consists in the following (see Fig. 2): ATP energizes the *oxidized* form of the MoFe protein and thereby raises the redox potential. This process creates the effective electron acceptor pair viz. *energized* oxidized MoFe protein – reduced (super-reduced) MoFe protein, onto which the Fe protein now donates its electron *isopotentially*. The term "oxidized MoFe protein" is used here synonymously for the dithionite-reduced state (also called semi-reduced) or the dye-oxidized state, because the concept would hold for either form which can be shown to be involved in the physiological redox cycle. Energization of the MoFe protein can be envisaged as a transient phosphorylation [phosphate exchange into ATP by the MoFe protein is of interest in this context (Eady et al., 1979), as well as the presence of phosphorus in the clostridial Mo-complex (Zumft, 1978)], as a conformational change, or as a combination of both effects. The difficulty in subjecting this or the reductive hypothesis to experimental proof is due to the presumed short lifetimes of the super-reduced and energized species of the MoFe protein. More evidence for an interaction of ATP at the MoFe protein could further substantiate the role of ATP.

6 Nitrogenase and Glutamine Synthetase of Phototrophic Bacteria

Nitrogenase from photosynthetic diazotrophs merits a separate discussion since it requires an activating component to achieve full catalytic activity. The additional component was on various occasions misleadingly referred to as "third nitrogenase component." Once an activation has been achieved no further action of this component is required to maintain the catalytic activity of nitrogenase. It does not participate in each enzyme turnover like the two other nitrogenase components, and is therefore not a necessary third component for the reduction of dinitrogen.

Molecular parameters are available so far only for the nitrogenase from *Rhodospirillum rubrum* (Ludden and Burris, 1978; Nordlund et al., 1978). The MoFe protein again is a tetrameric molecule with 230,000 molecular weight (determined by gel filtration). Different, however, from all other Fe proteins, two subunits with molecular weights of 30,000 and 31,500 were observed by detergent electrophoresis of the Fe protein. The ultraviolet absorption spectrum of the inactive Fe protein had besides the protein maximum a second absorption maximum at 268 nm which was attributed to a nonamino acid group.

The first hint of the distinctive nature of nitrogenase from phototrophic bacteria came from the observation of nonlinear reaction kinetics and demonstration of a manganese requirement for activity (Ludden and Burris, 1976). On cell-breakage the nitrogenase from phototrophic bacteria is mostly found in an inactive form. By washing the chromatophore membranes with salt solution, a factor can be solubilized which acts on the Fe protein and converts it to a catalytically active form (Ludden and Burris, 1976, 1978; Nordlund et al., 1977). The criteria for the activating component mentioned above provided circumstantial evidence for its presence also in *Rhodopseudomonas palustris* (Zumft and Castillo, 1978), *Rhodopseudomonas capsulata* (Yoch, 1979), and in the nonphotosynthetic bacterium *A. lipoferum* (Ludden et al., 1978).

The activating component has not yet been purified to reveal its molecular properties, but from the work of Ludden and Burris it appears to be an enzyme which removes an adenine-like molecule from the Fe protein (Ludden and Burris, 1979). ATP which is required for, but not hydrolyzed during the activation process, is believed to open the site of the modifying group via a conformational change to allow its removal. Phosphate and a pentose are also part of the nonproteinaceous moiety on the Fe protein; they constitute, however, no AMP or ADP residue. The actuation of the activating component implicitly asks for its counterpart that would add the modifying group to the Fe protein for the inactivation process. This postulate has been incorporated in a scheme for nitrogenase regulation in phototrophic bacteria, although physical evidence does not yet exist for this further system (Fig. 3).

The factors in response to which nitrogenase activity is being modulated are neither fully understood nor all identified. The enzyme is found inactive on cell breakage from glutamate-grown (Ludden and Burris, 1978), or N_2-grown cells (Nordlund et al., 1978; Zumft and Castillo, 1978). In contrast, from "nitrogen-starved" cells (i. e., cells that were derepressed for nitrogenase but nonfixing and nonassimilating, apart from internal ammonium turnover) a fully active, nonregulatory nitrogenase was obtained (Carithers et al., 1979). Whether light, indirectly via the shut-down of photosynthesis, or perhaps more directly plays an essential role in the inactivation process continues to be an intriguing possibility for investigation (Meyer et al., 1978).

Low concentrations of ammonium (<5 m*M*) cause an immediate cessation of nitrogenase activity (Zumft and Castillo, 1978, and references therein). Again, this is a peculiarity of nitrogenase from phototrophic bacteria, with the possible exception of *Azotobacter* (see below). We have termed the phenomenon of nitrogenase inactivation

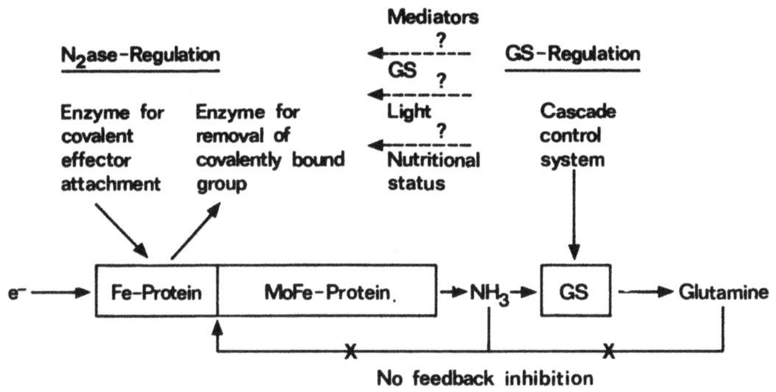

Fig. 3. Hypothetical scheme for the regulation of nitrogenase from phototrophic bacteria. *GS* glutamine synthetase

upon addition of ammonium, and nitrogenase reactivation on ammonium depletion a *switch-off* and *switch-on* effect (Zumft and Castillo, 1978), to underline the fact that nitrogenase of photosynthetic bacteria belongs to the group of enzymes which undergo a reversible inactivation process (Switzer, 1977). To explain this phenomenon, both the activating factor discussed above (Nordlund and Eriksson, 1979), and the key enzyme for ammonium assimilation, i. e., glutamine synthetase, are being considered as possible mediators between ammonium and the regulation of nitrogenase activity (Zumft and Castillo, 1978; Hillmer and Fahlbusch, 1979). The regulation of glutamine synthetase and that of nitrogenase might be interlaced in such a way that from the cascade control system of glutamine synthetase a signal is being transmitted to nitrogenase, causing its inactivation mechanism to shut it down, whereas the activating component under the proper control signal will reactivate nitrogenase. This scheme is still speculative because many pieces of this puzzle remain to be found and put into proper place (Fig. 3). When nitrogenase was inactivated in vivo by ammonium, the enzyme in vitro was found active (Zumft and Castillo, 1978), required the activating component to achieve full activity (Nordlund and Eriksson, 1979), or could not be reactivated by the activating component (Carithers et al., 1979).

For an understanding of a possible connection between the two regulatory systems, it is necessary to expand our knowledge on the molecular and regulatory properties of glutamine synthetase of phototrophic bacteria. Glutamine synthetase from *R. palustris* is a dodecameric molecule of 670,000 molecular weight (Alef et al., 1980). The 12 subunits are spatially arranged in form of two stacked hexagons, and thus resemble the structure described for *Escherichia coli* and other bacterial glutamine synthetases. The enzyme from *R. palustris* is covalently modified by adenylylation, which was shown by removal of AMP from a pure enzyme sample. Adenylylated and deadenylylated forms of glutamine synthetase had different pH profiles, metal requirements, and kinetic properties. The adenylylated form showed maximal activity in the presence of Mn^{2+}, whereas the deadenylylated form was Mg^{2+}-dependent. The apparent K_m for glutamate increased about fivefold when the enzyme became adenylylated. The deadenylylated form in turn had two apparent K_m values for both ammonium and ATP, indicating the existence of two conformational states of the enzyme in response to these substrates. In addition to covalent modification, the glutamine synthetase from *R. palustris* was under negative feedback control by alanine, serine, and glycine (Alef and Zumft, unpubl.). Activation of a partially purified glutamine synthetase from *R. capsulata* by treatment with phosphodiesterase and demonstration of differ- and adenylylation/deadenylylation mechanism also in this species (Johansson and Gest, 1977).

Outside the phototrophic bacteria a short-term inactivation process for nitrogenase in response to external ammonium was shown only for *A. vinelandii* (Kleiner, 1975). Nitrogenase activity was found there to respond antagonistically to the activity of glutamine synthetase. Recently, ammonium was shown to affect slightly the electrical component ($\Delta\psi$) of the protonmotive force (Laane et al., 1980). A decrease of about 25 mV (of initially 106 mV) inhibited nitrogenase activity completely by abolishing the flow of reducing equivalents. If this is the only mechanism operative in *Azotobacter* in response to ammonium, it would demonstrate the unique status of the more complex regulatory system in the phototrophic bacteria. It can be rationalized that regulation of nitrogenase is special insofar as the cell cannot regulate the entry of substrate, i. e., dinitrogen, and therefore must use as control signal the output of nitrogenase, i. e., ammonium. However, rather than using a simple feedback mechanism with ammonium or even glutamine [which was erroneously stated as feedback inhibitor (Jones and Monty, 1979)], that would not allow an integration of the status of other metabolic routes requiring reduced nitrogen, the regulation was devised to operate via the assimilatory system. Control of glutamine synthetase integrates herein particularly an extraordinary number of metabolic signals. Due to the importance of nitrogen for the organism, it is not surprising that alternatively or simultaneously the two other requirements for dinitrogen reduction, energy and reductant, can provide means for regulation.

Acknowledgment. I would like to thank Prof. M. Losada for many helpful discussions.

References

Alef K, Burkardt JH, Horstmann HJ, Zumft WG (1980) unpublished
Benson DR, Arp DJ, Burris RH (1979) Science 205:688–689
Berndt H, Wölfle D (1978) In: Schlegel HG, Schneider K (eds) Hydrogenases: their catalytic activity, structure and function. Erich Goltze KG, Göttingen, pp 327–351
Berndt H, Lowe DJ, Yates MG (1978) Eur J Biochem 86:133–142
Bishop EO, Lambert MD, Orchard D, Smith BE (1977) Biochim Biophys Acta 482:286–300
Bont De JAM, Leijten MWM (1976) Arch Microbiol 107:235–240
Burgess BK, Stiefel EI, Newton WE (1980a) J Biol Chem 255:353–356
Burgess BK, Wherland S, Stiefel EI, Newton WE (1980b) In: Newton WE, Otsuka S, eds (1980), pp 73–84
Burns RC, Hardy RWF (1975) Nitrogen fixation in bacteria and higher plants. Springer, Berlin Heidelberg New York
Carithers RP, Yoch DC, Arnon DI (1979) J Bacteriol 137:779–789
Christou G, Garner CD, Mabbs FE, King TJ (1978) J Chem Soc Chem Commun 740
Coucouvanis D, Baenziger NC, Simhon ED, Stremple P, Swenson D, Simopoulos A, Kostikas A, Petrouleas V, Papaefthymiou V (1980a) J Am Chem Soc 102:1732–1734
Coucouvanis D, Baenziger NC, Simhon ED, Stremple P, Swenson D, Kostikas A, Simopoulus A, Petrouleas V, Papaefthymiou V (1980b) J Am Chem Soc 102:1730–1732
Cramer SP, Hodgson KO, Gillum WO, Mortenson LE (1978a) J Am Chem Soc 100:3398–3407

Cramer SP, Gillum WO, Hodgson KO, Mortenson LE, Stiefel EI, Chisnell JR, Brill WJ, Shah VK (1978b) J Am Chem Soc 100:3814–3819
Davis LC, Shah VK, Brill WJ (1975) Biochim Biophys Acta 403:67–78
Davis LC, Henzl MT, Burris RH, Orme-Johnson WH (1979) Biochemistry 18:4860–4869
Eady RR, Lowe DL, Thorneley RNF (1978) FEBS Lett 95:211–213
Eady RR, Imam S, Lowe DJ, Miller RW, Smith BE, Thorneley RNF (1979) In: Proc Phytochem Soc Symp Nitrogen Fixation Sussex, pp 17–19
Emerich DW, Burris RH (1976) Proc Natl Acad Sci USA 73:4369–4373
Emerich DW, Burris RH (1978a) Biochim Biophys Acta 536:172–183
Emerich DW, Burris RH (1978b) J Bacteriol 134:936–943
Emerich DW, Ljones T, Burris RH (1978) Biochim Biophys Acta 527:359–369
Evans MCW, Albrechts SL (1975) Biochem Biophys Res Commun 61:1187–1193
Ferrante JV, Nicholas DJD (1976) FEBS Lett 66:187–190
Gillum WO, Mortenson LE, Chen J-S, Holm RH (1977) J Am Chem Soc 99:584–595
Haaker H, Veeger C (1977) Eur J Biochem 77:1–10
Hageman RV, Burris RH (1978a) Proc Natl Acad Sci USA 75:2699–2702
Hageman RV, Burris RH (1978b) Biochemistry 17:4117–4124
Hageman RV, Burris RH (1979) J Biol Chem 254:11189–11192
Hallenbeck PC, Kostel PJ, Benemann JR (1979) Eur J Biochem 98:275–284
Hardy RWF, Bottomley F, Burns RC, eds (1979) A treatise on dinitrogen fixation, Section I and II: Inorganic and physical chemistry and biochemistry. John Wiley & Sons, New York
Hillmer P, Fahlbusch K (1979) Arch Microbiol 122:213–218
Huynh BH, Münck E, Orme-Johnson WH (1979) Biochim Biophys Acta 527:192–203
Imam S, Eady RR (1980) FEBS Lett 110:35–38
Johansson BC, Gest H (1977) Eur J Biochem 81:365–371
Johnson JL. Hainline BE, Rajagopalan KV (1980) J Biol Chem 255:1783–1786
Jones BL, Monty KJ (1979) J Bacteriol 139:1007–1013
Kennedy C, Eady RR, Kondorosi E, Rekosh DK (1976) Biochem J 155:383–389
Kleiner D (1975) Arch Microbiol 104:163–169
Kurtz DM, McMillan RS, Burgess BK, Mortenson LE, Holm RH (1979) Proc Natl Acad Sci USA 76:4986–4989
Laane C, Krone W, Konings W, Haaker H, Veeger C (1980) Eur J Biochem 103:39–46
Ljones T (1979) FEBS Lett 98:1–8
Ljones T, Burris RH (1978a) Biochemistry 17:1866–1872
Ljones T, Burris RH (1978b) Biochem Biophys Res Commun 80:22–25
Losada M (1978) Bioelectrochem Bioenerg 5:296–310
Lowe DJ (1978) Biochem J 175:955–957
Ludden PW, Burris RH (1976) Science 194:424–426
Ludden PW, Burris RH (1978) Biochem J 175:251–259
Ludden PW, Burris RH (1979) Proc Natl Acad Sci USA 76:6201–6205
Ludden PW, Okon Y, Burris RH (1978) Biochem J 173:1001–1003
Lundell DJ, Howard JB (1978) J Biol Chem 253:3422–3426
McKenna CE, Jones JB, Eran H, Huang CW (1979) Nature (London) 280:611–612
McKenna CE, Nakajima T, Jones JB, Huang C, McKenna MC, Eran H, Osumi A (1980) In: Newton WE, Otsuka S, eds (1980), pp 39–57
Meyer J, Kelley BC, Vignais PM (1978) J Bacteriol 136:201–208
Mortenson LE, Thorneley RNF (1979) Annu Rev Biochem 48:387–418
Münck E, Rhodes H, Orme-Johnson WH, Davis LC, Brill WJ, Shah VK (1975) Biochim Biophys Acta 400:32–53
Newton WE, Nyman CJ, eds (1976) Proc 1st Int Symp Nitrogen Fixation, vol I, II. Washington State University Press, Pullman
Newton WE, Orme-Johnson WH, eds (1980) Nitrogen fixation, vol I: Free-living systems and chemical models, University Park Press, Baltimore
Newton WE, Otsuka S, eds (1980) Molybdenum chemistry of biological significance. Plenum Press, New York London

Newton WE, Postgate JR, Rodriguez-Barrueco C, eds (1977) Recent developments in nitrogen fixation. Academic Press, London New York
Nordlund W, Eriksson U (1979) Biochim Biophys Acta 547:429–437
Nordlund S, Eriksson U, Baltscheffsky H (1977) Biochim Biophys Acta 462:187–195
Nordlund S, Eriksson U, Baltscheffsky H (1978) Biochim Biophys Acta 504:248–254
O'Donnell MJ, Smith BE (1978) Biochem J 173:831–839
Orme-Johnson WH, Davis LC, Henzl MT, Averill BA, Orme-Johnson NR, Münck E, Zimmermann R (1977) In: Newton WE, Postgate JR, Rodriguez-Barrueco C, eds (1977), pp 131–178
Pienkos PT, Shah VK, Brill WJ (1977) Proc Natl Acad Sci USA 74:5468–5471
Postgate JR, ed (1971) The chemistry and biochemistry of nitrogen fixation. Plenum Press, New York London
Pratt JM (1979) Inorg Perspect Biol Med 2:357–386
Rawlings J, Shah VK, Chisnell JR, Brill WJ, Zimmermann R, Münck E, Orme-Johnson WH (1978) J Biol Chem 253:1001–1004
Robson RL (1979) Biochem J 181:569–575
Shah VK, Brill WJ (1977) Proc Natl Acad Sci USA 74:3249–3253
Shah VK, Davis LC, Brill WJ (1975) Biochim Biophys Acta 384:353–359
Shah VK, Chisnell JR, Brill WJ (1978) Biochem Biophys Res Commun 81:232–236
Smith BE, Thorneley RNF, Eady RR, Mortenson LE (1976) Biochem J 157:439–447
Stephens PJ, McKenna CE, Smith BE, Nguygen HT, McKenna M-C, Thomson AJ, Devlin F, Jones JB (1979) Proc Natl Acad Sci USA 76:2585–2589
Stiefel EI (1977) In: Newton WE, Postgate JR, Rodriguez-Barrueco C, eds (1977), pp 69–108
Switzer RL (1977) Annu Rev Microbiol 31:135–157
Tanaka M, Haniu M, Yasunobu KT, Mortenson LE (1977) J Biol Chem 252:7093–7100
Teo B-K, Averill BA (1979) Biochem Biophys Res Commun 88:1454–1461
Thorneley RNF, Eady RR, Yates MG (1975) Biochim Biophys Acta 403:269–284
Thorneley RNF, Yates MG, Lowe DJ (1976) Biochem J 155:137–144
Thorneley RNF, Eady RR, Lowe DJ (1978) Nature (London) 272:557–558
Tsai LB, Mortenson LE (1978) Biochem Biophys Res Commun 81:280–287
Tsopanakis AD, Bray RC, Smith BE (1979) Biochem J 181:509–512
Watt GD (1980) In: Newton WE, Otsuka S, eds (1980), pp 3–21
Watt GD, Bulen WA (1976) In: Newton WE, Nyman CJ, eds (1976), vol I, pp 248–256
Watt GD, Burns A (1977) Biochemistry 16:264–270
Watt GD, Bulen WA, Burns A, Hadfield KL (1975) Biochemistry 14:4266–4272
Wolff TE, Berg JM, Hodgson KO, Frankel RB, Holm RH (1979) J Am Chem Soc 101:4140–4150
Yamada T, Sakaguchi K (1980) Arch Microbiol 124:161–167
Yates MG, Lowe DJ (1976) FEBS Lett 72:121–126
Yates MG, Planqué K (1975) Eur J Biochem 60:467–476
Yates MG, Thorneley RNF, Lowe DJ (1975) FEBS Lett 60:89–93
Yoch DC (1979) J Bacteriol 140:987–995
Zimmermann R, Münck E, Brill WJ, Shah VK, Henzl MT, Rawlings J, Orme-Johnson WH (1978) Biochim Biophys Acta 537:185–207
Zolg W, Ottow JCG (1975) Z Allg Mikrobiol 15:287–299
Zumft WG (1974) Ber D Bot Ges 87:135–145
Zumft WG (1976) Struct Bonding (Berlin) 29:1–65
Zumft WG (1978) Eur J Biochem 91:345–350
Zumft WG, Castillo F (1978) Arch Microbiol 117:53–60
Zumft WG, Mortenson LE (1975) Biochim Biophys Acta 416:1–52
Zumft WG, Mortenson LE, Palmer G (1974) Eur J Biochem 46:525–535
Zumft WG, Hase T, Matsubara H (1980) In: Newton WE, Otsuka S, eds (1980), pp 59–72

Pathways and Regulatory Aspects of N_2 and NH_4^+ Assimilation in N_2-Fixing Bacteria

D. KLEINER[1], S, PHILLIPS[2], and E. FITZKE[2]

1 Introduction

N_2-fixing organisms are able to use a variety of nitrogen sources, notably the inorganic compounds NH_4^+, NO_3^- and N_2. Although the N_2-fixing strains discovered so far are scattered among many species differing widely in their physiology (see Stewart, 1977a), the major pathways for the assimilation of inorganic nitrogen – leading to glutamine, glutamate, and carbamyl phosphate – are rather similar in all species. A general pattern of major and minor pathways will be outlined in the first part.

The second part will be devoted to the discussion of some regulatory aspects. An important regulation occurs via catabolite repression. All species prefer assimilation of NH_4^+ (or sometimes NO_3^-) over fixation of N_2. Generally the availability of NH_4^+ leads to repression of the nitrogenase synthesis.

2 Pathways

2.1 Free-Living Heterotrophic Organisms

The primary pathways for the assimilation of inorganic nitrogen are summarized in Fig. 1. A brief description of some of these enzymes is found in Table 1, while Table 2 gives the intracellular concentrations of some metabolites during growth on N_2 (nitrogenase derepressed) and NH_4^+ (nitrogenase repressed). Results are mainly derived from studies with three species, which differ widely in their energy metabolism: *Azotobacter vinelandii* (obligate aerobe), *Klebsiella pneumoniae* (faculative, but fixing N_2 almost exclusively under anaerobic conditions), and *Clostridium pasteurianum* (obligate anaerobe).

The best nitrogen source for all organisms is NH_4^+, which is transported across the membrane by a specific carrier, that has been discovered in *C. pasteurianum* (Kleiner and Fitzke, 1979), *A. vinelandii* (Laane et al., 1980), and *K. pneumoniae* (Kleiner, to be published). NH_4^+ uptake is concentrative. Up to 100-fold gradients across the membrane have been observed (Kleiner, 1975, 1976; Kleiner and Fitzke, 1979).

Extracellular NO_3^- is converted to NH_4^+ by the nitrate/nitrite reductase couple, which has been found in *A. vinelandii* and *K. pneumoniae*, but is absent in *C. pasteuria-*

1 Universität Bayreuth, Mikrobiologie, 8580 Bayreuth/FRG
2 Chemisches Laboratorium der Universität, Albertstr. 21, 7800 Freiburg i.Br./FRG

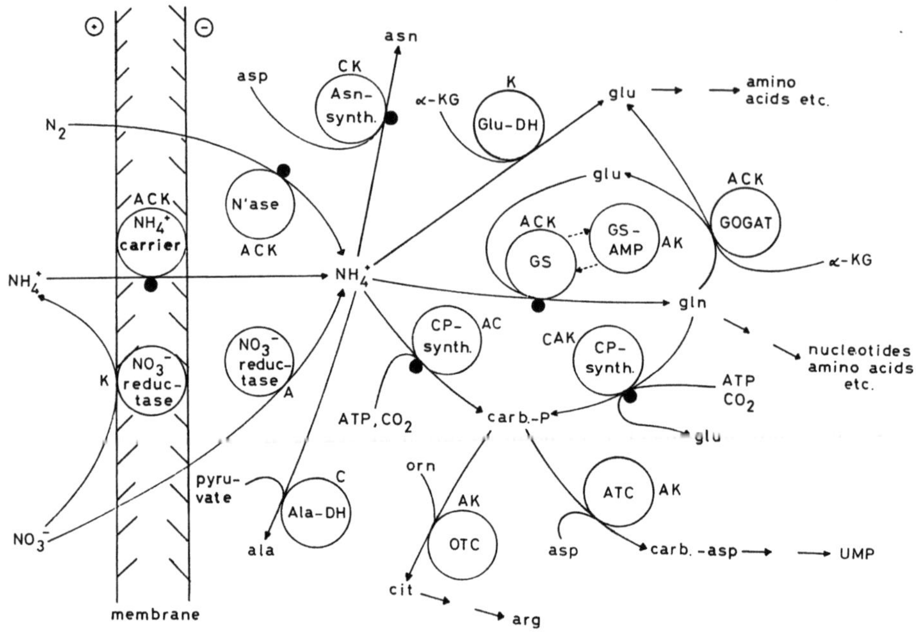

Fig. 1. Major and minor pathways for the assimilation of N_2, NH_4^+, and NO_3^- in *Azotobacter vinelandii (A)*, *Klebsiella pneumoniae (K)*, and *Clostridium pasteurianum (C)*. *Ala-DH* alanine dehydrogenase; *Asn-synth.* asparagine synthetase; *ATC* aspartate transcarbamylase; *CP-synth.* carbamyl phosphate synthetase; *Glu-DH* glutamate dehydrogenase; *GOGAT* glutamate synthase; *GS, GS-AMP* unadenylylated and adenylylated glutamine synthetase; *N'ase* nitrogenase; *OTC* ornithine transcarbamylase; *α-KG* α-ketoglutarate; *carb.-P* carbamyl phosphate. Reducing equivalents are omitted. Reactions marked with a *full circle* are energy-dependent

num. Under N_2 fixing conditions – which are referred to in Fig. 1 – *K. pneumoniae* (anaerobic growth) contains a dissimilatory, membrane-bound complex, which reduces NO_3^- to extracellular NH_4^+, while *A. vinelandii* (aerobic growth) contains an assimilatory, cytoplasmic complex, which yields intracellular NH_4^+ directly. A detailed discussion is found in the chapters of Losada and Whatley in this volume.

N_2 very likely passes the membrane by diffusion and is then reduced to N_2 by the nitrogenase complex.

The product of the action of the first assimilatory enzymes thus is intracellular NH_4^+. Until 1970 it was generally assumed that its further metabolization occurs mainly via direct incorporation into the α-amino group of glutamate, a reaction mediated by glutamate dehydrogenase (Glu-DH). This enzyme, however, has only been detected in *K. pneumoniae,* where it is induced by high supplies of NH_4^+ and NO_3^-. It is generally absent in *A. vinelandii* and *C. pasteurianum,* and repressed in *K. pneumoniae* under N_2-fixing conditions (Nagatani et al., 1971; Kleiner, 1976). After the discovery of the glutamate cycle (Meers et al., 1970) it became evident, that in *C. pasteurianum* and *A. vinelandii* under all growth conditions and in *K. pneumoniae* under N_2-fixing conditions the bulk of NH_4^+ is first incorporated into the amide group of glutamine. This ATP-dependent reaction is catalyzed by the glutamine synthetase (GS), which has a low K_m value for NH_4^+ (see Table 1). In contrast, the high K_m value of Glu-DH renders this enzyme unsuitable for the assimilation of the low NH_4^+ concentrations (see Table 2) produced by the nitrogenase action. Furthermore the amide group of glutamine can be regarded as "activated" NH_4^+, since its group transfer potential is 14.2 kJ/mol higher than in the free state, providing an additional thermodynamic driving force for its subsequent utilization. For further nitrogen metabolization analogous enzymes often exist, which use either free or activated NH_4^+ and transfer it to the same acceptor molecule. The most important enzyme for the transfer of the amide group, the glutamate synthase (GOGAT, from *g*lutamine:2-*o*xo*g*lutarate *a*mido *t*ransferase) is an enzyme analogous to Glu-DH. This enzyme has been discovered in all heterotrophic N_2-fixing species. It is NAD-specific in *C. pasteurianum*, but NADP-specific in *K. pneumoniae* and *A. vinelandii* (Nagatani et al., 1971). The transfer of the amide group to α-ketoglutarate completes the glutamate cycle (GS/GOGAT pathway).

Another important pool for the distribution of nitrogen equivalents is carbamyl phosphate, which especially serves the routes to arginine and UMP. Two carbamyl phosphate synthetases can exist, using either glutamine or NH_4^+ as donors. We were able to demonstrate both activities (one enzyme with dual specificity?) in crude extracts of *A. vinelandii*, but only the glutamine-dependent one in *K. pneumoniae* (unpubl. results). Mortenson reported the discovery of these enzymes in *C. pasteurianum* (1978).

The alanine dehydrogenase-catalyzed amination of pyruvate (Ala-DH) has been detected in *C. pasteurianum* (Mortenson, 1978).

Investigations in our laboratory revealed the existence of an NH_4^+-dependent asparagine synthetase in *C. pasteurianum* (Kleiner, 1979) and *K. pneumoniae* (unpubl. data), but not in *A. vinelandii*, whose enzyme may be glutamine-dependent.

2.2 Phototrophic Bacteria

Munson and Burris (1969) observed that N_2-fixing cultures of *Rhodospirillum rubrum* absorb extracellular NH_4^+ down to traces below 1 μM. If we assume that the intracellar NH_4^+ concentration in this organism is close to the one determined in nonphotosynthetic species and thus in the range of the K_m of the GS (see Tables 1 and 2), the existence of active NH_4^+ transport must be inferred.

Comparative studies by Brown and Herbert (1977 a, b) with several strains of phototrophic bacteria revealed that all species contain high GS levels under N_2-fixing conditions. GOGAT and Glu-DH were detected in most strains too, but their nucleotide specificity varied from one strain to the other.

2.3 Cyanobacteria

No investigations about NH_4^+ uptake by cyanobacteria have been reported. Clarification of this point might help to understand the mechanism by which NH_4^+ is released in associations (see Sect. 2.4.2).

The GS/GOGAT pathway is now well established as the main route of NH_4^+ assimilation in almost all cyanobacteria (see Stewart, 1977b; Wolk, 1980). At variance with the bacterial enzyme, GOGAT from cyanobacteria requires reduced ferredoxin as reductant (Lea and Miflin, 1975). The levels of (mostly NADP-specific)Glu-DH are generally low, but some strains contain high levels of Ala-DH (see Stewart, 1977b), which under conditions of high NH_4^+ supply may provide a major route for NH_4^+ assimilation. Both the NH_4^+ and the glutamine-dependent carbamyl phosphate synthetase activities have been detected in *Anabaena cylindrica* (Lawrie, 1979).

2.4 Symbiotic Organisms

Since nothing is known about NH_4^+ assimilation by the endosymbionts causing nodulation of nonleguminous plants, only the studies with rhizobia and cyanobacteria will be briefly summarized.

2.4.1 *Rhizobia*

Free-living rhizobia are unable to cover their entire nitrogen demand by N_2 fixation and require an additional nitrogen source. Few strains are able to use NH_4^+, while the majority require glutamate (see Dilworth, 1980). This demand is probably due to the high permeability of rhizobial membranes towards NH_3 (Laane et al., 1980), which leads to excretion of newly fixed nitrogen (Bergersen and Turner, 1967). Since rhizobia (Laane et al., 1980) like other bacteria (see Harold, 1978) maintain a pH gradient with an alkaline interior across the membrane, NH_4^+ should accumulate outside the cell according to

$$[NH_4^+]_{outside}/[H^+]_{outside} = [NH_4^+]_{inside}/[H^+]_{inside}$$

$$([NH_3]_{outside} = [NH_3]_{inside})$$

When growing in symbiosis with the host plant this leakage is of physiological importance, and is probably the reason for the maintenance of a high NH_4^+ concentration in the cytosol (Kennedy, 1966).

All free-living rhizobia contain GS and GOGAT. In nodules only low bacterial GS levels have been found (see Dilworth, 1980), while the levels of GOGAT and Glu-DH show no consistent patterns. It is generally assumed that the lack of GS in bacteroids is the cause of excretion of unassimilable NH_3 into the plant cytosol. But as discussed above, the high permeability of the rhizobial membrane towards NH_3 in conjunction with a suitably directed pH gradient must result in considerable NH_4^+ depletion of the bacteroid. This loss is partially compensated by an important of amino acids (Shanmugam et al., 1977).

Ala-DH has been detected in one rhizobial strain (Werner and Stripf, 1978). No data have been published about carbamyl phosphate metabolism.

2.4.2 *Cyanobacteria*

Free-living cyanobacteria frequently release fixed nitrogen, mostly NH_4^+ (see Stewart, 1977b). This excretion is of importance for species living in symbiotic associations. The best-investigated system are the *Anabaena-Azolla* association (see Peters et al., 1980), some lichens (see Millbank, 1977; Stewart, 1977b), and liverworts (see Millbank, 1977; Stewart, 1977b). Evidence has been obtained for an interaction of the NH_4^+ metabolism in the endosymbiont by the host, which results in enhanced NH_4^+ secretion. The target of control is mostly assumed to be the GS, but if NH_4^+ transport across the cyanobacterial membrane should occur in response to the membrane potential $\Delta\psi$ as in free living heterotrophic bacteria (see Sect. 3.1), a decrease of $\Delta\psi$ (e. g., by a compound acting like colicin K, see Weiss and Luria, 1978) should lead to a considerable loss of NH_4^+.

3 Regulatory Aspects

Most enzymes of the metabolic pathways outlined in Fig. 1 are regulated on several levels. Some metabolites play multiple roles in these regulations, indicating the existence of a complicated biocybernetic network. For example, all reactions of Fig. 1 marked with a full circle are energy-dependent, and thus are putative sites for control by the energy charge or by the individual nucleotide levels, providing a link between nitrogen and energy metabolism.

The regulatory properties of some enzymes are briefly summarized and discussed.

Table 1. Selected properties of some enzymes

Enzyme	EC number	Source	K_m (m*M*) for NH_4^+ or gln	Inducible + repressible – unaffected O by NH_4^+	Ref.
NH_4^+ carrier	–	*C. pasteurianum*	0.009	0	Kleiner and Fitzke (submitted)
		K. pneumoniae	0.007	–	Kleiner (to be published)
Nitrogenase	–	All strains (wild type)	0.002–0.012	–	see Zumft and Mortenson (1975)
GS	6.3.1.2	*A. vinelandii*	0.15	0	Kleinschmidt and Kleiner (1978)
		R. palustris	0.2–0.5	0	Alef and Zumft (unpubl.)
		A. cylindrica	1	(–)	Stewart et al. (1977)
		K. pneumoniae	0.33	–	Kleiner (1976)
GOGAT	1.4.1.13	*K. pneumoniae*	1.0	(0)	Kleiner (1976)
Glu-DH	1.4.1.2	*K. pneumoniae*	12	+	Kleiner (1976); Nagatani et al. (1971)

Table 2. Intracellular concentrations (m*M*) of some metabolites during growth on N_2 and NH_4^+ (from Kleiner, 1975, 1976; Kleiner and Fitzke, 1979, 1979; Kleinschmidt, 1979; Mortenson, 1978)

Metabolite	*C. pasteurianum*		*K. pneumoniae*		*A. vinelandii*	
	N_2	NH_4^+	N_2	NH_4^+	N_2	NH_4^+
NH_4^+	0.2–0.4	–	0.3–0.4	–	0.6	(0.6)
α-KG	0.7	0.7	–	–	–	–
gln	0.8	6.2	0.2	0.6	0.45	0.5–2.0
glu	5.2	8.0	4.5	4.0	6.6	9.2
Carbamyl phosphate	0.060	0.104	–	–	–	–

3.1 Ammonium Transport

Good evidence has been obtained for concentrative NH_4^+ uptake being driven by the membrane potential $\Delta\psi$ (negative inside) in *A. vinelandii* (Laane et al., 1980), *C. pasteurianum* (Kleiner and Fitzke, in press), and *K. pneumoniae* (Kleiner, in prep.). This implies that a) NH_4^+ (and not NH_3) is the transported molecule, b) the size of $\Delta\psi$ governs extent and probably rate of uptake (and release).

In *K. pneumoniae,* but not in *C. pasteurianum,* the NH_4^+ carrier is repressible by high NH_4^+ concentrations, similarly as in lower fungi (see Silver, 1978). The NH_4^+ carrier is absent in rhizobia and may be lacking in some cyanobacteria.

3.2 Glutamine Synthetase

Controls of this enzyme include repression by NH_4^+, feed-back inhibition by nitrogenous metabolites, and chemical modification by adenylylation-deadenylylation.

3.2.1 Repression by NH_4^+

Repression of GS is mainly found in strains that contain alternative routes for NH_4^+ incorporation, especially by Glu-DH. The free-living organisms *A. vinelandii* (Kleinschmidt, 1979) and *C. pasteurianum* (Kleiner, 1979) do not synthesize Glu-DH, and conversely their GS is not repressible. In contrast, the GS of *K. pneumoniae* is repressed by NH_4^+, while Glu-DH is induced (Nagatani et al., 1971; Kleiner, 1976). Other organisms with repressible GS include some phototrophic organisms (Brown and Herbert, 1977 a,b), all rhizobia (see Dilworth, 1980) and some cyanobacteria (see Stewart et al., 1977; Peters et al., 1980).

3.2.2 Feed-Back Inhibition

Feed-back inhibition by some metabolites has been reported for the GS from several species (Kleinschmidt and Kleiner, 1978; Orr et al., 1978; Stacey et al., 1979). A detailed kinetic analysis with the *A. vinelandii* enzyme revealed rather complicated relations between the inhibitory power of some metabolites and the availability of substrates, thus providing a very efficient regulation of the flow of nitrogen equivalents along this route.

3.2.3 Chemical Modification

Extensive studies with *E. coli* (see Tyler, 1978) demonstrated that the GS of this organism can be enzymatically modified by the attachment of AMP, leading to a rapid decrease in the GS activity. The modifying enzyme system is under strict metabolic control by several metabolites, notably glutamine and α-ketoglutarate. This mechanism ensures a rapid and stable change in activity.

Adenylylation of GS has been demonstrated in *K. pneumoniae* (see Tyler, 1978), *A. vinelandii* (Kleinschmidt and Kleiner, 1978), and *Rhodopseudomonas palustris* (Alef and Zumft, unpubl.). Good evidence for this control has been obtained for *R. capsulata* (Johannsson and Gest, 1977), *Azospirillum brasiliense* (Okon et al., 1976), and rhizobia (see Dilworth, 1980). No evidence has been found for *C. pasteurianum* (see Mortenson, 1978; Kleiner, 1979) and cyanobacteria (Stewart et al., 1977; Orr et al., 1978; Stacey et al., 1979).

3.3 Glutamate Synthase

Repressibility of GOGAT by NH_4^+ or glutamate has been reported to occur in rhizobia (Upchurch and Elkan, 1978). So far no conclusive data are available for other organisms. Rhizobial GOGAT is very sensitive towards inhibition by amino acids (Dilworth, 1980).

3.4 Nitrogenase

Nitrogenase synthesis and activity are regulated on several levels, indicating a very tight control of N_2 fixation, probably due to the high energy requirement of this reaction (see Zumft and Mortenson, 1975). Regulation of nitrogenase activity includes feed-back inhibition by ADP and carbamyl phosphate (see Mortenson, 1978), enzymatic modification of one nitrogenase component (Ludden and Burris, 1979), reversible "switch-off" by O_2 (Drozd and Postgate, 1970) and control of reductant flow (see Veeger et al., 1980). Nitrogenase synthesis may be controlled by the concentrations of O_2 and molybdenum (see Brill, 1975; Mortenson, 1978), by the temperature (Hennecke and Shanmugam, 1979) and by the NH_4^+ supply (see Mortenson, 1978). Under some circumstances enhanced turnover seems to occur (Kleiner and Kleinschmidt, 1976; Vallespinos and Kleiner, 1980).

Only a few aspects of the repression by NH_4^+ will be discussed here. This control, which occurs on the level of transcription (Tubb and Postgate, 1973), has been observed in all wild-type strains, at least under certain growth conditions. Rhizobia generally require higher NH_4^+ concentrations for full repression than free-living species. This is probably due to the lack of an NH_4^+ carrier, which amplifies the extracellular NH_4^+ concentration in the cell, and thus the concentration of the regulatory compound(s). The nature of the latter thus far has eluded identification. An attractive theory suggested direct involvement of the unadenylylated GS in the transcription of *nif* (nitrogen fixation) genes (Tubb, 1974; Streicher et al., 1974), but in the light of recent findings this hypothesis had to be abandoned (see Tyler, 1978; Ausubel et al., 1979). Possibly a compound involved in the regulation of the synthesis and adenylylation of GS plays a major role in the expression of *nif* genes. Recent investigations concentrate on the yet unidentified *gln F* gene product (see Tyler, 1978). Ausubel et al. (1979) even consider that GS may not be absolutely required for *nif* derepression.

At this point it may be fruitful to discuss two questions, which thus far have attracted little attention:

1. Is the mechanism of *nif* repression by NH_4^+ the same for all organisms, or has it changed during evolution?

2. Is there only one regulatory circuit or does control of nitrogenase synthesis by NH_4^+ occur on several levels?

The first question results from the fact that the hypothesis which postulates involvement of the *gln F* product (or another factor of the GS modification system) has been almost exclusively derived from genetic investigations with *K. pneumoniae,* which possesses a modifiable GS, whereas *C. pasteurianum* or cyanobacteria do not. This leaves four possibilities:

a) the mechanism is different in at least two groups;

b) the hypothesis has to be revised;

c) the *gln F* product is of more general importance and occurrence, and plays a role in many regulatory systems, and is also present in cyanobacteria and *C. pasteurianum;*

d) there is one type of control common in all species, which sometimes is superimposed by a second regulatory mechanism that varies from one group to the other.

Some support for a multiple control hypothesis has been forwarded by Shanmugam et al. (1977) and Ausubel et al. (1979). Comparative investigations with several strains should be useful to elucidate the nature of the common regulatory compound. The level of this regulator should in all species vary fast and correlate strictly with the rate of nitrogenase synthesis. Thus possible condidates are glutamine and carbamyl phosphate (see Table 2). NH_4^+ itself has been ruled out, because it does not repress nitrogenase synthesis when its further metabolization is blocked (Gordon and Brill, 1974; Stewart, 1977b). Recently we turned our attention to postclassical modes of regulation and investigated whether guanosine 3′diphosphate 5′diphosphate (ppGpp) might not be involved. In enterobacteria high levels of this compound have been found under conditions of amino acid starvation (see Nierlich, 1978; Gallant, 1979). ppGpp can interact with RNA polymerase and change its initiation specificity, thus influencing the transcription rates of several operons. Generally the transcription of those operons seems to be stimulated whose enzymes might circumvent the cause of amino acid starvation. This could also apply to the *nif* operon. Analysis of the total amino acid pools of several N_2-fixing bacteria showed generally higher levels under growth on NH_4^+ than on N_2 (Drozd et al., 1972; Ohmori and Hattori, 1974; Kleiner, 1979; Kleinschmidt, 1979). Thus it may be that nitrogenase synthesis is derepressed in response to continued amino acid depletion. In accordance with these reasonings preliminary experiments in our laboratory revealed rather high ppGpp levels in *C. pasteurianum, A. vinelandii,* and *K. pneuomiae* under N_2-fixing conditions, which rapidly decreased 3- to 10-fold after addition of NH_4^+. These results are in accordance with the hypothesis that nitrogenase synthesis is generally stimulated by ppGpp, the level of which is regulated by the type and supply of the nitrogen source.

This hypothesis would also explain the repression of nitrogenase synthesis by some amino acids in mutants of *K. pneumoniae*, that do not synthesize GS (see Shanmugam et al., 1977).

References

Ausubel FM, Bird SC, Durbin KJ, Janssen KA, Margolskee RF, Peskin AP (1979) J Bacteriol 140:597–606
Bergensen FJ, Turner GL (1967) Biochim Biophys Acta 141:507–515
Brill WJ (1975) Annu Rev Microbiol 29:109–129
Brown CM, Herbert RA (1977a) FEMS Lett 1:39–42
Brown CM, Herbert RA (1977b) FEMS Lett 1:43–46
Dilworth MJ (1980) In: Newton WE, Orme-Johnson WH (eds) Nitrogen fixation, voll II. University Park, Baltimore, pp 3–31
Drozd JW, Postgate JR (1970) J Gen Microbiol 63:63–73
Drozd JW, Tubb RS, Postgate JR (1972) J Gen Microbiol 73:221–232
Gallant JA (1979) Annu Rev Genet 13:393–415
Gordon JK, Brill WJ (1974) Biochem Biophys Res Commun 59:967–971
Harold FM (1978) Bacteria 6:463–521
Hennecke H, Shanmugam KT (1979) Arch Microbiol 123:259–265
Johannsson BC, Gest H (1977) Eur J Biochem 81:365–371
Kennedy IR (1966) Biochim Biophys Acta 130:285–294
Kleiner D (1975) Arch Microbiol 104:163–169
Kleiner D (1976) Arch Microbiol 111:85–91
Kleiner D (1979) Arch Microbiol 120:263–270
Kleiner D, Fitzke E (1979) Biochem Biophys Res Commun 86:211–217
Kleiner D, Kleinschmidt JA (1976) J Bacteriol 128:117–122
Kleinschmidt JA (1979) Ph D Thesis, Freiburg
Kleinschmidt JA, Kleiner D (1978) Eur J Biochem 89:51–60
Laane C, Krone W, Haaker H, Veeger C (1980) Eur J Biochem 103:39–46
Lawrie AC (1979) J Bacteriol 139:115–119
Lea PJ, Miflin BJ (1975) Biochem Soc Trans 3:381–384
Ludden PW, Burris RH (1979) Proc Natl Acad Sci USA 76:6201–6205
Meers JL, Tempest DW, Brown CM (1970) J Gen Microbiol 64:187–194
Millbank (1977) In: Hardy RWF, Silver WS (eds) A treatise on dinitrogen fixation, vol III. Wiley, New York London Sydney Toronto, pp 125–151
Mortenson LE (1978) Curr Top Cell Regul 13:179–232
Munson TO, Burris RH (1969) J Bacteriol 97:1093–1098
Nagatani H, Shimizu M, Valentine RC (1971) Arch Microbiol 79:164–175
Nierlich DP (1978) Annu Rev Microbiol 32:393–432
Ohmori M, Hattori A (1974) Plant Cell Physiol 15:131–142
Okon Y, Albrecht SL, Burris RH (1976) J Bacteriol 128:592–597
Orr J, Toan ND, Ohtsuki M, Wellems T, Haselkorn R (1978) Fed Proc 37:1431
Peters GA, Ray TB, Mayne BC, Toia RE (1980) In: Newton WE, Orme-Johnson WH (eds) Nitrogen fixation, vol II. University Park, Baltimore, pp 293–309
Rowell P, Stewart WDP (1976) Arch Microbiol 107:115–124
Shanmugam KT, O'Gara F, Andersen K, Morandi C, Valentine RC (1977) In: Newton W, Postgate JR, Rodriguez-Barrueco (eds) Recent developments in nitrogen fixation. Academic Press, London New York, pp 321–330
Silver S (1978) In: Rosen BP (ed) Bacterial transport. Marcel Dekker, New York Basel, pp 281–284

Stacey G, Van Baalen C, Tabita FR (1979) Arch Biochem Biophys 194:457–467
Stewart WDP (1977a) Ambio 6:166–173
Stewart WDP (1977b) In: Hardy RWF, Silver WS (eds) A treatise on dinitrogen fixation, vol III. Wiley, New York London Sydney Toronto, pp 63–123
Stewart WDP, Rowell P, Apte SK (1977) In: Newton W, Postgate JR, Rodriguez-Barrueco C (eds) Recent developments in nitrogen fixation. Academic Press, London New York, pp 287–307
Streicher SL, Shanmugam KT, Ausubel F, Morandi C, Goldberg R (1974) J Bacteriol 120:815–821
Tubb RS (1974) Nature (London) 251:481–485
Tubb RS, Postgate JR (1973) J Gen Microbiol 79:103–117
Tyler B (1978) Annu Rev Biochem 47:1127–1162
Upchurch RG, Elkan GH (1978) Biochim Biophys Acta 538:244–248
Vallespinos F, Kleiner D (1980) J Gen Microbiol 117:543–545
Veeger C, Laane C, Scherings G, Matz L, Haaker H, Van Zeeland-Wolbers (1980) In: Newton WE, Orme-Johnson WH (eds) Nitrogen fixation, vol I. University Park, Baltimore, pp 111–137
Weiss MG, Luria SE (1978) Proc Natl Acad Sci USA 75:2483–2487
Werner D, Stripf R (1978) Z Naturforsch 33c:245–252
Wolk CP (1980) In: Newton WE, Orme-Johnson WH (eds) Nitrogen fixation, vol II. University Park, Baltimore, pp 279–292
Zumft WG, Mortenson LE (1975) Biochim Biophys Acta 416:1–52

The Hydrogenase-Nitrogenase Relationship in Nitrogen-Fixing Organisms

H. BOTHE and G. EISBRENNER[1]

1 Introduction

Historically, an enzyme catalyzing the reaction $2H^{+}+2e^{-}\rightarrow H_2$ was first demonstrated by Stephenson and Stickland (1931) using extracts from *Escherichia coli.* A relationship between H_2-metabolism and N_2-fixation was recognized in the early 1940's by P.W. Wilson and his coworkers (Phelps and Wilson, 1941; Wilson et al., 1942; Lee and Wilson, 1943; Green and Wilson, 1953; Hyndman et al., 1953). They detected a hydrogenase in *Azotobacter* and *Rhizobium* bacteroids which catalyzed the uptake of molecular hydrogen and which was particularly active under nitrogen-fixing conditions. In the past few years, the hydrogenase-nitrogenase relationship has been investigated with increasing efforts. Such renewed interest mainly comes from considerations to use H_2-formation by photosynthetic organisms in projects of solar energy conversion. Consequently, since several agencies gave financial support to investigations in this field, the output of publications has recently been increased explosively. It is not the scope of this communication to review the vast literature, particularly since articles on H_2-metabolisms by different groups of microorganisms have recently become available. The reader is referred to the review articles by Schlegel and Schneider (1978) or Mortenson and Chen (1974) for fermentative bacteria, by Meyer et al. (1978) or Schlegel and Schneider (1978) for photosynthetic bacteria, by Dixon (1978), Albrecht et al. (1979) or Walker and Yates (1978) for *Rhizobium* and *Azotobacter,* by Hallenbeck and Benemann (1979) or Bothe et al. (1978) for blue-green algae and by Kessler (1974), Bishop et al. (1977) or Ben-Amotz (1979) for green algae. This communication will summarize aspects of the hydrogenase-nitrogenase relationship with special emphasis on blue-green algae which are in the centre of our own interest.

2 The H_2-Metabolizing Enzymes in Nitrogen-Fixing Microorganisms

Nitrogen-fixing organisms may contain three H_2-metabolizing enzymes which are clearly distinguishable by their physiological and biochemical properties.

a) Reversible Hydrogenase. Clostridium pasteurianum contains a soluble hydrogenase which couples to ferredoxin, flavodoxin and artificial electron carriers with an electro-

1 Botanisches Institut, Universität Köln, Gyrhofstr. 15, 5000 Köln 41/FRG

negative redox potential. Both uptake and evolution of H_2 by this enzyme are independent of ATP and are affected by carbon monoxide. The enzyme has a molecular weight of about 60,000, mediates a D_2O-exchange reaction and contains three 4Fe:4S centres among which only one is believed to undergo oxidation/reduction during catalysis. Physiologically, the enzyme functions to remove excess of reducing equivalents generated during fermentation. The occurrence of this soluble hydrogenase has clearly been established for saccharolytic *Clostridia* and for facultative anaerobic bacteria, e. g., *Bacillus polymyxa* (see Mortenson and Chen, 1974; Schlegel and Schneider, 1978). Hydrogenase of the Enterobacteriaceae (e. g., *Klebsiella pneumoniae, Escherichia coli)* is part of the membrane-bound formate: hydrogen lyase complex which has not yet been characterized in full detail. The same complex apparently also occurs in photosynthetic bacteria (e. g. *Rhodospirillum rubrum*) which ferment pyruvate to formate catalyzed by pyruvate: formate lyase in the dark (see Voelskow and Schön, 1980). A hydrogenase, catalyzing the evolution of H_2 at the expense of carbohydrates, is apparently soluble in the non-N_2-fixing green algae (e. g., *Chlamydomonas, Chlorella* strains; see Bishop et al., 1977) but it is uncertain at present which other properties is shares with the "classical" hydrogenase from *Clostridium.* A soluble, reversible hydrogenase was suggested to occur also in blue-green algae (Tel-Or et al., 1978; Hallenbeck and Benemann, 1978) which however, was questioned by us (Bothe et al., 1978, see Sect. 5).

b) Nitrogenase. This enzyme catalyzes the formation of H_2 from H_3O^+ in a unidirectional, ferredoxin, ATP-dependent and CO-insensitive reaction. These features clearly allow to differentiate whether a H_2 formation is catalyzed by hydrogenase or by nitrogenase. The mechanisms by which H_2 is formed from nitrogenase have not yet been resolved, although an attractive working hypothesis has recently been forwarded (Chatt, 1980; see Bothe et al., 1980b). In the presence of N_2, H_2 is released concomittantly with the ammonia produced, presumably according to the following equation:

$$8H^+ + 8e^- + N_2 \rightarrow 2NH_3 + H_2 .$$

In the absence of any other substrate of nitrogenase, substantial amounts of H_2 are still evolved despite the low H_3O^+-concentrations available in such assays normally performed around pH 7–8. It must be emphasized that H_2 formation catalyzed by nitrogenase is substantial with nitrogenases isolated from all organisms but is often marginal in intact cells.

c) Uptake Hydrogenase. Although this membrane-bound enzyme has been known since the early 1940's, as mentioned earlier, its physiological role and its biochemical properties have not been fully elucidated at present. This uptake hydrogenase is obviously an integral protein of membranes and therefore difficult to characterize. The enzyme appears to be more or less ubiquitous in aerobic nitrogen-fixing organisms possibly only with the exception of some *Rhizobium* strains (Albrecht et al., 1979). It is, however, not restricted to N_2-fixing cells, though activity levels are generally enhanced under such conditions. It can be demonstrated in non-N_2-fixing blue-green algae and is also present in aerobic hydrogen oxidizing bacteria (*Alcaligenes, Paracoccus* and the N_2-fixing *Xanthobacter;* see Schlegel and Schneider, 1978). A biochemical characterization and a direct comparison of the properties of the uptake hydrogenase from various sources is lacking at present.

3 The Evolution of Molecular Hydrogen

Contrary to anaerobic nitrogen-fixing bacteria (e.g., *Clostridium pasteurianum*) which evolve H_2 in large quantities, aerobic nitrogen-fixing microorganisms (*Azotobacter, Rhizobium, Xanthobacter,* blue-green algae) can form the gas at best in moderate amounts. There is agreement among the different experimentators that virtually all H_2-formation in aerobes is catalyzed by nitrogenase. The gas is only evolved under nitrogen-fixing conditions, and the addition of combined nitrogen to the cultures suppresses its formation. The reaction requires the presence of optimal concentrations of oxygen to allow respiration, or, in the phototrophs, depends on light under anaerobic conditions, indicating the requirement for ATP. Addition of uncouplers to the assays abolishes the activities. C_2H_2 and N_2 which compete with H_3O^+ for electrons in nitrogenase reduce H_2-formations in all aerobic nitrogen-fixing organisms. The quantities formed are rather variable, depending on the strain used, the culture conditions employed and on the experimentator. According to Albrecht et al. (1979) H_2 was evolved in larger quantities in such *Rhizobium* strains which were unable to utilize the gas. Walker and Yates (1978) noted that batch-grown cultures of *Azotobacter vinelandii* did not evolve significant amounts of H_2, whereas O_2-, N_2-, as well as C-limited continuous cultures showed considerable H_2-productions.

Discrepancies about the quantities of H_2 produced by blue-green algae have arisen in the literature (for a detailed review see Bothe et al., 1978). It has recently become clear that variable rates are not so much due to the strain used but more to the culture conditions employed. Factors influencing H_2-formation include **during growth**: The age of the cultures, the light intensities, the temperature, the supply with iron and CO_2, the oxygen tensions and **during incubation in the assay flasks**: the duration of the experiment, the assay temperature, the concentrations of O_2, N_2, or CO_2 in the vessels, the nutritional status of the cells, particularly their reserves of endogenous carbohydrates or nitrogen compounds (see Benemann and Weare, 1974 a,b; Weissman and Benemann, 1977; Jeffries et al., 1978; Spiller et al., 1978; Ernst et al., 1979; Miyamoto et al., 1979; Lambert et al., 1979 a,b).

H_2-formation rates can be enhanced to some extent by adding low levels of carbon monoxide and high amounts of C_2H_2 to the assays (Bothe et al., 1977b; Lambert and Smith, 1978) or by increasing the number of heterocysts as the site of nitrogenase which can be achieved by supplementing the cultures with 7-azatryptophan (Bothe and Eisbrenner, 1977). Despite this, in our own assays with *Anabaena* species or *Nostoc muscorum* routinely performed under oxygen exclusion, the activities were always marginal as compared to the rates of N_2-fixation (C_2H_2-reduction) and CO_2-fixation (see also Tel-Or et al., 1978). Likewise, H_2-formation capabilities were low in a whole series of blue-green algae newly isolated from the Zürich area (Berchthold and Bachofen, 1979). A more neutral position is taken by Bishop and coworkers, who, in their recent publication (Tetley and Bishop, 1979, Fig. 1), gave a valuable comparison of the rates of H_2-formation, CO_2-fixation, and C_2H_2-reduction by *Anabaena*. Other authors reported extraordinarily high and long lasting H_2-formations (Weissman and Benemann, 1977; Mitsui and Kumazawa, 1977; Jeffries et al., 1978; Lambert et al., 1979b). The absolute maximum probably are the 7,500 μl/mg chl·h^{-1} with *Anabaena cylindrica* (Berchthold and Bachofen, 1979), although the experimental procedure for obtaining such high values was unfortunately omitted in this publication.

Since H_2-formation is catalyzed by nitrogenase, C_2H_2-reduction activities are an indication for the maximal achievable H_2-evolutions. There is general consensus that C_2H_2-reductions are generally 5–20 μmol C_2H_4/h·mg chlorophyll with various blue-green algal strains or 1/5–1/10 of the rates of photosynthetic CO_2-fixation, in accord with the requirements of the cells for carbon and nitrogen. A rate of 20 μmol H_2 formed/h·mg chlorophyll would mean that all the electrons flowing to nitrogenase must reduce H_3O^+ and that the organisms do not possess a regulatory mechanism to switch off the energy-consuming H_2-formations. Such high activities sustained over a longer period would necessitate the supply of the cells with fixed nitrogen, which, however, represses nitrogenase biosynthesis. Activities exceeding 20 μmol H_2/h · mg chlorophyll would imply an additional source of reductant and ATP for nitrogenase, which can hardly be provided by the cells over a longer period. We are, therefore, currently not inclined to give up our "pessimism about eventual application of the biophotolytic process by blue-green algae" (see Lambert et al., 1979b).

Having said that all H_2-formations by aerobic nitrogen-fixing microorganisms are catalyzed by nitrogenase, a marked exception must be mentioned: The halophytic *Oscillatoria limnetica* evolves H_2 in a hydrogenase-dependent reaction under anaerobisis and in the absence of any reducible substrate (Belkin and Padan, 1978). Under these conditions, the alga needs a valve to remove excess of reductants similar as the anaerobic, fermentative bacteria. With other blue-green algae, H_2-formation catalyzed by hydrogenase is observed only in the presence of methyl viologen and of high concentrations of $Na_2S_2O_4$. This latter reaction is also observed with intact algae (Daday et al., 1979; Eisbrenner and Bothe, unpubl.); it is, however, questionable whether the organisms survive the drastic treatment by the reducing chemicals.

4 The Utilization of Molecular Hydrogen

Aerobic nitrogen-fixing microorganisms consume H_2 in an oxygen-dependent reaction. This Knallgas reaction is mediated by a membrane-bound hydrogenase, proceeds via the respiratory electron transport chain and supplies the organisms with extra energy. Direct ATP-formation by the Knallgas reaction was demonstrated with particles from *Azotobacter* (Hyndman et al., 1953), *Rhizobium* bacteroids (Dixon 1972) or heterocysts from *Anabaena* (Peterson and Burris, 1978). The ATP/2e ratio and the electron acceptor of uptake hydrogenase are unknown at present. In the isolated state, the enzyme does not couple with ferredoxin or flavodoxin but only with natural carriers of a redox potential of O volt or even more positive (see Bothe et al., 1978). This reflects the unidirectional nature of the enzyme under physiological conditions. Studies with inhibitors revealed that hydrogenase feeds in electrons close to the quinone site of respiration. According to Laane et al. (1979), site I and flavoproteins are not involved in H_2-utilization by *Azotobacter* indicating that hydrogenase feeds in at a particular proton-translocating site of the respiratory chain. Hydrogenase of crude preparations is virtually insensitive to exposure to oxygen regardless of whether obtained from *Azotobacter, Rhizobium,* or blue-green algae. In the latter group of organisms, H_2 is much faster utilized than all other respiratory substrates (Eisbrenner et al., 1978; Peschek, 1979a). The oxyhydrogen reaction proceeds via the terminal cytochrome oxidase complex and not via the CN^- insensitive branch of the respiratory chain.

Phototrophs can also consume molecular hydrogen in a light-dependent reaction. For bacteria, a H_2-dependent CO_2-fixation in the light was first demonstrated by Klemme (1968) for *Rhodopseudomonas* strains. It has recently been established for blue-green algae that the major pathway of H_2-utilization is the oxyhydrogen reaction (Bothe et al., 1977a). Additionally, the algae utilize the gas in a photosystem I-dependent pathway proceeding under strict O_2 exclusion (Bothe et al., 1977a; Eisbrenner et al., 1978). This was demonstrated in *Anabaena* depleted of carbohydrate reserves and therefore limited for reductants (Benemann and Weare, 1974b; Bothe et al., 1977a) or in heterocysts isolated from the same organism (Eisbrenner et al., 1978; Peterson and Wolk, 1978). In heterocyst preparations, H_2 turned out to be the most effective electron donor for C_2H_2-reduction in a reaction strictly dependent on illumination and oxygen exclusion.

The two pathways of H_2-utilization presumably do not proceed independently from each other in blue-green algae (see Fig. 1). Recent studies with the inhibitors dibromothymoquinone and tetramethylphenylenediamine (Eisbrenner and Bothe, 1979) suggest that plastoquinone is a common intermediate of both H_2-utilizations. Since plastoquinone is the major quinone and ubiquinone as well as menaquinone are absent in blue-green algae, plastoquinone could be shared by the H_2-utilizations via the

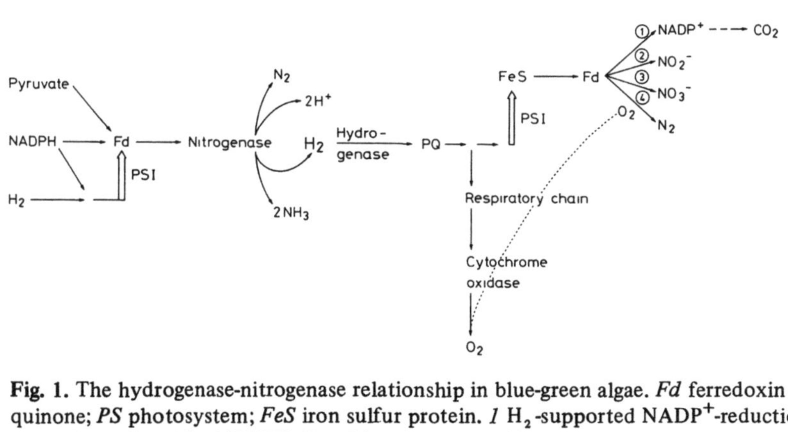

Fig. 1. The hydrogenase-nitrogenase relationship in blue-green algae. *Fd* ferredoxin; *PQ* plastoquinone; *PS* photosystem; *FeS* iron sulfur protein. *1* H_2-supported $NADP^+$-reduction. The reaction is mediated by NADPH:fd oxidoreductase and is active in heterocysts but weak in vegetative cells (see text); *2* H_2-supported NO_2^--reduction, mediated by nitrite reductase; *3* H_2-supported nitrate reduction, mediated by nitrate and nitrite reductases, demonstrated for *Anacystis nidulans;* *4* H_2-supported C_2H_2-reduction of heterocysts; *5* the oxyhydrogen reaction

photosynthetic and respiratory electron transport. This implies that the respiratory chain is, at least partly, located on the thylakoid membranes; a concept which is substantiated by several experimental observations (see Eisbrenner and Bothe, 1979). Alternatively, phylloquinone could be the respiratory quinone involved in the oxyhydrogen reaction of the cytoplasmic membrane, as may be concluded from experiments with particles from the non-N_2-fixing *Anacystis nidulans* (Peschek, 1980).

Four physiological roles, mainly first suggested by Dixon (1972), can be envisaged for the uptake hydrogenase in the context of nitrogen fixation:

a) ATP supplied by the oxyhydrogen reaction could minimize the loss of energy caused by the formation of the gas catalyzed by nitrogenase. Such a function appears to be particularly important under carbon-limiting conditions, as shown for *Azotobacter* (Walker and Yates, 1978) and *Anabaena* (Bothe et al., 1977 a). In the latter alga, significant C_2H_2-reduction activities (25%–40% of the light-dependent rate) were obtained in the dark in a reaction strictly dependent on optimal concentrations of H_2 and O_2 in the vessels.

b) All purified nitrogenases are extremely sensitive to exposure to O_2. In vivo, the Knallgas reaction could remove the deleterious O_2 from the nitrogenase site, thereby helping to protect nitrogenase from damage. Indeed, C_2H_2-reduction was found to be more resistant to O_2 in experiments with intact *Azotobacter vinelandii* (Walker and Yates, 1978) or *Anabaena variabilis* (Bothe et al., 1978) or with heterocysts isolated from the same alga (Wolk, 1979). Such a function requires a close spatial association between nitrogenase and hydrogenase and a high H_2-formation rate catalyzed by nitrogenase which is difficult to verify experimentally with intact organisms.

c) High concentrations of molecular hydrogen are known to inhibit N_2-fixation. H_2-utilization catalyzed by the uptake hydrogenase could remove the deleterious effects of the gas. However, experimental evidence for such a role is lacking at present. It is also likely that H_2 disperses too rapidly to the environment to reach the critical concentrations.

d) H_2 could supply reducing equivalents to nitrogenase. This has been verified experimentally in photosynthetic bacteria (see Meyer et al., 1978) and blue-green algae (see Bothe et al., 1978) where the transfer of electrons from H_2 to nitrogenase requires light activation. There is also experimental evidence that H_2 is electron donor to nitrogenase in carbon-limited cultures of *Azotobacter* (Walker and Yates, 1978). If the electron acceptor of hydrogenase is, indeed, a quinone or a carrier with a redox potential of about O volt, the transfer of electrons to nitrogenase via ferredoxin or flavodoxin must proceed uphill and would necessitate an ATP-dependent reversed electron flow in *Azotobacter.*

All these functions imply a close interrelationship between the biosynthesis of uptake hydrogenase and nitrogenase. In all aerobes, activity levels of hydrogenase are enhanced in cells grown under nitrogen-fixing conditions. The enzyme is, however, not restricted to nitrogen-fixing cells. It is demonstrable in *A. cylindrica, A. variabilis,* and *N. muscorum* grown with NH_4^+ in the media and in the non-N_2-fixing *Anacystis nidulans* (Eisbrenner et al., 1978; Peschek, 1979 a,b). Growth with H_2 in the gas phase drastically increases the enzyme activities in all cases investigated (Tel-Or et al., 1977; Eisbrenner et al., 1978; Peschek, 1979 a,b; Partridge et al., 1980).

Clearly, uptake hydrogenase must have functions not connected to nitrogen fixation. H_2 could act as electron donor in photosynthetic CO_2-fixation, similarly as in the photoreduction experiments of green algae (Gaffron, 1940). Such a reaction has been demonstrated with significant activities in *Oscillatoria limnetica*, a blue-green alga capable of performing anoxygenic photosynthesis with high rates (Belkin and Padan, 1978). In the blue-green algae routinely used in the laboratory, H_2 only poorly supports CO_2-fixation with activities of less than 5% of the control, even when grown under H_2 (Jones and Bishop, 1976; Eisbrenner et al., 1978; Peschek, 1979 a,c). Recent measurements with particles from vegetative cells of *N. muscorum* (Eisbrenner et al., 1981) indicate that H_2 virtually does not support $NADP^+$-reduction, as compared to the control with dichlorophenolindophenol/ascorbate as the electron donor, explaining the low values in the H_2-dependent CO_2-fixations. In contrast, H_2 supported $NADP^+$-reductions are readily demonstrable in particles from heterocysts, with rates exceeding those of the C_2H_2-reductions. Even after growth in the presence of H_2 in the gas phase, the activity levels of hydrogenase in vegetative cells are too low to allow growth at the expense of H_2 as the electron donor. In *Anacystis* grown under H_2, activity levels are, however, sufficient for H_2-supported and light-dependent nitrate- and nitrite-reductions reaching approximately 50% of the control activities with H_2O as the electron donor (Bothe et al., 1980a).

Autotrophic growth at the expense of molecular H_2 as the sole hydrogen donor and energy substrate is a property common to a whole group of gram-negative genera (*Alcaligenes, Pseudomonas, Paracoccus, Xanthobacter;* see Schlegel and Schneider, 1978). Simultaneous fixation of N_2 and CO_2 at the expense of H_2 and O_2 under autotrophic conditions was reported for *Xanthobacter autotrophicus* (Berndt et al., 1976). H_2-dependent CO_2-fixation has recently been demonstrated in *Rhizobium japonicum* (Simpson et al., 1979) and *Derxia gummosa* (Pedrosa et al., 1980) which are able to form ribulose 1-5 bisphosphate carboxylase under chemoautotrophic growth conditions. Similar investigations failed with *Azotobacter* (Pedrosa et al., 1980).

Besides supporting CO_2-fixation, nitrate assimilation or supplying ATP via respiration for cell metabolism, H_2 utilized by the uptake hydrogenase may have other functions. Incubation under H_2 was recently reported to stimulate akinete (spore) formation in the blue-green alga *Cylindrospermum licheniforme* (Hirosawa and Wolk, 1979). H_2 may also affect the photosynthetic electron transport chain at a site between photosystem II and I in a special strain of *Anabaena* (Antarikanonda et al., 1980). Surprisingly, this effect appears to be nonenzymatic and is found in N_2-fixing but not in NH_4^+-grown cells.

The occurrence of an enzyme catalyzing H_2-utilization is somewhat mysterious for aerobic, non-N_2-fixing microorganisms. Recent measurements (Conrad and Seiler, 1979), however, indicate that soils may evolve H_2 to an appreciable extent, due to the action of microorganisms, though with large seasonal fluctuations. The uptake hydrogenase has a high affinity for H_2 in all aerobes and may partly mop up the H_2 available. The enzyme may have had better times in early earth's history where reducing conditions with high levels of H_2 prevailed. Reminiscent of that period may be the capability to enhance the enzyme concentration with the growth under H_2.

5 The Number of Hydrogenases in the Organisms

According to a recent publication (Chen and Blanchard, 1978), *Clostridium pasteurianum* contains an uptake hydrogenase with high activities in addition to the reversible enzyme. In regard to aerobic organisms, two hydrogenases catalyzing the utilization of H_2 have unambiguously been demonstrated in *Alcaligenes eutrophus* (see (Schlegel and Schneider, 1978). In addition to the membrane-bound enzyme, the organism possesses a soluble protein which contains FMN in the prosthetic group and which catalyzes the reduction of NAD^+. Otherwise there is no clear-cut demonstration for the occurrence of two different uptake hydrogenases in one organism. The properties of the one enzyme may, however, partly change upon solubilization from the membranes, as shown for *Alcaligenes* (Schneider and Schlegel, 1977), *Paracoccus denitrificans* (Sim and Vignais, 1979), *Thiocapsa roseopersicina* (Gogotov et al., 1978), *Rhizobium* bacteroids (Arp and Burris, 1979), *Desulfovibrio vulgaris* (Van der Westen et al., 1978), and *Proteus mirabilis* (Schoenmaker et al., 1979). Two hydrogenases were repeatedly reported to occur in blue-green algae (see Hallenbeck and Benemann, 1979).

The existence of two hydrogenases is concluded from experiments to determine the distributions of the enzymes among heterocysts and vegetative cells. A reversible, apparently soluble enzyme, catalyzing the evolution of H_2 in a methyl viologen and $Na_2S_2O_4$ dependent reaction, is present in extracts from both heterocysts and vegetative cells with essentially the same specific activity levels (Tel-Or et al., 1978; see also Bothe et al., 1978). The latter authors, however, noted large variations in the distribution of the specific activity levels with different batches from *Anabaena* or *Nostoc*. As determined by the H_2-dependent $NADP^+$-reduction, the uptake hydrogenase is almost completely confined to heterocysts in aerobically grown cultures of *Nostoc muscorum* (Eisbrenner et al., 1980; see also Section 4 and Peterson and Wolk, 1979). Despite this, we are reluctant to conclude the occurrence of two hydrogenases. Recent experiments (Eisbrenner et al., 1980) indicated that the more drastic the treatment is to break the cells, the higher are the activity levels of the reversible, methyl viologen and $Na_2S_2O_4$-dependent H_2-evolution. In contrast, the O_2-dependent H_2-utilization activity is rapidly destroyed with the breakage of the cells in the case of *Anacystis nidulans* grown under H_2 or of heterocysts from aerobically grown *A. cylindrica* and *N. muscorum.* Hydrogenase obviously becomes the more readily accessible to $Na_2S_2O_4$ and methyl viologen the more solubilized from the membranes. Any enzyme should catalyze the reversible reaction provided the proper concentrations of substrates are given in the assay vessel. A membrane-bound hydrogenase, catalyzing H_2-utilization under physiological conditions, must be able to evolve the gas in vitro. The enzyme may be partly loosely and partly tightly attached to the membranes of blue-green algae. Difficulties have to be encountered to find a physiological role for a reversible, soluble hydrogenase catalyzing H_2-evolution in the obligate autotrophic *Anabaena, Anacystis* or *Nostoc* which can remove excess of reducing equivalents by respiration. In conclusion, the reversible, soluble hydrogenase is probably an artifact of cell-free preparations (Eisbrenner et al., 1980). Such soluble hydrogenases from blue-green algae have recently been isolated and characterized from *Spirulina maxima* (Llama et al., 1979) and *Anabaena cylindrica* (Hallenbeck and Benemann, 1979) with remarkably different properties.

Claims have also been forwarded for the occurrence of two different uptake hydrogenases in blue-green algae (Tetley and Bishop, 1979; Peschek, 1979 c). All their experimental findings can easily be reconciled with our concept of two pathways of H_2-utilization. These are probably not catalyzed by two different enzymes if the respiratory chain is located on the thylakoid membranes, which, however, remains to be established. At present it is more neutral to use the term pathways instead of enzymes before the experimental evidence for two hydrogenases with different physiological and biochemical properties is convincing.

References

Albrecht SL, Maier RJ, Hanus FJ, Russel SA, Emerich DW, Evans HJ (1979) Science 203:1255–1257
Antarikanonda P, Berndt H, Mayer F, Lorenzen H (1980) Arch Microbiol 126:1–10
Arp DI, Burris RH (1979) Biochim Biophys Acta 570:221–230
Belkin S, Padan E (1978) FEBS Lett 94:291–294
Ben-Amotz A (1979) In: Gibbs M, Latzko E (eds) Encyclopedia of plant physiol, new series, vol VI. Springer, Berlin Heidelberg New York, pp 497–506
Benemann JR, Weare NM (1974a) Science 184:174–175
Benemann JR, Weare NM (1974b) Arch Microbiol 104:401–408
Berchthold M, Bachofen R (1979) Arch Microbiol 123:227–232
Berndt H, Ostwal K-P, Laculat J, Schumann C, Mayer F, Schlegel HG (1976) Arch Microbiol 108: 17–26
Bishop NI, Frick M, Jones LW (1977) In: Mitsui A, Miyachi S, San Pietro A, Tamura S (eds) Biological solar energy conversion. Academic Press, London New York, pp 3–22
Bothe H, Eisbrenner G (1977) Biochem Physiol Pflanz 171:323–332
Bothe H, Tennigkeit J, Eisbrenner G (1977a) Arch Microbiol 114:43–49
Bothe H, Tennigkeit J, Eisbrenner G, Yates MG (1977b) Planta 133:237–242
Bothe H, Distler E, Eisbrenner G (1978) Biochimie 60:227–289
Bothe H, Neuer G, Kalbe I, Eisbrenner G (1980a) In: Stewart WDP, Gallon JR (eds) Nitrogen fixation. Proc Phytochem Soc Eur. Academic Press, London New York, pp 83–112
Bothe H, Yates MG, Cannon FC (1980b) In: Läuchli A, Bieleski R (eds) Encyclopedia of plant physiology, new series, vol XII. Springer, Berlin Heidelberg New York, in press
Chatt J (1980) In: Stewart WDP, Gallon JR (eds) Nitrogen fixation. Proc Phytochem Soc Eur. Academic Press, London New York, in press
Chen JS, Blanchard DK (1978) Biochem Biophys Res Commun 84:1144–1150
Conrad R, Seiler W (1979) Soil Biol Biochem 11:689–690
Daday A, Platz RA, Smith GD (1977) Appl Environ Microbiol 34:478–483
Daday A, Lambert GR, Smith GD (1979) Biochem J 177:139–144
Dixon ROD (1972) Arch Microbiol 85:193–201
Dixon ROD (1978) Biochimie 60:233–236
Eisbrenner G, Bothe H (1979) Arch Microbiol 123:37–45
Eisbrenner G, Distler E, Floener L, Bothe H (1978) Arch Microbiol 118:177–184
Eisbrenner G, Roos P, Bothe H (1981) submitted
Ernst A, Kerfin W, Spiller H, Böger P (1979) Z Naturforsch 34c:820–825
Gaffron H (1940) Am J Bot 27:273–283
Gogotov IN, Zorin NA, Serebriakova LT, Kondratieva EN (1978) Biochim Biophys Acta 523: 335–343

Green M, Wilson PW (1953) J Bacteriol 65:511–517
Hallenbeck PC, Benemann JR (1978) FEBS Lett 94:261–264
Hallenbeck PC, Benemann JR (1979) In: Barber J (ed) Photosynthesis in relation to model systems. Elsevier/North-Holland Biomedical Press, Amsterdam, pp 333–364
Hirosawa T, Wolk CP (1979) J Gen Microbiol 114:423–432
Hyndman LA, Burris RH, Wilson PW (1953) J Bacteriol 65:522–531
Jeffries TW, Timourian H, Ward RJ (1978) Appl Environ Microbiol 35:704–710
Jones LW, Bishop NI (1976) Plant Physiol 57:659–665
Kessler E (1974) In: Algal physiology and biochemistry, botanical monograph, vol X. Blackwell, Oxford, pp 456–473
Klemme H (1968) Arch Microbiol 64:29–42
Laane C, Haaker H, Veeger C (1979) Eur J Biochem 97:369–377
Lambert GR, Smith GD (1978) FEBS Lett 83:159–162
Lambert GR, Daday A, Smith GD (1979a) Appl Environ Microbiol 38:521–529
Lambert GR, Daday A, Smith GD (1979b) Appl Environ Microbiol 38:530–536
Lee SB, Wilson PW (1943) J Biol Chem 151:377–385
Llama MJ, Serra JL, Rao KK, Hall DO (1979) FEBS Lett 98:342–345
Meyer J, Kelley BC, Vignais PM (1978) Biochimie 60:245–270
Mitsui A, Kumazawa S (1977) In: Mitsui A et al (eds) Biological solar energy conversion. Academic Press, London New York, pp 23–43
Miyamoto K, Hallenbeck PC, Benemann JR (1979) Appl Environ Microbiol 38:440–446
Mortenson LE, Chen JS (1974) In: Neilands JB (ed) Microbiol iron metabolism. Academic Press, London New York, pp 231–282
Partridge CDP, Walker CC, Yates MG, Postgate JR (1980) J Gen Microbiol 119:313–319
Pedrosa FO, Döbereiner J, Yates MG (1980) J Gen Microbiol 119:547–551
Peschek GA (1979a) Biochim Biophys Acta 548:187–202
Peschek GA (1979b) Biochim Biophys Acta 548:203–215
Peschek GA (1979c) Arch Microbiol 123:81–92
Peschek GA (1980) Biochem J 186:515–523
Peterson RB, Burris RH (1978) Arch Microbiol 116:125–132
Peterson RB, Wolk CP (1978) Proc Natl Acad Sci USA 75:6271–6275
Peterson RB, Wolk CP (1979) Plant Physiol 61:688–691
Phelps AS, Wilson PW (1941) Proc Soc Exp Biol Med 47:473–476
Schlegel HG, Schneider K (1978) In: Schlegel HG, Schneider K (eds) Hydrogenases, their catalytic activity, structure and function. Goltze E KG, Göttingen, pp 15–44
Schneider K, Schlegel HG (1977) Arch Microbiol 112:229–238
Schoenmaker GS, Oltmann LF, Stouthammer AH (1979) Biochim Biophys Acta 567:511–521
Sim E, Vignais PM (1979) Biochim Biophys Acta 570:43–55
Simpson FB, Maier RJ, Evans HJ (1979) Arch Microbiol 123:1–8
Spiller H, Ernst A, Kerfin W, Böger P (1978) Z Naturforsch 33c:541–547
Stephenson M, Stickland LH (1931) Biochem J 25:205–214
Tel-Or E, Luijk LW, Packer L (1977) FEBS Lett 78:49–52
Tel-Or E, Luijk LW, Packer L (1978) Arch Biochem Biophys 185:185–194
Tetley RM, Bishop NE (1979) Biochim Biophys Acta 546:43–53
Van der Westen HM, Mayhew SG, Veeger C (1978) FEBS Lett 86:122–126
Voelskow H, Schön G (1980) Arch Microbiol 125:245–249
Walker CC, Yates MG (1978) Biochimie 60:225–231
Weissman JC, Benemann JR (1977) Appl Environ Microbiol 33:123–131
Wilson JR, Lee SB, Wilson PW (1942) J Biol Chem 144:265–271
Wolk CP (1979) In: Nichols JM (ed) Int Symp Photosynthetic Prokaryotes Abstract Book. The University of Liverpool, Abstract D-1

Sulfur Metabolism in General

Assimilatory Sulfate Reduction

J.A. SCHIFF and H. FANKHAUSER[1]

1 Introduction

The roles of sulfur and nitrogen in cell nutrition show many similarities. Both must be reduced by living systems to form proteins and other macromolecules and both, therefore, appear among the macroelements required by plants and plant-like microorganisms. Nutritional interactions between the two elements have been described (Friedrich and Schrader, 1978) and both elements show a rather uniform distribution throughout the tissues and organs of higher plants and among cells of microorganisms. Sulfate reduction occurs in mitochondria and microbodies in *Euglena* cells (Brunold and Schiff, 1976) and is localized in the plastids of higher plants (Schmidt and Schwenn, 1971; Schwenn and Trebst, 1976). Since cell, tissue, and organ cultures can be grown on sulfate as a sole sulfur source (White, 1963; White and Grove, 1965; Butenko, 1968; Thomas and Davey, 1975), and since nongreen tissues such as beet discs (Ellis, 1963) and roots (Pate, 1965) reduce sulfate, assimilatory sulfate reduction must be widely distributed throughout the plant; studies of the localization of certain enzymes of sulfate reduction are in agreement (Mayer, 1967; Ellis, 1969; Onajobi et al., 1973; Schmidt, 1976; Tamura et al., 1976; Cacco et al., 1977; Fankhauser and Brunold, 1978, 1979). In this paper we will try to summarize what is known of the process of assimilatory sulfate reduction and how it might have evolved; some recent work on an enzymatic reaction of adenosine 5′ phosphosulfate (APS) and ammonia (Fankhauser et al., 1979; Fankhauser and Schiff, 1980; Fankhauser et al., 1980) will also be described which may provide a connection between sulfur and nitrogen metabolism.

2 The Place of Sulfate Reduction in the Sulfur Cycle

Unlike most of the other elements required by living systems, carbon, nitrogen, and sulfur undergo extensive metabolic transformations. Sulfur, unlike carbon and nitrogen, can be utilized in its most highly oxidized naturally occurring form, sulfate; sulfate reduction is necessary for the formation of sulfur-containing amino acids and proteins (Schiff and Hodson, 1973; Roy and Trudinger, 1970; Siegal in Greenberg, 1977).

---

1 Photobiology Institute, Brandeis University, Waltham, MA 02254/USA

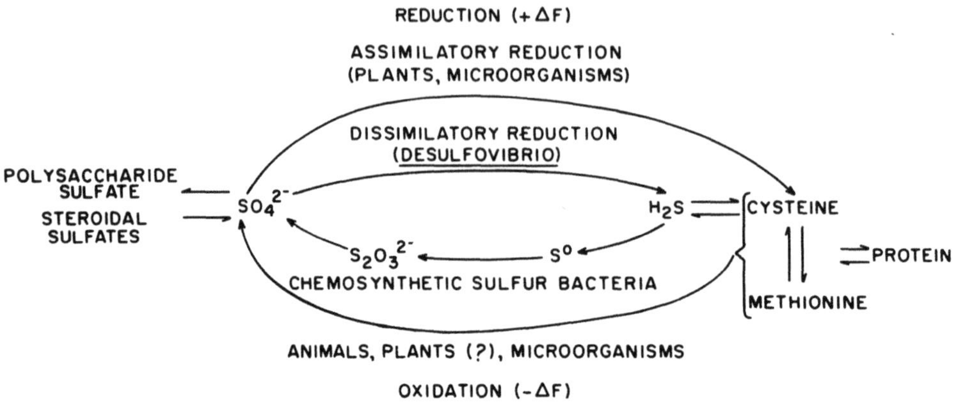

Fig. 1. Reactions of sulfur in the biosphere

Figure 1 shows the relation of sulfate to the many reactions which sulfur undergoes in the biosphere. Sulfate is used by living systems to form sulfate esters of polysaccharides, phenols, steroids and other organic compounds through a series of activation and transfer reactions. Sulfate is reduced by certain anaerobic organisms such as the bacterium *Desulfovibrio* in their energy metabolism where, during anaerobic respiration, sulfate is used in place of oxygen to oxidize substrates thereby releasing energy and hydrogen sulfide. Many organisms reduce sulfate to the thiol level found in many coenzymes, and in the amino acids cysteine and methionine which serve as building blocks for peptides and proteins; this process is called assimilatory sulfate reduction. Most organisms (including higher animals and higher plants) oxidize reduced sulfur to sulfate although the chemosynthetic bacteria are the only organisms which have been observed to couple the energy released (some 180 kcals per mol) to the reduction of carbon dioxide. The anoxygenic photosynthetic bacteria use reduced sulfur compounds as photosynthetic electron donors, thereby oxidizing them to sulfate.

3 Phylogenetic Distribution of Reactions Involving Sulfate Transfer and Reduction

Reactions resulting in the esterification of organic compounds by sulfate are widely distributed (Schiff and Hodson, 1973; DeMeio in Greenberg, 1977). Sulfate esters of organic compounds are formed by higher animals in the form of steroid sulfates, phenol sulfates, and polysaccharide sulfates such as chondroitin sulfate. The algae produce sulfated polysaccharides as wall constituents, frequently in copious amounts. The better-known sulfated polysaccharides of commercial importance such as carrageenan and agar are produced by marine members of the red algae or Rhodophyta. There is at least one reliable report of sulfated polysaccharides in a halophilic bacterium (Steber and Schleifer, 1975) among the prokaryotes (bacteria and cyanobacteria or blue-green algae), but none from higher plants. These gelatinous materials may have evolved in the algae as adaptations to periodic wetting and drying as water conservation mechanisms, a particular problem for benthic and terrestrial

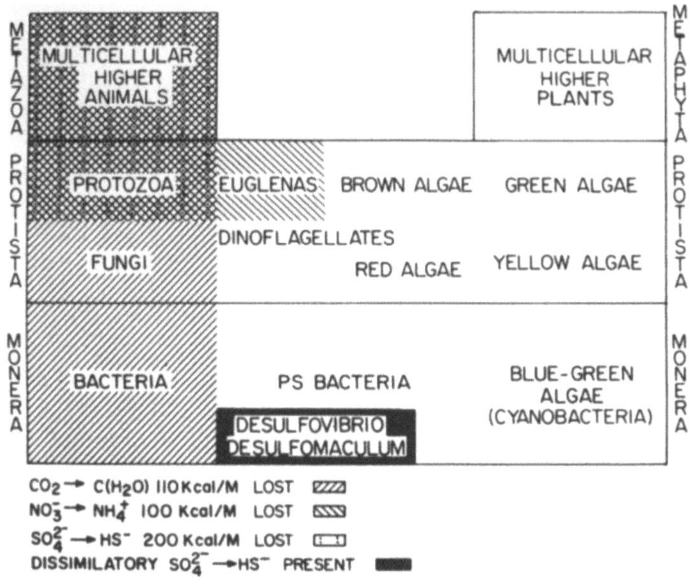

Fig. 2. Loss of energetically expensive reduction reactions in various contemporary groups of organisms. Dissimilatory sulfate reduction is restricted to two small genera of anaerobic bacteria *(Desulfovibrio* and *Desulfomaculum).* The *clear areas* indicate organisms that can carry out the reduction of sulfate, nitrate and carbon dioxide [a small group of aerobic chemosynthetic bacteria (not shown) are capable of carrying out carbon dioxide reduction given oxidizable inorganic substrates as energy sources]. The various patterns indicate the loss of these reduction reactions in various groups. The loss of all three reduction pathways has occurred in the evolution of the protozoa and higher animals

organisms. In the higher animals, the sulfated polysaccharides are used to provide a matrix for cellular structures such as cartilage and skin. From our present knowledge the occurrence of sulfated polysaccharides seems to represent the adaptation of particular groups or organisms to special needs.

The distribution of reduction reactions, however, seems to follow a more regular evolutionary pattern (Fig. 2) (Schiff and Hodson, 1973). Reduction reactions are expensive, energetically, and require the investment of hundreds of kcal per mol in order to take place (Fig. 2). The reduction of nitrate, carbon dioxide, and sulfate seems to have been lost in the evolution of the primitive animals, the protozoa, and this deficiency persists in the evolution of the multicellular animals. These reactions may have been lost when photosynthetic reactions trapping light energy were lost. Animals require energy for movement and, having lost photosynthetic energy, found themselves with a tight energy budget. Being predators, they could fulfill their architectural and energy requirements for reduced sulfur, nitrogen, and carbon by eating organisms which could perform these reductions such as microorganisms or plants, thereby originating the food chains we know today. Contemporary animals obtain their reduced sulfur compounds by eating or harboring sulfate-reducing plants or microorganisms, or by eating other animals which do.

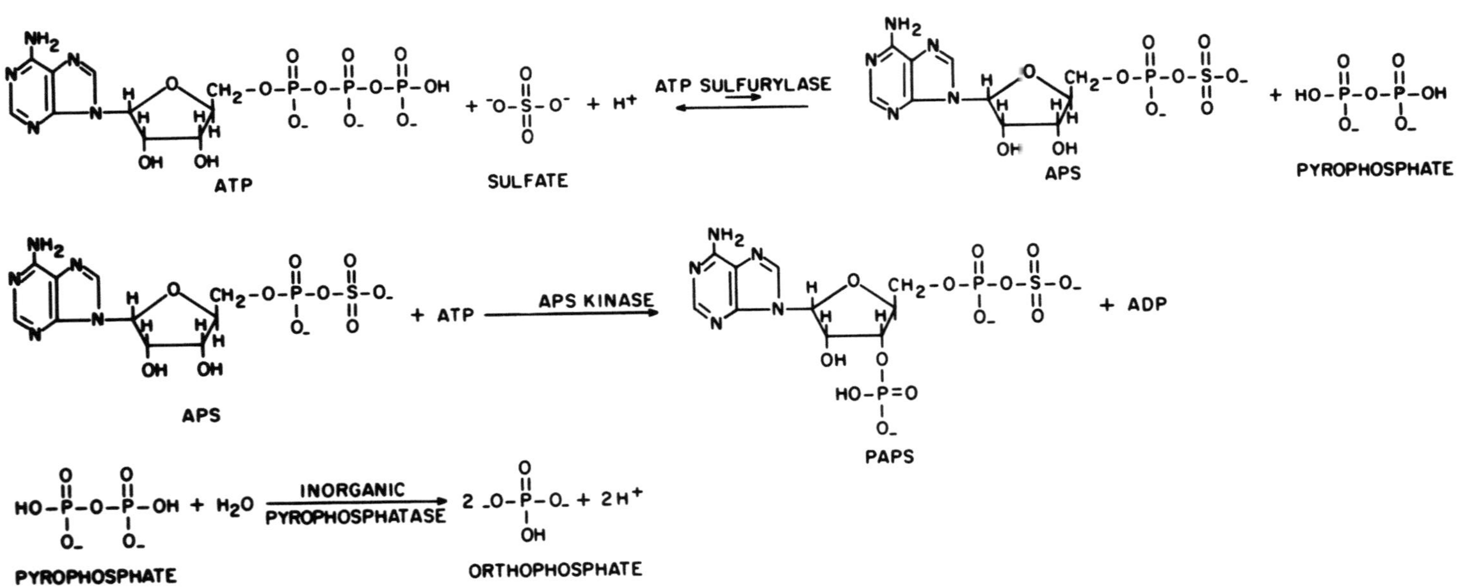

Fig. 3. Enzymatic reactions involved in sulfate activation. The enzymes shown are ATP sulfurylase (ATP:sulfate adenylyl transferase 2.7.7.4), APS kinase (ATP:adenylyl sulfate 3′ phosphotransferase 2.7.1.25), and inorganic pyrophosphatase (pyrophosphate phosphohydrolase 3.6.1.1)

4 Sulfate Uptake, Activation, and Transfer

Sulfate uptake and its control has been well studied in many systems (Schiff and Hodson, 1973; Siegel in Greenberg, 1977). Uptake seems to be accomplished through active transport mediated by a carrier enzyme system.

As far as is known, the cellular metabolism of sulfate begins with a series of activation reactions with ATP to form the nucleoside phosphosulfates adenosine 5′ phosphosulfate (APS) and adenosine 3′ phosphate 5′ phosphosulfate (PAPS) (Fig. 3) (Schiff and Hodson, 1973; DeMeio in Greenberg, 1977). APS is formed against an unfavorable equilibrium since the free energy of hydrolysis of the phosphosulfate bond is high than that of the pyrophosphate linkage (Roy and Trudinger, 1970). To offset this unfavorable equilibrium, pyrophosphate removal through inorganic pyrophosphatase activity and/or reaction of APS with another molecule of ATP to form PAPS are used to allow an accumulation of nucleoside phosphosulfates. A 3′(2′), 5′ diphosphonucleoside 3′(2′) phosphohydrolase (DPNPase) is widely distributed which converts PAPS to APS and may facilitate APS formation despite the unfavorable equilibrium of the first reaction forming APS (Fig. 5) (Tsang and Schiff, 1976c). This enzyme may also provide control of APS and PAPS levels in organisms where both are required for different purposes.

All sulfate transfer reactions resulting in sulfate esterification appear to use PAPS as the sulfate donor as far as is known (DeMeio in Greenberg, 1977; Schiff and Hodson, 1973). APS is a substrate for hydrolases that release sulfate (Tsang and Schiff, 1976c) and for a novel enzyme reaction [adenylyl sulfate:NH_3 adenylyl transferase (APSAT)] which will be discussed subsequently. Sulfonic acids which contain sulfur at the redox level of sulfite occur in many living systems, the plant sulfolipid of chloroplast thylakoids being an important example (Schiff and Hodson, 1973). Although the formation of the sulfonic acid group of compounds such as taurine can arise through oxidation of reduced sulfur in animals (Siegel in Greenberg, 1977), the formation of the sulfonic acid group of the plant sulfolipid appears to come more readily from sulfate via as yet unknown reactions. The formation of sulfonic acids is an area requiring more study.

5 Enzymatic Reaction of Adenosine 5′ Phosphosulfate (APS) with Ammonia

Cells of *Chlorella, Euglena,* higher plants such as barley and spinach, *Dictyostelium discoideum,* and *Escherichia coli* contain an enzyme activity called adenylyl sulfate (APS):ammonia adenylyl transferase (APSAT) which catalyzes the reaction shown in Fig. 4 (Fankhauser et al., 1979; Fankhauser and Schiff, 1980; Fankhauser et al., 1980) (the reaction is written on the assumption that NH_3 rather than NH_4^+ is the substrate). This enzyme activity from *Chlorella* has been purified to homogeneity and is found to be a protein of 64 KD with subunits of 26, 21, and 17 KD. The enzyme is specific for APS and ammonia and will not use PAPS, ATP, ADP or AMP, or a large variety of amines, amides, and other ammonia analogs. APSAT does not catalyze any

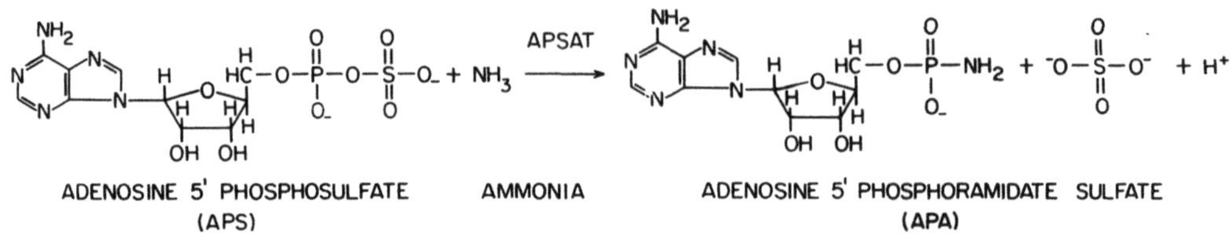

Fig. 4. Reaction of adenosine 5′ phosphosulfate (APS) and ammonia to yield adenosine 5′ phosphoramidate (APA) and sulfate catalyzed by adenylyl sulfate (APS):ammonia adenylyl transferase (APSAT). The reaction is written on the assumption that NH_3 rather than NH_4^+ is the reactive form of ammonia since the reaction is promoted by increasing pH

of the known reactions of APS catalyzed by other enzymes such as APS kinase, ATP sulfurylase, ADP sulfurylase, APS sulfotransferase, and APS sulfohydrolase. The only products of the enzyme reaction are adenosine 5′ phosphoramidate (APA) and sulfate and, perhaps, a hydrogen ion. If the reaction is carried out nonenzymatically by boiling APS and ammonia together, both APA and AMP are formed, indicating that the enzyme prevents the hydrolytic reaction from occurring.

Enzymatic reactions are known which catalyze the formation of phosphoramide bonds between amino acids and phosphate as in the phosphagens, arginine, and creatine phosphates. An enzyme-bound phosphoramide intermediate involving an AMP residue bound to the epsilon amino group of lysine has been described in the DNA ligase reaction (Gumport and Lehman, 1971). To our knowledge, however, there is no precedent for the direct enzymatic activation of amino groups or ammonia through reaction with high-energy nucleotides to form high-energy adenylyl adducts of the nitrogen atom. The formation of APA described here may be the first of such reactions and may also be an example of a type of activation reaction requiring a higher free energy than that provided by the pyrophosphate group of ATP. Other reactions of APS (or even PAPS) may be found in which they supply a higher free energy than ATP can supply. The role of APA in cellular metabolism remains to be elucidated. APA, a high energy compound, is used in the chemical synthesis of nucleotides (Moffat and Khorana, 1961) and will replace AMP in the activation of certain enzymes (Okazaki et al., 1968; Black and Wang, 1970; Nakazawa et al., 1967; Hizuka et al., 1976); perhaps this interesting compound serves as a regulatory link between nitrogen and sulfur metabolism. We have very recently been able to isolate APA from *Chlorella* cells (Berkowitz et al., 1981); thus APA appears to be a normally-occuring constituent of the cellular nucleotide pool.

6 Sulfate Reduction

6.1 The Two Assimilatory Pathways

Dissimilatory sulfate reduction occurs in certain anaerobic bacteria such as *Desulfovibrio* which use sulfate as an oxidant in place of molecular oxygen, and in this process APS is the nucleotide sulfate donor for reduction, which accumulates hydrogen sulfide and results in oxidative phosphorylation (Roy and Trudinger, 1970; Siegel in Greenberg, 1977; Peck, 1970).

Assimilatory sulfate reduction, where sulfate is reduced to form the thiol groups of sulfur-containing amino acids, coenzymes and other organic compounds, now appears to exhibit two distinct patterns (Tsang and Schiff, 1975; Schiff and Hodson, 1973:

a) In most oxygen-evolving photosynthesizers (including all eukaryotic algae and some prokaryotic blue-green algae (Tsang and Schiff, 1975; Schmidt, 1977), all higher plants tested (Schmidt, 1975 a), and spinach chloroplasts (Schmidt and Schwenn, 1971; Schmidt, 1975 b), the nucleotide sulfate donor for reduction is APS via a highly specific APS sulfotransferase (adenylyl sulfate:thiol sulfotransferase) which will not use PAPS (Fig. 5) (Tsang and Schiff, 1976 a). Where this form of reduction has been studied in detail; e. g., *Chlorella, Euglena,* and spinach chloroplasts (Schmidt and Schwenn, 1971; Brunold and Schiff, 1976; Schmidt et al., 1974; Abrams and Schiff, 1973) a low molecular weight thiol carrier [which appears to be glutathione in *Chlorella* (Tsang and Schiff, 1976 a)] is also involved but the APS sulfotransferase will transfer promiscuously to other added thiols in the presence or absence of the carrier (Tsang and Schiff, 1976 a). Ferredoxin (in the *Chlorella* and higher plant system) or NADPH (in *Euglena* extracts) must be added to achieve reduction of the sulfo group of APS to the thiol level of amino acids.

b) In organisms which lack oxygen-evolving photosynthesis [besides animals, which do not reduce sulfate to the thiol level (Schiff and Hodson, 1973)], such as yeast and *E. coli* and other bacteria (Tsang and Schiff, 1975; Torii and Bandurski, 1967; Tsang and Schiff, 1978 b; Dreyfuss and Monty, 1963; Fujimoto and Ishimoto, 1961; Tsang and Schiff, 1976 b; Wilson and Bierer, 1976) and a few oxygen-evolving blue-green algae (Schmidt, 1977), PAPS in the nucleoside phosphosulfate donor for reduction via a specific PAPS sulfotransferase (3′ phosphoadenylyl sulfate:thiol sulfotransferase) which will not use APS (Tsang and Schiff, 1975; Peck, 1961) (Fig. 6). The transfer of the sulfo group of PAPS and further reduction requires thioredoxin in yeast and *E. coli* (Porqué et al., 1970; Tsang and Schiff, 1976 b, 1978 b). Only reduced pyridine nucleotide need be added to obtain reduction, but the PAPS sulfotransferase will only function in the presence of thioredoxin (Tsang and Schiff, 1976 b) or an analogous molecule (Tsang and Schiff, 1978 b); other added thiols in the absence of thioredoxin are inactive.

6.2 Detailed Reactions of the Two Assimilatory Pathways

6.2.1 The APS Pathway

Our most detailed information comes from work on the enzymes from *Chlorella, Euglena,* spinach chloroplasts (Schmidt and Schwenn, 1971; Schmidt, 1975 b; Brunold and Schiff, 1976; Abrams and Schiff, 1973; Tsang and Schiff, 1976 a; Tsang and Schiff, 1978 a) and duckweed *(Lemna)* (Brunold and Schmidt, 1978) (Fig. 5). It has taken some time to sort out the reactions of APS sulfotransferase because the enzyme transfers the sulfo group of APS not only to the normal carrier [glutathione (Tsang and Schiff, 1978 a) in *Chlorella,* a somewhat larger molecule in spinach chloroplasts (Schmidt and Schwenn, 1971)] but to practically any thiol added (Tsang and Schiff, 1976 a). Thiols forming rings on oxidation (such as dithiothreitol) form sulfite in this

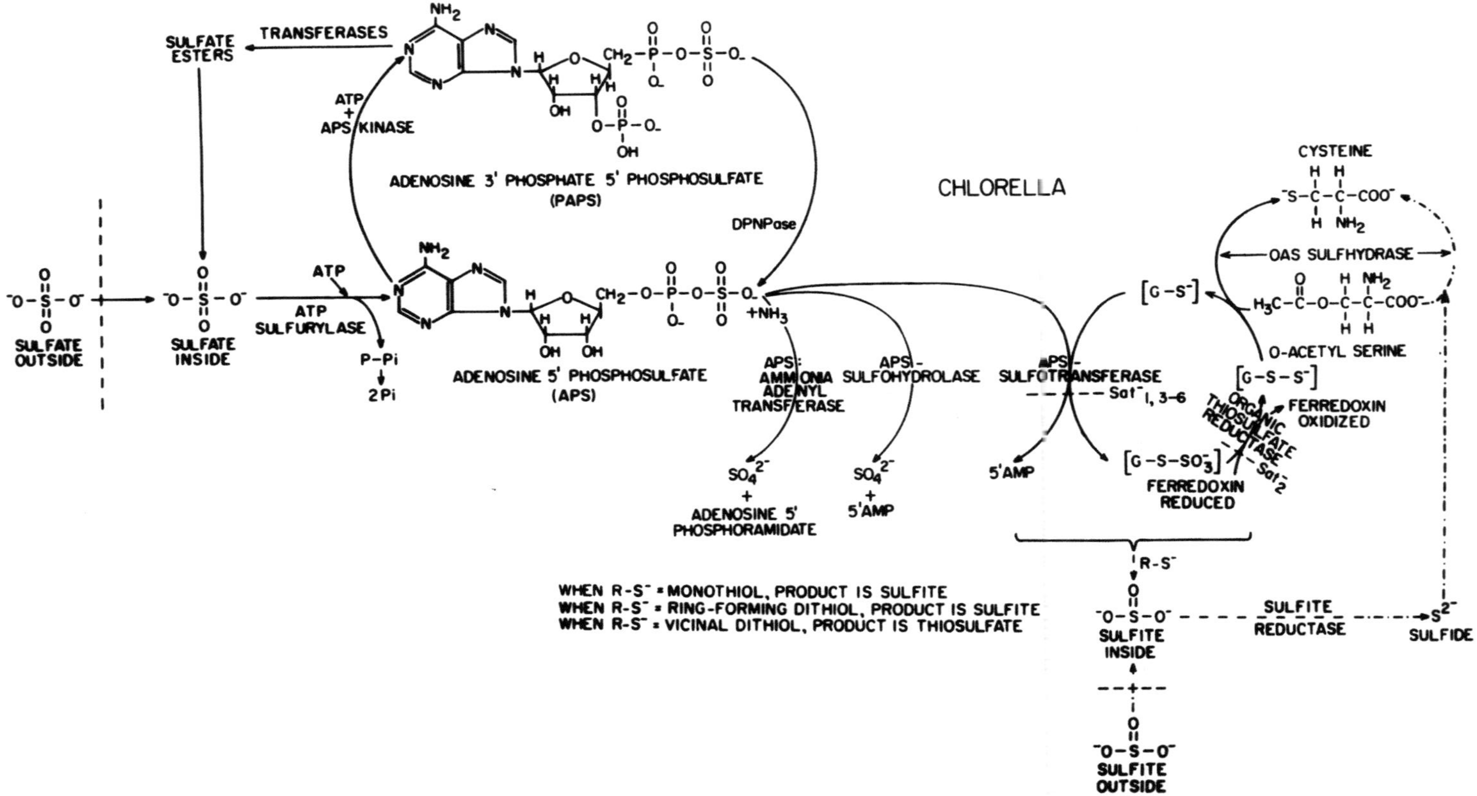

Fig. 5. APS pathway of sulfate reduction in *Chlorella*. Reactions thought to be on the main path of sulfate reduction in vivo are shown as *solid lines*. Reactions blocked in various mutants that cannot grow on sulfate (Sat⁻ mutants) are shown by *dashes* through the reactions deleted. Side reactions are shown by *alternate dashes and dots*. Sulfite (or thiosulfate) is only produced through chemical reactions requiring thiols and probably only occurs in vivo when sulfite (or thiosulfate) is available to the cells from outside. G-S⁻ designates reduced glutathione

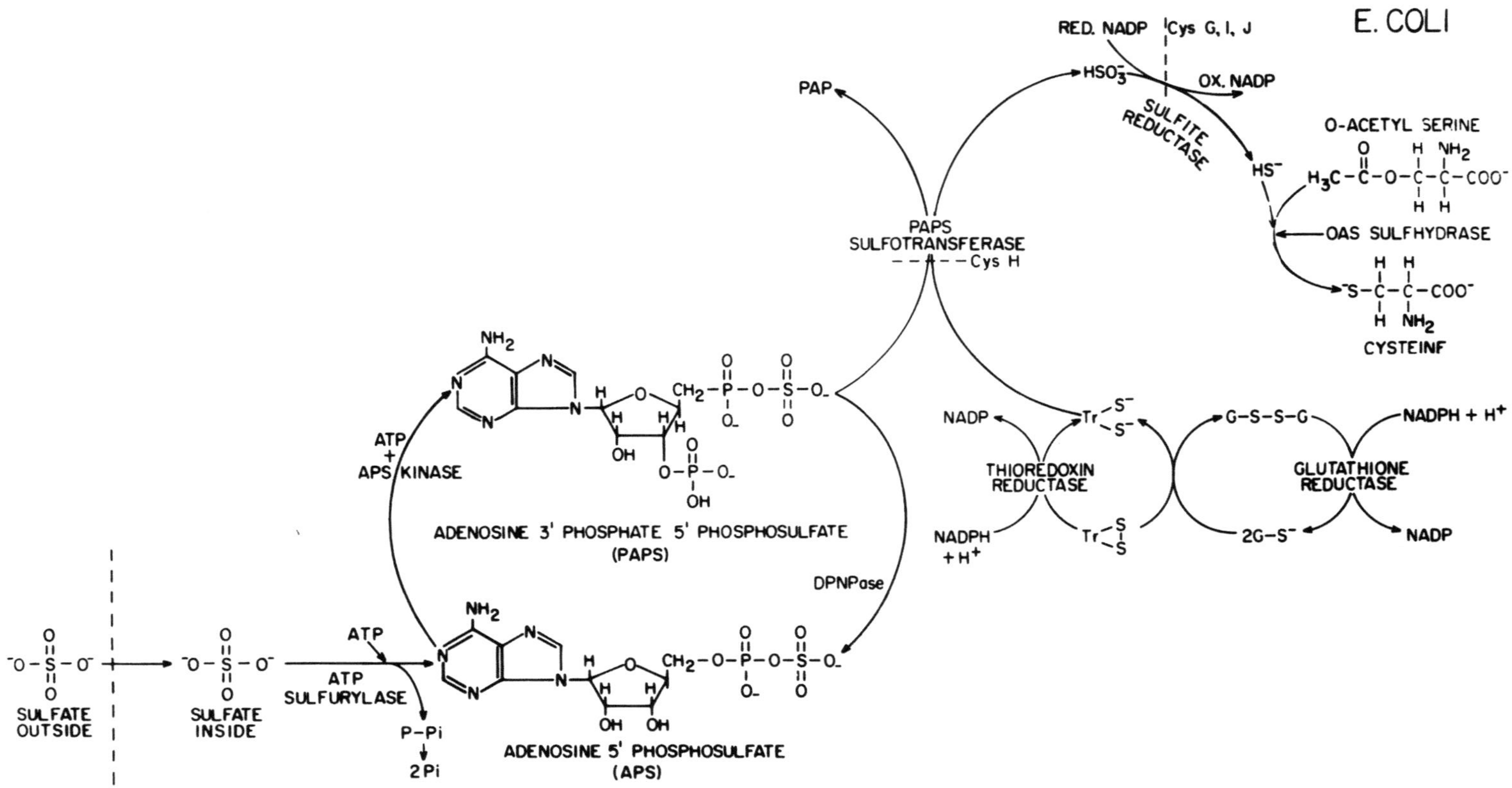

Fig. 6. PAPS pathway of sulfate reduction in *E. coli.* Thioredoxin is required in the PAPS sulfotransferase reaction and formation of sulfite, but the reactions it undergoes have not been elucidated fully. Mutants (Cys) blocked in various enzyme reactions are shown by *dashes* through the reactions deleted. G-S$^-$ designates reduced glutathione. $Tr{<}^{S^-}_{S^-}$ designates reduced thioredoxin

reaction, monothiols (such as glutathione and thioethanolamine) form the Bunte salt ($R\text{-}S\text{-}SO_3^-$) of the thiol and in the presence of excess thiol, sulfite. Vicinal dithiols (such as 2,3 dithiopropanol, BAL) form thiosulfate ($^-S\text{-}SO_3^-$) where the oxidized sulfur comes from APS and the reduced sulfur from the thiol. Of all the thiols, glutathione stands out because it has a much higher activity with the enzyme than any other monothiol and shows kinetics suggestive of a regulatory interaction. Glutathione seems to be the only small molecule in *Chlorella* extracts which is active in the APS sulfotransferase reaction (Tsang and Schiff, 1978 a). If thiols are omitted from the reaction, it is possible to demonstrate that the enzyme transfers the sulfo group of APS to an enzyme-bound acceptor in the crude extracts (Abrams and Schiff, 1973). This labeled material shows the properties of a Bunte salt or organic thiosulfate ($R\text{-}S\text{-}SO_3^-$) and is thought to be the sulfo form of the thiol carrier. Reconstruction experiments using glutathione-S-SO_3^- show that sulfite and thiosulfate can be formed chemically by a nonenzymatic reaction with the appropriate thiol (Tsang and Schiff, 1976a). While the transfer of the sulfate group from APS to the enzyme-bound carrier is heat-labile, the subsequent reaction of the enzyme-bound Bunte salt with added thiols is nonenzymatic and the same products are formed with each thiol as in the model experiments (Tsang and Schiff, 1976 a; Abrams and Schiff, 1973).

Thus we view the reaction catalyzed by APS sulfotransferase as the transfer of the sulfo group of APS to glutathione ($G\text{-}S^-$) to form the Bunte salt ($G\text{-}S\text{-}SO_3^-$). As will be discussed shortly, further physiological reduction appears to involve this bound sulfo group. The reductive release of sulfite or thiosulfate by addition of thiols results from chemical side reactions of this carrier-bound sulfo group.

There are two candidates for enzymes catalyzing reduction to the thiol level. One is sulfite reductase (hydrogen-sulfide:$NADP^+$ oxidoreductase 1.8.1.2) which is present in these systems and will bring about the reduction of free sulfite to free sulfide with reduced pyridine nucleotides (Siegel in Greenberg, 1977; Schmidt et al., 1974; Tsang and Schiff, 1976b). The other enzyme is an organic thiosulfate reductase (ferredoxin: sulfoglutathione oxidoreductase) (formerly called „thiosulfonate reductase") which reduces glutathione-S-SO_3^- to bound sulfide (probably glutathione-S-S^-) or dithionite to sulfide, with reduced ferredoxin (Schmidt, 1973; Schmidt et al., 1974). Studies of a mutant of *Chlorella* (Sat_2^-) which cannot grow on or reduce sulfate have shown that organic thiosulfate reductase activity is extremely low or absent while normal levels of sulfite reductase activity are present (Schmidt et al., 1976). Thus the presence of sulfite reductase alone does not allow sulfate reduction to proceed in vivo and shows that organic thiosulfate reductase is the preferred enzyme for sulfate reduction in this system. We view the physiological reaction as the reduction of $G\text{-}S\text{-}SO_3^-$ to $G\text{-}S\text{-}S^-$ which seems to be the primary reduction reaction in vivo in the APS system. Sulfite reductase would probably only act on free sulfite if it arose through side-reactions in vivo or if it was taken up from outside by the cell.

These systems contain a very active O-acetyl serine sulfhydrase [O-acetyl-L-serine acetate-lyase (adding H_2S) 4.2.99.8] which forms cysteine from free sulfide or from carrier-bound sulfide (Brunold and Schiff, 1976; Schmidt et al., 1974). Two electrons would be required to convert $G\text{-}S\text{-}S^-$ to the thiol group of cysteine in this reaction.

An important advance in studies of this system was finding conditions for the reduction of APS to form cysteine in cell-free extracts without the addition of thiols

(Schmidt et al., 1974) which form free sulfite and sulfide through chemical side reactions (Tsang and Schiff, 1976a). Studies of this cell-free system from wild-type *Chlorella* and mutants blocked for organic thiosulfate reductase and APS sulfotransferase have proven to be consistent with the interpretations offered above and have led to a scheme in which what are thought to be the normal reactions in vivo are shown in solid lines and the side reactions in broken lines (Fig. 5).

Studies of the cellular location and synthesis of the various enzymes of this pathway have been undertaken. The activating enzymes for sulfate, APS sulfotransferase, organic thiosulfate reductase (reductant, ferredoxin) and O-acetyl serine sulfhydrase have been found in spinach chloroplasts (Schmidt and Schwenn, 1971; Schwenn and Trebst, 1976; Fankhauser and Brunold, 1979). It is likely that this is also the case in *Chlorella* judging from the close similarity of green algal and higher plant systems. In *Euglena,* however, which is an animal-like cell resembling the protozoa but containing chloroplasts, APS sulfotransferase is found in the mitochondria and microbodies and the organic thiosulfate reductase (reductant, NADPH) and OAS sulfhydrase are found in the mitochondria (Brunold and Schiff, 1976); thus the mitochondria of *Euglena* contain all of the presently known enzymes of the bound pathway of APS reduction. It is interesting that green algal and higher plant cells cannot be induced to lose their chloroplasts, perhaps because certain enzyme systems (such as sulfate reduction) which are essential for viability are plastid-localized. In *Euglena,* however, chloroplasts and plastid DNA can be lost (Schiff, 1972). *Euglena* does not seem to localize essential reactions (other than those essential for photosynthesis) in the plastids.

In the course of studies on the APS sulfotransferase two other activities were found which act on APS (Tsang and Schiff, 1976c). One is an APS sulfohydrolase (adenylyl sulfate sulfohydrolase 3.6.2.1) which forms sulfate and AMP from APS. PAPS is much more stable than APS in extracts; perhaps this is why the DPNPase reaction forming APS from PAPS is advantageous as an APS source. The second activity, already discussed above, forms adenosine 5′ phosphoramidate (Fankhauser et al., 1979).

6.2.2 The PAPS Pathway

This system has been studied in yeast and *Escherichia coli* (Tsang and Schiff, 1975; Torii and Bandurski, 1967; Fujimoto and Ishimoto, 1961; Dreyfuss and Monty, 1963; Tsang and Schiff, 1976b; Wilson and Bierer, 1976) (Fig. 6). PAPS sulfotransferase (also called „PAPS reductase" before thioredoxin was identified as a component) is specific for PAPS as the sulfate donor and is also completely specific for thioredoxin; other thiols will not replace thioredoxin (Tsang and Schiff, 1976b) or an analogous molecule (Tsang and Schiff, 1978b). As in *Chlorella,* an enzyme-bound intermediate with the properties of a Bunte salt ($R\text{-}S\text{-}SO_3^-$) can be detected as a product of the PAPS sulfotransferase reaction (Tsang and Schiff, 1976b; Torii and Bandurski, 1967). The formation of free sulfite requires thioredoxin reductase (NADPH:oxidized-thioredoxin oxidoreductase 1.6.4.5) and reduced pyridine nucleotide (Torii and Bandurski, 1967; Porqué et al., 1970) but these can be replaced by appropriate thiols (Tsang and Schiff, 1976b) to yield sulfite or thiosulfate as in the APS system. Under certain conditions this system can accept electrons from the glutathione-glutathione reductase system [(NAD(P)H:oxidized-glutathione oxidoreductase 1.6.4.2)]; (Tsang and Schiff, 1978b; Holmgren et al., 1978). The product of this PAPS sulfotransferase-thioredoxin system is sulfite when NADPH is used as the reductant. Free sulfite is reduced to sulfide with reduced pyridine nucleotides via sulfite reductase (Siegel in Greenberg, 1977). In *Chlorella*, two reductase activities are present, one that acts on free sulfite and one that acts on a bound intermediate in the form of a Bunte salt or organic thiosulfate, the latter being the physiological reaction in sulfate reduction (Schmidt et al., 1974). However, *E. coli* mutants blocked for sulfate reduction lack sulfite reductase activity, suggesting that this enzyme system is the physiological reaction in these organisms (Tsang and Schiff, 1978b). O-acetyl serine sulfhydrase is also present, (Kredich and Tomkins, 1966) which can form cysteine from O-acetyl serine and sulfide. Taking together the information presented here, and various suggestions made previously by various workers, it is possible to write a scheme consistent with current knowledge (Fig. 6).

The role of thioredoxin in this system is still under study. Reduced thioredoxin $\left(\mathrm{Tr}{<}^{S^-}_{S^-}\right)$ could act as the thiol carrier acceptor for the sulfo group of PAPS in the PAPS sulfotransferase reaction forming $\mathrm{Tr}{<}^{S-SO_3^-}_{S^-}$ (sulfothioredoxin). Elimination of sulfite would yield oxidized thioredoxin $\left(\mathrm{Tr}{<}\begin{array}{c}S\\ |\\ S\end{array}\right)$ which could then be reduced as shown in Fig. 6. In view of the ready elimination of sulfite from Bunte salts of nonvicinal dithiols (Tsang and Schiff, 1976a), some mechanism would be necessary to prevent the immediate release of sulfite from sulfothioredoxin, since the bound intermediate formed from PAPS in *E. coli* is rather stable (Tsang and Schiff, 1976b). Perhaps linkage of one of the thiol groups of thioredoxin to a thiol group of the transferase via an S-S bond accomplishes this. Another possibility is suggested by recent work which shows that a large proportion of the thioredoxin has a thiol group substituted with phosphate (Pigiet and Conley, 1978; Conley and Pigiet, 1978). Perhaps this substitution of one thiol group by phosphate prevents spontaneous elimination of sulfite from sulfothioredoxin until timely enzymatic removal of the phosphate. It is possible, of course, that thioredoxin is not the carrier and that the bound intermediate contains the sulfo group bound to a different acceptor. In this case, thioredoxin would act as the immediate reductant.

It is worth noting that although the bound intermediates in both the APS and PAPS systems show many reactions of Bunte salts, frequently not all of the activity is exchangeable with sulfite (Tsang and Schiff, 1976b; Torii and Bandurski, 1967) and further work is needed to prove that these are definitely organic thiosulfates and that they account for all of the activity in both types of systems.

7 Speculations on the Origin and Evolution of Pathways of Sulfate Reduction

It is likely that the most primitive form of sulfate reduction is the dissimilatory pathway of the anaerobes (Roy and Trudinger, 1970; Siegel in Greenberg, 1977; Peck,

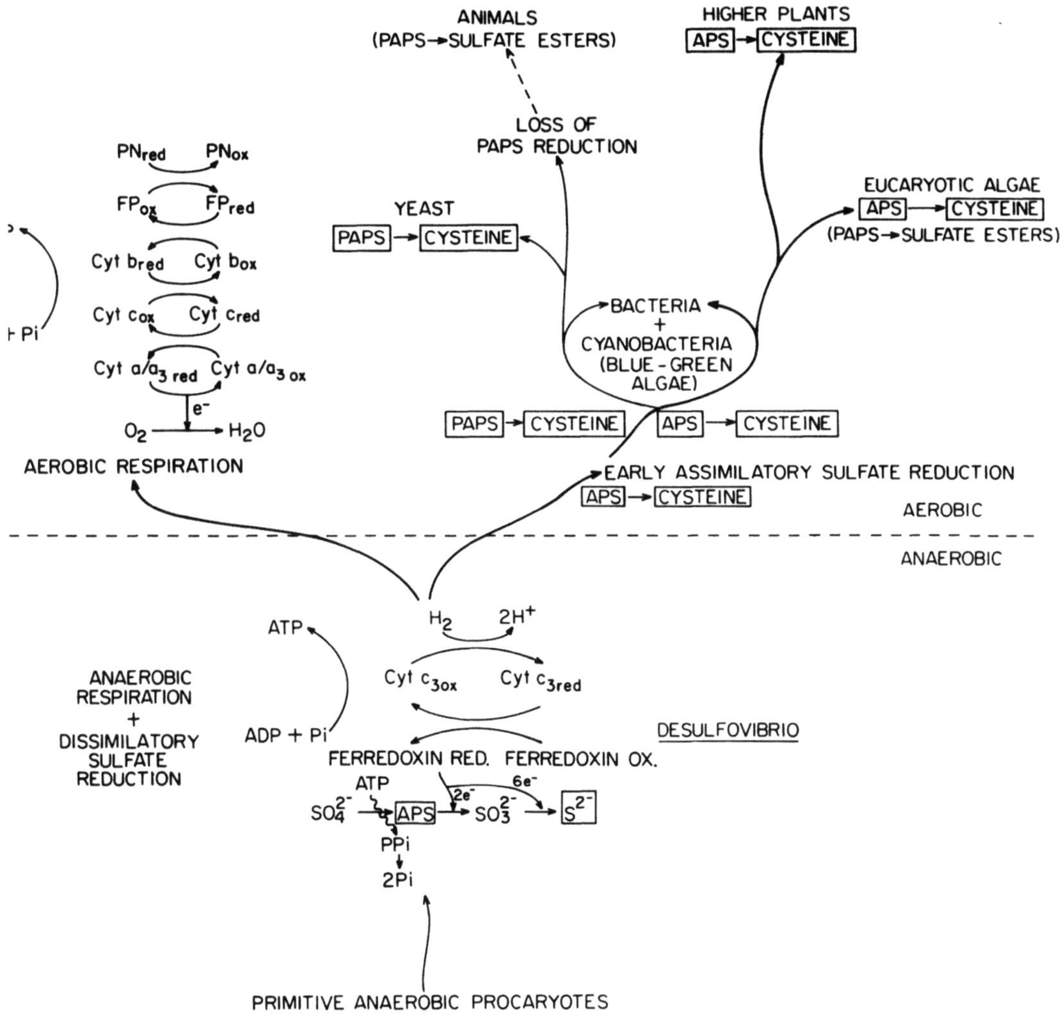

Fig. 7. Suggested evolutionary relationships of sulfate-reducing pathways in various groups. Assimilatory sulfate reduction, as part of anaerobic respiration, probably arose in the anaerobic phase of the origin of life and has persisted in the modern prokaryotic anaerobes *Desulfovibrio* and *Desulfomaculum* which use sulfate as an oxidant in respiration forming ATP and sulfide. This pathway begins with APS. With the release of oxygen into the atmosphere, oxygen became the electron acceptor in respiration and sulfate reduction occurred leading to the establishment of assimilatory sulfate reduction, still using APS, as the activated sulfate for reduction. For as yet unknown reasons, the PAPS pathway for reduction evolved from the APS pathway and modern prokaryotes (bacteria and blue-green algae or cyanobacteria) have one or the other. The APS pathway has persisted in the evolution of the eukaryotic photosynthetic oxygen evolvers including the eukaryotic algae and higher plants. The PAPS pathway leads to yeast, among the fungi. The animals have lost the ability to reduce PAPS to the thiol level of the amino acids but retain PAPS as a donor of the sulfo group in esterification reactions via specific sulfotransferases. The eukaryotic algae also use PAPS as the donor of the sulfo group in esterification reactions although they use APS as the substrate for reduction

1970) (Fig. 7) which may have survived with little modification from the anaerobic phase of the origin of life and evolution. In the absence of molecular oxygen, other oxidants were pressed into service to oxidize substrates. The result was anaerobic respiration employing sulfate or nitrate as oxidants with the consequent formation of their reduced forms, sulfide and nitrogen gas. In organisms such as *Desulfovibrio* and *Desulfomaculum* the formation of ATP from the oxidation of substrates is coupled to dissimilatory sulfate reduction which serves as a sink for the electrons removed in the oxidation of the substrate (hydrogen in the present example). This dissimilatory pathway uses APS as the activated form of sulfate for reduction and the pathway proceeds through the formation of free sulfite and sulfide. In an anaerobic environment, there is no danger from chemical oxidation of these intermediates by molecular oxygen. Anaerobic conditions also allow the undisturbed participation of autoxidizable low potential reductants such as ferredoxin in the sulfate reduction pathway. Multielectron reductions utilizing low potential reductants such as those involved here, in sulfite reduction, in nitrite reduction and in nitrogen fixation very likely evolved under the early anaerobic conditions in which life originated.

When oxygen came into the atmosphere through the activities of the first oxygenic photosynthesizers [very likely early prokaryotes related to the cyanobacteria or blue-green algae (see Schiff, 1972)] aerobic respiration using oxygen as the electron acceptor evolved and the organisms carrying out anaerobic respiration were forced into anaerobic niches (such as deep mud) to survive. With the availability of oxygen, we suppose, came the separation of respiration and sulfate reduction in the aerobic organisms. Respiration now employed the reduction of oxygen to water (Fig. 7) in the formation of ATP. Sulfate reduction, although no longer required for respiration, was still necessary for the formation of reduced sulfur compounds including cysteine and methionine of protein. This separation of sulfate reduction from respiration led to the establishment of assimilatory sulfate reduction as a separate process to provide these reduced sulfur compounds.

This assimilatory sulfate reduction process was, we suppose, at first similar to the process of dissimilatory sulfate reduction in the anaerobes in that it employed APS as the activated intermediate and reduced ferredoxin as the reductant. Free sulfite and sulfide react with oxygen, however, and this may have been the point at which carrier-bound intermediates such as glutathione-S-sulfonate ($G\text{-}S\text{-}SO_3^-$) and glutathione persulfide ($G\text{-}S\text{-}S^-$) which are not autoxidizable, were introduced into the APS pathway. This process would have been present in the early aerobic prokaryotes and is found among contemporary prokaryotes such as the cyanobacteria or blue-green algae (Tsang and Schiff, 1974; Schmidt, 1977). It is the conviction of many modern biologists, including ourselves, that the chloroplasts and mitochondria of modern eukaryotes originated from the invasion of nonphotosynthetic, nonrespiratory eukaryotes by appropriate prokaryotic endosymbionts such as primitive blue-green algae and bacteria (Schiff, 1972). In view of this, it is not surprising that the APS pathway of assimilatory sulfate reduction with ferredoxin as reductant is found in eukaryotic algae such as *Chlorella* (Tsang and Schiff, 1975; Schmidt et al., 1974) and higher plants (Schmidt, 1975a; Brunold and Schmidt, 1978) and in spinach has been shown to be localized in the chloroplasts (Schmidt and Schwenn, 1971; Schwenn and Trebst, 1976). This is probably one of the reasons why chloroplast function can never be eliminated completely from green algae and higher plants even when reduced carbon is available; essential functions such as sulfate reduction (and nitrite reduction) are localized in the plastids. Organisms more closely related to the animals such as *Euglena* do not put essential functions other than carbon dioxide fixation in their plastids and when provided with a reduced carbon source plastid function can be eliminated completely. In *Euglena*, sulfate reduction is localized in the mitochondria and microbodies (Brunold and Schiff, 1976). In this organism, the APS pathway characteristic of oxygenic eukaryotes is present but the reductant has become NADPH rather than reduced ferredoxin, which in eucaryotes is available only in the chloroplasts as far as we know.

Although the APS pathway is pervasive among eukaryotes having oxygenic photosynthesis, another pathway appears to have evolved at the same time. While some prokaryotes use the APS pathway of assimilatory sulfate reduction, others use a PAPS pathway (Tsang and Schiff, 1975; Schmidt, 1977), suggesting that the PAPS pathway evolved from the APS pathway at or after the establishment of assimilatory sulfate reduction (Fig. 6). This pathway is found in bacteria such as *E. coli* (Tsang and Schiff, 1975, 1976b, 1978b; Dreyfuss and Monty, 1963; Fujimoto and Ishimoto, 1961) and in fungi such as yeast (Torii and Bandurski, 1967; Wilson and Bierer, 1976) where the reductant is NADPH. Since PAPS is formed from APS it presumably arose later in evolution than APS but why PAPS was selected for assimilatory sulfate in these organisms is not readily explained. As we have already noted (Fig. 2) sulfate reduction was lost in the evolution of the protozoa and higher animals; the pathway from sulfate to PAPS formation has been conserved, however, and PAPS serves as the source of sulfate for esterification reactions via PAPS sulfotransferases of various specificities (DeMeio in Greenberg, 1977). PAPS also serves as the donor for esterification reactions in organisms that use APS for sulfate reduction, suggesting that PAPS formation was fairly common among primitive prokaryotes but was selected for use in assimilatory sulfate reduction only in certain prokaryotes and in nonphotosynthetic eukaryotes such as the fungi.

Acknowledgments. Support from grants from the National Science Foundation (PCM 76-21486) is gratefully acknowledged. Jerome A. Schiff is the Abraham and Etta Goodman Professor of Biology.

References

Abrams WR, Schiff JA (1973) Arch Mikrobiol 94:1–10
Berkowitz GA, Fankhauser H, Schiff JA (1981) Plant Physiol (Supp) in press
Black WJ, Wang JH-C (1970) Biochim Biophys Acta 212:257–268
Brunold C, Schiff JA (1976) Plant Physiol 57:430–436
Brunold C, Schmidt A (1978) Plant Physiol 61:342–347
Butenko RG (1968) Plant tissue culture and plant morphogenesis. Israel Program for Scientific Translation. US Dep Commerce, Springfield, VA
Cacco G, Saccomani M, Ferrari G (1977) Plant Physiol 60:582–584
Conley RR, Pigiet V (1978) J Biol Chem 253:5568–5572
Dreyfuss J, Monty KJ (1963) J Biol Chem 238:3781–3783
Ellis RJ (1963) Phytochemistry 2:129–136
Ellis RJ (1969) Planta 88:34–42
Fankhauser H, Brunold C (1978) Planta 143:285–289

Fankhauser H, Brunold C (1979) Plant Sci Lett 14:185–192
Fankhauser H, Schiff JA (1980) Plant Physiol S–65:17
Fankhauser H, Garber LJ, Schiff JA (1979) Plant Physiol S-63:162
Fankhauser H, Schiff JA, Garber LJ (1980) submitted for publication
Friedrich JW, Schrader LE (1978) Plant Physiol 61:900–903
Fujimoto D, Ishimoto M (1961) Sulfate reduction in *E. coli.* J Biochem (Tokyo) 50:533–537
Greenberg DM (ed) (1977) Metabolic pathways, vol VII. Metabolism of sulfur compounds. Academic Press, London New York
Gumport RI, Lehman IA (1971) Proc Natl Acad Sci USA 68:2559–2563
Hizuka H, Adachi K, Halprin KM, Levine V (1976) Biochim Biophys Acta 444:685–693
Holmgren A, Ohlsson I, Crankvist ML (1978) J Biol Chem 253:430–436
Kredich N, Tomkins G (1966) J Biol Chem 241:4955–4965
Mayer AM (1967) Plant Physiol 42:324–326
Moffatt JG, Khorana HG (1961) J Am Chem Soc 83:649–658
Nakazawa A, Tokushige M, Hayaishi O, Ikehara M, Mizuno Y (1967) J Biol Chem 242:3868–3872
Okazaki T, Nakazawa A, Hayaishi O (1968) J Biol Chem 243:5266–5271
Onajobi FD, Cole CV, Ross C (1973) Plant Physiol 52:580–584
Pate JS (1965) Science 143:547–548
Peck HD Jr (1961) J Bacterol 82:933–939
Peck HD Jr (1970) In: Muth OH, Oldfield JE (eds) Symp Sulfur in Nutrition. Avi, Westport, CT, pp 61–79
Pigiet V, Conley RR (1978) J Biol Chem 253:1910–1920
Porqué PG, Baldestein A, Reichard P (1970) J Biol Chem 245:2371–2374
Roy AB, Trudinger PA (1970) The biochemistry of inorganic compounds of sulfur. University Press, Cambridge
Schiff JA (1972) In: Abercrombie M et al. (eds) Advances in morphogenesis, vol X. Academic Press, London New York, pp 265–309
Schiff JA, Hodson RC (1973) Annu Rev Plant Physiol 24:381–414
Schmidt A (1973) Arch Mikrobiol 93:29–52
Schmidt A (1975a) Plant Sci Lett 5:407–415
Schmidt A (1975b) Planta 124:267–275
Schmidt A (1976) Z. Pflanzenphysiol 78:164–168
Schmidt A (1977) FEMS Microbiol Lett 1:137–140
Schmidt A, Schwenn JD (1971) Proc 2nd Int Congr Photosynth. Junk, The Hague, pp 507:514
Schmidt A, Abrams WR, Schiff JA (1974) Eur J Biochem 47:423–434
Schwenn JD, Trebst A (1976) In: Barber J (ed) The intact chloroplast. Elsevier, Amsterdam, pp 315–334
Steber J, Schleifer KH (1975) Arch Microbiol 105:173–177
Tamura G, Iwasawa T, Masada M, Iukushima K (1976) Agric Biol Chem 40:637–638
Thomas E, Davey MR (1975) From single cells to plants. Springer, Berlin Heidelberg New York
Torii K, Bandurski RS (1967) Biochim Biophys Acta 136:286–295
Tsang M L-S, Schiff JA (1975) Plant Sci Lett 4:301–307
Tsang M L-S, Schiff JA (1976a) Plant Cell Physiol 17:1209–1220
Tsang M L-S, Schiff JA (1976b) J Bacteriol 125:923–933
Tsang M L-S, Schiff JA (1976c) Eur J Biochem 65:113–121
Tsang M L-S, Schiff JA (1978a) Plant Sci Lett 11:177–183
Tsang M L-S, Schiff JA (1978b) J Bacteriol 134:131–138
White PR (1963) Ronald Press, NY
White PR, Grove AR (eds) (1965) Proc Conf Plant Tissue Culture. McCutchan, Berkeley, CA
Wilson LG, Bierer D (1976) Biochem J 158:255–270

Ecology and Physiology of Some Anaerobic Bacteria from the Microbial Sulfur Cycle

N. PFENNIG and F. WIDDEL[1]

1 Introduction

Considering the conditions under which the different groups of microorganisms involved in the sulfur cycle are active, it is advisable to differentiate an aerobic and an anaerobic part. Under aerobic conditions, the reductive part of the cycle is the assimilatory sulfate reduction of microorganisms and green plants. The reduced organic sulfur compounds which are ultimately liberated during aerobic decomposition of organic substances from plants and animals become reoxidized to sulfate by the aerobic chemoautotrophic and mixotrophic colorless sulfur bacteria (e. g., Thiobacilli, Beggiatoa). Under anaerobic conditions, the sulfur cycle is entirely microbial. Sulfate is used as the electron acceptor in the energy-conserving, dissimilatory sulfate reduction by the various groups of sulfate-reducing bacteria. The sulfide arising from this process is either reoxidized to sulfate under anaerobic conditions in the light by the phototrophic green or purple sulfur bacteria, or the hydrogen sulfide reaches aerobic regions and is then chemically or biologically oxidized via sulfur and thiosulfate to sulfate (Fig. 1).

During times of the earth's history when an oxygen-containing atmosphere was not yet developing, the anaerobic phototrophic formation of sulfate from reduced sulfur compounds could well have been an important biological process. It allowed the development of microbial processes by which organic substances could be completely degraded to CO_2 when sulfate or elemental sulfur rather than organic carbon compounds served as electron acceptors.

In illuminated anaerobic aquatic environments with degradable organic matter, the so-called sulfureta, sulfate-reducing and sulfide-oxidizing groups of bacteria are found in more or less close association; this has been well known for a long time (Baas-Becking, 1925). It was, however, only recently that true dissimilatory sulfur-reducing bacteria were discovered which preferentially develop together with the sulfur-excreting phototrophic green sulfur bacteria (Pfennig and Biebl, 1976; Biebl and Pfennig, 1978).

2 True Dissimilatory Sulfur-Reducing Bacteria

The first strains of anaerobic acetate-oxidizing, sulfur-reducing bacteria were isolated from anaerobic, sulfide-containing marine sediments and acetate-grown cultures of

1 Fakultät für Biologie, Universität Konstanz, 7750 Konstanz/FRG

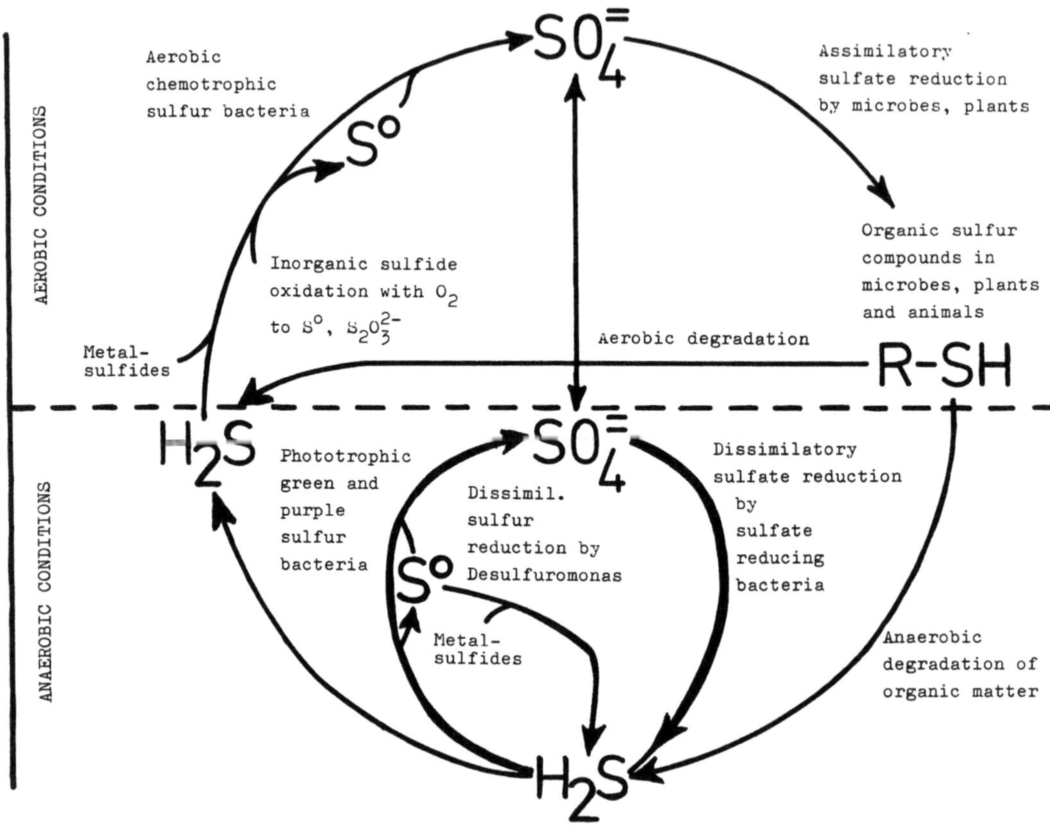

Fig. 1. The biological sulfur cycle

phototrophic green sulfur bacteria from saline environments. The cells of these strains are rod-shaped, 0.4–0.7 μm wide, 1–4 μm long; they possess a single flagellum in lateral to subpolar position and exhibit a characteristic propeller-like movement. The strains are gram-negativ and do not form spores. The new species and genus *Desulfuromonas acetoxidans* was established for this type of bacteria (Pfennig and Biebl, 1976).

The selective enrichment culture was achieved by the use of a sulfate-free, bicarbonate-containing mineral salts medium with acetate as carbon source and electron donor and sulfur flower as electron acceptor. These enrichment conditions proved to be useful in all subsequent studies. Pure cultures were obtained by the repeated application of the agar shake culture method. To obtain a fine and homogeneous distribution of elemental sulfur in agar deeps, 3–4 drops of a clear, orange-colored, autoclaved polysulfide solution (10 g $Na_2S \cdot 9\ H_2O$ and 3 g of sulfur flower dissolved in 15 ml dist. water) were added to 50 ml of medium (final pH: 7.8–8.0). Clearing of the sulfur turbidity in the „polysulfide" tubes and appearance of pink colonies indicated growth of sulfur reducers. With pure cultures it was shown that the dissimilatory, energy-conserving reduction of elemental sulfur to hydrogen sulfide is indeed stoichiometrically linked to the oxidation of acetate to CO_2: $CH_3COOH + 2H_2O + 4S^o \rightarrow 2CO_2 + 4H_2S$; $\Delta G_0' = -5.7$ kcal/reaction.

The precise demonstration of this stoichiometry appeared to be of particular importance because the free energy change of the reaction was considered to be too low to support growth of bacteria. The process was even more surprising since acetate and CO_2 were the only carbon sources required for growth. With the exception of the acetate cleavage to methane and CO_2 by methanogenic bacteria (Mah et al., 1977; Zehnder et al., 1980), no anaerobic nonphototrophic, dissimilatory oxidation of acetate had been demonstrated with pure cultures of bacteria until 1976. The argument that not S-S bonds of elemental sulfur but more oxidized sulfur compounds, present as impurities in sulfur flower, might be used as electron acceptors, is invalidated by the observed stoichiometry of the process. In addition it was shown that the only other sulfur compounds used as electron acceptors by *Desulfuromonas* apart from elemental sulfur and polysulfide sulfur were compounds with disulfide bonds such as cystine and oxidized glutathione. Not even thiosulfate with its S-S bond between a sulfane and a sulfone-sulfur was reduced, nor sulfite or sulfate (Pfennig and Biebl, 1976). It is very likely that in anaerobic natural habitats, particularly marine sediments, oligosulfide and FeS_2 play a significant role as electron acceptors for sulfur reducers. The extensive anaerobic formation of sulfur and polysulfides may well be restricted to habitats which obtain sufficient light to allow the development of green sulfur bacteria or Ectothiorhodospira.

The salt-requiring strains (2% NaCl and 0,3% $MgCl \cdot 6H_2O$) grouped in the species *Desulfuromonas acetoxidans* were all able to utilize ethanol or propanol instead of acetate; and elemental sulfur could be replaced by the nonselective electron acceptors L-malate or fumarate which were quantitatively reduced to succinate. With the culture medium originally used no growth was observed within three weeks on malate or fumarate alone. Later studies revealed, however, that in a bicarbonate-free medium supplemented with yeast extract, malate or fumarate could be fermented after a lag period in the absence of electron donors like acetate, ethanol, or propanol (A. Kümmel, personal communication).

In the meantime more than 20 new strains were isolated in pure culture from marine and freshwater environments (Opitz et al., 1980; Bache, personal communication). The results obtained with these strains can be summarized as follows:

1. Samples from anaerobic marine environments regularly yielded well-growing enrichment cultures of the *D. acetoxidans* type, while only a low percentage of all anaerobic freshwater samples gave rise to positive enrichments. This indicates clearly that the physiological type of the true dissimilatory sulfur-reducing bacteria has its adequate ecological niche in anaerobic marine or salt water habitats.

2. A number of isolates were unable to utilize malate or fumarate as an alternative electron acceptor instead of sulfur or as a fermentable substrate.

3. Most strains isolated from freshwater habitats differed morphologically from *Desulfuromonas acetoxidans*. Their cells were ovoid to elongated ovoid, polarly flagellated and unable to use any electron donor other than acetate. In order to classify this new type of obligate dissimilatory sulfur-reducing bacteria as a new species of the genus *Desulfuromonas*, the genus description had to be emended. The new group of strains was named *Desulfuromonas acetexigens* (Opitz et al., 1980).

4. All *Desulfuromonas* strains form pink to peach-colored cell material exhibiting characteristic absorption spectra of c-type cytochromes. A major component of the

cytochromes from *Desulfuromonas acetoxidans* strain 5,071 is the low potential, three heme cytochrome c-551,5 (c-7); its amino acid sequence is identical with that of cytochrome c-7 from "Chloropseudomonas" cultures (Probst et al., 1977). The study of further cytochromes and their function as a sulfur reductase is in progress (Bache, 1978).

5. In combination with pure cultures of phototrophic green sulfur bacteria, all strains of sulfur-reducing bacteria formed well-growing syntrophic cultures on acetate (Biebl and Pfennig, 1978). Combined cultures of marine or salt lake isolates with *Prosthecochloris aestuarii* as the green bacterium grow particularly efficiently at all light intensities between 5 and 1,000 lx. In contrast, maximum yields with combined cultures of freshwater isolates were obtained only under light-limiting growth conditions (5–200 lx). This was readily attributed to the fact that at high light intensities freshwater green sulfur bacteria oxidize H_2S largely directly to sulfate, while saltwater isolates form and excrete elemental sulfur almost exclusively.

3 Facultative Sulfur-Reducing Bacteria

Enrichment cultures for *Desulfuromonas*, particularly from freshwater habitats, almost regularly contained small spirilla which were identified as saprophytic *Campylobacter* sp. strains (Wolfe and Pfennig, 1977; Laanbroek et al., 1978; Winter, 1978). Pure cultures of these bacteria resemble *Vibrio succinogenes* in most respects including the possession of c-type cytochromes, except that the latter bacterium is unable to utilize malate or fumarate without hydrogen or formate as electron donor. In the presence of hydrogen or formate, all these bacteria can be selectively enriched when the bicarbonate-containing mineral medium is supplemented with elemental sulfur and acetate. It is very likely, therefore, that Pelsh's spirilloid „hydrogenthiobacteria" (Pelsh, 1936) which he for the first time enriched with hydrogen, sulfur, and mud, were in fact saprophytic Campylobacters. In contrast to *Desulfuromonas*, the acetate utilization of *Campylobacter* is entirely assimilatory and hydrogen or formate are required as electron donor for the dissimilatory, energy-conserving sulfur reduction. The saprophytic *Campylobacter* strains exhibit a great metabolic diversity. Under microaerobic conditions, oxygen may serve as electron acceptor; under anaerobic conditions nitrate, sulfite, thiosulfate, elemental sulfur or malate and fumarate may be utilized in the presence of a suitable carbon source (e. g., acetate, pyruvate, or succinate). Like *Desulfuromonas*, *Vibrio succinogenes* and *Campylobacter* strains form well-growing syntrophic cultures when combined with pure cultures of phototrophic green sulfur bacteria. The reduction of elemental sulfur formed from sulfide by the green bacteria is in this case dependent on the presence of formate as electron donor and acetate as carbon source for *Campylobacter* (Wolfe and Pfennig, 1977).

Finally, some strains of sulfate-reducing bacteria have to be mentioned as facultative sulfur reducers in the absence of sulfate. From the well-known *Desulfovibrio* species and strains only the rod-shaped strain Norway 4 was able to grow with elemental sulfur as electron acceptor. Further strains were isolated from different enrichment cultures for sulfur-reducing bacteria (Biebl and Pfennig, 1977). When ethanol was used as the electron donor, these strains grew well together with phototrophic green sulfur bacteria; the fact that no globules of elemental sulfur appeared in such combined cultures showed that the potential sulfate reducers were effectively reducing the sulfur formed by the green bacteria in the light.

4 Dissimilatory Sulfate-Reducing Bacteria as Terminal Oxidizers of Organic Substances Under Anaerobic Conditions

The dissimilatory sulfate-reducing bacteria known until a few years ago, the species of *Desulfovibrio* and *Desulfotomaculum*, all oxidize organic substrates incompletely, acetate being the regular end product when either lactate, pyruvate or ethanol served as electron donor and carbon source (Postgate, 1979). These bacteria were often considered to be essentially fermentative with indispensable energy conservation by substrate phosphorylation. With a *Desulfovibrio vulgaris* strain that can be grown with hydrogen as electron donor and acetate plus CO_2 as carbon source, recently Badziong et al. (1978) provided unequivocal evidence that the reduction of sulfate to sulfide is linked to sufficient energy conservation for growth. Therefore, dissimilatory sulfate reduction should be considered as an energy-conserving process analogous to nitrate reduction. This is an important extension of our concepts, because it allows us to think of possible substrates for sulfate-reducing bacteria, the anaerobic oxidation of which is not coupled to substrate phosphorylation. Only one year after the isolation of the acetate-oxidizing, sulfur-reducing *Desulfuromonas acetoxidans*, still other anaerobic bacteria were obtained in which the oxidation of acetate is coupled to the reduction of sulfate. Here again, energy conservation must be connected with electron-transport. The first type, a clostridium-shaped, spore-forming bacterium, *Desulfotomaculum acetoxidans*, was isolated from manure-contaminated anaerobic mud samples; it could not be enriched from nonpolluted anaerobic sediments. This fact, and the observed optimum growth temperature of around 36 °C, make it likely that the bacterium belongs to the intestinal microflora of livestock (Widdel and Pfennig, 1977). Later isolations with acetate from anaerobic marine and brackish water habitats yielded short rod-shaped, nonsporing bacteria, *Desulfobacter postgatei*, as well as filamentous gliding bacteria, *Desulfonema limicola* (Widdel, 1980). With these and perhaps further metabolic types, the sulfate-reducing bacteria appeared more and more capable of functioning as terminal oxidizers of organic substances in sulfate-rich anaerobic environments. This idea was strongly supported by quantitative ecological studies on the sulfur cycle and the degradation of plant material in anaerobic marine sediments (Jørgensen and Fenchel, 1974; Jørgensen, 1977). Such measurements showed that in the course of dissimilatory sulfate reduction more than 50% of organic matter were completely degraded to CO_2. Therefore, in sulfate-sufficient habitats the sulfate-reducing bacteria do function at the end of the anaerobic food chain, as do the methanogenic bacteria in sulfate-poor or -free environments. All these results encouraged further enrichment experiments with substances originating in the anaerobic degradation of organic residues like lower and higher fatty acids or benzoate. The extensive studies led to the isolation in pure culture of several new types of sulfate-reducing bacteria. A brief summary of the results is given in Table 1. The strains of all new species were strict anaerobes and required highly reducing conditions for growth

Table 1. Selected characteristics of new species and genera of fatty acid oxidizing sulfate-reducing bacteria. (From Widdel, 1980)

Species	Cell shape width × length (μm)	Enrichment substrate	Electron donors and carbon sources utilized with sulfate					
			H_2, CO_2	HCOOH	Acetate	Propionate	Fatty acids $C_nH_{2n}O_2$ n=	Benzoate, phenylacetate, phenylpropionate
1. *Incomplete oxidation:*								
Desulfobulbus propionicus	Ovoid, lemon 1–1.3x1.8–2	Propionate	–	–	–	+	3	–
Desulfovibrio sapovorans	Spirillum 0.5–1.5x3–6	Palmitate	–	–	–	–	4–18	–
2. *Complete oxidation:*								
Desulfobacter postgatei	Rod 1–1.5x1.7–3.5	Acetate	–	–	+	–	2	–
Desulfotomaculum acetoxidans	Rod 1–1.5x3.5–9	Acetate	–	–	+	–	2,4,5	–
Desulfococcus multivorans	Sphere 1.5–2.2	Benzoate	–	+	+	+	1–14	+
Desulfosarcina variabilis	Ovoid, Sarcina 1–1.5x1.5–2.5	Benzoate	+	+	+	+	1–14	+
Desulfonema limicola	Filament 2.5–3xSome mm	Acetate	+	+	+	+	1–12	–
Desulfonema magnum	filament 7.0xSeveral mm	Benzoate	–	–	+	+	1–10	+
Desulfovibrio baarsii	Vibrio, spirillum 0.5–1.0x2.0–4	Stearate	–	+	+	+	1–18	–

Utilization for growth = +; no growth = –

(e. g., sulfide plus dithionite-reduced media). With the exception of *Desulfobulbus propionicus,* capable of reducing nitrate to ammonia in the absence of sulfate, all other species were obligately dependent on oxidized sulfur compounds as electron acceptors. On the basis of physiological criteria, two main groups of sulfate reducers were differentiated: the first group includes those strains and species that carry out an incomplete oxidation of their substrate. All species of the classical sulfate reducers and two new species belong to this group. The second group includes all species that are able to completely oxidize their organic substrates to CO_2. The way of overall utilization of the organic electron donors and carbon sources was established by quantitative determinations of the substrates and products and calculation of the stoichiometries. The experimental results agreed very well with the stoichiometric equations given in Table 2 for the most important incomplete and complete substrate oxidations.

Enrichment cultures with propionate or butyrate all yielded strains that incompletely oxidized their substrates. Apparently, under conditions of competition such strains have a selective advantage at high concentrations of these substrates. Pure cultures of strains enriched with other substrates were even capable of completely oxidizing propionate and butyrate; however, such strains could not be selectively enriched with these substrates. Another interesting result is that those species of sulfate reducers which are most restricted in their substrate utilization: *Desulfobacter postgatei* and *Desulfobulbus propionicus,* are specialized to those fatty acids which occur in largest amounts in anaerobic waters and sediments: acetate and propionate (Ansbaek and Blackburn, 1980). Several of the species capable of completely oxidizing acetate and propionate were at the same time those which can utilize C_1-compounds, CO_2

Table 2. Examples of experimentally established stoichiometric relationships for substrate conversions of sulfate-reducing bacteria listed in Table 1

1. *Incomplete oxidation*	
Propionate:	$4CH_3CH_2COO^- + 3SO_4^{2-} \rightarrow 4CH_3COO^- + 3HS^- + H^+$
Straight chain fatty acids with uneven numbers of C-atoms:	$CH_3(CH_2)_{2n+1}COO^- + \frac{n}{2}SO_4^{2-} \rightarrow CH_3CH_2COO^- + nCH_3COO^- + \frac{n}{2}HS^- + \frac{n}{2}H^+$
Straight chain fatty acids with even numbers of C-atoms:	$CH_3(CH_2)_{2n}COO^- + \frac{n}{2}SO_4^{2-} \rightarrow (n+1)CH_3COO^- + \frac{n}{2}HS^- + \frac{n}{2}H^+$
2. *Complete oxidation*	
Formiate:	$4HCOO^- + SO_4^{2-} + H_2O \rightarrow 4HCO_3^- + HS^- + OH^-$
Acetate:	$CH_3COO^- + SO_4^{2-} \rightarrow 2HCO_3^- + HS^-$
Palmitate:	$2CH_3(CH_2)_{14}COO^- + 23SO_4^{2-} \rightarrow 32HCO_3^- + 23HS^- 7H^+$
Fatty acids, general:	$4H(CH_2)_nCOO^- + (3n+1)SO_4^{2-} + H_2O \rightarrow (4n+4)HCO_3^- + (3n+1)HS^- + HO^- + nH^+$
Benzoate:	$4C_6H_5COO^- + 15SO_4^{2-} + 16H_2O \rightarrow 28HCO_3^- + 15HS^- + 9H^+$

and/or formate, as sole carbon sources. This means that the chemoautotrophic mode of life is now for the first time shown to exist in the group of bacteria conserving energy by sulfate reduction. Finally it is worth emphasizing that three of the new species completely oxidize aromatic compounds like benzoate, phenylacetate, and 3-phenylpropionate. For some of the benzoate oxidizing strains a specific requirement for selenium (10^{-8} mol/l) and molybdenum (10^{-7} mol /l) was established when benzoate was substrate. Further experiments led to the assumption that the first attack on the benzoate ring is a selenium-enzyme-dependent hydroxylation yielding 2-hydroxybenzoate (Widdel, 1980). The filamentous gliding strains of the new genus *Desulfonema* were found to be true sediment inhabitants. Pure cultures of these bacteria could only be grown reproducibly in liquid media when some kind of sediment was provided. A precipitate of aluminum phosphate allowed very good growth; since this precipitate is dissoluble in citrate it did not prevent the harvest of pure cell material.

With the morphologically different new species and genera, the sulfate-reducing bacteria became a very heterogeneous group from a taxonomic point of view. Considering their outstanding dissimilatory metabolism it appears justified, however, to treat the different genera together as a physiological-ecological group. The true genetic relationships of the morphological types to other groups of bacteria can only be established by genetic analyses.

In conclusion it can be stated: with pure cultures of new morphological and physiological types of sulfate-reducing bacteria further evidence is presented that energy conservation connected to sulfate reduction in itself yields sufficient energy to support growth of the bacteria. Substrate utilization spectra of the new species demonstrate that ecologically significant anaerobic degradation products of organic residues can be completely oxidized to CO_2 in the course of sulfate reduction. The results fully corroborate quantitative ecological studies on the significance of sulfate reduction for the terminal oxidation of organic substances in sulfate-sufficient anaerobic environments.

Acknowledgment. The cooperation of R. Bache, G. Kirchner, and R. Opitz in some of the experimental results of sulfur-reducing bacteria is kindly acknowledged.

References

Ansbaek J, Blackburn TH (1980) Microb Ecol 5:253–264
Baas-Becking LGM (1925) Ann Bot 39:613–650
Bache R (1978) Dipl thesis. Univ Göttingen
Badziong W, Thauer RK, Zeikus JG (1978) Arch Microbiol 116:41–49
Biebl H, Pfennig N (1977) Arch Microbiol 112:115–117
Biebl H, Pfennig N (1978) Arch Microbiol 117:9–16
Jørgensen BB (1977) Limnol Oceanogr 22:814–832
Jørgensen BB, Fenchel T (1974) Mar Biol 24:189–201
Laanbroek HJ, Stal LJ, Veldkamp H (1978) Arch Microbiol 119:99–102
Mah RA, Ward DM, Baresi L, Glass TL (1977) Annu Rev Microbiol 31:309–341
Opitz R, Kirchner G, Pfennig N (1980) Arch Microbiol, in press
Pelsh AD (1936) Isdatelstvo Akademii Nauk SSSR, Moskau Leningrad, pp 109–126
Pfennig N, Biebl H (1976) Arch Microbiol 110:3–12

Postgate JR (1979) The sulphate-reducing bacteria. Cambridge University Press, Cambridge
Probst J, Bruschi M, Pfennig N, LeGall J (1977) Biochim Biophys Acta 460:58–64
Widdel F (1980) Doct thesis. Univ Göttingen
Widdel F, Pfennig N (1977) Arch Microbiol 112:119–122
Winter M (1978) Dipl thesis. Univ Göttingen
Wolfe RS, Pfennig N (1977) Appl Environ Microbiol 33:427–433
Zehnder AJB, Huser BA, Brock TD, Wuhrmann K (1980) Arch Microbiol 124:1–11

Dissimilatory Sulfate Reduction, Mechanistic Aspects

J.M. AKAGI[1]

1 Introduction

Dissimilatory sulfate-reducing bacteria, belonging to the genera *Desulfovibrio* and *Desulfotomaculum* can carry out the process of anaerobic respiration by utilizing inorganic sulfate as a terminal electron acceptor. The overall process of dissimilatory sulfate reduction may be considered to occur in two phases. They are (1) the reduction of sulfate to bisulfite through the intermediate compound, adenylyl sulfate (APS), and (2) the reduction of bisulfite to sulfide. The biochemistry of (1), with respect to the pathway by which sulfate is reduced to bisulfite, has been firmly established. The sequence of events and the intermediate involved were unequivocally determined. In contrast, the issue concerning the pathway of bisulfite reduction to sulfide is still unsettled. One view advocates a direct reduction of bisulfite to sulfide comparable to the mechanism ascribed to the assimilatory sulfite reductase. The other hypothesis supports the view that bisulfite is reduced to sulfide sequentially through a pathway containing trithionate and thiosulfate as intermediates. It is the purpose of this review to discuss the current status of the mechanistic aspects of sulfate reduction with emphasis placed on the bisulfite reduction process catalyzed by dissimilatory sulfate-reducing bacteria.

2 Sulfate Activation and Reduction to Bisulfite

2.1 Adenosine Triphosphate Sulfurylase

The primary step of sulfate metabolism is catalyzed by ATP sulfurylase (EC 2.7.7.4; ATP:sulfate adenylytransferase), which forms APS and inorganic pyrophosphate from ATP and sulfate. Although the reaction lies in the direction of ATP and sulfate, the hydrolysis of pyrophosphate by inorganic pyrophosphatase (EC 3.6.1.1, pyrophosphate phosphohydrolase) probably is responsible for accumulating APS inside the cell (Akagi and Campbell, 1962; Ware and Postgate, 1971).

---

1 Department of Microbiology, University of Kansas, Lawrence, KS 66045/USA

2.2 Adenylyl Sulfate Reductase

Peck (1959) and Ishimoto and Fujimoto (1959) independently discovered that APS was the form of active sulfate used by *Desulfovibrio vulgaris* in the dissimilatory process of sulfate reduction. Furthermore, these investigators discovered an enzyme, APS reductase, which reduced APS to AMP and sulfite (Peck, 1961; Ishimoto and Fujimoto, 1961). Although the natural electron carrier for this reaction has not been elucidated, several physicochemical properties of APS reductase have been characterized. This enzyme has a MW of 220,000 with-one flavin adenine dinucleotide (FAD), 12 nonheme irons and 12 acid-labile sulfides per molecule of enzyme (Bramlett and Peck, 1975). The mechanism of action of APS reductase catalysis was proposed by Michaels et al. (1970) to consist of an adduct formation between sulfite and the N-5 position of the isoalloxazine ring of the enzyme-bound FAD. Their mechanism is shown below:

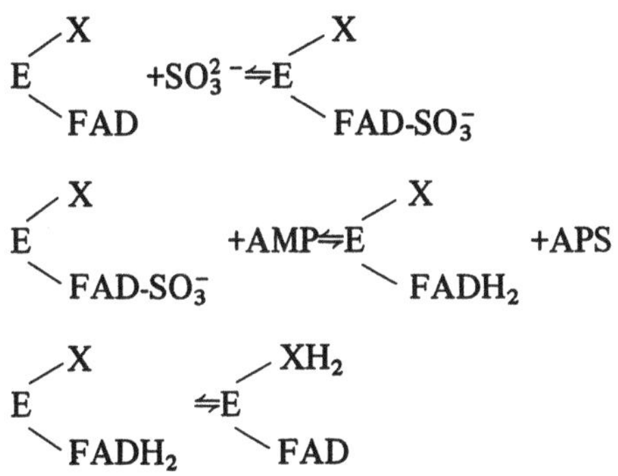

Although this mechanism was proposed for sulfite oxidation to APS, the reversal of this sequence is applicable to APS reduction.

3 Dissimilatory Bisulfite Reductase System

The sulfite, arising from APS reduction, is subsequently reduced to sulfide through a pathway which is still unresolved at this time. When Ishimoto and Koyama (1957) isolated thiosulfate reductase activity from *D. vulgaris* (Miyazaki) extracts, it suggested that thiosulfate might be formed as an intermediate during dissimilatory sulfate reduction. Kobayashi et al. (1969) described the sequential formation of trithionate and thiosulfate during sulfite reduction by *D. vulgaris* extracts. They succeeded in obtaining two fractions which were required to reduce sulfite to sulfide. One fraction reduced sulfite to trithionate and thiosulfate, while the second fraction, which did not reduce sulfite, rapidly reduced trithionate and thiosulfate to sulfide. On the basis of their finding Kobayashi et al. (1969) proposed the following pathway for dissimilatory sulfite reduction:

$$3SO_3^{2-} \xrightarrow{H_2} S_3O_6^{2-} \underset{SO_3^{2-}}{\xrightarrow{H_2}} S_2O_3^{2-} \underset{SO_3^{2-}}{\xrightarrow{H_2}} S^{2-}.$$

In the same year Suh and Akagi (1969) reported that two separate proteins, isolated from *D. vulgaris* extracts, formed thiosulfate from sulfite. One fraction was identified as the green pigment, desulfoviridin, while the other protein, only partially purified, was designated FII. Both desulfoviridin and FII were required for thiosulfate formation from sulfite; this activity was described as the "thiosulfate-forming system." In addition, from absorption spectra, pH optimum and substrate concentration studies, Suh and Akagi (1969) concluded that bisulfite, rather than sulfite, was the ionic species which was reduced to thiosulfate. Their isolation of thiosulfate from reaction mixtures during bisulfite reduction was in accordance with the dissimilatory pathway proposed by Kobayashi et al. (1969).

3.1 Bisulfite Reductase

3.1.1 Products of Bisulfite Reductase

Additional evidence for the existence of the trithionate pathway was provided by Lee and Peck (1971) when they isolated an enzyme from *Desulfovibrio gigas* extracts which catalyzed the reduction of bisulfite to trithionate. They identified this protein as the green pigment, desulfoviridin, and proposed the name bisulfite reductase for this enzyme. Under the conditions of their assay Lee and Peck (1971) reported that trithionate was the sole product of bisulfite reduction. Subsequent studies by Kobayashi et al. (1972, 1974) and Jones and Skyring (1975) on bisulfite reductase (desulfoviridin) demonstrated that, in addition to trithionate, thiosulfate, and sulfide were formed during bisulfite reduction. Furthermore, trithionate and thiosulfate were not reduced by bisulfite reductase, indicating that they were products of bisulfite reduction. These workers reported that the pattern of product formation was dependent upon the assay conditions. The principle of the assay for bisulfite reduction involves the reduction of the artificial dye, methyl viologen, (either chemically or enzymatically), which supplies the electrons to bisulfite reductase for bisulfite reduction. Kobayashi et al. (1972, 1974) and Jones and Skyring (1975) found that a high bisulfite concentration with low methyl viologen or hydrogenase concentration resulted in trithionate being the major product. Under opposite conditions, or at a higher pH value, lower trithionate and higher sulfide levels were observed (Table 1). On the basis of these observations, Kobayashi et al. (1974) proposed that sulfide was the true product of bisulfite reductase with possible enzyme-bound intermediates being sulfoxylate and elemental sulfur. Trithionate and thiosulfate were postulated as being nonenzymic products formed during bisulfite reduction to sulfide. This was in agreement with the hypothesis by Trudinger (1970), who isolated a carbon monoxide-binding pigment (P582) from *Desulfotomaculum nigrificans.* He demonstrated that P582 reduced sulfite to sulfide and suggested that *Dm. nigrificans* reduced sulfite directly to sulfide in a manner similar to assimilatory sulfite reductase. However, Akagi and Adams (1973) reported that P582 formed trithionate as the major product of bisulfite reduction with thiosulfate and sulfide being formed in lesser amounts. They concluded that P582 from *Dm. nigrificans* was analogous to bisulfite reductase (desulfoviridin) isolated from *D. vulgaris* and *D. gigas.* Another protein, functioning in the same capacity as desulfoviridin and P582 is desulforubidin (Lee et al., 1973a). This

Table 1. Effect of assay conditions on bisulfite reductase activity

pH	H_2ase mg/ml	mM Methyl viologen	SO_3^{2-}	μ mol formed $S_3O_6^{2-}$	$S_2O_3^{2-}$	H_2S	$S_3O_6^{2-}$ / H_2S
6.0[a]	3.5	1	5	1.35	0.13	0.1	13.5
6.0	3.5	10	5	1.13	0.43	0.27	4.2
6.0	0.7	1	5	1.49	0.14	0.07	21
6.0	14	10	5	0.85	0.35	0.2	4.25
6.0	7	20	1	0.03	0	0.12	0.21
7.0	3.5	1	5	1.35	0.05	0.23	5.9
7.0	3.5	10	5	1.08	0.1	0.59	1.82
7.0	3.5	10	1	0.05	0.08	0.35	0.15
6.0[b]	0.5 U.	2	2.5	1.62	0.46	0.6	2.7
6.0	0.5 U.	2	5	3.27	0.86	1.02	3.2
6.0	0.5 U.	2	10	6.78	1.42	1.39	4.8

Data from: [a] Jones and Skyring (1975) [b] Kobayashi et al. (1974)
U. = units

reddish brown pigment was isolated from *Desulfovibrio desulfuricans* (Norway), which lacks desulfoviridin, and was found to catalyze the reduction of bisulfite to trithionate and sulfide.

3.1.2 Proposed Model for Catalysis

Further studies on bisulfite reductase of *D. vulgaris* by Drake and Akagi (1977 a) showed that the pattern of product formation depended on the strength of the reducing system relative to the concentration of bisulfite in the reaction mixture. They observed that when the pyruvic phosphoroclastic system was used as the reductant for bisulfite reductase, the relative proportion of the products was dependent upon pyruvate concentration (Table 2). A higher pyruvate concentration (stronger reducing system) caused a lower trithionate, higher thiosulfate pattern whereas trithionate was the only product when a weaker reducing system (low pyruvate concentration) was used. Sulfide formation was detected when the reaction was allowed to proceed to completion. Using a weaker reducing system (lower pyruvate concentration), which favored trithionate formation, the effect of time on product formation was tested. Trithionate was observed to be the initial product formed. When trithionate formation reached a plateau, the amount of thiosulfate increased until its formation ceased. Sulfide formation became evident only when the bisulfite concentration was significantly decreased. On the basis of these findings a model for bisulfite reductase catalysis was proposed (Fig. 1). The active site is postulated to consist of three adjacent sites for bisulfite binding. A bisulfite ion binds to one site (C) and is reduced to the level of sulfoxylate. If the bisulfite concentration is high relative to the reducing power of the electron-donating system, a bisulfite molecule, occupying site B, reacts

Table 2. Coupling bisulfite reductase with the phosphoroclastic reaction[a]

Pyruvate added (μmol)	Product formed (μmol)				HSO_3^- consumed (μmol)
	$S_3O_6^{2-}$	$S_2O_3^{2-}$	S^{2-}	Acetyl phosphate	
50	0.66	1.81	Trace	5.0	5.60
25	1.51	1.12	0	4.9	6.77
10	2.50	0.29	0	4.6	8.08
8	2.70	0	0	3.2	8.10
5	1.28	0	0	1.3	3.84
50, TC[b]	1.25	2.27	0.28	8.7	8.57
10, TC	2.60	0.52	0.53	6.5	9.37

[a] Data from Drake and Akagi (1977a)
[b] Reaction allowed to go to completion (150 min)
TC = to completion

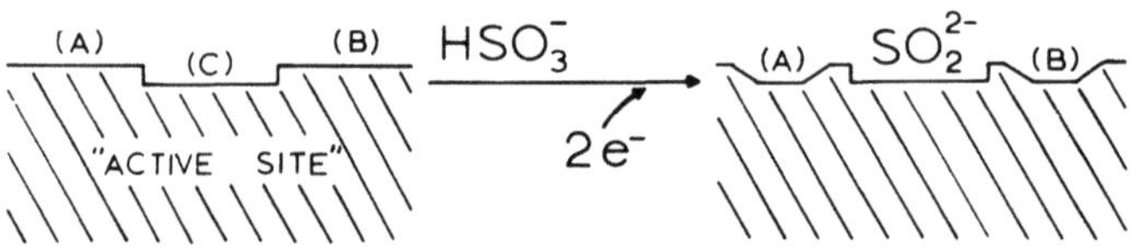

Fig. 1. Model for bisulfite reductase catalysis

with the sulfoxylate to form a two-sulfur intermediate. A third bisulfite molecule (site A) reacts with the two-sulfur intermediate to form trithionate. As the reaction proceeds, the availability of bisulfite ions decreases to a level where they cannot effectively compete with the reducing system to form trithionate and the two-sulfur intermediate is reduced to thiosulfate. When the bisulfite ion concentration is depleted to a point where sites A and B are no longer occupied, the sulfoxylate is reduced to sulfide.

3.2 Trithionate Reduction

3.2.1 "Thiosulfate-Forming Enzyme" (TF)

While reinvestigating the "thiosulfate-forming system" of *D. vulgaris,* Drake and Akagi (1977b) isolated an enzyme which formed thiosulfate from trithionate. Since the "thiosulfate-forming system" consisted of bisulfite reductase and fraction FII (Suh and Akagi, 1969), and since bisulfite reductase was shown to reduce bisulfite to trithionate (Lee and Peck, 1971), it seemed logical to conclude that FII contained trithionate reductase activity. This proved not to be so, instead, an enzyme was purified from the FII fraction which required both trithionate and bisulfite as substrates for thiosulfate

Table 3. Product formation by the thiosulfate-forming system[a]

System	Product (μmol)		
	$S_3O_6^{2-}$	$S_2O_3^{2-}$	S^{2-}
Bisulfite reductase $+HSO_3^-$	2.37	0	0.23
Bisulfite reductase $+TF+HSO_3^-$	0.27	3.68	0.20
Bisulfite reductase $+S_3O_6^{2-}$		0	0.10
$TF+S_3O_6^{2-}$		0	0
$TF+HSO_3^-$	0	0	0
$TF+HSO_3^-+S_3O_6^{2-}$		5.53	0
Bioled TF[b] $+HSO_3^-+S_3O_6^{2-}$		0	0

[a] Data from Drake and Akagi (1977b)
[b] TF placed in boiling water bath for 10 min
TF = thiosulfate-forming enzyme

formation (Table 3). This enzyme was called the "thiosulfate-forming" (TF) enzyme. To elucidate the mechanism of the TF-catalyzed reaction, kinetic and ^{35}S-labeled substrate utilization studies were conducted. The labeling pattern of thiosulfate, obtained by chemical degradation studies, indicated the source from which the sulfane and sulfonate sulfur atoms of thiosulfate were derived. The mechanism for TF-catalysis, proposed by Drake and Akagi (1977b) is shown below:

$$H\overset{*}{S}O_3^- + O_3S\text{-}\overset{*}{S}\text{-}SO_3^{2-} \xrightarrow{2e^-} \overset{*}{S}\text{-}\overset{*}{S}O_3^{2-} + SO_3^{2-} + HSO_3^-$$

The mechanism is analogous to the cyanolysis reaction of trithionate. In this case the cyanide anion makes a nucleophilic attack on the inner sulfane atom of trithionate, forming thiocyanate, with the concomitant release of sulfite and sulfate anions. The reductive TF-catalyzed reaction is postulated to involve a similar displacement reaction where the nucleophile, bisulfite, combines with the trithionate sulfane atom, forming thiosulfate, with the simultaneous release of the two sulfonate groups (as bisulfite ions). Although the TF enzyme is not a typical trithionate reductase, it is the first instance where a purified enzyme was shown to form thiosulfate reductively from trithionate.

3.2.2 Bisulfite Reductase-Dependent Trithionate Reductase

Another trithionate-reducing enzyme has been isolated from the thermophilic sulfate reducer, *Dm. nigrificans* (unpubl. results). This enzyme has been partially purified and it reduces trithionate to thiosulfate in the presence of bisulfite reductase (P582). This bisulfite reductase-dependent trithionate reductase can utilize reduced methylviologen as an electron donor and, unlike TF, does not require bisulfite to form thiosulfate (unpubl. results). The roles of the two enzyme fractions in this reaction have not been elucidated at this time.

3.3 Thiosulfate Reductase

The dissimilatory process of bisulfite reduction terminates with the reduction of thiosulfate to sulfide by thiosulfate reductase. Ishimoto et al. (1955) and Ishimoto and Koyama (1957) first described the presence of this enzyme in sulfate-reducing bacteria and demonstrated that the products of the reaction were sulfide and sulfite. Nakatsukasa and Akagi (1969) partially purified thiosulfate reductase from *Dm. nigrificans.* Using ^{35}S-labeled sulfane and sulfonate thiosulfate, they determined that the sulfane atom was reduced to sulfide, while the sulfonate moiety accumulated as sulfite. Sulfite was not reduced by this enzyme, however, it was a strong inhibitor of thiosulfate reduction. In contrast, Haschke and Campbell (1971) found that the thiosulfate reductase from *D. vulgaris* was capable of reducing both thiosulfate and sulfite to sulfide. The enzyme from *Dm. nigrificans* (Nakatsukasa and Akagi, 1969) and *D. gigas* (Hatchikian, 1975) was inhibited by sulfhydryl reagents and reversed by the addition of cysteine. This suggested that one or more sulfhydryl groups are involved in the catalysis.

3.4 Does a Dissimilatory Pathway Exist?

3.4.1 Studies in Growing and Resting Cells

The existence of the dissimilatory pathway containing trithionate and thiosulfate is still under consideration by several investigators. Chambers and Trudinger (1975) questioned the occurrence of these intermediates from their studies on growing and resting cells of *Desulfovibrio desulfuricans* and *D. gigas.* Their evidence included (1) no intermediate compound accumulating during $^{35}SO_4{}^{2-}$ reduction by resting cells of *D. desulfuricans* in the presence of lactate (2) both sulfur atoms of thiosulfate were reduced to sulfide at equal rates and (3) ^{35}S was incorporated mainly into the sulfonate sulfur atom of thiosulfate when $^{35}SO_4{}^{2-}$ plus unlabeled thiosulfate were incubated with whole cells of *D. desulfuricans.* The authors attributed this to an exchange reaction between $^{35}SO_3{}^{2-}$ and the sulfonate group of the unlabeled thiosulfate. Theoretically, the sulfur atoms of the thiosulfate molecule should have been equally labeled if the dissimilatory pathway was operating.

3.4.2 Crude Extract Studies

In light of these results Drake and Akagi (1978) incubated crude extracts of *D. vulgaris* with bisulfite under conditions for bisulfite reduction to sulfide. They observed sulfide formation with no detectable intermediate such as trithionate or thiosulfate. The crude extracts were centrifuged at 100,000 *g* for 90 min, to separate the soluble and particulate fractions. When the soluble fraction was tested for bisulfite-reducing capacity under identical conditions as performed with crude extracts, thiosulfate accumulated and decreased concomitant with sulfide formation. The addition of the particulate fraction to the soluble fraction restored the activity exhibited by crude extracts, i. e., no intermediate was detected. Membranes were suspected to be responsible for the observed phenomenon so they were purified from *D. vulgaris* by differential and sucrose density centrifugations. The purified membranes plus the soluble fraction behaved exactly as the crude extract and the soluble plus particulate fractions in reducing bisulfite. Crude extracts reduced both the sulfane and sulfonate sulfur atoms of thiosulfate at equal rates. The soluble fraction reduced the sulfane atom significantly faster than the sulfonate group. Addition of membranes to this system caused the sulfane and sulfonate sulfur atoms to be reduced to sulfide at equal rates. To demonstrate that the soluble fraction recycled the sulfonate group of thiosulfate through the dissimilatory pathway, ^{35}S-sulfonate thiosulfate was incubated with soluble fraction, under reducing conditions, and the experiment was halted at various time intervals. Degradation of thiosulfate in the reaction mixture showed that doubly labeled thiosulfate increased with time. The recycling was inhibited when antiserum directed against purified bisulfite reductase was added to the soluble fraction. In contrast, the recycling phenomenon could not be demonstrated with crude extracts. These results were interpreted to mean that the sulfonate moiety of thiosulfate was reduced to trithionate which was subsequently reduced to thiosulfate. The presence of membranes somehow masks the detection of the recycling process.

3.4.3 Role of Membranes in Bisulfite Reduction

These studies strongly suggest that a dissimilatory pathway for bisulfite reduction, as proposed by Kobayashi et al. (1969), exists in sulfate-reducing bacteria and that it is intimately associated with membranes. The role of membranes in this process is postulated to be as in Fig. 2. This shows that the sequential arrangement of bisulfite reductase (A ase), trithionate reductase or TF (B ase) and thiosulfate reductase (C ase) on the membrane surface. This arrangement would result in bisulfite reduction without any release of intermediates. It also can explain how the sulfane and sulfonate sulfur atoms of thiosulfate can apparently be reduced at equal rates.

4 Assimilatory (Bi)Sulfite Reductase

Another pathway for bisulfite reduction is through a direct route involving an assimilatory sulfite reductase. This enzyme was isolated from *D. vulgaris* by Lee et al. (1973b) and was shown to reduce sulfite to sulfide without the formation of any intermediates. It was identified as an assimilatory sulfite reductase by the absence of intermediates during bisulfite reduction and by the similarity of its absorption spectrum to assimilatory sulfite reductase from plants and sulfate-assimilating bacteria.

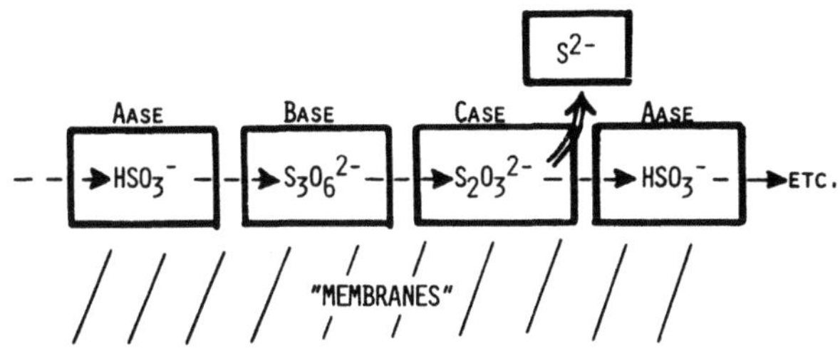

Fig. 2. Role of membranes in bisulfite reduction

It was postulated that this enzyme is present in sulfate-reducing bacteria under some regulatory control in case the organisms are unable to utilize the dissimilatory pathway involving trithionate and thiosulfate (Lee et al., 1973b). Presumably the same enzyme was isolated from *D. vulgaris* by Drake and Akagi (1976). The absorption specta of the oxidized and reduced forms of the enzyme were identical to those reported by Lee et al. (1973x). However, the enzyme purified by Drake and Akagi formed thiosulfate in addition to sulfide. High concentrations of bisulfite inhibited sulfide but not thiosulfate formation. Because of the optimum pH for this reaction, Drake and Akagi (1976) named this enzyme bisulfite reductase II. The reason for the apparent discrepancy in the results obtained by the two laboratories is not known at this time.

5 Summary and Conclusion

Dissimilatory sulfate-reducing bacteria metabolize inorganic sulfate by initially forming the activated compound APS, followed by its reduction to bisulfite. The bisulfite is subsequently reduced to sulfide which accumulates as one of the end products of the respiratory process. The reactions involved in bisulfite formation are fairly well understood, however, the pathway for its reduction to sulfide is currently a matter of dispute.

The enzyme responsible for the controversy is bisulfite reductase. The fundamental question concerning this enzyme is "what is the true product of bisulfite reduction?" Depending upon the assay conditions the pattern of products formation has been observed to vary. Consequently, the true enzymic product arising from bisulfite reduction has been a matter of interpretation. Studies by Kobayashi et al. (1972, 1974), Akagi et al. (1974), Jones and Skyring (1975), and Drake and Akagi (1977a) demonstrated that trithionate was the major product of bisulfite reductase catalysis under most of the assay conditions tested. In these cases the difference in the pattern under varying assay conditions was reflected in the trithionate:sulfide ratio. If trithionate is assumed to be the product of bisulfite reduction, it should be utilized as a substrate by another enzyme. The "thiosulfate-forming" (TF) enzyme is such an example. In addition, when bisulfite reductase and TF were incubated with bisulfite, thiosulfate was formed (Table 3). This lends support to the hypothesis that bisulfite is sequentially reduced to thiosulfate through trithionate.

One puzzling fact concerning bisulfite reductase activity is that some investigators observed the formation of fairly substantial amounts of sulfide. Murphy and Siegel (1973) found that bisulfite reductase from *Desulfovibrio* and *Desulfotomaculum* contained a tetrahydroporphyrin compound in common with assimilatory sulfite reductase. The name sirohydrochlorin was given to this class of compounds. This finding was interesting since it suggested that bisulfite reductase was capable of reducing bisulfite directly to sulfide. In fact, if bisulfite reductase (desulfoviridin) were treated with heat or chemicals to extract the sirohydrochlorin chromophore, the chromophore groups can reduce bisulfite to thiosulfate and sulfide as the major products (Seki and Ishimoto, 1979). It is possible that discrepancies in the literature pertaining to the ratio of trithionate:sulfide may have been due to the methods by which bisulfite reductase was prepared. If varying degrees of denaturation occurred during bisulfite reductase purification, it may account for the differences observed in the pattern of products formed.

The three enzyme activities necessary to operate the dissimilatory pathway of bisulfite reduction have been purified from sulfate-reducing bacteria. These are bisulfite reductase, thiosulfate-forming enzyme (TF) and thiosulfate reductase. In addition, an assimilatory sulfite reductasetype of enzyme has been purified from *D. vulgaris* (Lee et al., 1973; Drake and Akagi, 1976). It does not seem unreasonable to assume that the capacity for reducing bisulfite by the dissimilatory *and* assimilatory routes are available to sulfate-reducing bacteria. The extent of usage of either pathway may be under control of a regulatory system.

References

Akagi JM, Adams V (1973) J Bacteriol 116:392–396
Akagi JM, Campbell LL (1962) J Bacteriol 86:563–568
Akagi JM, Chan M, Adams V (1974) J Bacteriol 120:240–244
Bramlett RN, Peck HD (1975) J Biol Chem 250:2979–2986
Chambers LA, Trudinger PA (1975) J Bacteriol 123:36–40
Drake HL, Akagi JM (1976) Biochem Biophys Res Commun 71:1214–1219
Drake HL, Akagi JM (1977a) J Bacteriol 132:129–143
Drake HL, Akagi JM (1977b) J Bacteriol 132:132–138
Drake HL, Akagi JM (1978) J Bacteriol 136:916–923
Haschke RH, Campbell LL (1971) J Bacteriol 106:603–607
Hatchikian EC (1975) Arch Microbiol 105:249–256
Ishimoto M, Koyama J (1957) Proc Jpn Acad Sci 35:243–245
Ishimoto M, Fujimoto D (1961) J Biochem (Tokyo) 50:299–304
Ishimoto M, Koyama J (1957) J Biochem (Tokyo) 44:233–242
Ishimoto M, Koyama J, Nagai Y (1955) J Biochem (Tokyo) 42:41–53
Jones HE, Skyring GW (1975) Biochim Biophys Acta 377:52–60
Kobayashi K, Tachibana S, Ishimoto M (196) J Biochem (Tokyo) 65:155–157
Kobayashi K, Takahashi E, Ishimoto M (1972) J Biochem (Tokyo) 72:879–887
Kobayashi K, Seki Y, Ishimoto M (1974) J Biochem (Tokyo) 75:519–529
Lee JP, Peck HD (1971) Biochem Biophys Res Commun 45:583–589
Lee JP, Yi CS, LeGall J, Peck HD (1973a) J Bacteriol 115:453–455
Lee JP, LeGall J, Peck HD (1973b) J Bacteriol 115:529–542
Michaels GB, Davidson JT, Peck HD (1970) Biochem Biophys Res Commun 39:321–328
Murphy MJ, Siegel LM (1973) J Biol Chem 248:6911–6919
Nakatsukasa W, Akagi JM (1969) J Bacteriol 98:429–433
Peck HD (1959) Proc Natl Acad Sci USA 45:701–708
Peck HD (1961) Biochim Biophys Acta 49:621–624
Seki Y, Ishimoto M (1979) J Biochem (Tokyo) 86:273–276
Suh B, Akagi JM (1969) J Bacteriol 99:210–215
Trudinger PA (1970) J Bacteriol 104:158–170
Ware D, Postgate JR (1971) Nature (London) 226:1250–1251

Dissimilatory Sulfate Reduction, Energetic Aspects

R.K. THAUER and W. BADZIONG[1]

1 Introduction

Several strict anaerobic prokaryotes are capable of using sulfate as electron acceptor for catabolic oxidation processes. In doing so sulfate is reduced to H_2S (Fig. 1). Respiration with sulfate as acceptor is termed dissimilatory sulfate reduction, bacteria capable of sulfate respiration are classified as sulfate-reducing bacteria. Seven genera have been described: Desulfovibrio (Postgate and Campbell, 1966), Desulfotomaculum (Campbell and Postgate, 1965), Desulfobulbus, Desulfobacter, Desulfococcus, Desulfosarcina, and Desulfonema (Widdel, 1980). In this paper it will be discussed whether the dissimilatory reduction of sulfate to H_2S is coupled with the net synthesis of ATP from ADP and P_i, or whether the reduction of sulfate serves only as an electron sink, the function of which is to enable ATP synthesis via substrate level phosphorylation (SLP). It is shown that dissimilatory sulfate reduction is coupled with the net synthesis of ATP. Results are presented from which the stoichiometry of coupling and the mechanism of coupling can be deduced. The treatise begins with a short summary of what was known when the investigation was started. Detailed reviews on the subject have appeared by Postgate (1979), LeGall and Postgate (1973), Peck (1974), Siegel (1975), Thauer et al. (1977), and Thauer and Badziong (1980). For original literature the reader is referred to these reviews.

2 Existing Knowledge

Dissimilatory sulfate reduction to H_2S proceeds via adenosine phosphosulfate (APS) and bisulfite (HSO_3^-) as free intermediates:

$$SO_4^{2-} \xrightarrow[\uparrow\ 2ATP]{} APS \xrightarrow{2e^-\ \downarrow} HSO_3^- \xrightarrow[\downarrow\ xATP]{6e^-\ \downarrow} H_2S.$$

1 Fachbereich Biologie, Mikrobiologie, Philipps-Universität, 3550 Marburg/FRG

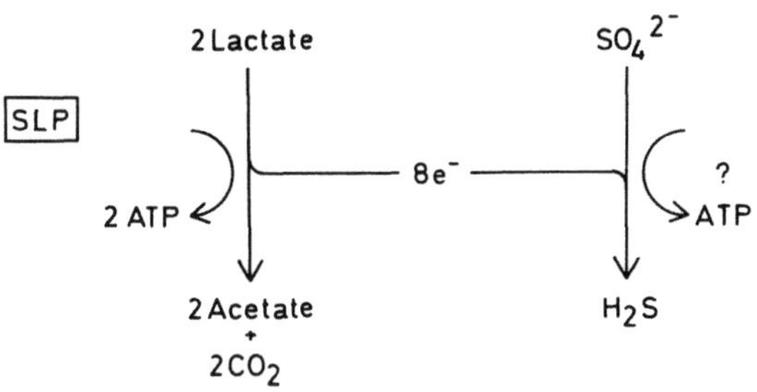

Fig. 1. Energy metabolism of sulfate-reducing bacteria growing on lactate plus sulfate as energy source. The oxidation of lactate to acetate and CO_2 is coupled with the synthesis of two mol of ATP via substrate level phosphorylation (SLP). Before the experiments were started it was unclear whether the reduction of sulfate to H_2S is additionally coupled with a net formation of ATP, or whether the reduction of sulfate serves only as an electron sink

The formation of APS from sulfate is catalyzed by ATP sulfurylase; pyrophosphate is formed as the second product. The pyrophosphate is probably quantitatively hydrolyzed to inorganic phosphate since sulfate reducers contain very high activities of inorganic pyrophosphatase. It is therefore generally assumed that two equivalents of ATP are consumed in the ATP sulfurylase reaction, which is also supported by thermodynamic considerations: The formation of APS from sulfate and ATP is exergonic only if the pyrophosphate formed is hydrolyzed (Thauer et al., 1977). APS is reduced to bisulfite in a two electron consuming reaction, which is catalyzed by APS reductase. The pathway from bisulfite to H_2S is still a matter of considerable controversy. It is uncertain whether bisulfite is reduced to H_2S via one or three consecutive steps. What is sure, however, is that six electrons are required for the reduction, and that bisulfite reductase is involved as enzyme. Bisulfite reductase is present in sulfate-reducing bacteria in very high concentrations, contains siroheme as prosthetic group, and is, dependent on the absorption spectrum, named desulfoviridin, desulforubidin, or P 582. The physiological electron donors for both the APS reductase and the bisulfite reductase are not known. Evidence is available that membrane components are involved in the reduction reactions. For details see the chapter by Akagi on page 178 of this volume.

The reduction of sulfate to bisulfite (sulfate/bisulfite: $E_0' = -516$ mV) with most electron donors (e. g., NAD/NADH: $E_0' = -320$ mV) is an endergonic reaction. The reduction of sulfate to bisulfite can therefore proceed only if coupled with the hydrolysis of ATP. The reduction of bisulfite to H_2S (bisulfite/H_2S: $E_0' = -116$ mV) with most electron donors, however, is strongly exergonic. It is known by the work of Peck (1960, 1966) that the latter reaction is coupled with the formation of ATP in sulfate-reducing bacteria. From growth yield data is was deduced that two mol of ATP are generated per mol bisulfite reduced to H_2S. If a stoichiometry of two were correct, then not more ATP would be formed between bisulfite and H_2S than required for the activation of sulfate to APS, and dissimilatory sulfate reduction would not be coupled with a net synthesis of ATP. In the following this is shown not to be the case.

3 Coupling of Dissimilatory Sulfate Reduction with the net Synthesis of ATP

The experimental approach was based on the following idea: If the dissimilatory reduction of sulfate to H_2S is coupled with a net synthesis of ATP bacteria should exist that can grow on H_2 and sulfate as sole energy source.

$$4H_2 + SO_4^{2-} + 2H^+ \rightarrow H_2S + 4H_2O \Delta G_0' = -36.4 \text{ kcal/mol.}$$

This is an exergonic reaction which theoretically could be coupled with the net synthesis of up to 3 mol of ATP; approximately 12 kcal/mol are required in the bacterial cell to drive the synthesis of ATP (Thauer et al., 1977). Evidence was available from the literature that such chemolithotrophic sulfate reducers do exist (Sorokin, 1966). We therefore tried to isolate a bacterium that can grow on H_2 plus sulfate as sole energy source. A scheme of the growth arrangement is given in Fig. 2. The bacteria were enriched and grown in 400 ml fermenters gassed with a mixture of 80%H_2/20%CO_2 at a rate of 100 ml per min. The medium contained besides sulfate and other mineral salts acetate as the only organic component. The pH of the culture was kept constant at 6.55 via an automatic pH stat. Using this arrangement sulfate-reducing bacteria were enriched from various sources and then isolated. The following experiments were performed with a strain isolated from anaerobic sludge of the sewage treatment plant in Marburg. The Marburg strain was classified as subspecies of *Desulfovibrio vulgaris.* It is not identical with the ATCC strain, since the latter did not grow under the experimental conditions described above. A typical growth curve of *D. vulgaris* (Marburg) on H_2 plus sulfate is given in Fig. 3. Under optimal conditions (pH

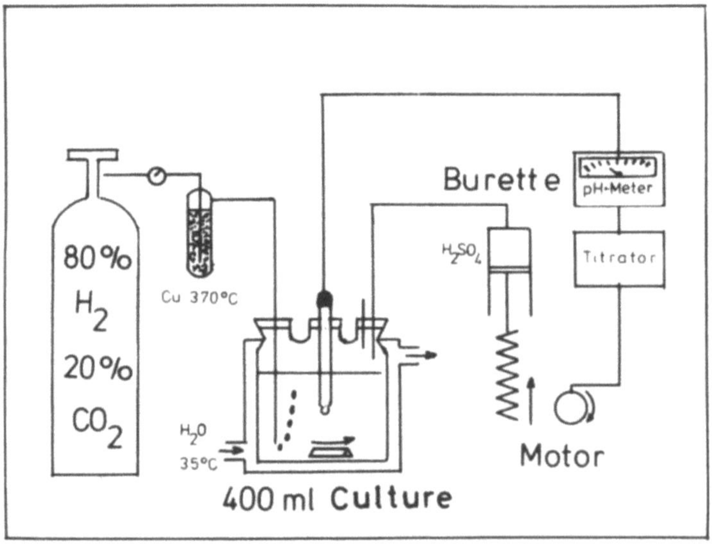

Fig. 2. Arrangement for growth of sulfate reducers on H_2 plus sulfate as sole energy source (Badziong, 1979)

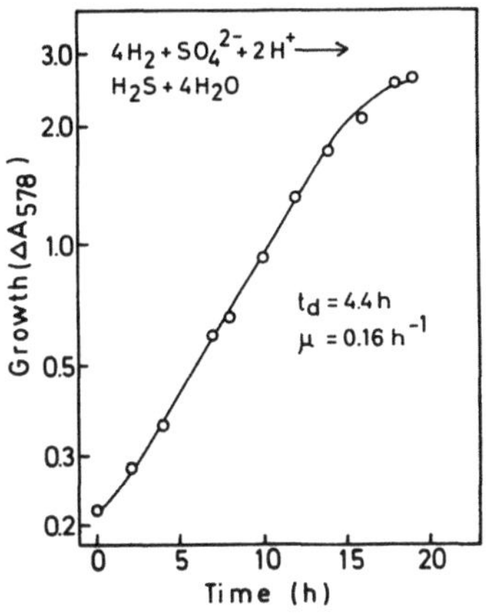

Fig. 3. Growth of *Desulfovibrio vulgaris* (Marburg) on H_2 plus sulfate as sole energy source at pH 6.55 and 35 °C. The 400 ml culture was gassed with 80%H_2/20%CO_2 at rate of 100 ml/min (Fig. 2). The medium contained 50 m*M* sulfate and 30 m*M* acetate. The pH was kept constant by automatic titration with 9 N H_2SO_4. At the end of exponential growth the cell density was 5 g wet cells/1=1 g dried cells/1 (Badziong and Thauer, 1978)

6.55; 35 °C) the bacterium grew exponentially with a doubling time of 4.4 h up to a cell density of 5 g wet cells per 1(=1 g dried cells per 1). Per mol of sulfate utilized one mol of H_2S was formed and 4.1 mol of H_2 was consumed. Acetate in the medium was shown to be required for cell carbon synthesis only rather than as electron donor; the monocarboxylic acid was not oxidized to CO_2. These findings proved without doubt that *D. vulgaris* (Marburg) can grow on H_2 plus sulfate as sole energy source and that therefore dissimilatory sulfate reduction must be coupled with the net formation of ATP (Badziong and Thauer, 1978; Badziong et al., 1978).

4 Stoichiometry of Coupling

The net amount of ATP formed per mol sulfate reduced to H_2S was determined from Y_{ATP}^{max} and $Y_{SO_4^{2-}}^{max}$ by using the following equation:

$$Y_{SO_4^{2-}}^{max} = n\, Y_{ATP}^{max})$$

where $Y_{SO_4^{2-}}^{max}$ is the growth yield in g cells per mol sulfate obtained at infinite growth rate (μ=∞), Y_{ATP}^{max} the theoretical yield in g cells per mol ATP at μ=∞, and n the number of mol ATP formed per mol sulfate. Y_{ATP}^{max} can be caldulated if the anabolism of the organism is known, $Y_{SO_4^{2-}}^{max}$ can be determined experimentally. Critical revies on the method are by Stouthamer (1976) and by Pirt (1975).

4.1 Calculation of Y_{ATP}^{max}

Desulfovibrio vulgaris (Marburg) grows on CO_2 and acetate as sole carbon sources, and with NH_3 and H_2S as N and S sources, respectively. Seventy percent of the cell carbon is derived from acetate, 30% from CO_2. The anabolism of the organism can be represented by the following equation:

$$1 \text{ Acetate} + 1\ CO_2 + 2\ H_2 \xrightarrow[\uparrow \atop ATP]{+NH_3\ +H_2S \atop \downarrow} \text{Cells.}$$

The pathway of acetate and CO_2 assimilation in *D. vulgaris* (Marburg) is depicted in Fig. 4. The scheme is based on long-term labeling data and on enzymatic studies (Badziong et al., 1979). By adding up the ATP consuming reactions it was calculated that between 11.5 and 14.5 g cells should be formed per mol ATP, provided that ATP is used for biosynthetic purposes only, which theoretically is the case when the bacterium grows at infinite growth rate. The calculated value is thus equal to Y_{ATP}^{max}. Some assumptions concerning the ATP consumption have to be made in the calculation. 11.5 g/mol was obtained if in all reactions the highest possible ATP consumption was assumed, 14.5 g/mol if minimal assumptions were made (Badziong, 1979).

4.2 Determination of $Y_{SO_4^{2-}}^{max}$

The growth yield Y is dependent on the growth rate μ via the Pirt equation (Pirt, 1965):

$$1/Y = m/\mu + 1/Y^{max},$$

where m is the maintenance coefficient (mol substrate or mol ATP consumed per hour to maintain 1 g cells). m is a constant provided the growth temperature and the growth medium composition is kept constant. The equation predicts that plots of $1/Y$ versus $1/\mu$ are linear with the intercept on the ordinate equal to $1/Y^{max}$. In order to determine $Y_{SO_4^{2-}}^{max}$ the growth yield per mol sulfate was determined in dependence of the growth rate. Via extrapolation to infinite growth rate a $Y_{SO_4^{2-}}^{max}$ of 12.2 g/mol was obtained (Badziong and Thauer, 1978) (Fig. 5). A comparison of $Y_{SO_4^{2-}}^{max}$ with the calculated Y_{ATP}^{max} of 11.5–14.5 g/mol shows that per mol sulfate as many cells were

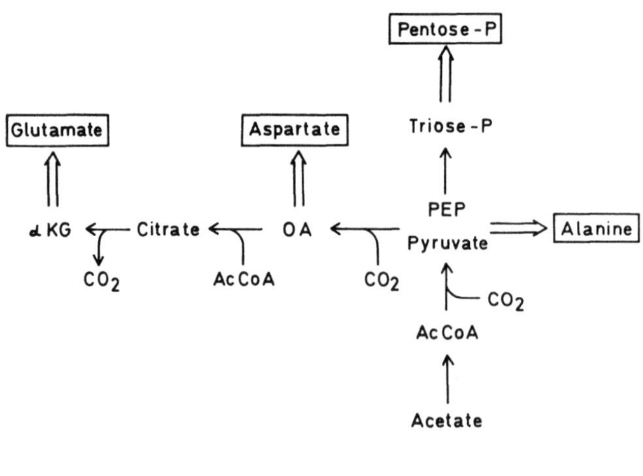

Fig. 4. Assimilation of acetate and CO_2 by *Desulfovibrio vulgaris* (Marburg) during growth on H_2 plus sulfate as sole energy source (Badziong et al., 1979)

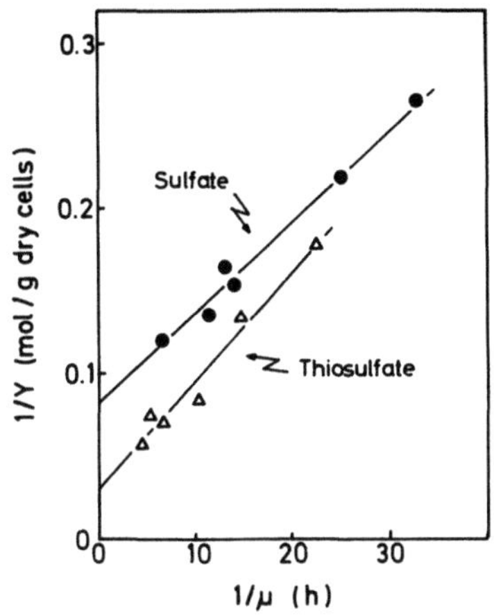

Fig. 5. Growth yields on H_2 plus sulfate and H_2 plus thiosulfate of *Desulfovibrio vulgaris* (Marburg) in dependence of the growth rate. Double reciprocal plot of Y versus μ (Badziong and Thauer, 1978)

formed as predicted per mol ATP. The finding thus indicates that dissimilatory sulfate reduction to H_2S is coupled with a net synthesis of 1 mol of ATP.

$$4H_2 + SO_4^{2-} + 2H^+ \xrightarrow[\downarrow\ 1\ ATP]{} H_2S + 4H_2O.$$

4.3 Determination of $Y^{max}_{S_2O_3^{2-}}$

The above findings indicate that the reduction of bisulfite to H_2S must be associated with the generation of 3 ATP, since 2 ATP are required for the activation of sulfate to APS. This conclusion was checked by determining the amount of ATP formed between bisulfite and H_2S directly. *D. vulgaris* (Marburg) was grown on H_2 plus thiosulfate as sole energy source and $Y_{S_2O_3^{2-}}$ was determined in dependence of the growth rate. Thiosulfate is reduced in sulfate-reducing bacteria to H_2S via bisulfite:

$$S_2O_3^{2-} \xrightarrow[\downarrow\ H_2S]{\downarrow\ 2e^-} HSO_3^- \xrightarrow[\downarrow\ ATP]{\downarrow\ 6e^-} H_2S.$$

The reduction of thiosulfate to bisulfite plus H_2S with H_2 is not associated with a change in free energy; $\Delta G_0'$ of the reaction is 0 kcal/mol (Thauer et al., 1977). The thiosulfate reductase reaction thus operates near or at equilibrium, i. e., ATP is not required for the reaction to proceed nor can the reaction be coupled with the synthesis of ATP. $Y_{S_2O_3^{2-}}$ is therefore equal to $Y_{HSO_3^-}$, which cannot directly be determined, because bisulfite at higher concentrations is toxic for the organism. From the data given in Fig. 5 a $Y^{max}_{S_2O_3^{2-}}$ of 33.5 g/mol was extrapolated, which is almost three times the growth yield obtained with sulfate. Dissimilatory reduction of bisulfite to H_2S

is thus associated with the net formation of 3 ATP, as was predicted from the stoichiometry of one to one for the coupling of sulfate reduction with ATP synthesis.

$$3H_2 + HSO_3^- + H^+ \xrightarrow[\downarrow \; 3ATP]{} H_2S + 3H_2O.$$

5 Mechanism of Coupling

It is generally assumed that coupling of a reductive process with the synthesis of ATP proceeds via a chemiosmotic mechanism. This implies that the redox process is vectorially organized over a proton-impermeable membrane (in bacteria over the cytoplasmic membrane) and that the electron chain is oriented in the membrane such that a protonmotive force is generated consisting of a pH gradient (inside alkaline) and of an electrical potential (inside negative). The protonmotive force is then used to drive the synthesis of ATP via a membrane-bound ATP synthetase (Mitchell, 1961). Having this hypothesis in mind it was determined whether the reduction of sulfate to H_2 in *D. vulgaris* (Marburg) is a vectorial redox process. For this purpose the topography of enzymes and electron carriers known to be involved in dissimilatory sulfate reduction was investigated (Badziong and Thauer, 1980).

5.1 Enzymes and Electron Carriers Involved

The following enzymes and electron carriers are involved in dissimilatory sulfate reduction with H_2 :Hydrogenase for the activation of H_2 ; cytochrome c_3 as coenzyme of hydrogenase; ATP sulfurylase for the activation of sulfate and APS reductase plus bi-

Table 1. Electron transport components determined in *Desulfovibrio vulgaris* (Marburg) grown on hydrogen plus sulfate as sole energy source in a medium containing excess iron (From Badziong and Thauer, 1980)

Component	Content per mg cell (dry weight)[a]
Hydrogenase	30 U
APS reductase	0.6 U
Desulfoviridin (= bisulfite reductase)	35 μg
Thiosulfate reductase	30 mU
Cytochrome c_3	0.3 nmol
Cytochrome *b*	0.04 nmol
Menaquinone	0.85 nmol
Ferredoxin	0.4 nmol

1 U = 1 μmol substrate oxidized or reduced per min

[a] The content of the electron transport components per mg cells (dry weight) is the sum of the content found in the periplasm-, cytoplasm-, and membrane fraction

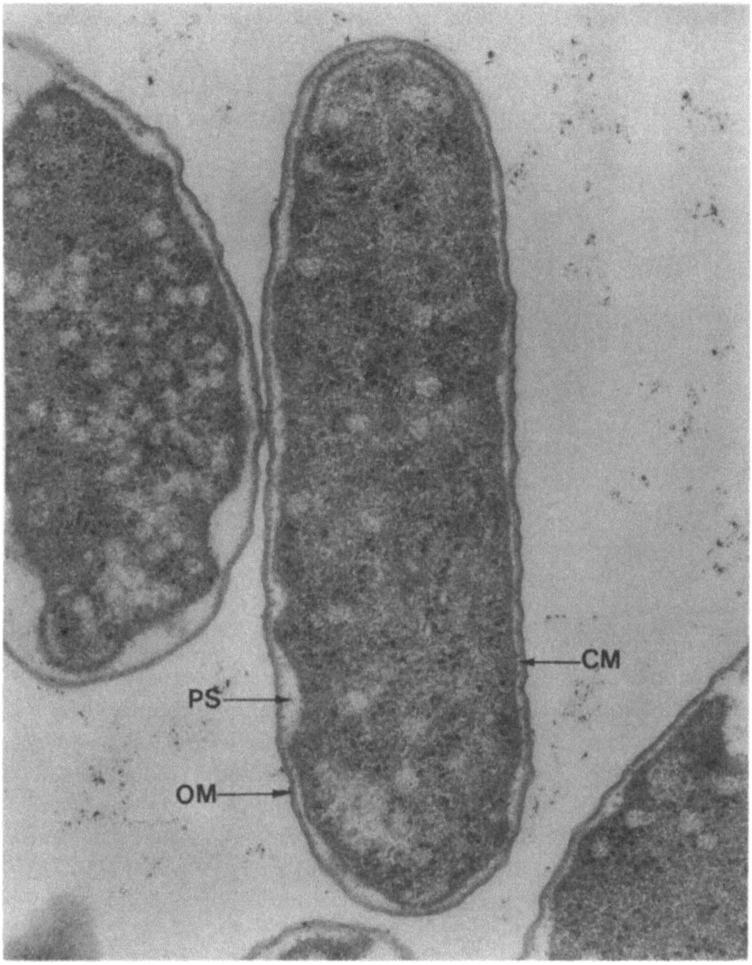

Fig. 6. Grazing section of *Desulfovibrio vulgaris* revealing outer membrane of cell wall *(OM)*, cytoplasmic membrane *(CM)* and periplasmic space *(PS)*. x43,700. The bacterium was grown on H_2 plus sulfate as sole energy source (Badziong et al., 1978)

sulfite reductase for the reduction of APS to H_2S. The membrane fraction of *D. vulgaris* (Marburg) contains cytochrome b and meanaquinone, the soluble fraction ferredoxin. The concentrations or specific activities of electron transport components in *D. vulgaris* (Marburg) are summarized in Table 1 (Badziong and Thauer, 1980).

5.2 Topography of Electron Transport Components

Desulfovibrio vulgaris (Marburg) is enveloped by an outer membrane, a murein sacculus and a cytoplasmic membrane (Fig. 6). Intracytoplasmatic membranes are missing (Badziong et al., 1978). If the reduction of sulfate to H_2S is a vectorial redox process

Table 2. Location of electron transport components in *Desulfovibrio vulgaris* (Marburg) grown on hydrogen plus sulfate as the sole energy source (from Badziong and Thauer, 1980)

Compound	Periplasm fraction %	Membrane fraction %	Cytoplasm fraction %
Hydrogenase	>95	<4	1
Cytochrome c_3	82	10	8
Cytochrome *b*	–	100	–
Menaquinone	–	100	–
Desulfoviridin (= bisulfite reductase)	4	>1	96
APS reductase	2	11	87
Thiosulfate reductase	3	23	74
Ferredoxin	29	–	71
Adenylates	51	3	46

– = not determined

then the electron transport components involved must be vectorially organized over the cytoplasmic membrane.

Hydrogenase and cytochrome c_3 were found to be located on the periplasmic site of the cytoplasmic membrane, APS-reductase, bisulfite reductase and thiosulfate on the cytoplasmic site. Cytochrome b and menaquinone are membrane components (Table 2; Fig. 7). This topography was easy to demonstrate since hydrogenase and cytochrome c_3 could quantitatively be washed off from intact cells after incubation in 50 m*M* EDTA at pH=9, whereas APS reductase, bisulfite reductase and thiosulfate reductase were only released after lysis of the cells with lysozyme. A detailed discussion of the results and of the methods employed is found in the original paper (Badziong and Thauer, 1980).

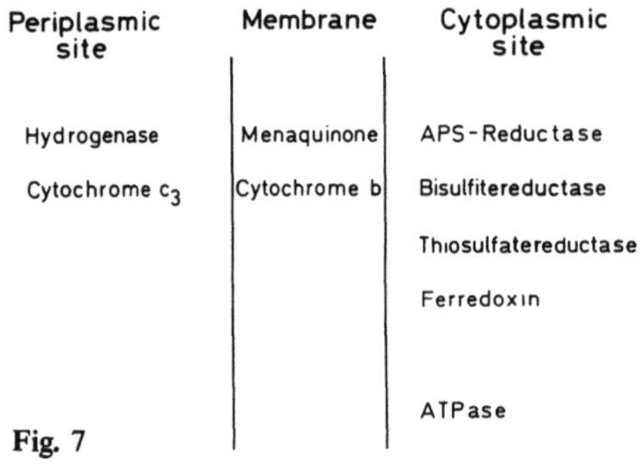

Fig. 7

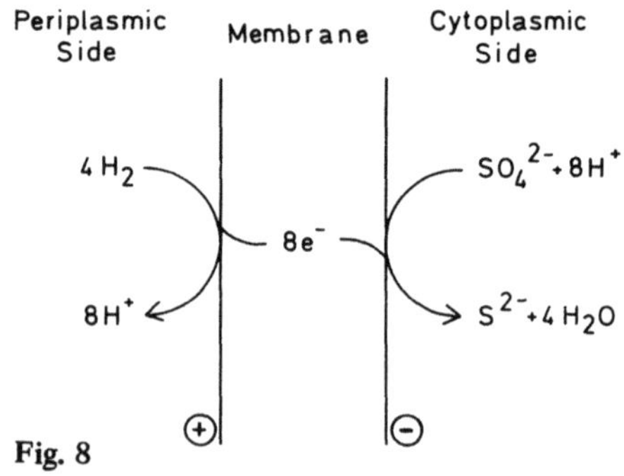

Fig. 8

Fig. 7. Topography of electron transport components in *Desulfovibrio vulgaris* (Marburg) grown on H_2 plus sulfate as sole energy source (Badziong and Thauer, 1980)

Fig. 8. Vectorial reduction of sulfate with H_2 in *Desulfovibrio vulgaris* (Marburg) (Badziong and Thauer, 1980)

5.3 Chemiosmotic Model of Coupling

From the topography of electron transport components it can be deduced that dissimilatory sulfate reduction with H_2 is a vectorial redox process with H_2 being oxidized on the periplasmic site and sulfate reduced to H_2S on the cytoplasmic site of the membrane (Fig. 8). It can therefore be assumed that the energy liberated in the redox reaction is conserved in a pH gradient and an electrical potential. This still remains to be shown.

A hypothetical scheme of the mechanism of coupling of dissimilatory sulfate reduction with the synthesis of ATP is given in Fig. 9. The scheme is based on the found stoichiometry of coupling and accounts for the observed topography of electron transport components. It is assumed that sulfate is taken up by the cell together with two protons. This could also be two sodium ions, which subsequently would have to be transported out again in an anti-port system consuming two protons. In the cell, sulfate is activated to APS consuming 2 equivalents of ATP. APS is reduced to bisulfite. This step must involve a membrane-bound redox carrier mediating both proton and electron transport, since the reduction of APS to bisulfite with H_2 is not associated with the formation of ATP and thus should not be coupled with the generation of a protonmotive force. The reduction of bisulfite to H_2S should involve redox carrier(s) mediating the transport of electrons rather than protons. The reduction of bisulfite to H_2S is thus associated with the generation of 6 protons on the periplasmic site of the membrane. The model implies that the six protons are used to drive the synthesis of 3 mol of ATP via the membrane-bound ATP synthetase assuming that 2 protons are required per mol ATP to poise the ATP synthetase. A membrane-bound ATPase with the properties of an ATP synthetase has been demonstrated in sulfate-reducing bacteria (Guarraia and Peck, 1971). The scheme contains only reactions which can be checked experimentally. This will be the subject of future investigation.

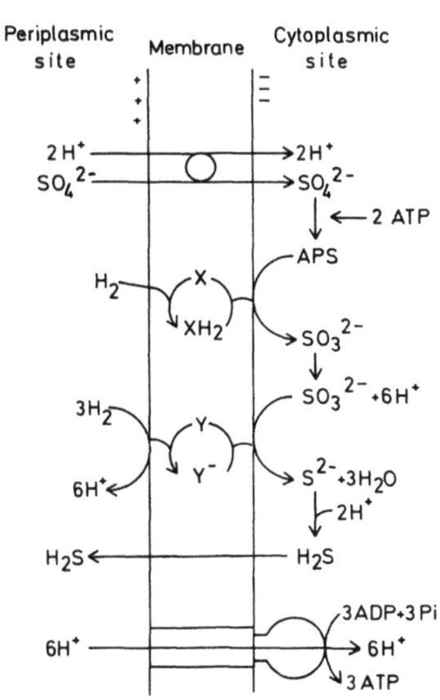

Fig. 9. A chemiosmotic model of sulfate respiration. x electron plus proton carrier; y electron carrier (Thauer and Badziong, 1980)

A chemiosmotic model for the coupling of dissimilatory sulfate reduction with the synthesis of ATP has been published by Wood (1978). It assumes that H_2 is oxidized on the cytoplasmic site of the membrane and involves a proton pump. This model can be dismissed since more than 99% of the hydrogenase can be localized on the periplasmic site rather than on the cytoplasmic site of the cytoplasmic membrane (Table 2).

References

Badziong W (1979) Dissertation Universität Marburg
Badziong W, Thauer RK (1978) Arch Microbiol 117:209–214
Badziong W, Thauer RK (1980) Arch Microbiol 125:167–174
Badziong W, Ditter B, Thauer RK (1979) Arch Microbiol 123:301–305
Badziong W, Thauer RK, Zeikus JG (1978) Arch Microbiol 116:41–49
Campbell LL, Postgate JR (1965) Bacteriol Rev 29:359–363
Guarraia LJ, Peck HD (1971) J Bacteriol 106:890–895
LeGall J, Postgate JR (1973) In: Rose AH, Tempest DW (eds) Advances in microbial physiology, vol X. Academic Press, London New York, pp 81–133
LeGall, J, DerVartanian DV, Peck HD (1979) In: Sanadi DR (ed) Current topics in bioenergetics, vol IX. Academic Press, London New York, pp 237–266
Mitchell P (1961) Nature (London) 191:144–148
Peck HD (1960) J Biol Chem 235:2734–2738
Peck HD (1966) Biochem Biophys Res Commun 22:112–118
Peck HD (1974) Symp Soc Gen Microbiol 24:241–262
Pirt SJ (1965) Proc R Soc London ser B 163:224–231
Pirt SJ (1975) Principles of microbe and cell cultivation. Blackwell Scientific Publications, Oxford London Edinburgh Melbourne
Postgate JR (1979) The sulfate-reducing bacteria. University Press, Cambridge London New York Melbourne
Postgate JR, Campbell LL (1966) Bacteriol Rev 30:732–738
Siegel LM (1975) In: Greenberg DM (ed) Metabolism of sulfur compounds, vol VII. Academic Press, London New York, pp 217–286
Sorokin YU (1966) Nature (London) 210:551–552
Stouthamer AH (1976) Yield studies in microorganisms. Meadowfield Press Ltd, Durham, England
Thauer RK, Badziong W (1980) In: Knowles CJ (ed) Diversity of bacterial respiratory systems. CRC Press Inc West Palm Beach, in press
Thauer RK, Jungermann K, Decker K (1977) Bacteriol Rev 41:100–180
Widdel F (1980) Dissertation Universität Göttingen
Wood PM (1978) FEBS Lett 95:12–18

Photolithotrophic Sulfur Oxidation

H.G. TRÜPER[1]

1 Introduction

The utilization of sulfur compounds (at oxidation levels below that of sulfate) as electron donors in anoxygenic photosynthesis is – though to a different extent – common to most groups of phototrophic prokaryotes (cf. Tables 1 and 2). Classical in this respect are the purple and green sulfur bacteria (Chromatiaceae and Chlorobiaceae) all of which utilize reduced sulfur compounds as electron donors. But also a number of classical "purple-nonsulfur" bacteria (Rhodospirillaceae) and the filamentous flexible green bacteria (Chloroflexaceae) are able to oxidize reduced sulfur compounds during photosynthesis. Even certain species of the cyanobacteria can perform anoxygenic photosynthesis at the expense of sulfide as electron donor.

Recent studies on phylogenetic relationships of phototrophic bacteria have revealed some surprising data (Stackebrandt and Woese, 1979; Gibson et al., 1979; Dickerson, 1980; Woese et al., 1980). From these studies one may derive that while the cyanobacteria and the green bacteria (Chlorobiales) each form a phylogenetic cluster different from the purple bacteria (Rhodospirillales) the latter constitute a major phylogenetic unit – apparently as extensive as the gram-positive bacteria – that is genealogically intermixed with nonphototrophic gram-negative organisms (Gibson et al., 1979). The relationships found between nonphototrophic species and phototrophic ones were often closer than the relationships within the Rhodospirillales. These findings are apt to throw doubt upon the opinion that, e. g., sulfur metabolism in phototrophic prokaryotes may follow a uniform pattern – an opinion based on the observations of sulfide and sulfur metabolism by whole cells.

Therefore I shall try in this review to point out the specific differences that so far have been revealed by research of photolithotrophic sulfur metabolism. A comprehensive presentation of sulfur metabolism in phototrophic bacteria was published in 1978 (Trüper, 1978).

1 Institut für Mikrobiologie, Rheinische Friedrich-Wilhelm-Universität, 5300 Bonn 1/FRG

Table 1. Utilization of reduced sulfur compounds as photosynthetic electron donors by Rhodospirillales

Organisms	S^{2-}	S^{0}	$S_2O_3^{2-}$	SO_3^{2-}
Chromatiaceae				
Chromatium okenii	+	+	–	–
Chromatium weissei	+	+	–	–
Chromatium warmingii	+	+	–	–
Chromatium buderi	+	+	–	–
Chromatium minus	+	+	+	o
Chromatium vinosum	+	+	+	+
Chromatium violascens	+	+	+	o
Chromatium minutissimum	+	+	+	o
Chromatium gracile	+	+	+	o
Chromatium purpuratum[a]	+	+	+	o
Thiospirillum jenense	+	+	–	–
Thiocystis violacea	+	+	+/–	+/–
Thiocystis gelatinosa	+	+	–	o
Thiocapsa roseopersicina	+	+	+	–
Thiocapsa pfennigii	+	+	–	o
Lamprocystis roseopersicina	+	+	o	o
Thiodictyon elegans	+	+	o	o
Thiodictyon bacillosum	+	+	o	o
Thiopedia rosea	+	+	o	o
Amoebobacter roseus	+	+	+	+
Amoebobacter pendens	+	+	+	+
Ectothiorhodospira mobilis	+	+	+	+
Ectothiorhodospira shaposhnikovii	+	+	+	+
Ectothiorhodospira halophila	+	+	+	o
Ectothiorhodospira halochloris[b]	+	+	–	o
Ectothiorhodospira abdelmalekii[c]	+	+	–	o
Rhodospirillaceae				
Rhodospirillum rubrum	(+)	–	–	–
Rhodospirillum tenue	o	–	–	–
Rhodospirillum fulvum	o	–	–	–
Rhodospirillum molischianum	o	–	–	–
Rhodospirillum photometricum	o	–	–	–
Rhodopseudomonas palustris	(+)	–	+	o
Rhodopseudomonas viridis	–	–	–	o
Rhodopseudomonas sulfoviridis	+	–	+	o
Rhodopseudomonas acidophila	–	–	–	o
Rhodopseudomonas gelatinosa	o	o	–	o
Rhodopseudomonas capsulata	(+)	–	–	o
Rhodopseudomonas sulfidophila	+	–	+	o
Rhodopseudomonas sphaeroides	o	o	–	o
Rhodopseudomonas globiformis	o	o	+	o
Rhodocyclus purpureus[d]	–	o	–	o
Rhodomicrobium vannielii	(+)	o	–	o

+, utilized; –, not utilized; o, not tested; +/–, strain differences; (+), in continuous culture at sulfide limitation

If not otherwise indicated, data were taken from Pfennig and Trüper (1974, 1978)

[a] Imhoff and Trüper (1980)

[b] Imhoff and Trüper (1977)

[c] Imhoff and Trüper, description of new species in press

[d] Pfennig (1978)

Table 2. Utilization of reduced sulfur compounds as photosynthetic electron donors by Chlorobiales and Cyanobacteria

Organisms	S^{2-}	S^{o}	$S_2O_3^{2-}$
Chlorobiaceae			
Chlorobium limicola	+	+	–
– f. *thiosulfatophilum*	+	+	+
Chlorobium vibrioforme	+	+	–
– f. *thiosulfatophilum*	+	+	+
Chlorobium chlorovibrioides[a]	+	+	–
Chlorobium phaeovibrioides	+	+	–
Chlorobium phaeobacteroides	+	+	–
Prosthecochloris aestuarii	+	+	–
Prosthecochloris phaeoasteroides[b]	+	+	–
Pelodictyon chlathratiforme	+	+	–
Pelodictyon luteolum	+	+	–
Pelodictyon phaeum	+	+	–
Ancalochloris perfilievii	+	+	–
Chloroflexaceae			
Chloroflexus aurantiacus	+	o	o
Chloronema giganteum[c]	+	o	o
Oscillochloris chrysea[d]	+	o	o
Cyanobacteria			
Oscillatoria limnetica[e]	+	–	o
Oscillatoria salina[f]	+	o	o
Oscillatoria sp. (2)[f]	+	o	o
Oscillatoria amphigranulata[g]	+	o	o
Aphanothece halophytica[f]	+	–	o
Lyngbya sp. (2)[f]	+	o	o
Microcoleus sp.[f]	+	o	o
Phormidium sp. (3)[f]	+	o	o
Synechococcus sp.[h]	+	o	o
Spirulina labyrinthiformis[i]	+	o	o
Anacystis nidulans[k]	+	o	+

+, utilized; –, not utilized; o, not tested. If not otherwise indicated, data were taken from Pfennig and Trüper (1974, 1978).

[a] Gorlenko et al. (1974)
[b] Puchkova and Gorlenko (1976)
[c] Dubinina and Gorlenko (1975)
[d] Gorlenko and Pivovarova (1977)
[e] Cohen et al. (1975)
[f] Garlick et al. (1977)
[g] Castenholz (1976)
[h] Imhoff et al. (1978)
[i] Castenholz (1977)
[k] Peschek (1978)

2 Anaerobic Sulfur Transformations Observed with Whole Cells in the Light

2.1 Oxidation of Sulfide

During phototrophic growth with sulfide different oxidation patterns have been found with respect to the detectable intermediate and end products (Table 3).

Where elemental sulfur occurs as an intermediate it is deposited as "plastic sulfur" (Hageage et al., 1970) either inside or outside the cells. The assumption of Puchkova et al. (1974) that the Chlorobiaceae excrete elemental sulfur globules by means of cytoplasmic membrane sacs still appears rather doubtful, especially in comparison with Remsen's (1978) hypothesis of sulfur globule formation in *Chromatium*. Remsen (1978) reported that the nonunit membrane surrounding the sulfur globules (Schmidt et al., 1971) may originate from the outer layer of the (unit) cytoplasmic membrane.

A new pattern of sulfide utilization was found by Schedel (1977): When cell suspensions of *Chlorobium limicola* f. *thiosulfatophilum* were (anaerobically) supplied with sulfide, first thiosulfate was formed, later also elemental sulfur appeared. Both

Table 3. Detectable products of anaerobic photooxidation of sulfide by whole cells of phototrophic bacteria

Organisms in groups	Intermediate	End products
1 Ectothiorhodospira (all species)	S^o (extr.)	SO_4^{2-}
2 All other Chromatiaceae species	S^o (intr.)	SO_4^{2-}
3 *Rhodospirillum rubrum*	–	S^o (extr.)
Rhodopseudomonas capsulata	–	S^o (extr.)
Rhodopseudomonas sphaeroides	–	S^o (extr.)
4 *Rhodopseudomonas palustris*	–	SO_4^{2-}
Rhodopseudomonas sulfidophila	–	SO_4^{2-}
5 *Rhodomicrobium vannielli*	{ –	$S_4O_6^{2-}$
		$(S_2O_3^{2-} + S^o)$[a]
6 Chlorobiaceae (species lacking thiosulfate utilization)	S^o (extr.)	SO_4^{2-}
7 *Chlorobium limicola* f. *thiosulfatophilum*[b]	$S^o + S_2O_3^{2-}$	SO_4^{2-}
8 Chloroflexaceae[c]	{ [S^o (extr.)]	S^o (extr.)
		(SO_4^{2-})
9 *Oscillatoria limnetica*	–	S^o (extr.)
Aphanothece halophytica	–	S^o
10 *Anacystis nidulans*	–	S^o (intr.)

For the references of data cf. Table 1 and 2

[a] In batch culture only; cf. Hansen (1974)

[b] Schedel (1977)

[c] As yet sulfate has not been clearly proven to be the end product

continued to be formed until the sulfide was exhausted. Then thiosulfate and elemental sulfur were oxidized to sulfate, the latter at a much slower rate. Tetrathionate was not formed.

2.2 Oxidation of Elemental Sulfur

If a phototrophic bacterium is able to utilize elemental sulfur as electron donor for photosynthesis (cf. Tables 1 and 2) it will oxidize it to sulfate. Other sulfur compounds have never been detected under these conditions. All species that store sulfur inside the cells are also able to oxidize it to sulfate and, in addition, are also capable of oxidizing externally supplied elemental sulfur. Organisms with intermediate extracellular sulfur formation such as the Chlorobiaceae and *Ectothiorhodospira* always completely utilize sulfide from their nutrient solution before they start to further oxidize the elemental sulfur formed (Trüper, 1978). Organisms with intracellular sulfur storage, such as *Chromatium,* start elemental sulfur oxidation before the external sulfide is totally oxidized (Trüper, 1964, 1978). The oxidation of elemental sulfur therefore appears to be inhibited by external sulfide. The site of sulfur oxidation in the Chlorobiaceae and Ectothiorhodospira is obviously the cytoplasmic membrane whereas in *Chromatium*-like organisms the sulfur-oxidizing enzyme(s) must be localized further inside the cells in order to reach the stored sulfur globules. The consequence is that in these bacteria the sulfur-oxidizing enzyme is not completely blocked by extracellular sulfide.

When extracellular sulfur is oxidized by *Chlorobium limicola* f. *thiosulfatophilum,* only sulfate appears as the product, whereas neither thiosulfate nor sulfide could be found (Schedel, 1977).

2.3 Disproportionation of Elemental Sulfur

Paschinger et al. (1974) found that *Chlorobium limicola* f. *thiosulfatophilum* formed sulfide and sulfate simultaneously when the cells were incubated in the light but without CO_2 under an inert atmosphere. Further investigation of this phenomenon in our laboratory (Lorenz and Trüper, unpubl.) showed that the capability of sulfur disproportionation of elemental sulfur is restricted to the thiosulfate-utilizing *formae* of *Chlorobium limicola* and *Chl. vibrioforme,* whereas all other Chlorobiaceae as well as *Ectothiorhodospira mobilis* and *Chromatium vinosum* are incapable of this process. The reaction occurs only anaerobically in the light in the absence of CO_2 and only as long as hydrogen sulfide is purged from the suspensions by argon or nitrogen. It was shown that not sulfate but thiosulfate is the oxidized product of the reaction(s) that follow an overall equation of

$$4S + 3H_2O \rightarrow 2H_2S + H_2S_2O_3.$$

Rather small but measurable amounts of sulfite indicated that the thiosulfate might have been formed by nonenzymatic condensation of sulfite with sulfur (Trüper et al., 1976). As the disproportionation ceases immediately upon the addition of CO_2

it seems that elemental sulfur may mimic the role of CO_2 as an electron acceptor in these bacteria.

2.4 Oxidation of Thiosulfate

The utilization of thiosulfate as a photosynthetic electron donor is more often found in Chromatiaceae and Rhodospirillaceae (Table 1) than in Chlorobiaceae and cyanobacteria (Table 2) although for the latter group only few data are available. Under anaerobic conditions in the light thiosulfate is either split into the sulfane- and the sulfone-group or oxidized to tetrathionate (Table 4).

In thiosulfate-utilizing species of the Chromatiaceae thiosulfate utilization is constitutive (Hashwa, 1972) whereas about *Rps. palustris* contradictory results have been reported (inducible: Rolls and Lindstrom, 1967; Rodova and Pedan, 1980; constitutive: Hashwa, 1972). In this bacterium tetrathionate is formed from thiosulfate only when the initial concentration of the latter is 10–20 m*M*; at 2–5 m*M* thiosulfate only sulfate is found (Rodova and Pedan, 1980).

Smith (1966) found that in *Chromatium vinosum* the splitting of thiosulfate, i. e., the formation of visible intracellular sulfur, is inhibited by tetrathionate.

The oxidation of tetrathionate as a photosynthetic electron donor has so far only been reported for *Chl. limicola* f. *thiosulfatophilum* (Larsen, 1952).

Table 4. Detectable products of anaerobic photooxidation of thiosulfate by whole cells of phototrophic bacteria

Organisms	Intermediate	End products
1 *Chromatium vinosum*[a]	S^o (intr.)	SO_4^{2-}
Chromatium vinosum[b]	–	$S_4O_6^{2-}$
2 *Rhodopseudomonas palustris*[c]	–	SO_4^{2-}
Rhodopseudomonas palustris[d]	–	$S_4O_6^{2-}+SO_4^{2-}$
3 *Rhodopseudomonas globiformis*[e]	–	$S_4O_6^{2-}$
4 *Chlorobium limicola* f. *thiosulfatophilum*[f]	–	SO_4^{2-}
5 *Chlorobium vibrioforme* f. *thiosulfatophilum*[g]	S^o (extr.)	SO_4^{2-}

[a] At pH 7.3 (Smith, 1966)
[b] At pH 6.25 (Smith, 1966)
[c] At thiosulfate concentration below 5 m*M* (Rodova and Pedan, 1980)
[d] At thiosulfate concentration above 10 m*M* (Rodova and Pedan, 1980)
[e] Then and Trüper (unpubl.)
[f] Schedel (1977)
[g] Trüper (unpubl.)

2.5 Oxidation of Sulfite

Only very few phototrophic bacteria have been reported to utilize sulfite as electron donor in photosynthesis (Table 1). So far no quantitative studies have been reported. Because of the toxicity and instability in nutrient solutions of sulfite such studies would have to be made in chemostat cultures.

Chl. limicola f. *thiosulfatophilum* is unable to utilize sulfite (Larsen, 1952).

3 Enzymatic Steps of Anaerobic Oxidative Sulfur Transformations

3.1 Sulfide to Sulfur (S^o)

While Petushkova and Ivanovskii (1976a) found that the formation of S^o from sulfide in *Thiocapsa roseopersicina* is catalyzed in the soluble fraction and is a heat-stabile process, Fischer and Trüper (1977) isolated a heat-stabile cytochrome c-550 from the soluble fraction. Oxidized cytochrome c-550 easily oxidizes H_2S to S^o. The molecular weight is 34,000. In this connection it is interesting that Fauque et al. (1979) identified cytochrome c_3 of *Desulfovibrio desulfuricans,* strain Norway (molecular weight 9,200) as sulfur reductase.

Schedel (1977) studied anaerobic sulfur metabolism in *Thiobacillus denitrificans* and found that here sulfide is oxidized by cytochrome c-552.

Therefore in our opinion there is a good chance of cytochromes being involved in the step $S^{2-} \rightleftharpoons S^o$ in phototrophic bacteria.

Also the flavocytochromes c-552 of *C. vinosum* and c-553 of *Chl. limicola* f. *thiosulfatophilum* are likely candidates to participate in in vivo sulfide oxidation. Kusai and Yamanaka (1973 a,c) reported that flavocytochrome c-553 of *Chl. limicola* f. *thiosulfatophilum* functions as a sulfide : cytochrome c reductase. The reaction is strongly inhibited by cyanide. Flavocytochrome c-553, as well as flavocytochrome c-552 from *C. vinosum,* are heat-labile and are reduced by sulfide under formation of thiosulfate (under strict oxygen-free conditions) (Fischer, 1977). As mentioned above, cells of *Chl. limicola* f. *thiosulfatophilum* form thiosulfate (and S^o) from sulfide during photosynthesis.

Fukumori and Yamanaka (1979a) found that flavocytochrome c-552 from *C. vinosum* also has sulfide : cytochrome c reductase activity and also catalyzes the reduction of elemental sulfur to sulfide with reduced benzyl viologen as the electron donor. Their method of measuring the latter function, however, is – as we have found – much less reliable than that of Fauque et al. (1979).

It is not yet clear whether the sulfide-oxidizing system in *Chl. limicola* f. *thiosulfatophilum* in vivo is particle-bound. Evans and Buchanan (1965) studied light-dependent reductive carboxylation of acetyl-CoA to pyruvate with a particulate fraction and found that sulfide was utilized as the electron donor. From studies with "chromatophores" from *Chl. limicola* f. *thiosulfatophilum* and *C. vinosum,* Knaff and Buchanan (1975) suggested that bound cytochrome b might be involved in noncyclic electron flow from sulfide to NADP in *Chlorobium* "chromatophores," i. e., pieces of the disrupted cytoplasmic membrane. As so far no cytochrome b has been isolated from *Chlorobium,* the significance of these suggestions remains open.

3.2 Thiosulfate Splitting

The enzyme rhodanese (thiosulfate : cyanide sulfur transferase) has been shown to occur in *C. vinosum, Rps. sphaeroides, R. rubrum* (Smith and Lascelles, 1966), *Rps. capsulata, Rps. palustris, Ectothiorhodospira mobilis, Chl. limicola* f. *thiosulfatophilum* (Yoch and Lindstrom, 1971), *C. minus* (Hashwa, 1972), *Rps. sulfidophila, Rps. globiformis, Rps. viridis, R. fulvum, Thiocapsa roseopersicina, Thiocystis gelatinosa, C. warmingii* (Prangenberg, 1976). Because cyanide is unlikely to participate in an in vivo reaction the question remains which in vivo reaction is catalyzed by the activity measurable as rhodanese.

Another thiosulfate-splitting enzyme found in phototrophic bacteria is thiosulfate reductase (Hashwa, 1972; Hashwa and Pfennig, 1972). This enzyme has so far been found in *C. vinosum, C. minus, Rps. palustris* (Hashwa, 1972, 1975), *Rps. sulfidophila, Rps. globiformis, Rps. capsulata, Rps. viridis, R. rubrum, R. fulvum, Thiocapsa roseopersicina, Thiocystis gelatinosa, C. warmingii, Chl. limicola* f. *thiosulfatophilum* (Prangenberg, 1976). Neither rhodanese nor thiosulfate reductase were found in *Chl. limicola* and *R. photometricum* (Prangenberg, 1976).

Thiosulfate reductase reacts with reduced benzyl viologen, cysteine, reduced glutathione, dihydrolipoic acid, reduced lipoamide, and dithioerythrol as electron donors (Hashwa, 1972, 1975; Prangenberg, 1976).

Experiments to prove clearly either identity or nonidentity of the two enzymes have failed so far. Ammonium sulfate fractionations, gel electrophoresis, gel filtrations, and ion exchange chromatography (Hashwa, 1975; Prangenberg, 1976; Giani, 1976) only revealed that both enzyme activities may be found in a variety of fractions and different subunits.

Rhodanese of *C. vinosum* occurs in three fractions of different molecular weights (10,000, 35,000–40,000, 150,000–200,000) (Giani, 1976) and in this respect is similar to rhodanese from the nonphototrophic *Thiobacillus denitrificans* (Bowen et al., 1965: subunit 9,000, monomer 36,000, polymer 120,000).

For the latter organism it was clearly demonstrated by Schedel (1977) and Schedel and Trüper (1980) that the enzyme responsible for thiosulfate-splitting is identical with the protein that exhibits rhodanese activity.

3.3 Sulfide/Sulfur to Sulfite

In *C. vinosum* Schedel (1977) found a siroheme sulfite reductase; the properties of this enzyme indicate that in vivo it is responsible for the oxidation of sulfide and elemental sulfur to sulfite. The enzyme occurred in the cells only after autotrophic growth in a sulfide/bicarbonate medium, not, however, after heterotrophic (malate/sulfate) growth. It is located in the soluble fraction. The molecular weight (in phosphate buffer) was determined as 280,000 consisting of two different subunits, α (MW 37,000) and β (MW 42,000), arranged in an $\alpha_4\beta_4$ structure (Schedel, 1977; Schedel et al., 1979).

Kobayashi et al. (1978) also isolated a sulfite reductase from *C. vinosum* and estimated a molecular weight of 180,000. As they worked in tris buffer the difference in molecular weights may be explained by a different aggregation behavior of the subunits in different buffers. Other properties have been found to be rather similar to Schedel's (1977) results. The enzyme contains siroheme as prosthetic group, and 51 Fe atoms plus 47 labile S atoms per molecule (Schedel, unpubl.).

Its general properties render it rather similar to the dissimilatory siroheme sulfite reductases of *Desulfovibrio vulgaris* (Lee et al., 1973; Kobayashi et al., 1972, 1974) and *Thiobacillus denitrificans* (Schedel and Trüper, 1980), where the subunits are arranged in an $\alpha_2\beta_2$ structure and the molecular weight is between 160,000 and 226,000. As in *Thiobacillus denitrificans* (Schedel and Trüper, 1979) in *C. vinosum* the physiological role of sulfite reductase must be seen in the oxidation of sulfide (and elemental sulfur, which in the outer layers of the globules is not present as S_8 but as polysulfane) to sulfite. Although this function has not yet been proven experimentally because no test system is available, the following arguments support this thesis: (1) reversibility of the sulfite reductase reaction; (2) high content of the enzyme based on total cellular protein which indicates a dissimilatory function; (3) analogy to APS reductase which in sulfate-reducing bacteria catalyzes APS reduction and in phototrophic sulfur bacteria and thiobacilli catalyzes the oxidation of sulfite to APS; (4) in *C. vinosum* the enzyme only occurred in autotrophic cells.

In *Chl. limicola* f. *thiosulfatophilum* (Schedel, 1977) and in *Thiocapsa roseopersicina* (Jorzig, 1979) no siroheme-containing sulfite reductase could be found in spite of numerous attempts. In the latter organism sulfite reductase activities were, however, found in particulate as well as in soluble fractions, which might indicate a different cellular localization of a (probably nonsiroheme) sulfite reductase.

Theoretically, sulfite formation could also occur simply by the combined action of a flavocytochrome c (cf. above) and a thiosulfate-splitting enzyme:

$$H_2S \rightarrow S_2O_3^{2-} \rightarrow SO_3^{2-}$$

(with an arrow returning from the second step to H_2S)

This would allow phototrophic sulfur bacteria to exist without a reversed sulfite reductase.

3.4 Sulfite to Adenylylsulfate (APS)

APS reductase, well known as an essential enzyme in sulfate-reducing bacteria and *Thiobacillus* species, also occurs in phototrophic sulfur bacteria (Thiele, 1968; Trüper and Peck, 1970). The enzyme was purified so far from *Thiocapsa roseopersicina* (Trüper and Rogers, 1971) and *Chl. limicola* f. *thiosulfatophilum* (Kirchhoff and Trüper, 1974). In *C. vinosum,* it is firmly membrane-bound and could not be solubilized so far although kinetic studies could be done (Schwenn and Biere, 1979). It was partly purified from *Thiocystis violacea* (Tatzki, 1979). Although Trüper and Peck (1970) reported the formation of APS in crude extracts of *Ectothiorhodospira mobilis,* we have not found APS reductase in this and the other *Ectothiorhodospira* species. Also *Rps. sulfidophila* lacks APS reductase.

APS reductases have molecular weights of 180,000–220,000. The enzymes from *Thiocapsa roseopersicina* and *Thiocystis violacea* contain heme groups, flavin, and nonheme iron sulfur clusters (Trüper and Rogers, 1971; Tatzki, 1979), the enzyme from *Chlorobium* lacks heme groups.

In *C. vinosum* (and, perhaps, other phototrophic bacteria) APS reductase activity depends on the growth phase. Schwenn and Biere (1979) observed high enzyme activity only at the late exponential phase and a drastic drop of activity in the stationary phase.

In the Rhodospirillaceae, APS reductase is absent.

3.5 Adenylylsulfate to Sulfate

This step is catalyzed by ADP sulfurylase or – theoretically possible – ATP sulfurylase. All organisms that assimilate sulfate for amino acid or sulfolipid biosyntheses possess ATP sulfurylase which is considered an assimilatory enzyme. In phototrophic bacteria ADP sulfurylase was first observed in crude extracts of *Thiocapsa roseopersicina* by Thiele (1968). Enzyme activity was further found in particle-free extracts of *C. vinosum, Thiocystis violacea* and *Chl. limicola* f. *thiosulfatophilum*, not, however, in *Ectothiorhodospira mobilis* and *E. shaposhnikovii* (Schug, 1979).

Also *Rps. sulfidophila* lacks ADP sulfurylase, and under photolithotrophic conditions ATP sulfurylase is fully repressed in this organism (Cooper, 1979).

Although studies with the purified enzyme are still lacking, the present status allows the conclusion that species containing APS reductase also contain ADP sulfurylase, (e. g., *C. vinosum, Thiocapsa roseopersicina*) whereas species lacking APS reductase also lack ADP sulfurylase (e. g., *E. mobilis, Rps. sulfidophila*).

3.6 Sulfite to Sulfate

When photosynthetic sulfur bacteria exist that do not contain the APS reductase pathway the most probable – though less energy-conserving – way of anaerobic sulfite oxidation in such bacteria would be a sulfite : acceptor oxidoreductase. In crude extracts of *Thiocapsa roseopersicina* Petushkova and Ivanovskii (1976 a,b) measured an AMP-independent sulfite-oxidizing enzyme (electron acceptor : cytochrome c) besides (AMP-dependent) APS reductase.

Brückenhaus (1977) partially purified such an enzyme from the soluble fraction of *C. vinosum* and determined the molecular weight as 70,000. The absorption spectrum of the enzyme reveals features of a cytochrome-like heme; the activity was maximal in the stationary growth phase. Also *Thiocystis violacea* (Tatzki, 1979), *Ectothiorhodospira mobilis, E. shaposhnikovii* (Imhoff, unpubl.), and *C. purpuratum* (Ulbricht and Trüper, unpubl.) contain sulfite : acceptor oxidoreductases.

3.7 Thiosulfate to Tetrathionate

Chl. limicola f. *thiosulfatophilum* contains a thiosulfate : cytochrome c oxidoreductase that was detected by Mathewson et al. (1968) and first purified by Kusai and Yamanaka (1973 b,c). This enzyme catalyzes the reduction of *Chlorobium* cytochrome c-551 by thiosulfate. Reduced cytochrome c-551 then reduces cytochrome c-555. The latter was found to stimulate the reduction of cytochrome c-551 by thiosulfate (Kusai and Yamanaka, 1973 c).

In *C. vinosum* the formation of tetrathionate from thiosulfate was discovered by Smith (1966), who partly purified the responsible enzyme using ferricyanide as the electron acceptor. The pH optimum was found to be 5.0. Recently Fukumori and Yamanaka (1979 b) reported that the partly purified enzyme rapidly reduced HiPIP (high-potential nonheme iron protein) from the same organism in the presence of thiosulfate. Also cytochromes c of yeast and tuna were suitable electron acceptors like ferricyanide, whereas horse cytochrome c was a poor acceptor, and cytochromes c-552, c′ (from *C. vinosum*) and c-553 (from *Chl. limicola* f. *thiosulfatophilum*) did not act as electron acceptors. The enzyme was inhibited by cyanide and sulfite. The authors inferred that tetrathionate was the oxidized product of thiosulfate.

In *Rps. palustris* thiosulfate is oxidized anaerobically to tetrathionate (Rolls and Lindstrom, 1967). Eley et al. (1971) and Knobloch et al. (1971) reported the presence of a thiosulfate : cytochrome c oxidoreductase in a cytochrome-containing fraction isolated from thiosulfate-grown *Rps. palustris.* Further study of this enzyme has been communicated by Appelt and Knobloch (1977) and Appelt et al. (1979). Also in *Rps. globiformis* a thiosulfate: acceptor oxidoreductase has been found to catalyze the formation of tetrathionate from thiosulfate (Then and Trüper, unpubl.).

4 Conclusion

Photolithotrophic sulfur metabolism is one aspect of sulfur metabolism in phototrophic prokaryotes. The others are assimilatory sulfur metabolism, dark fermentative sulfur metabolism, and chemolithotrophic sulfur metabolism.

The work of Gibson (1967), Uspenskaya and Kondratieva (1972), Gorlenko (1974), Kondratieva et al. (1975, 1976 a,b), Petushkova and Ivanovskii (1976 a), Krassilnikova et al. (1980), and finally the big survey of Kämpf and Pfennig (1980) has shown that many of the Chromatiaceae but none of the Chlorobiaceae are able to grow in the dark at more or less reduced oxygen tensions, oxidizing reduced sulfur compounds and fixing CO_2, i. e., they are facultative chemolithoautotrophic bacteria and in their aerobic sulfur metabolism probably resemble thiobacilli.

From the results reviewed above the following picture is emerging:

1. The four aspects of sulfur metabolism cannot be viewed separately from each other. There are pathways that may be used both in aerobic and anaerobic, or both assimilatory and dissimilatory metabolism or that may function in both directions, oxidative and reductive.

2. Cytochromes and other heme-containing enzymes are centrally involved in photolithotrophic anaerobic oxidation of reduced sulfur compounds.

3. Principally different dissimilatory pathways exist in
   a) oxidation of sulfite to sulfate (direct or via APS)
   b) utilization of thiosulfate (splitting or oxidation to tetrathionate)
   c) sulfide/sulfur oxidation (siroheme sulfite reductase or other mechanisms)

4. Different dissimilatory pathways may be found in one organism, and may depend upon different physiological conditions and ecological situations.

References

Appelt N, Knobloch K (1977) Hoppe Seiler's Z Physiol Chem 358:1174–1175
Appelt N, Weber H, Wieluck S, Knobloch K (1979) Ber Dtsch Bot Ges 92:365–378
Bowen TJ, Butler PJ, Happold FC (1965) Biochem J 97:651–657
Brückenhaus I (1977) Sulfitoxidation bei einigen Arten phototropher Schwefelbakterien. Dipl thesis, Univ Bonn
Castenholz RW (1976) J Phycol 12:54–68
Castenholz RW (1977) Microb Ecol 3:79–105
Cohen Y, Padan E, Shilo M (1975) J Bacteriol 123:855–861
Cooper BP (1979) Der enzymatische Mechanismus der Sulfataktivierung bei Rhodospirillaceae. Doc thesis, Univ Bonn
Dickerson RE (1980) Nature (London) 283:210–212
Dubinina GA, Gorlenko VM (1975) Mikrobiologiya 44:511–517
Eley JH, Knobloch K, Aleem MIH (1971) Arch Biochem Biophys 14:419–429
Evans MCW, Buchanan BB (1965) Proc Natl Acad Sci USA 53:1420
Fauque G, Herve D, LeGall J (1979) Arch Microbiol 121:261–264
Fischer U (1977) Die Rolle von Cytochromen im Schwefelstoffwechsel phototropher Schwefelbakterien. Doc thesis, Univ Bonn
Fischer U, Trüper HG (1977) FEMS Microbiol Lett 1:87–90
Fukumori Y, Yamanaka T (1979a) J Biochem (Tokyo) 85:1405–1414
Fukumori Y, Yamanaka T (1979b) Curr Microbiol 3:117–120
Garlick S, Oren A, Padan E (1977) J Bacteriol 129:623–629
Giani D (1976) Rhodanese- und Thiosulfatreduktase-Aktivitäten in Extrakten phototropher Bakterien: Gel-Elektrophorese-Verhalten und Molekulargewichte. Dipl thesis, Univ Bonn
Gibson J (1967) Arch Mikrobiol 59:104–112
Gibson J, Stackebrandt E, Zablen LB, Gupta R, Woese CR (1979) Curr Microbiol 3:59–64
Gorlenko VM (1974) Mikrobiologiya 43:729–731
Gorlenko VM, Pivovarova TA (1977) Izv Akad Nauk SSSR Ser Biol 23:396–409
Gorlenko VM, Chebotarev EN, Kachalkin VI (1974) Mikrobiologiya 43:908–914
Hageage GJ, Eanes ED, Gherna RL (1970) J Bacteriol 101:464–469
Hansen TA (1974) Sulfide als electrondonor voor Rhodospirillaceae. Doc thesis, Univ Groningen, Netherlands
Hashwa F (1972) Die enzymatische Thiosulfatspaltung bei phototrophen Bakterien. Doct thesis, Univ Göttingen
Hashwa F (1975) Plant Soil 43:41–47
Hashwa F, Pfennig N (1972) Arch Mikrobiol 81:36–44
Imhoff JF, Trüper HG (1977) Arch Microbiol 114:115–121
Imhoff JF, Trüper HG (1980) Zentralbl Bakteriol Abt 1 Orig Reihe C1:61–69
Imhoff JF, Hashwa F, Trüper HG (1978) Arch Hydrobiol 84:381–388
Jorzig EB (1979) Sulfitreduktion in *Thiocapsa roseropersicina*. Dipl thesis, Univ Bonn
Kämpf C, Pfennig N (1980) Arch Microbiol 127:125–137
Kirchhoff J, Trüper HG (1974) Arch Microbiol 100:115–120
Knaff DB, Buchanan BB (1975) Biochim Biophys Acta 376:549–560
Knobloch K, Eley JH, Aleem MIH (1971) Arch Mikrobiol 80:97–114
Kobayashi K, Takahashi E, Ishimoto M (1972) J Biochem (Tokyo) 72:879–887
Kobayashi K, Seki Y, Ishimoto M (1974) J Biochem (Tokyo) 75:519–529
Kobayashi K, Katsura E, Kondo T, Ishimoto M (1978) J Biochem (Tokyo) 84:1209–1215
Kondratieva EN, Petushkova YP, Zhukov VG (1975) Mikrobiologiya 44:389–394
Kondratieva EN, Krassilnikova EN, Pedan LV (1976a) Mikrobiologiya 45:172–173

Kondratieva EN, Zhukov VG, Ivanovskii RN, Petushkova YP, Monosov EZ (1976b) Arch Microbiol 108:287–292
Krassilnikova EN, Zakharchuk LM, Limnik LM (1980) Mikrobiologiya 49:244–248
Kusai A, Yamanaka T (1973a) FEBS Lett 34:235–237
Kusai A, Yamanaka T (1973b) Biochem Biophys Res Commun 51:107–112
Kusai A, Yamanaka T (1973c) Biochim Biophys Acta 325:304–314
Larsen H (1952) J Bacteriol 64:187–196
Lee JP, LeGall J, Peck HD (1973) J Bacteriol 115:529–542
Mathewson JH, Burger LJ, Milstone HG (1968) Fed Proc 27:774
Paschinger H, Paschinger J, Gaffron H (1974) Arch Microbiol 96:341–352
Peschek GA (1978) Arch Microbiol 119:313–322
Petushkova YP, Ivanovskii RN (1976a) Mikrobiologiya 45:9–14
Petushkova YP, Ivanovskii RN (1976b) Mikrobiologiya 45:592–597
Pfennig N (1978) Int J Syst Bacteriol 28:283–288
Pfennig N, Trüper HG (1974) The phototrophic bacteria. In: Buchanan RE, Gibbons NE (eds) Bergey's manual of determinative bacteriology, 8th edn. Williams & Wilkins, Baltimore, pp 24–64
Pfennig N, Trüper HG (1978) Taxonomy of the Rhodospirillales. In: Clayton RK, Sistrom WR (eds) The photosynthetic bacteria, Plenum Press, New York London, pp 19–27
Prangenberg B (1976) Thiosulfatspaltende Enzyme in einigen phototrophen Bakterien. Dipl thesis, Univ Bonn
Puchkova NN, Gorlenko VM (1976) Mikrobiologiya 45:655–659
Puchkova NN, Gorlenko VM, Pivovarova TA (1974) Mikrobiologiya 44:108–114
Remsen CC (1978) Comparative subcellular architecture of photosynthetic bacteria. In: Clayton RK, Sistrom WR (eds) The photosynthetic bacteria. Plenum Press, New York London, pp 31–60
Rodova NA, Pedan LV (1980) Mikrobiologiya 49:221–226
Rolls JP, Lindstrom FS (1967) J Bacteriol 94:860–866
Schedel M (1977) Untersuchungen zur anaeroben Oxidation reduzierter Schwefelverbindungen durch *Thiobacillus denitrificans, Chromatium vinosum* und *Chlorobium limicola.* Doct thesis, Univ Bonn
Schedel M, Trüper HG (1980) Arch Microbiol 124:205–210
Schedel M, Vanselow M, Trüper HG (1979) Arch Microbiol 121:29–36
Schmidt GL, Nicolson GL, Kamen MD (1971) J Bacteriol 105:1137–1141
Schug A (1979) Anreicherung und Charakterisierung der ADP-Sulfurylase aus *Thiobacillus denitrificans* RT. Dipl thesis, Univ Bonn
Schwenn JD, Biere M (1979) FEMS Microbiol Lett 6:19–22
Smith AJ (1966) J Gen Microbiol 42:371–380
Smith AJ, Lascelles J (1966) J Gen Microbiol 42:357–370
Stackebrandt E, Woese CR (1979) Forum Mikrobiol 2:183–190
Tatzki G (1979) Sulfitoxidierende Enzyme bei Thiocystis violacea St. 2311. Dipl thesis, Univ Bonn
Thiele HH (1968) Antonie van Leeuwenhoek J Microbiol Serol 34:350–356
Trüper HG (1964) Antonie van Leeuwenhoek J Microbiol Serol 30:385–394
Trüper HG (1978) Sulfur metabolism. In: Clayton RK, Sistrom WR (eds) The photosynthetic bacteria. Plenum Press, New York London, pp 677–690
Trüper HG, Peck HG (1970) Arch Mikrobiol 73:125–142
Trüper HG, Rogers LA (1971) J Bacteriol 108:1112–1121
Trüper HG, Lorenz C, Fischer U (1976) In: Codd GA, Stewart WDP (eds) Proc 2nd Symp Photosynth Prokaryotes, Dundee, pp 41–42
Uspenskaya VE, Kondratieva EN (1972) Mikrobiologiya 41:449–456
Woese CR, Gibson J, Fox GE (1980) Nature (London) 283:212–214
Yoch D, Lindstrom E (1971) J Bacteriol 106:700–701

Oxidation of Ammonia by Nitrosomonas and of Inorganic Sulfur by Thiobacilli

I. SUZUKI, S.-CHN. KWOK, D.C.Y. TSANG, J.K. OH, and R.S. BHELLA[1]

1 Introduction

Inorganic nitrogen and sulfur compounds exist in nature in various oxidation states and the reduced compounds are used by chemolithotrophic bacteria as sole source of energy for growth. Thus ammonia is oxidized to nitrite by *Nitrosomonas* and further to nitrate by *Nitrobacter*. Sulfide, sulfur, and thiosulfate are oxidized by thiobacilli to sulfate. Interestingly the oxidation of both nitrogen and sulfur compounds starts with the oxygenation of reduced substrates, NH_3 and S^o, in *Nitrosomonas europaea* and *Thiobacillus thiooxidans*, respectively. A possible interaction between nitrogen and sulfur compounds takes place in *Thiobacillus denitrificans* where reduced sulfur compounds are oxidized with nitrate.

This article limits the discussion to the mechanism of ammonia oxidation by *Nitrosomonas europaea* and of sulfur oxidation by thiobacilli.

2 Oxidation of Ammonia by Nitrosomonas europaea

Ammonia is oxidized by *N. europaea* to nitrite: $NH_3 + 1^1/_2 O_2 \rightarrow HNO_2 + H_2O$. Hydroxylamine is the intermediate of oxidation (Less, 1952) and the initial conversion of ammonia to hydroxylamine is through hydroxylation: $NH_3 + AH_2 + O_2 \rightarrow NH_2OH + A + H_2O$ (Suzuki, 1974; Dua et al., 1979). NH_3 rather than NH_4^+ seems to be the actual substrate for ammonia oxidation either by cells or cell-free extracts, (Suzuki et al., 1974; Drozd, 1976) although ammonia exists mainly as NH_4^+ at pH 7.5–8.0, the optimum for growth and ammonia oxidation.

A further oxidation of hydroxylamine is catalyzed by hydroxylamine oxidoreductase which has been extensively purified and studied (Falcone et al., 1963; Aleem and Lees, 1963; Hooper and Nason, 1965; Rees, 1968; Ritchie and Nicholas, 1974; Hooper et al., 1978; Yamanaka et al., 1979). The enzyme can use either cytochrome c or phenazine methosulfate as electron acceptor and carries out the removal of at least two electrons from hydroxylamine: $NH_2OH \rightarrow (NOH) + 2\bar{e} + 2H^+$. Although *Nitrosomonas* cells can oxidize hydroxylamine to nitrite stoichiometrically ($NH_2OH + O_2 \rightarrow HNO_2 + H_2O$), the nitrite yield by the enzyme system is less than 50% even aerobically, the remainder being nitrate (NHO_3), nitric oxide (NO) and nitrous oxide (N_2O) (Hooper et al., 1977; Hooper and Terry, 1979).

---

1 Department of Microbiology, University of Manitoba, Winnipeg, Mantioba, Canada R3T 2N2

Table 1. Cytochromes purified from *N. europaea*

Cytochromes	Reduced γ	β	Spectra (nm) α	CO	Isoelectric point	Molecular weight 10^3	E_0 (v)
Cytochromes							
c-552[a]	416	523	552		3.7	10	0.25
c-554[a]	421	524	554		10.7	22	Autoxid.
c-553[b]	418	523	553	417	4.7	52	0.46
a_1 [c]	444		595	428			Autoxid.
p-460[d]	460			446		36	Autoxid.
Hydroxylamine oxidoreductase[e]	418 (463)	521	553	(no 463)	4.7–5.3	180–200	Autoxid.

[a] Yamanaka and Shinra (1974)
[b] Tronson et al. (1973). Cytochrome c-552 and a c-p-460 complex also reported
[c] Erickson et al. (1972)
[d] Erickson and Hooper (1972)
[e] Ritchie and Nicholas (1974); Hooper et al. (1978); Yamanaka et al. (1979). A 463 nm shoulder disappears with CO

A number of cytochromes have been purified from *N. europaea* and their characteristics are shown in Table 1. Cytochrome c-552 is reduced by NH_2OH in the presence of catalytic quantities of hydroxylamine oxidoreductase and cytochrome c-554, and reduced cytochrome c-552 is oxidized by cytochrome a_1 in air. Thus the electron flow from NH_2OH is: $NH_2OH \rightarrow$ hydroxylamine oxidoreductase $\rightarrow$ c-554 $\rightarrow$ c-552 $\rightarrow a_1 \rightarrow O_2$ (Yamanaka and Shinra, 1974). The relationship of c-553 to c-552 and c-554 is unclear. So is the relationship between hydroxylamine oxidoreductase (463 nm shoulder) and cytochrome P-460 in spite of their spectral similarities.

Ammonia is oxidized by *Nitrosomonas* cells stoichiometrically to nitrite. Cell-free extracts prepared by a French Pressure Cell carry out the stoichiometric oxidation of ammonia at a rate as high as 10%–20% of that by intact cells, but only in the presence of bovine serum albumin, spermine, or Mg^{2+} (Suzuki and Kwok, 1970). The relative effectiveness of these activators on this membrane system is influenced by the concentration of phosphate as shown in Table 2. Thus BSA is most effective in 0.1 *M* potassium phosphate and spermine at lower phosphate concentrations. Bovine serum albumin has an additional effect of stabilizing the ammonia-oxidizing activity of extracts and therefore extracts are routinely prepared from a cell suspension (30 mg wet cells/ml) in 0.1 *M* potassium phosphate (pH 7.5) with BSA (20 mg/ml), unless otherwise indicated.

Extracts prepared without BSA can also oxidize ammonia when catalytic quantities of NH_2OH and NADH are added as electron donor (Suzuki et al., 1976). Thus the hydroxylation of ammonia is initiated by electrons supplied either through the externally added NH_2OH (NADH) or through the endogenous metabolism in the pres-

Table 2. Effect of phosphate on ammonia oxidation by *N. europaea* extracts

Potassium phosphate (m*M*)	Ammonia-oxidizing activity (nmol O_2/min) with the following additions			
	None	BSA (20 mg/ml)	Spermine (2 m*M*)	$MgCl_2$ (10 m*M*)
Extracts without BSA				
100	3	42	30	27
50	0	5	30	
Extracts with BSA				
100	50	50	56	56
50	5	5	74	
20	5	5	75	56

Nitrosomonas extracts were prepared in 0.1 *M* potassium phosphate (pH 7.5) with or without 20 mg/ml bovine serum albumin (BSA). Ammonia oxidation was measured in 1.5 ml reaction mixtures at pH 7.5 with 0.3 ml extracts and 1.7 m*M* ammonium sulfate as described in Suzuki and Kwok (1970). Potassium phosphate concentrations in the reaction mixtures were varied as indicated

ence of BSA. The presence of BSA seems to help the partial reduction of cytochromes in the extracts upon addition of ammonia. As shown in Fig. 1 the difference sprectrum of BSA-extracts with and without ammonia indicates a partial reduction of *Nitrosomonas* c-type and a-type cytochromes. In the absence of BSA such a reduction is not observed, but with a further addition of NADH the cytochromes are reduced similarly (Fig. 1).

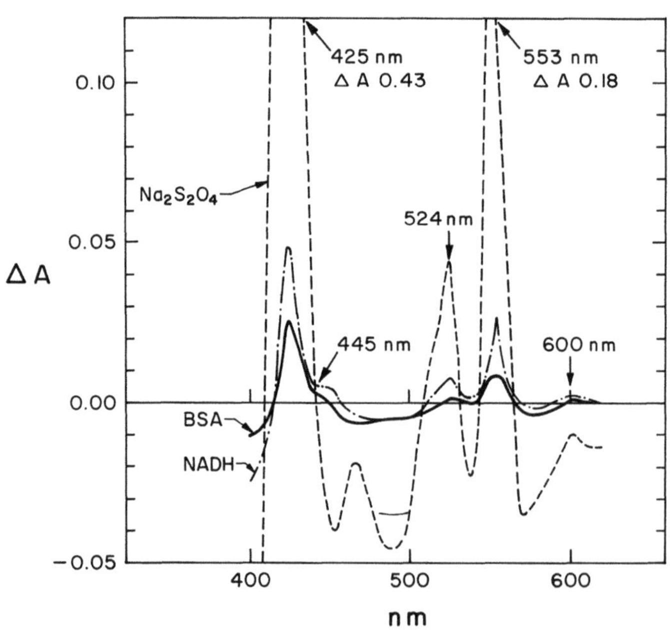

Fig. 1. Reduction of cytochromes by ammonia in *N. europaea* extracts. Both reference and sample cuvettes (1 cm light path) contained 0.5 ml extracts in a total volume of 1 ml in 0.1 *M* potassium phosphate buffer (pH 7.5). After drawing a baseline in a Shimadzu MPS-50L spectrophotometer ammonium sulfate was added to the sample cuvette (2 m*M*) and the difference spectrum was taken. *BSA:* extracts with BSA (20 mg/ml). *NADH:* extracts without BSA, after addition of ammonia plus NADH (50 µ*M*). $Na_2S_2O_4$: extracts with BSA reduced with dithionite in the sample cuvette

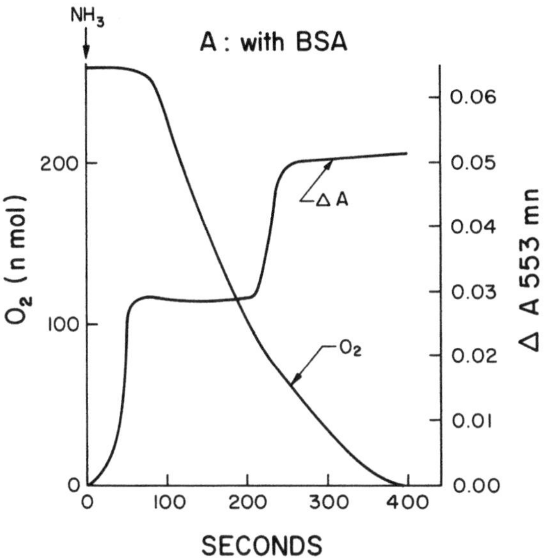

Fig. 2A, B. Kinetics of cytochrome c and ammonia oxidation in *N. europaea* extracts. Reaction mixtures contained 0.5 ml extracts with BSA (A) or without BSA (B) in a total volume of 1 ml in 0.1 *M* potassium phosphate (pH 7.5). The reaction was initiated with the addition of ammonium sulfate (2 m*M*). The reduction of c-type cytochrome was followed in a Shimadzu spectrophotometer as absorbance increase at 553 nm (ΔA553). The oxygen consumption was measured in a Gilson Oxygraph as described by Suzuki and Kwok (1970) in separate, but identical 1 ml reaction mixtures. With BSA (A) but not without BSA (B) a secondary reduction of cytochrome c was observed. *NADH* 10 *μM*

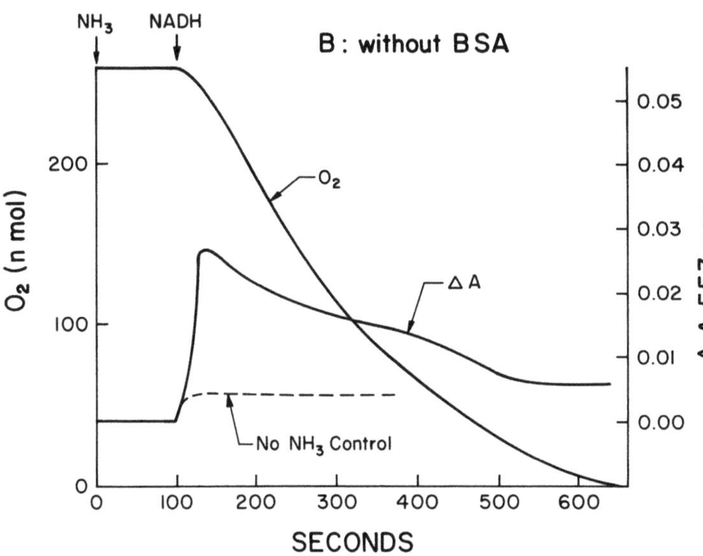

The kinetics of cytochrome c reduction and oxygen uptake is shown in Fig. 2. The reduction starts after the addition of ammonia and the partial reduction is almost complete before the O_2 consumption starts. With the addition of NADH or NH_2OH the partial reduction is achieved more quickly and the lag period required for O_2 consumption is shortened. Once the hydroxylation of ammonia (O_2 uptake) is initiated electrons from NH_2OH oxidation become available for further hydroxylation of ammonia for coupled reactions: $NH_3+O_2+NH_2OH \rightarrow NH_2OH+H_2O+(NOH)$ or $NH_3+O_2+(NOH) \rightarrow NH_2OH+HNO_2$.

Table 3. Inhibition of NH_3 oxidation by *N. europaea* extracts

Inhibitor	Type of inhibition	K_i
CH_2CH_2	Competitive	15 μM
CH_4	Competitive	50 μM
CO	Competitive	3 μM
CH_3OH	Competitive	120 μM
CHCH	Mixed, complex	5–10 μM

Table 4. NADH oxidation by a membrane fraction of *N. europaea* extracts

Additions		NADH oxidation (nmol NADH/min/mg protein)
None		1.0
$(NH_4)_2SO_4$	(0.5 m*M*)	2.5
CH_2CH_2	(0.5 m*M*)	2.2
CH_4	(0.5 m*M*)	2.0
CH	(0.5 m*M*)	2.3
CH_3OH	(17.0 m*M*)	2.5
CHCH	(0.5 m*M*)	0.3

NADH oxidation was measured spectrophotometrically at 340 nm. NADH concentration, 0.1 m*M*; pH 7.5 in 0.1 *M* potassium phosphate. A particulate or membrane fraction was prepared from extracts as in table 5

Hydroxylation of ammonia has some similarity to that of methane to methanol (Wilkinson, 1975; Ferenci et al., 1975; Suzuki et al., 1976) and CH_4 or CO, the substrate for oxidation by methane hydroxylase, competitively inhibits the oxidation of ammonia in *N. europaea* extracts and replaces ammonia in the ammonia-stimulated NADH oxidation (Suzuki et al., 1976, Tables 3 and 4). In addition to methane and carbon monoxide, methanol and ethylene (CH_2CH_2) seem to compete for the NH_3-binding site, while acetylene (CHCH) shows a mixed complex type of inhibition (uncompetitive at low changing to noncompetitive at high CHCH concentrations). The K_i values in Table 3, with the exception of CH_3OH are in a similar order of magnitude as the K_m value for NH_3 of 24 μM (Suzuki et al., 1974). All the competitive inhibitors of ammonia oxidation stimulated the NADH oxidation similarly to ammonia, but acetylene inhibited the oxidation (Table 4). Acetylene inhibition of ammonia oxidation has also been reported on intact cells (Hynes and Knowles, 1978). In spite of the similarity between *N. europaea* and methane-oxidizing bacteria, the *Nitrosomonas* system is distinct in that NADH has only a catalytic or priming effect, unlike in methane oxidation where the stoichiometry is: $CH_4 + O_2 + NADH + H^+ \rightarrow CH_3OH + NAD^+ + H_2O$ (Ferenci et al., 1975). It is thus different from *Methylococcus capsulatus* (Bath) extracts which oxidize ammonia to hydroxylamine in the presence of high concentrations of NADH and ammonia (K_m 43.5 m*M* ammonium sulfate) before its further oxidation to nitrite (Dalton, 1977). Ascorbate, an effective electron donor for methane hydroxylase from *Methylosinus trichosporium* OB36 (Tonge et al., 1977), is ineffective in the *Nitrosomonas* system. There seems to be an obligatory coupling between ammonia hydroxylation and hydroxylamine oxidation in *Nitrosomonas.*

Hydroxylamine nevertheless accumulates during ammonia oxidation by *N. europaea* cells in the presence of hydrazine (Less, 1952) and ammonia hydroxylation can be conclusively demonstrated by analyzing NH_2OH formed from $^{15}NH_4Cl$ and $^{18}O_2$ or $H_2{}^{18}O$ (Dua et al., 1979). In these experiments hydrazine (NH_2NH_2) presumably supplies electrons required for ammonia hydroxylation, since it is a substrate for hydroxylamine oxidoreductase (Hooper and Nason, 1965) and it can replace NADH or hydroxylamine as catalytic electron donor for ammonia oxidation in our *N. europaea* extracts without BSA.

Resolution of the cell-free ammonia-oxidizing system of *N. europaea* has been only partially successful. The membrane or particulate fraction (fraction I) prepared by passage of extracts through a column of Sepharose 6B retains activity for the ammonia-stimulated NADH oxidation (Suzuki et al., 1976), but does not oxidize ammonia. It has cytochrome a_1 and other cytochromes (largely c type), a strong cytochrome oxidase activity and very little NH_2OH-cytochrome c oxidoreductase activity. The ammonia-oxidizing activity is restored when a Sepharose fraction containing hydroxylamine oxidoreductase and a fraction containing a c-type cytochrome (spectrally similar to c-552 but not replaceable with the purified cytochrome c-552) are combined with the membrane fraction. Interestingly, however, the ammonia-oxidizing activity is also restored when cytochrome c-554 is added to the membrane fraction. The effect of increasing concentrations of cytochrome c-554 is shown in Table 5. The NH_2OH-cytochrome c (horse heart) oxidoreductase activity of the membrane fraction also increases with increasing cytochrome c-554 concentrations. The oxidation of NH_2OH to HNO_2, however, requires the addition of both cytochromes c-552 and c-554, the latter increasing the yield of nitrite.

Table 5. Reconstitution of ammonia oxidizing activity

Fractions	cyt. c-554 (μ*M*)	Activity (nmol O_2/min)
Extracts	–	80
1	–	2
1	1.0	20
1	2.0	32
1	4.0	48

Nitrosomonas extracts were prepared in 0.1 *M* potassium phosphate (pH 7.5) with BSA and passed through a Sepharose 6B column (Suzuki et al., 1976) to obtain the membrane fraction (fraction 1). Ammonia oxidation was measured as in Table 2 with BSA and with 0.5 ml extracts or fraction 1. Cytochrome c-554 was added as indicated

As mentioned earlier the oxidation of hydroxylamine by hydroxylamine oxidoreductase leads not only to nitrite but also to nitrate. Nitrate is not produced from nitrite and its production from hydroxylamine is inhibited by diethyldithiocarbamate, resulting in the formation of nitrite and oxidation of diethyldithiocarbamate (Hooper et al., 1977). Production of nitrate from ammonia by a *Methylococcus* is also mentioned (Drozd et al., 1978).

The formation of nitrate from hydroxylamine possibly supports the model of hydroxylamine oxidation involving peroxonitrite (ONO_2^-) advanced by Hughes and Nicklin (1970a) based on studies of hydroxylamine autoxidation in alkaline solution (Hughes and Nicklin, 1970b). In their model the nitroxyl ion, NO^- produced from hydroxylamine, forms peroxonitrite in combination with oxygen: $NO^- + O_2 \rightarrow ONO_2^-$. Copper ion catalyzes the reaction of peroxonitrite with hydroxylamine: $ONO_2^- + NH_2OH + OH^- \rightarrow NO_2^- + NO^- + 2H_2O$. The overall reaction is the oxidation of hydroxylamine to nitrite: $NH_2OH + O_2 + OH^- \rightarrow NO_2^- + 2H_2O$. Peroxonitrite, when protonated, is unstable and isomerizes to nitrate: $ONOO^- + H^+ \rightarrow ONOOH \rightarrow HNO_3$ (Hughes and Nicklin, 1968). If peroxonitrite is indeed the intermediate of biological hydroxylamine oxidation the production of nitrate as well as nitrite is understandable. The effect of diethyldithiocarbamate may then be explained as the reduction of peroxonitrite.

Peroxonitrite, although unstable at nonalkaline conditions, is an attractive candidate as an intermediate of ammonia oxidation because of its ability to hydroxylate benzene to phenols (Halfpenny and Robinson, 1952) and to transfer oxygen atom to nucleophiles such as cyanide: $ONO_2^- + CN^- \rightarrow NO_2^- + CNO^-$ (Hughes et al., 1971). If ammonia can be hydroxylated by peroxonitrite, then the following hypothetical reactions can be visualized: $NH_3 + ONOOH \rightarrow NH_2OH + HNO_2$, $NH_2OH \rightarrow NOH + 2\bar{e} + 2H^+$, $NOH + O_2 \rightarrow ONOOH$. If the first and last reactions are tightly coupled we have $NH_3 + NOH + O_2 \rightarrow NH_2OH + HNO_2$, a reaction compatible with the mechanism proposed earlier (Suzuki, 1974). The hypothetical scheme of ammonia oxidation involving peroxonitrite is shown in Fig. 3. In Fig. 3a ammonia is initially hydroxylated by oxygen plus electrons derived from endogenous metabolism or added NADH (hydroxylamine). Cytochrome c-554 very likely mediates these electrons. Once the hydroxylation is initiated, electrons from hydroxylamine oxidation become available for further hydroxylation of ammonia. At the steady state of ammonia oxidation the oxidation of nitroxyl (NOH) is coupled to ammonia hydroxylation with peroxonitrite as possible intermediate (Fig. 3b). The only electrons going through the electron transport system to molecular oxygen for possible energy generation are the two electrons derived from hydroxylamine. The nature of ammonia hydroxylase, the mechanism of ammonia interaction with peroxonitrite (the same hydroxylase involved?), the role of cytochrome p-460 all remain still unsolved.

3 Oxidation of Inorganic Sulfur Compounds by Thiobacilli

Elemental sulfur is oxidized by *Thiobacillus thiooxidans* to sulfuric acid: $S^0 + 1^1/_2 O_2 + H_2O \rightarrow SO_4^{2-} + 2H^+$. Other sulfur compounds (e. g., thiosulfate and sulfide) are also oxidized to sulfate by various thiobacilli. The mechanism of oxidation of inorganic sul-

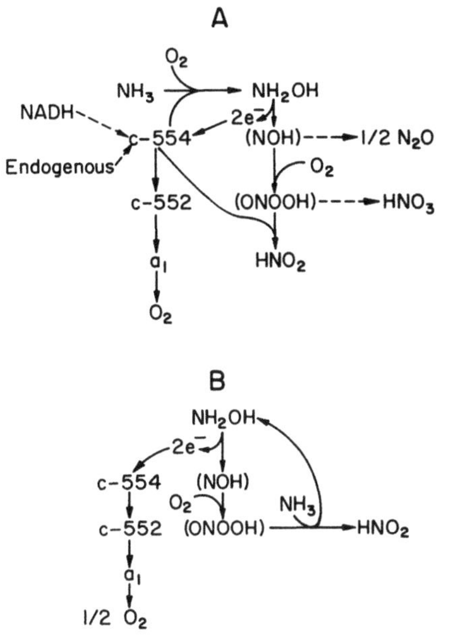

Fig. 3 A, B. Hypothetical scheme of ammonia oxidation in *N. europaea*. **A** represents the initial state of ammonia oxidation (see text) and explains also the possible production of nitrous oxide and nitrate from hydroxylamine. **B** the steady-state ammonia oxidation

fur compounds has been extensively reviewed recently (Suzuki, 1974; Oh and Suzuki, 1980).

Sulfite is the final intermediate in all these oxidations. The oxidation of sulfite (SO_3^{2-}) to sulfate (SO_4^{2-}) is achieved either by sulfite: cytochrome c oxidoreductase (Charles and Suzuki, 1966) or adenosine phosphosulfate (APS) reductase (Peck, 1960) with cytochrome oxidase in aerobic thiobacilli, and with nitrate reductase in *Thiobacillus denitrificans* (Sawhney and Nicholas, 1977).

Thiosulfate is oxidized either to tetrathionate by thiosulfate-oxidizing enzyme: $2S_2O_3^{2-} \rightarrow S_4O_6^{2-} + 2\bar{e}$ (Trudinger, 1961) or cleaved to sulfur and sulfite by thiosulfate-cleaving enzyme (rhodanese): $SSO_3^{2-} \rightarrow S + SO_3^{2-}$ followed by the oxidation of sulfur and sulfite (Charles and Suzuki, 1966).

Sulfide is oxidized first to polysulfide (sulfur): $S^{2-} \rightarrow S + 2\bar{e}$ by O_2 or nitrite before further oxidation (Aminuddin and Nicholas, 1974; Sawhney and Nicholas, 1978).

Elemental sulfur, after conversion to polysulfide, is oxidized to sulfite by the sulfur-oxidizing enzyme: $S_n + GSH \rightarrow GSS_nH$, $GSS_nH + O_2 + H_2O \rightarrow GSS_{n-1}H + SO_3^{2-} + 2H^+$ (Suzuki, 1965; Suzuki and Silver, 1966). The enzyme is supposedly an oxygenase in aerobic thiobacilli and if a similar enzyme were to operate in *T. denitrificans* an intermediate of denitrification could be a possible candidate as oxygen donor (Suzuki, 1974).

Recently a siroheme-containing sulfite reductase has been isolated and purified from *T. denitrificans* (Schedel et al., 1975; Schedel and Trüper, 1979). The physiological role of the enzyme is not likely to be the reduction of sulfite since only the sulfane sulfur (S-) of thiosulfate ($S\text{-}SO_3^{2-}$) is assimilated during growth (Schedel and Trüper, 1980). In fact sulfide is oxidized to thiosulfate with substrate amounts of oxidized enzyme (Schedel and Trüper, 1979). If indeed the sulfite reductase functions for the oxidation of sulfide (or polysulfide) to sulfite as proposed by Schedel and Trüper (1979), its relationship to the sulfide-linked nitrite reductase (Sawhney and Nicholas, 1978) and sulfite-linked nitrate reductase (Sawhney and Nicholas, 1977) should be interesting to study. The nitrite reductase which contains cytochromes c and d has also a sulfite reductase activity (Schedel and Trüper, 1979).

In our laboratory the sulfur-oxidizing enzyme of aerobic thiobacilli is being re-investigated. *Thiobacillus thioparus* contains a siroheme which is associated with the sulfur-oxidizing enzyme activity during purification. The sulfur-oxidizing enzyme of *Thiobacillus thiooxidans*, on the other hand, does not contain a siroheme. The possible role of siroheme on sulfur oxidation in aerobic thiobacilli is still under investigation.

Acknowledgment. The work was supported by a grant from the Natural Sciences and Engineering Research Council Canada.

References

Aleem MIH, Lees H (1963) Can J Biochem Physiol 41:763–778
Aminuddin M, Nicholas DJD (1974) J Gen Microbiol 82:115–123
Charles AM, Suzuki I (1966) Biochim Biophys Acta 128:510–521
Dalton H (1977) Arch Mikrobiol 114:273–279
Drozd JW (1976) Arch Mikrobiol 110:257–262
Drozd JW, Godley A, Bailey ML (1978) Proc Soc Gen Microbiol 5:66–67
Dua RD, Bhandari B, Nicholas DJD (1979) FEBS Lett 106:401–404
Erickson RH, Hooper AB (1972) Biochim Biophys Acta 275:231–244
Erickson RH, Hooper AB, Terry KR (1972) Biochim Biophys Acta 283:155–166
Falcone AB, Shug AL, Nicholas DJD (1963) Biochim Biophys Acta 77:199–208
Ferenci T, Strøm T, Quayle JR (1975) J Gen Microbiol 91:79–91
Halfpenny E, Robinson PL (1952) J Chem Soc 928:938
Hooper AB, Terry KR (1979) Biochim Biophys Acta 571:12–20
Hooper AB, Terry KR, Maxwell PC (1977) Biochim Biophys Acta 462:141–152
Hooper AB, Nason A (1965) J Biol Chem 240:4044–4057
Hooper AB, Maxwell PC, Terry KR (1978) Biochemistry 17:2984–2989
Hughes MN, Nicklin HG (1968) J Chem Soc (A) 450–452
Hughes MN, Nicklin HG (1970a) Biochim Biophys Acta 222:660–661
Hughes MN, Nicklin HG (1970b) J Chem Soc (A) 925–928
Hughes MN, Nicklin HG, Sackrule WAC (1971) J Chem Soc (A) 3722–3725
Hynes RK, Knowles R (1978) FEMS Lett 4:319–321
Lees H (1952) Nature (London) 169:156–158
Oh JK, Suzuki I (1980) In: Knowles CJ (ed) The diversity of bacterial respiratory systems, vol II. CRC Press, Boca Raton Florida, pp 113:137
Peck HD Jr (1960) Proc Natl Acad Sci USA 46:1053–1057
Rees MK (1968) Biochemistry 7:353–366
Ritchie GAF, Nicholas DJD (1974) Biochem J 138:471–480
Sawhney V, Nicholas DJD (1977) J Gen Microbiol 100:49–58
Sawhney V, Nicholas DJD (1978) J Gen Microbiol 106:119–128
Schedel M, Trüper HG (1979) Biochim Biophys Acta 568:454–467
Schedel M, Trüper HG (1980) Arch Microbiol 124:205–210
Schedel M, Legall J, Baldensperger J (1975) Arch Microbiol 105:339–341
Suzuki I (1965) Biochim Biophys Acta 104:359–371

Suzuki I (1974) Annu Rev Microbiol 28:85–101
Suzuki I, Kwok SC (1970) Biochem Biophys Res Commun 39:950–955
Suzuki I, Silver M (1966) Biochim Biophys Acta 122:22–33
Suzuki I, Dular U, Kwok SC (1974) J Bacteriol 120:556–558
Suzuki I, Kwok SC, Dular U (1976) FEBS Lett 72:117–120
Tonge GM, Harrison DEF, Higgins IJ (1977) Biochem J 161:333–344
Tronson DA, Ritchie GAF, Nicholas DJD (1973) Biochim Biophys Acta 310:331–343
Trudinger PA (1961) Biochem J 78:680–686
Wilkinson JF (1975) In: Microbial growth on C_1-compounds. Society of Fermentation Technology, Jpn, pp 45–57
Yamanaka T, Shinra M (1974) J Biochem (Tokyo) 75:1265–1273
Yamanaka T, Shinra M, Takahashi K, Shibasaka M (1979) J Biochem (Tokyo) 86:1101–1108

Contributed Papers: Nitrogen Metabolism

Enzymatic Mechanism and Regulation of Nitrate Reduction in Rhodopseudomonas capsulata

J.-H. KLEMME, J. CZICHOS, and R. WESCH[1]

1 Introduction

The ability to assimilate nitrate as N-source is widespread among prokaryotes. When studying a strain of the facultative phototrophic bacterium, *Rhodospirillum (R.) rubrum,* Katoh (1963a) found that the bacteria could be adapted to grow with nitrate as sole N-source under photosynthetic conditions or with nitrate as terminal electron acceptor under dark-anaerobic conditions. The nitrate reductase from such adapted cells was partially purified and characterized as a cytochrome b-containing enzyme complex (Katoh, 1963b). Taniguchi and Kamen (1963) obtained growth of the same *R. rubrum* strain (S1) with nitrate as N-source both under photosynthetic and aerobic conditions. Contrary to the results of the latter authors, we were unable to grow any of four different *R. rubrum* strains (including strain S1) with nitrate as sole N-source under photosynthetic conditions in defined media with D,L-malate as C-source. In our laboratory, among 35 strains of the more common species of the Rhodospirillaceae, the capacity of assimilatory nitrate reduction was restricted to 11 strains of the two taxonomically related species *Rhodopseudomonas (Rps.) capsulata* and *Rps. sphaeroides* (Klemme, 1979).

The nitrate-assimilating strains of *Rps. capsulata* and *Rps. sphaeroides* studied in our laboratory were unable to grow under dark-anaerobic conditions with nitrate as the terminal electron acceptor (Alef, 1978). On the other hand, Satoh et al. (1974, 1976) have described several newly isolated denitrifying strains of *Rps. sphaeroides* and purified the dissimilatory nitrite reductase from one strain (Sawada et al., 1978).

The present paper reports on the properties of assimilatory nitrate reductases isolated from two different strains of *Rps. capsulata.* The two enzymes differ in respect to intracellular localization and molecular properties (Alef and Klemme, 1979; Wesch and Klemme, 1980). In an approach to understanding the mechanisms underlying the regulation of the expression of the nitrate reductase gene as well as the regulation of the enzyme activity, mutant strains of *Rps. capsulata* unable to assimilate nitrate were studied.

---

1 Institut für Mikrobiologie der Universität, Meckenheimer Allee 168, 5300 Bonn 1/FRG

2 Material and Methods

The *Rps. capsulata* strains used in this study (AD2, BK5, and Kb1=DSM 155) are kept in the culture collection of the Institut für Mikrobiologie der Universität Bonn. If not otherwise stated, the bacteria were grown under photosynthetic conditions (2,500 lx, 30 °C) in completely filled screw-cap bottles containing a defined medium with 0.4% (w/v) Na-D,L-malate and 0.1% (w/v) KNO_3 (other constituents as specified by Weaver et al., 1975). For growth experiments with other N-sources, the KNO_3 of the basal medium was replaced by $(NH_4)_2SO_4$ (7.5 m*M*), $NaNO_2$ (2 m*M*), hypoxanthine (2.2 m*M*), uric acid (1.8 m*M*), or amino acids (10 m*M*). The cultures were started with an inoculum of 2% (v/v). For comparison, media without N-source were also inoculated. Growth with N_2 as sole N-source was tested by growing cells on N-free agar plates [N-free basal medium supplemented with 1.7% (w/v) of purified "Oxoid" agar] in the light (30 °C) under an atmosphere of H_2, N_2, and CO_2 in "GasPak" jars of Becton, Dickinson Co., Heidelberg. The growth of liquid cultures was monitored by measuring the OD_{660} in a "Spectronic 20" colorimeter (Bausch & Lomb, München).

The preparation and fractionation of cell-free extracts, the colorimetric assay of nitrate reductase and other analytical procedures were performed as described in a previous paper (Alef and Klemme, 1979).

For the isolation of *Rps. capsulata* AD2 mutants with a defective nitrate assimilation system, the following procedure was employed: 15 ml of a NH_4^+-malate-culture (OD_{660}=0.2, corresponding to about $2x10^8$ colony-forming units per ml) was transferred to a glass Petri dish and irradiated for 3 min by a UV-lamp (254 nm, 13 watt) at a distance of 26 cm and then incubated aerobically in the dark for one day. The cells were collected by centrifugation, resuspended in a malate medium without N-source and incubated photosynthetically for 24 h. The cells were then transferred to a KNO_3-malate-medium containing 10 units of penicillin G per ml and incubated for two days under photosynthetic conditions. The surviving cells were collected, washed twice in NH_4^+-malate-medium and incubated under photosynthetic conditions for 3–6 days in the same medium. The sequence of selective steps described above was repeated two to three times in order to increase the yield of mutants. To detect the mutant colonies, samples of the final penicillin-free NH_4^+-malate-medium were plated on KNO_3^- malate-agar and incubated aerobically in the dark. Wild-type cells gave rise to brown-red colonies of 1–2 mm diameter, while *nit*-mutants formed only very small, whitish colonies.

3 Properties of Assimilatory Nitrate Reductase from Different Strains of Rhodopseudomonas capsulata

3.1 Occurrence of Nitrate and Nitrite Reductases in Cell-Free Extracts of Different Rps. capsulata Strains

The three *Rps. capsulata* strains AD2, BK5, and Kb1 (=DSM 155) grow readily in the malate-KNO_3-medium both under photosynthetic and dark-aerobic conditions. They do not grow, however, in such media (supplemented with 0.05% yeast extract or 0.1% ammoniumsulfate as respective N-sources) under anaerobic-dark conditions, i. e., the three strains are capable only of assimilatory, but not of dissimilatory nitrate reduction. In ultrasonic extracts of nitrate-grown *Rps. capsulata* strains AD2, BK5, and Kb1, nitrate reductase activity was detected at the beginning only in cell-free extracts of strains AD2 and BK5. We were unable to find any nitrite reductase activity (tested according to Vega et al., 1973) in cell-free extracts of the three strains. Thus, nitrite reductase activity was not the cause of seemingly missing nitrate reductase in Kb1-extracts. Our recent experiments show that ultrasonic extracts of *Rps. capsulata* Kb1 grown photosynthetically in malate-KNO_3-medium (which appear to be devoid of nitrate reductase activity) strongly inhibit the nitrate reductase isolated from *Rps. capsulata* AD2. The inhibiting substance from strain Kb1 is heat-stable, of low molecular weight, and can be separated from the nitrate reductase by gelfiltration of the soluble fraction of Kb1-homogenates through Sephacryl S300 (M. Preuß, Diplomarbeit Universität Bonn, 1980). It may be noted in this connection that various inactivation systems (enzymatic or nonenzymatic) have been described for nitrate reductases from unicellular green algae (Pistorius et al., 1976; Diez et al., 1977), the yeast *Rhodotorula glutinis* (Ito and Suzuki, 1978), and the fungi *Aspergillus nidulans* (Dunn-Coleman and Pateman, 1977), and *Neurospora crassa* (Walls et al., 1978).

3.2 Purification of a Heme-Associated Nitrate Reductase from Rps. capsulata AD2

The assimilatory nitrate reductase of *Rps. capsulata* AD2 was a soluble enzyme in ultrasonic extracts of photosynthetically grown cells. Interestingly, the enzyme was not detected in ultrasonic extracts from cells grown aerobically in the dark in KNO_3-malate-medium. However, if such cells were treated with toluene in order to break the permeability barrier, nitrate reductase activity was measured having a specific activity of about 10 nmol/min·mg protein. In order to obtain high enzyme yields from photosynthetic cultures it was essential to harvest the cells before the OD_{660} attained a value of 1.

The soluble assimilatory nitrate reductase of *Rps. capsulata* AD2 was purified to electrophoretic homogeneity by ammoniumsulfate precipitation, DEAE-cellulose chromatography and isoelectric focusing (Alef and Klemme, 1979). The enzyme had a molecular weight of about 185,000, appeared to be composed of two identical subunits and contained about 0.8 g-atoms molybdenum and 1.3 mol heme per mol. On account of its absorption spectrum (α-band at 552 nm, β-band at 523 nm in the dithionite-reduced state), the heme group of the AD2-enzyme seemed to be of the c-type.

3.3 Solubilization of a Heme-Free Nitrate Reductase from the Chromatophores of Rps. capsulata BK5

Contrary to the enzymes from strains AD2 and Kb1, the assimilatory nitrate reductase of *Rps. capsulata* strain BK5 was firmly attached to the membraneous chromatophore fraction. In order to solubilize and purify the enzyme, the chromatophores were extracted at −15 °C with 80% (v/v) acetone and then treated with 1% (v/v) Triton X 100. The solubilized proteins were subjected to a mild heat treatment (55 °C, 5 min) and an ammoniumsulfate-fractionation. The protein fraction precipitating between 30% and 50% saturation with ammoniumsulfate contained most of the nitrate reductase activity. The enzyme was further purified by chromatography on DEAE-cellulose or by gel filtration through Sephacryl S 300. The nitrate reductase solubilized and purified from the chromatophore fraction of strain BK5 contained no heme group and had a molecular weight of about 360,000 (Wesch and Klemme, 1980). Similar to the enzyme from strain AD2, the purified nitrate reductase of strain BK5 was active only with dithionite as reductant in the presence of catalytic amounts of a viologen dye (benzyl- or methylviologen) or a flavin (FMN, FAD). The enzyme was competitively inhibited by chlorate, azide, thiocyanate, cyanate, and cyanide (Wesch and Klemme, 1980).

4 Isolation and Characterization of Rps. capsulata AD2 Mutants Unable to Assimilate Nitrate

4.1 Isolation of Chlorate-Resistant Mutants

Mutants of *Escherichia coli* – as well as of other Enterobacteriaceae – with defects in the nitrate reductase enzyme system can be obtained without treatment of the cells with a mutagenic agent by simply selecting chlorate-resistant colonies on agar plates (Piéchaud et al., 1967; Stouthamer et al., 1967). The selection procedure is based on the fact that the membrane-bound respiratory nitrate reductase (type A, according to the classification of Pichinoty, 1964) also catalyzes the reduction of chlorate to chlorite, the latter compound being toxic to the cells. The chlorate method was also successfully applied to isolate nitrate reductase-defective mutants of *Rhizobium* sp. (Pagan et al., 1977), *Azospirillum* sp. (Magalhaes et al., 1978), *Chlamydomonas reinhardii* (Nichols and Syrett, 1978), and other microorganisms. Since the assimilatory nitrate reductase of *Rps. capsulata* AD2 was also active with chlorate as substrate (Alef and Klemme, 1979), we expected that it should be possible to obtain nitrate reductase-defective mutants of that strain by isolating chlorate-resistant colonies.

The wild-type strain AD2 was markedly resistant to chlorate. Inhibition of photosynthetic growth in malate-NH_4^+-medium occurred only at chlorate concentrations greater than 50 m*M*. Growth was completely inhibited in the presence of 200 m*M* K-chlorate. By subculturing of *Rps. capsulata* strain AD2 in media containing 100 m*M* K-chlorate and subsequently plating cell suspensions on malate-NH_4^+-agar, 6 chlorate-resistant mutant strains were isolated in independent experiments. The chlorate resistance of these strains was not lost after several transfers in chlorate-free medium. This indicated a stable genetic mutation.

However, the chlorate-resistant mutants grew normally with nitrate as N-source and contained a soluble nitrate reductase at the same specific activity (20–30 nmol/-min·mg protein) as the wild type. *Rps. capsulata* strain C2.1, a mutant derived from the wild-type strain AD2, which is devoid of nitrate reductase activity (see Sect. 4.2), was also strongly inhibited by 100 m*M* K-chlorate. Thus, the growth inhibition of

Rps. capsulata AD2 by 100 m*M* K-chlorate was independent of the presence of an active nitrate reductase. The resistance to 100 m*M* chlorate by the above-mentioned mutant strains must therefore be related to the function of enzymes or enzyme systems different from nitrate reductase (see Newman et al., 1974, for a discussion of causes of chlorate inhibition in *Escherichia coli*).

4.2 Isolation and Characterization of Mutants Unable to Assimilate Nitrate

Following the method outlined in Section 2, several mutant strains of *Rps. capsulata* AD2 unable to grow with nitrate as the sole N-source were isolated. Note in this connection that, due to the activity of the N_2-fixing enzyme system under photosynthetic conditions, most of the mutant strains grew as well as the wild type on KNO_3-malate-agar plates in "GasPak" jars (H_2-N_2-CO_2 atmosphere) under photosynthetic conditions. However, the activity of nitrogenase was repressed under *aerobic* conditions, and then, growth of the *nit*-mutants was no longer supported by N_2.

Mutant strains C2.1 and W3.2, representing two different types of *nit*-mutants (see below), were selected for further studies (Table 1). Both mutants grew normally in malate-NH_4^+-medium. Mutant C2.1 did not grow with nitrate as N-source, even when incubated photosynthetically in liquid medium for more than 7 days. By contrast, mutant W3.2 grew readily with nitrate as N-source after an "adaptation" period of 4–5 days.

During the "adaptation" period observed in liquid KNO_3-malate cultures of mutant W3.2, the OD_{660} remained constant and no nitrate-assimilating colonies were detected on aerobic KNO_3-malate-agar plates. At the end of the "adaptation" period, the number of nitrate-assimilating colonies increased parallel to the increase of OD_{660}. Such nitrate-assimilating colonies, when transferred to liquid malate-NH_4^+-medium and cultivated with

Table 1. Utilization of various N-sources by the wild-type, mutant and revertant strains of *Rhodopseudomonas capsulata* AD2 during photosynthetic growth

N-source	Wild-type AD2	Mutant strains C2.1	Mutant strains W3.2	Revertant strain W3.2a
NH_4^+	+	+	+	+
NO_3^-	+	–	–	+
NO_2^-	+	–	–	+
N_2	+	+	–	+
Alanine	+	+	–	+
Glutamate	+	+	–	+
Hypoxanthine	+	+	–	n.d.
Uric acid	+	+	–	n.d.

+ = good growth after 1–2 days (growth with N_2 was checked after 5 days)
– = no growth after 3 days
n.d. = not determined

Satoh T, Hoshino Y, Kitamura H (1974) Agric Biol Chem 38:1749–1751
Satoh T, Hoshino Y, Kitamura H (1976) Arch Microbiol 108:265–270
Sawada E, Satoh T, Kitamura H (1978) Plant Cell Physiol 19:1339–1352
Stouthamer AH, Bettenhausen C, Hartingsveldt J (1967) Arch Microbiol 58:228–247
Taniguchi S, Kamen D (1963) In: Studies on microalgae and photosynthetic bacteria. University of Tokyo Press, Tokyo, p 465
Vega JM, Guerrero MG, Leadbetter E, Losada M (1973) Biochem J 133:701–708
Wall JD, Johansson BC, Gest H (1977) Arch Microbiol 115:259–263
Walls S, Sorger GJ, Gooden D, Klein V (1978) Biochim Biophys Acta 540:24–32
Weaver PF, Wall JD, Gest H (1975) Arch Microbiol 105:207–216
Wesch R, Klemme J-H (1980) FEMS Lett 8:37–41

HCN and the Control of Nitrate Reduction The Regulation of the Amount of Active Nitrate Reductase Present in Chlorella Cells

B. VENNESLAND[1]

1 Introduction

The control of the activity of the enzyme nitrate reductase is a subject which has received considerable attention. The processes seem to be quite complex. I will attempt in the present article to summarize our own current views on this subject. Attention will be confined mainly to the phenomenon of control as we see it in the unicellular green alga *Chlorella vulgaris,* though I hope that much of what is said can be applied more generally to other green algae. No attempt will be made to trace historical development in detail or to cover the problem of control in fungi or higher plants, though here also a certain amount of cross-fertilization is inevitable.

Our main experimental material, then, has been the nitrate reductase from *Chlorella vulgaris,* which has been studied extensively (Solomonson et al., 1975; Solomonson, 1975; Solomonson, 1979; Giri and Ramadoss, 1979). It is isolated mainly as a trimer of equal or very similar subunits of molecular weight of about 90,000. The active trimer contains between 2 and 3 molecules of each of the identified prosthetic groups, Mo, FAD, and heme, but we do not know exactly where they are located or distributed in the molecule. There is more recent evidence that the molecule can also exist as a dimer (Howard and Solomonson, 1980). NADH is the preferred reductant of the *C. vulgaris* enzyme, which can be classified as an NADH: nitrate oxidoreductase (EC.1.6.6.1).

Following Hewitt et al. (1979), we will distinguish here between two different types of inactivation-activation mechanisms for nitrate reductase in green algae. Since the chemistry of the reactions is not yet completely understood, it would be premature to exclude a certain level of overlap in the operation of these two mechanisms. Nevertheless, we feel that at present there is an advantage in making a clear distinction between them.

Mechanism I was the first mechanism which we encountered in enzyme studies with algae, but it is perhaps of secondary importance. This mechanism may be called the "switch" mechanism to indicate that it involves a reversible conversion of active enzyme to inactive enzyme by relatively simple chemical processes. Because mechanism I can sometimes be brought into play in vivo by addition of ammonia, we at first supposed that this was the mechanism whereby added ammonia or a product of ammonia metabolism stops the utilization of nitrate by algae (Syrett and Morris, 1963; Morris, 1973). Subsequent studies have shown, however, that the effect of ammonia

1 Forschungsstelle Vennesland der Max-Planck-Gesellschaft, Harnackstraße 23, 1000 Berlin 33/FRG

on mechanism I is quite indirect (Pistorius et al., 1978). We will propose later a working hypothesis regarding the possible relationship of the ammonia effect to mechanism II for controlling nitrate reductase levels. Mechanism II has to do with synthesis and degradation of the enzyme.

2 Mechanism I: the "Switch" Mechanism

2.1 Interaction of HCN with Nitrate Reductase

So far as I know, all nitrate reductases are inhibited by HCN. The nature of this inhibition has been studied intensively for the *Chlorella* enzymes (for details see Vennesland and Guerrero, 1979). It has been established that both *C. vulgaris* (Lorimer et al., 1974) and *C. fusca* (Gewitz et al., 1978) can form the HCN-inactivated enzyme in vivo.

The reduced form of nitrate reductase binds HCN quite firmly to give a protein-HCN complex which cannot activate nitrate, but which has full diaphorase activity. It is assumed that the HCN binds at the Mo site where the nitrate is activated. This is supported by the fact that the amount of HCN bound is equal to the Mo content of the enzyme, and also by the fact that the tungsten analog of the enzyme likewise has full diaphorase activity though it cannot activate nitrate. ESR studies of the enzyme also suggest that HCN binds to Mo (Solomonson, 1979).

The inactive HCN complex can readily be reactivated in vitro by treatment with ferricyanide. Note, however, that if ferricyanide is added to intact cells, the result is a rapid conversion of all intracellular nitrate reductase to the inactive HCN complex, since ferricyanide itself does not penetrate the cell wall, whereas the small amount of HCN which dissociates from ferricyanide will penetrate readily (Pistorius et al., 1978).

The conditions which favor formation of HCN-inactivated nitrate reductase in *C. vulgaris* have been found to include: absence of nitrate, high O_2 tension, high light intensity, and absence of CO_2 (Pistorius et al., 1976). Nitrate interferes with the formation of HCN-inactivated enzyme, presumably by keeping the enzyme in the oxidized form. The other variables probably act to increase HCN formation as discussed under Sect. 2.4.

2.2 Interaction of Vanadate with Nitrate Reductase

Buczek (1973) has reported that vanadium salts inhibit the nitrate reductase of tomato seedlings both in vitro and in vivo, and that the inhibited enzyme can be reactivated by treatment with EDTA. A more detailed study of the effect of vanadium salts on the nitrate reductase of *C. vulgaris* has shown that vanadate binds to the reduced purified enzyme in a manner rather similar to that of HCN (Ramadoss, 1979). The reaction is not instantaneous but requires time to go to completion, just as in the case of HCN. The binding is probably associated with the Mo site, since only nitrate activation is affected, not diaphorase activity. In both cases also, the inactive complex can be rapidly reactivated in vitro with ferricyanide. The vanadate binds less firmly to the enzyme than does HCN. Furthermore, the vanadate complex can be partially reactivated by

EDTA, and can thus be distinguished from the HCN complex, which does not respond to EDTA.

Addition of vanadate to intact *Chlorella* cells results in an increased reversible inactivation of nitrate reductase, but it is the HCN complex which is mainly formed rather than the vanadate complex (Ramadoss, 1979). This is ascribed to the fact that vanadate stimulates HCN formation. When vanadate and HCN are both present, the formation of the HCN complex is favored because of its greater stability.

Vanadium, a trace metal apparently essential for chlorophyll synthesis (Meisch and Bauer, 1978), has recently excited special interest because of the demonstration that vanadate is present in ATP derived from muscle, and that it is also a powerful inhibitor of (Na, K)-ATPase (Cantley et al., 1977; Simons, 1979).

2.3 Other Inactivation Reactions

The Seville group has long favored the view that the reversible inactivation of nitrate reductase in algae is an "overreduction" stimulated by ADP (Maldonado et al., 1973; Losada, 1974; Aparicio and Maldonado, 1979). Evidence has recently been presented with the enzyme from *C. fusca* that oxygen stimulates such inactivation, presumably through reduction by O_2^- (Chaparro et al., 1979). Because the enzyme thus inactivated can be reactivated in vitro by ferricyanide, it seems related to the HCN-inactivated enzyme or the vanadate-inactivated enzyme, but the chemical nature of the relationship is not yet clear. The authors have not shown that vanadate and/or HCN are absent from their preparations. It is curious also that a reaction previously reported to be favored by anaerobiosis, should now be reported to be O_2-dependent. There are other reagents also which have reversible effects on the activity of nitrate reductase. For a detailed discussion see Hewitt et al. (1979).

2.4 Sources of HCN

Very low levels of HCN ($\sim 10^{-9} M$) can account for the reversible inactivation of nitrate reductase in *C. vulgaris* (Solomonson, 1974; Lorimer et al., 1974). Two possible sources of this HCN have been described: (1) glyoxylate plus hydroxylamine (Solomonson and Spehar, 1977, 1979) and (2) aromatic amino acids, particularly histidine (Gewitz et al., 1976a, 1976b; Pistorius et al., 1977, 1979); Pistorius and Voss, 1977; Gewitz et al., 1980; Pistorius and Voss, 1980).

Figure 1 is a modification of a diagram originally presented by Solomonson and Spehar (1977) to show how HCN generation may interact with the reactions involved in nitrate utilization. The scheme shows hydroxylamine as an eventual reduction product of nitrite, reacting with glyoxylate to give HCN by way of glyoxylic oxime. The glyoxylate is represented as an eventual product of the oxygenase action of ribulose diphosphate carboxylase. The stimulatory effects of light and O_2, and the inhibitory effect of CO_2 on HCN generation (Pistorius et al., 1976) can be understood in such terms. This also applies to histidine, which is another possible source of HCN. Eventual metabolism of HCN as the β-carboxamide group of asparagine is likewise depicted. Sites of eventual inhibitory effects of HCN (indicated by the symbol ⊖→)

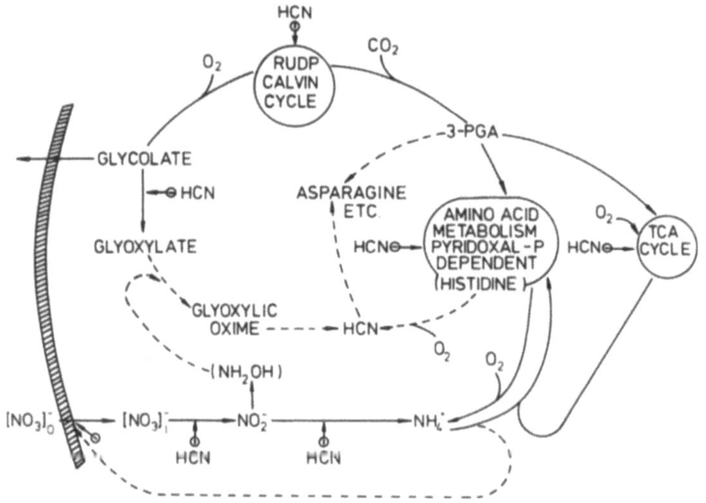

Fig. 1. Regulation of nitrate utilization and possible interrelationship with HCN metabolism in *Chlorella vulgaris.* Inhibition is indicated by the symbol ⊖→. The *hatched area* on the *left* represents the cell membrane. $[NO_3^-]_o \rightarrow [NO_3^-]_i$ represents nitrate uptake. This process is also inhibited by HCN, but the effect may be indirect, for example on energy supply

are manifold, and include not only nitrate and nitrite reduction, but also the oxidation of glycolate to glyoxylate, the carboxylation reaction of the Calvin cycle, the TCA cycle, and the pyridoxal phosphate-dependent reactions of amino acid metabolism.

The conversion of glyoxylate and hydroxylamine to HCN is catalyzed by *Chlorella* extracts. The reaction requires Mn^{2+} and is strongly stimulated by ADP. Added ATP potentiates the ADP effect (Solomonson and Spehar, 1979). The formation of HCN from histidine is a general reaction catalyzed by amino acid oxidases plus a cofactor. The latter may be peroxidase, cytochrome c, any one of a variety of redox metal ions (including vanadium) or a particulate fraction from *Chlorella* cells (Pistorius et al., 1977; Gewitz et al., 1980). In *Chlorella,* L-histidine serves as an HCN precursor in the light (Pistorius et al., 1977), but, among the amino acids, the best dark precursor of HCN is D-histidine (Pistorius and Voss, 1977). Nothing is known as yet about the formation or occurrence of D-histidine in *Chlorella* cells. A similar problem exists with regard to the occurrence of hydroxylamine, an essential reagent for the glyoxylic-oxime path of HCN formation.

Though it is our impression that small amounts of HCN can be formed by most if not all green algae, it is not possible yet to generalize about the significance of HCN formation in metabolic control, either of nitrate reduction or of anything else. More information is needed. The best algal HCN former which we have encountered to date is the prokaryote, *Anacystis nidulans* (Pistorius et al., 1979; Pistorius and Voss, 1980). In this case, a L-amino acid oxidase specific for basic amino acids is responsible for HCN generation from histidine. The enzyme seems to have a role in the dark respiration of these cells, and its interplay with the redox reactions of photosynthesis is being explored.

2.5 Reactivation in Vivo: Function

Relatively little work has been done as yet on the reactivation in vivo of the reversibly inhibited nitrate reductase of *Chlorella vulgaris,* though it has been demonstrated that the reaction can take place (Solomonson and Spehar, 1977). Reactivation of inactive enzyme in vivo, without de novo enzyme synthesis, has also been described in other algae (Aparicio and Maldonado, 1979). Blue light and flavin may have an important role in this reactivation, but this has so far only been demonstrated in vitro (Aparicio et al., 1976; Roldán et al., 1978; Aparicio and Maldonado, 1979).

Howard and Solomonson (1978) have shown that the HCN nitrate reductase complex is less susceptible to degradation by trypsin than the active enzyme. This suggests that the HCN-inactivated enzyme may be more stable in vivo than the active molecule. In fact, when nitrate-grown cells are transferred to NH_3, the level of active enzyme declines more rapidly with time than the level of reversibly inactivated enzyme (see for example Johnson, 1979). Thus, the HCN-inactivated complex may represent a relatively stable storage form of the enzyme which is degraded relatively slowly and is thus available for reactivation if the environment should suddenly provide nitrate instead of ammonia (see also Yamaya et al., 1980). In any case, however, the occurrence of ferricyanide-activatable enzyme in different unicellular algae is highly variable (Hipkin et al., 1979).

3 Mechanism II: Degradation and Assembly or Activation

The second mechanism for controlling the level of nitrate reductase has to do with the assembly and degradation of the enzyme. The partial catalytic activities of the intact protein (e. g., the reduction of nitrate by reduced viologens, and the oxidation of NADH by cytochrome c, i. e., diaphorase) have been shown to be subject to the control of different sets of genes in fungi and higher plants (Cove, 1979). Genetic studies of nitrate assimilation have also been undertaken with *Chlamydomonas rheinhardii* (Nichols and Syrett, 1978; Nichols et al., 1978). To my knowledge no such studies have been done with *Chlorella* species. We will show, however, that the synthesis of the enzyme in *C. vulgaris* can be separated in time into several steps in normal wild-type cells. The first steps involve a synthesis of protein and lead to diaphorase formation. The final step is concerned with the incorporation of molybdenum and can occur in the absence of new protein synthesis. The experimental evidence for this conclusion is only partly published and will be summarized in some detail.

3.1 Effect of Ammonia

On the bottom of Fig. 1 there is a dashed line leading from ammonia (or a metabolite of ammonia) to the point where nitrate is penetrating the cell membrane. This is intended to indicate that ammonia (or a metabolite of ammonia) very rapidly inhibits nitrate uptake by *Chlorella* (Morris, 1973; Pistorius et al., 1978; Tischner and Lorenzen, 1979).

Green algae such as *Chlorella* contain virtually no vacuoles and can accumulate no perceptible quantities of nitrate in the cytoplasm. The first locus for the control of nitrate utilization in such algae is undoubtedly the nitrate uptake system (Vennesland and Guerrero, 1979). Ullrich (1979) has concluded, mainly on the basis of experiments with *Ankistrodesmus braunii,* that external factors like CO_2, pH, high salt, and ammonium concentrations act directly on the nitrate carrier and not on the process of nitrate reduction or on the energy supply.

Algae grown on ammonia as a source of nitrogen contain only very low levels of nitrate reductase. The removal of ammonia from ammonia-grown cells often results in the appearance of considerable amounts of the enzyme, whether nitrate is added or not (Morris and Syrett, 1963, 1965; Hipkin and Syrett, 1977). Thus, nitrate is not specifically required for enzyme synthesis. *Chlorella vulgaris* requires nitrate to achieve the highest possible nitrate reductase levels, however. Funkhouser et al. (1980) and Funkhouser and Ramadoss (1980) have used the protein synthesis inhibitor, cyclo heximide, combined with labeling of nitrate reductase by radioactive arginine, and specific immunochemical techniques, to study the process of enzyme induction which occurs when ammonia-grown *Chlorella vulgaris* cells are transferred to a nitrate-containing growth medium. There is first a lag period of about an hour before the level of active enzyme rises rapidly. If cycloheximide is added at the beginning of the induction period, together with nitrate, then nitrate reductase activity never appears; the "induction" is completely inhibited. But if cycloheximide is added later, after the rise in enzyme activity has commenced, then the cycloheximide does not completely prevent the accumulation of additional enzyme activity. The rise in nitrate reductase level continues for a while, though growth stops eventually, of course. If labeled arginine is added after cycloheximide, there is no incorporation of label from added arginine into newly formed active enzyme during the "post-cycloheximide" rise in enzyme activity. If the labeled arginine is added prior to cycloheximide, for example at the time of nitrate addition, then the postcycloheximide rise in active enzyme is associated with movement of label into the active enzyme

These results show that the enzyme synthesis occurs in at least two steps, as outlined in Eq. (1). Only the second of these steps is not sensitive to cycloheximide.

$$\text{amino acids} \xrightarrow{\text{I}} \text{protein precursor} \xrightarrow{\text{II}} \text{active nitrate reductase} \tag{1}$$

Step I is a complex process, and can be separated in *Chlorella* into two phases, which we may call I a and I b. Phase I a is the situation in ammonia-grown cells. Funkhouser and Ramadoss (1980) have shown that such cells contain considerable amounts of protein which bind to antibodies to purified nitrate reductase and must therefore be immunologically related to the enzyme. This protein material contains a large number of components of different molecular size, mostly smaller than nitrate reductase. The implication is that most of them are degradation products. One gets an impression that nitrate reductase precursor is constantly being synthezised and degraded in ammonia-grown cells.

The other line of evidence has been provided by Johnson (1979). He used *Chlorella vulgaris,* and the same method of inducing nitrate reductase in ammonia-grown cells.

But he used density labeling with heavy water followed by gradient centrifugation to detect newly synthesized enzyme. With these different methods, Johnson concluded that at least 50% of the active enzyme which appears after 4 h of induction by nitrate represents activation (or assembly) of protein already synthesized and present in ammonia-grown cells.

Thus, different laboratories using different methods have come to the same conclusion. When we speak of assembly and Johnson speaks of activation, note that we are speaking of the same process.

The conclusion that the messenger for nitrate reductase is constantly being translated in ammonia-grown *Chlorella* cells is supported by the demonstration with the use of inhibitors (Hipkin and Syrett, 1977) that the messenger must be present in ammonia-grown *Chlorella fusca,* a different species. It should be noted, however, that nitrate is regarded as an inducer of de novo enzyme synthesis in the case of the fungal nitrate reductase (Cove, 1979; Bahns and Garrett, 1980). No material which cross-reacts with antibodies to nitrate reductase could be found in extracts of ammonia-grown *Neurospora crassa* (Amy and Garrett, 1979).

3.2 Effect of Nitrate and Molybdenum

Hewitt et al. (1979) in particular have discussed the assembly of nitrate reductase from its components in higher plants and proposed a model system for regulation of the enenzyme level by assembly and disassembly. They attribute an important role here to Mo. Serological techniques have also been used to show that Mo-deficient plants contain protein which binds antibodies to nitrate reductase but which is inactive enzymatically (Notton et al., 1974).

Phase Ib of nitrate reductase synthesis in *Chlorella vulgaris* is the situation brought about by addition of nitrate without Mo to ammonia-grown cells. The latter cells do not store Mo in a form which is readily available for the synthesis of nitrate reductase. Thus, addition of nitrate without Mo results only in a very slow synthesis of active nitrate reductase, from the trace amounts of Mo present in the other mineral salts. When molybdate is added together with nitrate, the rate of appearance of enzyme is strongly dependent on the concentration of added molybdate, as shown by the experiment of Fig. 2 (Gewitz and Piefke, unpubl.). The figure shows the level of nitrate reductase reached 3 h after nitrate addition, as a function of the negative log of the concentration of molybdate added with the nitrate.

When ammonia-grown cells are incubated with nitrate in the absence of added Mo, the diaphorase level rises to very high values. This diaphorase is related to nitrate reductase, since Ramadoss (unpubl.) has found that it is inhibited by antibodies raised to purified enzyme. Thus, the main effect of nitrate is to induce diaphorase and whatever else is needed for incorporation of Mo. If nitrate is added first, followed later by molybdenum, it can be shown that a considerable portion of the synthesis of active enzyme, after Mo addition, is not sensitive to cycloheximide. Thus, we regard the incorporation of Mo as the last step in the synthesis of active enzyme [step II of Eq. (1)]. Whatever protein components may be required can be preformed prior to Mo addition. If a peptide component synthesized on cytoplasmic ribosomes is part of the

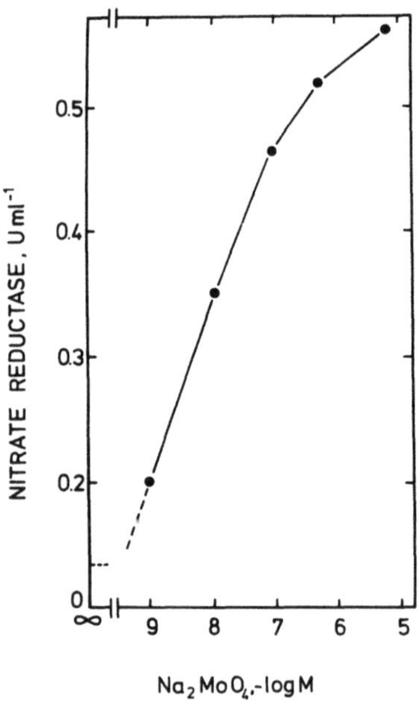

Fig. 2. Dependence of nitrate reductase induction rate on molybdate cencentration. *Chlorella* cells grown on ammonia were washed and added to nitrate-mineral salts medium under standard growth conditions, except that the added molybdate was adjusted to the indicated concentrations in separate flasks. Cells were harvested after 3 h and broken cell extracts were analyzed for nitrate reductase after ferricyanide activation. The [Mo] under our standard conditions is 10^{-7} *M*, but there is enough Mo impurity in the other salts to give a concentration of about 5×10^{-10} *M* when no Mo is added. This is sufficient to allow eventual cell growth that is almost as fast as that with added Mo, though the cells have a somewhat lower chlorophyll content than normal

Mo moiety, as studies of other Mo enzymes imply (Johnson, 1980; Johnson et al., 1980), then this peptide is preformed along with diaphorase. The effect of the absence of Mo on diaphorase synthesis is similar to the effect of W addition in the presence of Mo, and suggests that the W acts mainly by preventing Mo uptake (see also Hewitt et al., 1979 and Pienkos et al., 1980).

One of our working hypothesis has acquired the form shown in Fig. 3. Here we visualize the assembly for final activation of nitrate reductase by molybdenum incorporation as a process occurring in a membrane, perhaps the outer membrane of the *Chlorella* cell. P represents protein or a protein complex, which may contain diaphorase, and can also stand for molybdenum "permease," if you will. We think of the molybdenum as a substance which is taken up only when a suitable protein acceptor is present. Nitrate is necessary for the synthesis of P, and ammonia probably inhibits its function.

In the paper of Johnson (1979) previously cited, the conclusion was reached that nitrate stimulates the turnover of nitrate reductase and that ammonia inhibits the turnover virtually completely. Incorporation of label from heavy water was taken as an indicator of turnover, but the presence of enzyme was measured by determination of

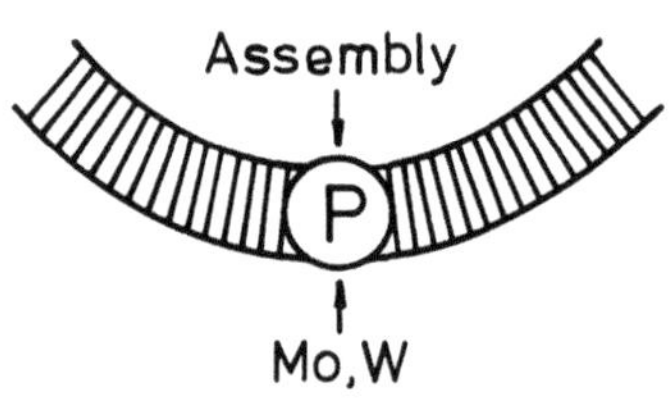

Fig. 3. The possible nature of the relationship between activation or assembly of nitrate reductase and Mo or W uptake. *P*, protein = assembler = permease. The *hatched area* represents *Chlorella* cell membrane

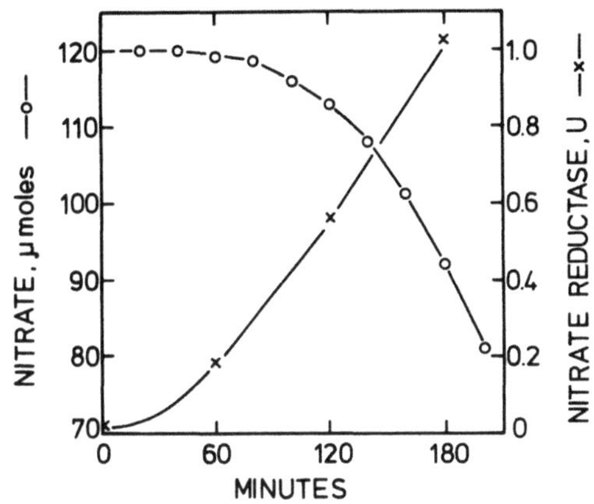

Fig. 4. Induction of nitrate reductase activity and nitrate uptake by *Chlorella* cells grown on ammonia. *Numbers* represent total amounts for 250 ml suspension containing 0.75 ml cells. Conditions are those used for growth, but with 0.5 m*M* KNO_3. Cells were washed and added to nitrate medium at t=0. Methods were as previously described (Pistorius et al., 1978). Nitrate reductase was measured after ferricyanide activation

its activity. At the same time, the conclusion was reached that protein present in ammonia-grown cells must contribute to the final active enzyme induced by nitrate. If ammonia inhibits the final process of Mo incorporation, then it would inhibit "turnover" as measured by Johnson's procedure.

Figure 3 contains no detail, and in this respect it is certain to be wrong. It is a rather primitive hypothesis which can easily be elaborated or abandoned. The suggestion that the enzyme may be assembled in the outer membrane is reminiscent of a different hypothesis previously presented by Butz and Jackson (1977). The latter authors were seeking to explain the apparent coordinate regulation of nitrate reductase and nitrate uptake. This phenomenon is striking in algae without vacuoles. During induction of nitrate reductase in ammonia-grown *Chlorella vulgaris*, uptake lags slightly behind the appearance of active enzyme, as shown in Fig. 4 (Gewitz and Piefke, unpubl., see also Tischner and Lorenzen, 1979). Butz and Jackson postulated that the nitrate reductase is embedded in the outer membrane of both cell and chloroplast and that it operates both as a transporter and as a reducer of nitrate. This is hard to reconcile with the fact that virtually all of the nitrate reductase of algae grown on nitrate is readily soluble. But it seems quite possible that the locus of enzyme assembly is in the outer membrane, and that this locus is also that of nitrate transport. The coordinated response of uptake and synthesis to ammonia and to nitrate and carbohydrate (light and CO_2) could thus be explained.

3.3 Enzyme Fluctuation in Synchronized Algae

Chlorella cells which are growing synchronously in alternating periods of light and dark show a striking "circadian" rhythm in the level of nitrate reductase. The enzyme level is quite low at the end of the dark period and then rises rapidly in the beginning of the light period. Tischner (1976) showed that these changes can not be explained by reversible inactivation and activation of active enzyme. He first concluded that de novo synthesis of enzyme was involved during rapid enzyme increase on illumination, but Tischner and Hüttermann (1979) later came to the conclusion on the basis of density labeling and isopycnic centrifugation that the increase in enzyme activity involved an activation of preexisting protein, at least in part. The procedure was analogous to that used in Johnson's (1979) studies of nitrate reductase induction in ammonia-grown *Chlorella,* and the conclusion was similar, though the synchronized *Chlorella* have nitrate available all the time. The rapid increase in enzyme activity of synchronized cells was shown to require light and CO_2 (Tischner and Lorenzen, 1979). Griffiths (1979) has shown that added glucose in the dark has the same effect as light and CO_2, provided the *Chlorella* cells can grow and divide on glucose and nitrate in the dark. Data on the diaphorase content of the synchronized *Chlorella* is needed to determine the nature of the process affected. It has been shown to be associated with nitrate uptake, and one can guess that it is associated with molybdate uptake as well.

References

Amy NK, Garret RH (1979) Anal Biochem 95:97–107
Aparicio PJ, Maldonado JM (1979) In: Hewitt EJ, Cutting CV (eds) Nitrogen assimilation of plants. Academic Press, London New York, pp 207–215
Aparicio PJ, Roldán JM, Calero F (1976) Biochem Biophys Res Commun 70:1071–1077
Bahns M, Garrett RH (1980) J Biol Chem 255:690–693
Buczek J (1973) Acta Soc Bot Pol 42:223–232
Butz RG, Jackson WA (1977) Phytochemistry 16:409–417
Cantley LC, Josephson L, Warner R, Yanagisawa M, Lechene C, Guidotti G (1977) J Biol Chem 252:7421–7423
Chaparro A, de la Rosa MA, Vega JM (1979) Z Pflanzenphysiol 95:77–85
Cove DJ (1979) In: Hewitt EJ, Cutting CV (eds) Nitrogen assimilation of plants. Academic Press, London New York, pp 289–305
Funkhouser EA, Ramadoss CS (1980) Plant Physiol 65:944–948
Funkhouser EA, Shen TC, Ackermann R (1980) Plant Physiol 65:939–943
Gewitz H-S, Pistorius EK, Voss H, Vennesland B (1976a) Planta 131:145–148
Gewitz H-S, Pistorius EK, Voss H, Vennesland B (1976b) Planta 131:149–153
Gewitz H-S, Piefke J, Vennesland B (1978) Planta 141:323–328
Gewitz H-S, Piefke J, Langowska K, Vennesland B (1980) Biochim Biophys Acta 611:11–26
Giri L, Ramadoss CS (1979) J Biol Chem 254:11703–11712
Griffiths DJ (1979) New Phytol 82:427–437
Hewitt EJ, Hucklesby DP, Mann AF, Notton BA, Rucklidge GJ (1979) In: Hewitt EJ, Cutting CV (eds) Nitrogen assimilation of plants. Academic Press, London New York, pp 255–287
Hipkin CR, Syrett PJ (1977) J Exp Bot 28:1270–1277
Hipkin CR, Syrett PJ, Al-Bassam BA (1979) In: Hewitt EJ, Cutting CV (eds) Nitrogen assimilation of plants. Academic Press, London New York, pp 309–312
Howard WD Jr, Solomonson LP (1978) Proc 176th Natl Am Chem Soc Meet, Abstr 89
Howard WD Jr, Solomonson LP (1980) Fed Proc Abstr
Johnson CB (1979) Planta 147:63–68
Johnson JL (1980) In: Coughlan M (ed) Molybdenum and molybdenum-containing enzymes. Pergamon Press, Oxford New York Toronto Sydney Paris Frankfurt, pp 345–383
Johnson JL, Hainline BE, Rajagopalan KV (1980) J Biol Chem 255:1783–1786
Lorimer GH, Gewitz H-S, Völker W, Solomonson LP, Vennesland B (1974) J Biol Chem 249: 6074–6079
Losada M (1974) In: 3rd Int Symp Metab Interconvers Enzymes, Seattle. Springer, Berlin Heidelberg New York, pp 257–270
Maldonado JM, Herrera J, Paneque A, Losada M (1973) Biochem Biophys Res Commun 51:27–33
Meisch H-U, Bauer J (1978) Arch Microbiol 117:49–52
Morris I, Syrett PJ (1963) Arch Mikrobiol 47:32–41
Morris I, Syrett PJ (1965) J Gen Microbiol 38:21–28

Morris J (1973) In: Stewart WDP (ed) Algal physiology and biochemistry, Blackwell, Oxford London Edinburgh Melbourne, pp 583–609
Nichols GL, Syrett PJ (1978) J Gen Microbiol 108:71–77
Nichols GL, Shehata SAM, Syrett PJ (1978) J Gen Microbiol 108:79–88
Notton BA, Graf L, Hewitt EJ, Povey RC (1974) Biochim Biophys Acta 364:45–58
Pienkos PT, Shah VK, Brill WJ (1980) In: Coughlan M (ed) Molybdenum and molybdenum-containing enzymes. Pergamon Press, Oxford New York Toronto Sydney Paris Frankfurt, pp 385–401
Pistorius EK, Voss H (1977) Biochim Biophys Acta 481:395–406
Pistorius EK, Voss H (1980) Biochim Biophys Acta 611:227–240
Pistorius EK, Gewitz H-S, Voss H, Vennesland B (1976) Planta 128:73–80
Pistorius EK, Gewitz H-S, Voss H, Vennesland B (1977) Biochim Biophys Acta 481:384–394
Pistorius EK, Funkhouser EA, Voss H (1978) Planta 141:279–282
Pistorius EK, Jetschmann K, Voss H, Vennesland B (1979) Biochim Biophys Acta 585:630–642
Ramadoss CS (1979) Planta 146:539–544
Roldán JM, Calero F, Aparicio PJ (1978) Z Pflanzenphysiol 90:467–474
Simons TJB (1979) Nature (London) 281:337–338
Solomonson LP (1974) Biochim Biophys Acta 334:297–308
Solomonson LP (1975) Plant Physiol 56:853–855
Solomonson LP (1979) In: Hewitt EJ, Cutting CV (eds) Nitrogen assimilation of plants. Academic Press, London New York, pp 199–205
Solomonson LP, Spehar AM (1977) Nature (London) 265:373–375
Solomonson LP, Spehar AM (1979) J Biol Chem 254:2176–2179
Syrett PJ, Morris I (1963) Biochim Biophys Acta 67:566–575
Tischner R (1976) Planta 132:285–290
Tischner R, Hüttermann A (1979) Plant Physiol 62:284–286
Tischner R, Lorenzen H (1979) Planta 146:287–292
Ullrich WR (1979) Ber Dtsch Bot Ges 92:273–284
Vennesland B, Guerrero MG (1979) In: Gibbs M, Latzko E (eds) Encyclopedia of plant physiology, new series, vol VI. Springer, Berlin Heidelberg New York, pp 425–444
Yamaya T, Solomonson LP, Oaks A (1980) Plant Physiol 65:146–150

Regulation of Nitrate Uptake in Green Algae and Duckweeds
Effect of Starvation and Induction

W. R. ULLRICH, H.-D. SCHMITT, and E. ARNTZ[1]

1 Introduction

Nitrate assimilation has been shown to be regulated in many cases by changes in the activity of nitrate reductase. This can occur by genetic induction and repression or via direct activity changes, i. e., by activation and inactivation of the enzyme (reviews of Beevers and Hageman, 1980; Losada and Guerrero, 1979; Vennesland and Guerrero, 1979). Nitrate reductase synthesis is mainly induced by nitrate and light and repressed by ammonia and amino acids. Compared with nitrate reduction, nitrate uptake as a process subject to regulation has been little investigated.

Previous papers reported on the regulation of nitrate uptake by external factors in the green microalgae *Ankistrodesmus braunii* and *Chlorella fusca.* Nitrate uptake was shown to be stimulated by CO_2 or other carbon sources, especially at low external pH, and by light, and inhibited by chloride and even by higher concentrations of nitrate (summarized in Ullrich, 1979). Monochromatic red or blue light was much less efficient in nitrate assimilation than white or red plus blue light (Calero et al., 1980), probably due to excess cyclic photophosphorylation supplying more energy for nitrate uptake in white light. Under respiratory conditions nitrate uptake is only enhanced by blue light, either by increased respiration or by direct blue light stimulation at the plasmalemma.

Membrane potential studies in *Lemna gibba* suggest the operation of a NO_3^-/H^+-cotransport mechanism, the extrusion of excess protons being the energy-requiring step (Novacky et al., 1978; Ullrich and Novacky, unpubl.).

The present paper deals with properties of nitrate uptake indicating the involvement of special transport proteins in both organisms, *Ankistrodesmus braunii* and *Lemna gibba,* and with various limitations by intracellular transport in *Lemna.*

2 Materials and Methods

2.1 Culture conditions

Ankistrodesmus braunii (Naeg.) Brunnth., strain 202-7d from the algal collection in Göttingen was cultured synchronously with a 14:10 h light-dark rhythm at 29 °C (cf. Eisele and Ullrich, 1977).

1 Institut für Botanik der Technischen Hochschule, Schnittspahnstr. 3, 6100 Darmstadt/FRG

Lemna gibba G1 (obtained from Prof. R. Kandeler, Wien) was grown under short-day conditions at 26 °/21 °C. Stock cultures were grown according to Pirson and Seidel (1950) under sterile conditions with 1% sucrose. Prior to the experiments the plants were transferred to an open culture with merely inorganic, nitrate-free medium for up to 5 days.

2.2 Experiments

One hour after the onset of the light period, i. e., after the first phase of light activation (cf. Tischner and Lorenzen, 1979), the *Ankistrodesmus* cells were collected on a membrane filter, washed with and resuspended in fresh medium with or without nitrate, and kept under culture conditions, from where samples were taken for assaying the uptake rates. Cycloheximide was added to one of these cultures at the time indicated in the figures. Nitrate uptake assays were performed in Plexiglas cuvettes placed in a water bath at 29 °C and illuminated from both sides (200 klx). Aeration was the same as in the culture (air+2% CO_2). At zero time 1 m*M* KNO_3 was added. Samples were removed and filtered at 10 s, 5, 10, 20, and 30 min for estimation of nitrate uptake rates. Disappearance of nitrate was determined from decreasing absorbance at 203 nm (Calero et al., 1980). The uptake rates were related to chlorophyll.

Experiments with *Lemna* were performed in 25 ml Erlenmeyer flasks placed on a rack in a photo-Warburg apparatus (Braun, Melsungen, W. Germany), using between 100 and 200 mg fresh weight. KNO_3 was applied at 0.1 m*M*, the pH was 5.7, if not otherwise stated. The incubation was started by adding the plants to the nitrate-containing medium (4–6 ml). Samples of the medium were removed at 10 s, 15 and 30 min, and nitrate was determined as in experiments with algae. Uptake rates were related to fresh weight.

3 Results

3.1 Stimulation of Nitrate Uptake in Ankistrodesmus by Nitrate Starvation

Comparable to the increasing phosphate uptake rate during the first few hours of phosphate starvation (Jeanjean, 1973), the rates of nitrate uptake increase with increasing nitrate starvation in *Ankistrodesmus.* According to the shorter half-life of the reaction systems in nitrate metabolism this increase is restricted to the first 2–3 h, reaching more than double rates, and is then followed by a decline (Fig. 1). The situation is principally the same at low or high CO_2 or O_2 concentrations (Eisele, 1976). In air (0.03% CO_2) the increase is similar to that in CO_2-saturated air, but the optimum appears only after 3–4 h, whereas in N_2 it is lower and already reached after 2 h starvation. The present experiments are restricted to saturating CO_2 (2%). The results are summarized in Fig. 1 and show the changing nitrate uptake rates during 6 h, together with inhibition experiments.

Cycloheximide (CHI) is currently used as a tool to show dependence on ribosomal protein synthesis in the cytoplasmic cell compartment. By adding 10 μ*M* of this in-

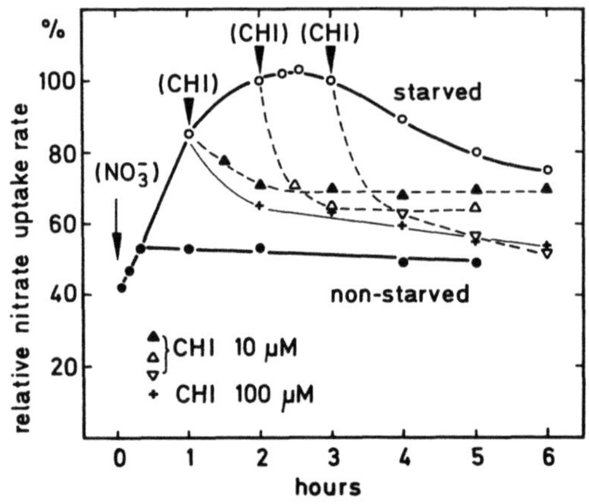

Fig. 1. Stimulation of nitrate uptake in *Ankistrodesmus braunii* by nitrate starvation, and inhibition by cycloheximide *(CHI)*. Light, pH 6.5, 29 °C, additions indicated by *arrows*. 20–30 µmol NO_3^- mg^{-1} chlorophyll h^{-1} = 100%

hibitor any further increase in nitrate uptake rates is stopped and the rates are even inhibited to a level that seems to be independent of the extent of induction before CHI addition. When CHI is added after 4 h starvation or at higher concentration, the inhibition is more severe and increases with time (Fig. 1). In nonstarved algae CHI is similarly inhibitory.

During investigation of the specificity of CHI, manometric experiments showed that 10 µ*M* CHI almost instantly also inhibit respiratory O_2 consumption by 20% and photosynthetic O_2 evolution by 54%, the data for 1 µ*M* being very similar.

3.2 Induction of Nitrate Uptake in Lemna gibba

Induction plays an even more important role in *Lemna* than in *Ankistrodesmus*. When *Lemna* plants are depleted of nitrate and then used immediately for the experiments, the uptake rates are low but show a nonlinear increase. Nonstarved and thus "induced" plants take up nitrate at medium rates of 0.5–0.7 µmol g^{-1} fr. wt. h^{-1}. Nitrate-starved plants, when induced by low nitrate concentrations for a short time, show uptake rates of more than 2.5 µmol NO_3^- g^{-1} fr. wt. h^{-1}.

The time course of induction and its dependence on nitrate is shown in Fig. 2. Plants starved for 3 days finally lose their ability to immediately absorb nitrate within 7 h in the light. Upon addition of 0.1 m*M* KNO_3 to plants pretreated in the same way, the first effect of induction appears after 1 h. The uptake rate then increases linearly with time up to saturation after 6 h of nitrate incubation. These induced rates remain constant for a fairly long time (10–20 h).

Thus for constant experimental conditions the plants were induced by addition of 0.1 m*M* KNO_3 (which is consumed in 2–3 h) already at the onset of the preceding dark period, and the starvation time before induction could be varied between 1 and 5 days. The diurnal time course of the nitrate uptake rates in scarcely and moderately starved plants is shown in Fig. 3 and is compared with plants whose starvation has been relieved by 0.1 m*M* KNO_3 during the whole experiment. In spite of the preceding induction, maximum uptake rates are reached only after 1.5 h in the light and only in

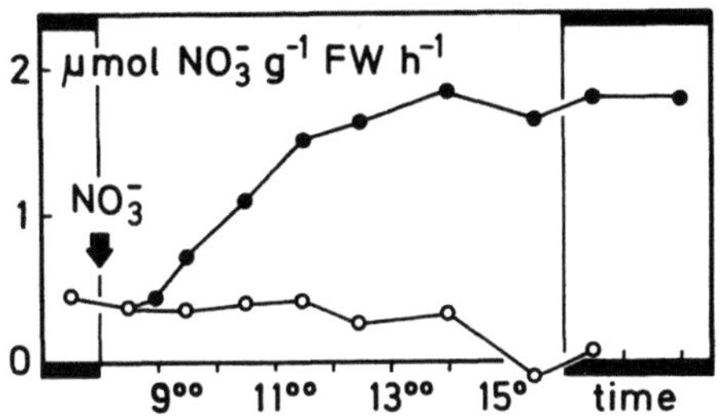

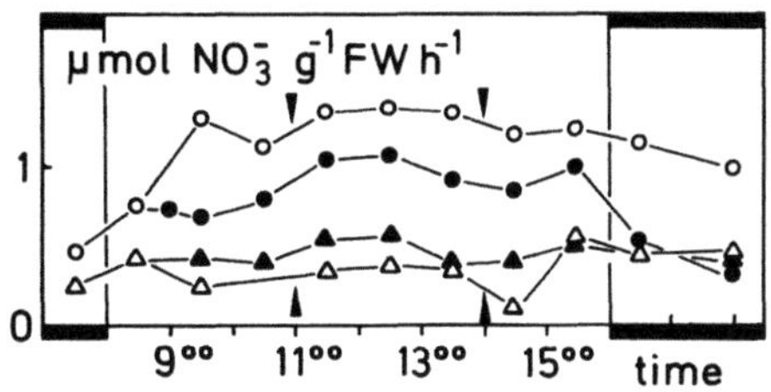

Fig. 2. Induction of nitrate uptake in *Lemna gibba* by nitrate. Plants nitrate-starved for 3 days, ○ without nitrate, ● 0.1 m*M* KNO_3 added at the onset of the light period *(arrow)*. Light, pH 5.7, 26 °C

Fig. 3. Diurnal time course of nitrate uptake in *Lemna gibba* induced during the preceding dark period (see text). *triangles* starvation for 1 day; *circles* starvation for 3 days; *black symbols* 0.15 m*M* KNO_3 added at onset of light period; *open symbols* no nitrate before uptake assay. Uptake assays in the light (light period) and dark (dark period, indicated by *black bars*). pH 5.7, 26 °C, 0.1 m*M* KNO_3 used in uptake assays

plants kept nitrate-free before the uptake assay. The following experiments were per-uration is reached already at 40 μ*M* NO_3^-, half-saturation in the order of 8–10 μ*M*, The uptake rates in the light (light period) are similar to those measured in the dark (following dark period).

3.3 Influence of Nitrate Starvation on Induced Nitrate Uptake in Lemna

Nitrate uptake is independent of the external concentration over a wide range. Saturation is reached already at 40 μ*M* NO_3^-, half-saturation in the order of 8–10 μ*M*, and there is still considerable uptake at concentrations below 1 μ*M*. The saturation rate, but not the half-saturation "constant" depends on the starvation of the plants (Fig. 4).

Nitrate is taken up by *Lemna gibba* in a wide pH range, i. e., between pH 3 and 9. After short nitrate starvation a plateau-like optimum is found between pH 4.5 and

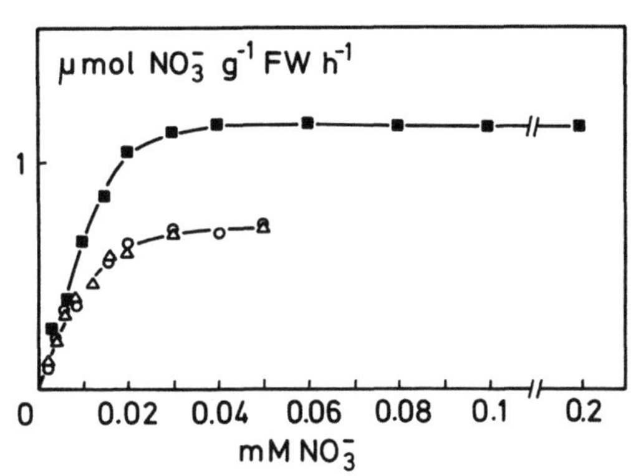

Fig. 4. Dependence of nitrate uptake on half initial nitrate concentrations in *Lemna gibba*. *Open symbols* plants starved for 2 days, *black symbols* plants starved 3 times for 2 days within 7 days. Light, pH 5.7, 26 °C

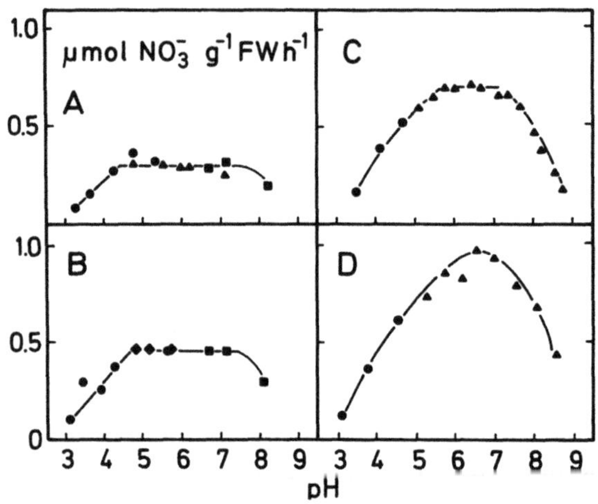

Fig. 5A–D. pH-dependence of nitrate uptake in *Lemna gibba* with increasing nitrate starvation. 0.1 m*M* KNO_3 used in cultures and experiments. **A** nitrate starvation for 2 days; **B**, **C**, and **D** 2x2 days starvation within 12, 8, and 5 days, respectively. Light, 26 °C

7.3, with low rates of only 0.3 µmol $NO_3^- g^{-1}$ fr. wt. h^{-1} (Fig. 5A). Increasing starvation produces increasing optimum rates and the plateau is more and more restricted to a medium pH (Fig. 5B,C). Only after several 2-day periods (Fig. 5D) or constant starvation for five or more days is a high and sharp peak found at pH 6.5. Additional carbon sources as glucose do not alter the pattern, but enhance the absolute uptake rates by a similar factor of 1.2–1.4 over the whole pH range. This is in agreement with *Chlorella fusca*, but not with *Ankistrodesmus*, where the pH dependence is abolished in a wide range by glucose or CO_2 at saturation (Eisele and Ullrich, 1977).

Also the temperature response of nitrate uptake in *Lemna gibba* depends very much on the preceding starvation period (Fig. 6). After starvation for 1 day the low uptake rates show one optimum at 35 °C and a shoulder at 28 °C. After 2 or 3 days a second optimum appears at 28 °C, which dominates after 4 days and still after

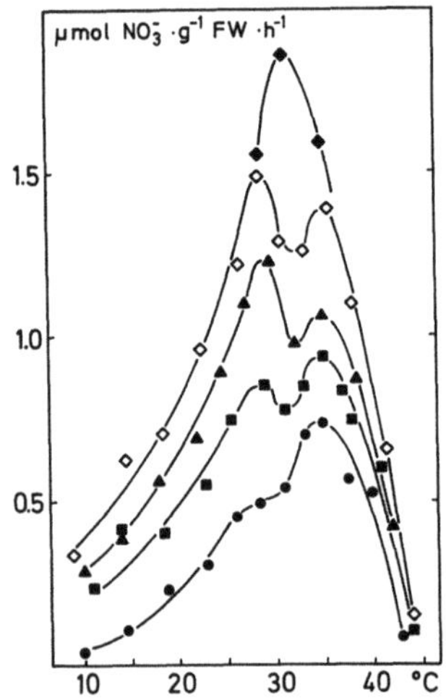

Fig. 6. Temperature dependence of nitrate uptake in *Lemna gibba*. Starvation: ● for 1 day, ■ for 2 or 3 days, ▲ for 4 days, ◇ for 5 days, but interrupted for 3 h (3rd day), ◆ for 5 days, Light, pH 6.5, 0.1 m*M* KNO_3

5 days starvation when interrupted for a short time. Only after 5 days of complete starvation does the valley at 31 °C disappear and is replaced by a sharp single peak.

This irregular pH and temperature dependence must be interpreted with limitations by other processes than the uptake in incompletely starved *Lemna* plants. Correspondingly, at pH 4.3, where the uptake step is expected to be limiting, only one temperature peak at 31 °C is found, independent of the starvation time (data not shown).

After application of tungstate (0.1 m*M*) one day before the experiments, the double peak of medium-starved plants remains unchanged, but the uptake rates are lower than in the control experiments (data not shown).

4 Discussion

4.1 Nitrate Uptake Stimulation in Ankistrodesmus by Starvation

When algae are transferred to a nitrate-free medium, their ability to take up nitrate increases during the first 3 h, then shows a decrease. This temporary increase is in contrast to the activity of nitrate reductase as observed in algae (Oesterheld, 1971) as well as in higher plants (Ferguson, 1969; Orebamjo and Stewart, 1974) and supports the view that nitrate uptake and nitrate reductase activity are induced independently of each other (cf. Tischner and Lorenzen, 1979). A very similar increase in the rates is seen in phosphate uptake by *Chlorella* (Jeanjean, 1973), but there is no decline after a few hours, suggesting a slower turnover of the system if no phosphate is taken up. Unknown inhibitors disturbing quantitative assays of nitrate reductase activity in extracts of *Ankistrodesmus* prevented a decision as yet, if the decrease after 3 h could also be attributed to the decline in nitrate reductase activity.

The experiments with cycloheximide (CHI) support the view of an induction phenomenon, but besides the inhibition of protein synthesis, CHI shows strong unspecific effects on nitrate uptake and on the rates of photosynthesis and respiration, as reported for various plant material (McDonald and Ellis, 1969; McMahon, 1975). The low specificity is less apparent in the inhibition of phosphate uptake induction (Jeanjean, 1973; Jeanjean and Ducet, 1974), in spite of much higher concentrations applied (0.4 m*M*), probably due to the fact that phosphate is not reduced and can be taken up well at lower rates of photosynthetic electron transport (Ullrich, 1971, 1972). Thus for a final interpretation a more detailed analysis is necessary.

4.2 Nitrate Uptake Induction in Lemna

Nitrate uptake rates in *Lemna gibba* are highly dependent on the induction by nitrate and on starvation. This is consistent with a similar dependence on induction of nitrate reductase activity in Lemnaceae (Ferguson, 1969; Joy, 1969; Orebamjo and Stewart, 1974; Humphrey et al., 1977), as in many higher plants and algae. The time course of induction seems also very similar for nitrate uptake in *Lemna gibba* as for nitrate reductase in *Lemna minor* (Orebamjo and Stewart, 1974; Humphrey et al., 1977), but unlike nitrate reductase the uptake shows saturation already after 6 h (Fig. 2).

In contrast to the microalga *Ankistrodesmus, Lemna* can accumulate nitrate in the vacuoles or transport it from the sites of uptake to other cells and tissues. There is good evidence for independence between nitrate uptake and reduction also in other higher plants with storage pools (Ferrari et al., 1973) and in nitrate reductase-free mutants (Doddema and Otten, 1979).

4.3 The Effect of Starvation on Induced Nitrate Uptake in Lemna

The present paper provides evidence that nitrate uptake can be limited by processes different from the uptake process itself. Nitrate affinity of *Lemna gibba* in induced plants is independent of the saturation rates (Fig. 4). Although this plant is generally found in natural habitats with high nitrate concentrations (Landolt and Wildi, 1977), the half saturation constant of 8–10 μM NO_3^- is lower than that of many higher land plants, even those of sandy and normally poor soils (Doddema and Otten, 1979).

In the pH experiments limitation by processes other than nitrate uptake is indicated by the independence of the external pH in a wide range after incomplete starvation (Fig. 5). The intracellular nitrate pools could be not depleted under these conditions. The same feed-back system may act in the temperature dependence, showing different optima with varying nitrate starvation (Fig. 6). The sharp peak at 31 °C after 5 days starvation is most likely the optimum of nitrate uptake itself. Apparently, the possible uptake rates cannot be reached, unless nitrate is removed from the compartment of its entrance, i. e., the cytosol of the lower epidermis. Thus the peaks at 28 ° and 35 °C in partly nitrate-depleted plants must represent optima of other processes. One of them could be nitrate reduction, but several arguments speak against this possibility: (1) Change from light to dark during the experiments did not change the rates or the peaks at 28 ° and 35 °C, although nitrate reduction is normally greatly enhanced by light. (2) Nitrate reductase shows rapid decay within a few hours after the onset of nitrate starvation in all Lemnaceae investigated as yet (Ferguson, 1969; Orebamjo and Stewart, 1974; Humphrey et al., 1977). Since the small quantities of nitrate used for induction in the dark period before the experiments are consumed within a few hours, the level of nitrate reductase activity must be very low again 16–20 h later. (3) Pretreatment with tungstate did not change the two peaks in partly depleted plants. Any nitrate reduction-dependent peak should be eliminated or reduced in its relative size, because tungstate forms inactive nitrate reductase (Heimer et al., 1969; Vega et al., 1971; Rao and Rains, 1976). The decrease in the uptake rates after tungstate treatment can be explained by direct or indirect inhibition of nitrate uptake (Rao and Rains, 1976).

For these reasons accumulation in the vacuoles and transport to other cells and tissues are regarded as the processes limiting nitrate uptake in scarcely or moderately starved *Lemna* plants, as long as there is little light stimulation (cf. Löppert and Kronberger, 1979). For a reliable association of each peak to a special transport process more detailed analysis is required.

Acknowledgments. The authors want to thank Mrs. Hedwig Conrad, Mrs. Gisela Reutlinger, and Mrs. Lieselotte Bartosch for able technical assistance.

References

Beevers L, Hageman RH (1980) In: Stumpf RK, Conn EE (eds) The biochemistry of plants, vol 5. Academic Press, New York

Calero F, Ullrich WR, Aparicio PJ (1980) In: Senger H (ed) The blue light syndrome. Springer, Berlin Heidelberg New York, pp 411–421

Doddema H, Otten H (1979) Phys Plant 45:339–346

Eisele R (1976) Thesis, Darmstadt

Eisele R, Ullrich WR (1977) Plant Physiol 59:18–21

Ferguson AR (1969) Planta 88:353–363

Ferrari TE, Yoder OC, Filner P (1973) Plant Physiol 51:423–431

Heimer YM, Wray JL, Filner P (1969) Plant Physiol 44:1197–1199

Humphrey TJ, Sarawek S, Davies DD (1977) Planta 137:259–264

Jeanjean R (1973) FEBS Lett 32:149–151

Jeanjean R, Ducet G (1974) In: Zimmermann U, Dainty J (eds) Membrane transport in plants. Springer, Berlin Heidelberg New York, pp 216–219

Joy KW (1969) Plant Physiol 44:849–853

Landolt E, Wildi O (1977) Ber Geobot Inst Eidg Tech Hochsch Stift Ruebel Zuerich 44:104–146

Löppert H, Kronberger W (1979) In: Marcelle R, Clijsters H, Van Poucke M (eds) Photosynthesis and plant development. Junk, The Hague

Losada M, Guerrero MG (1979) In: Barber J (ed) Photosynthesis in relation to model systems. Elsevier, Amsterdam New York Oxford, pp 365–408

McDonald IR, Ellis RJ (1969) Nature (London) 222:791–792

McMahon D (1975) Plant Physiol 55:815–821

Novacky A, Fischer E, Ullrich-Eberius CI, Lüttge U, Ullrich WR (1978) FEBS Lett 88:264–267

Oesterheld H (1971) Arch Mikrobiol 79:25–43

Orebamjo TO, Stewart GR (1974) Planta 117:1–10

Pirson A, Seidel F (1950) Planta 38:431–473

Rao KP, Rains DW (1976) Plant Physiol 57:59–62

Tischner R, Lorenzen H (1979) Planta 146:287–292

Ullrich WR (1971) Planta 100:18–30

Ullrich WR (1972) Arch Mikrobiol 87:323–339

Ullrich WR (1979) Ber Dtsch Bot Ges 92:273–284

Vega JM, Herrera J, Aparicio PJ, Paneque A, Losada M (1971) Plant Physiol 48:294–299

Vennesland B, Guerrero MG (1979) In: Pirson A, Zimmermann MH (eds) Encyclopedia of plant physiology, vol VI. Springer, Berlin Heidelberg New York, pp 425–444

Nitrate Uptake and Reduction in Chlorella – Characterisation of Nitrate Uptake in Nitrate-Grown and Nitrogen-Starved Chlorella sorokiniana

R. TISCHNER and H. LORENZEN[1]

1 Introduction

Investigations of the nitrate-reducing system (i. e., nitrate reductase and nitrite reductase) must include measurements of nitrate uptake. The latter is closely connected with nitrate reduction only in cells with small or no vacuoles. There exist only few experimental conditions to separate uptake and reduction from each other (Ullrich and Eisele, 1977). One of them is the availability of nitrogen-starved cells, which show nitrate uptake and only a quite low nitrate reductase activity. The data presented here will compare and characterize the kinetics of the nitrate uptake carrier of nitrate-grown and nitrogen-starved algae.

2 Materials and Methods

All experiments were carried out with *Chlorella sorokiniana* (strain 211 8k) from the Göttingen algae collection. The algae grow completely synchronously in a light-dark change of 7:5 h at 39 °C, producing eight autospores per cell (Lorenzen, 1961). Nitrogen-starved algae (one light-dark change) grew in a medium lacking nitrate. The ionic strength of the medium was adjusted by addition of KCl, NaCl, and $CaCl_2$ as described by Ssymank (1972). Measurement of nitrate uptake: The algae were harvested after 1 h of illumination, washed and resuspended in 30 ml phosphate buffer (5 m*M* pH 6.8) which was previously gassed with 2% CO_2 in air. The cell number in all experiments was 1.56×10^8 per ml, equivalent to 23.1 μg chlorophyll. Nitrate uptake rates were measured by using Orion ion-selective electrodes and continuously monitoring the signals from the electrodes on a recorder. All additional procedures are indicated in the text or were already described by Tischner and Lorenzen (1979).

1 Institut für Pflanzenphysiologie, Universität Göttingen, Untere Karspüle 2, 3400 Göttingen/FRG

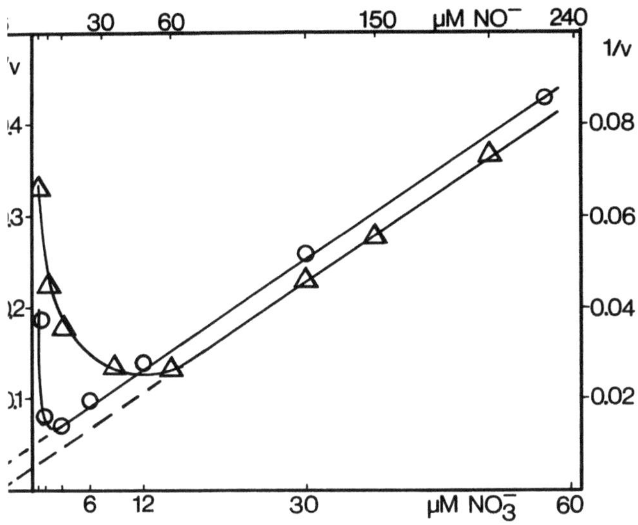

Fig. 1. Rate of nitrate uptake dependent on nitrate concentration in nitrogen-starved — o — *(left ordinate axis, lower abscissa)* and nitrate-grown cells — △ — *(right ordinate axis, upper abscissa)*. K'_s were calculated to be 7.5 *μM* and 15 *μM* respectively

3 Results

3.1 Rate of Nitrate Uptake into Nitrate-Grown and Nitrogen-Starved Algae Dependent on Nitrate Concentration, Temperature, and Inhibitors

The rate of nitrate uptake in nitrate-grown algae is dependent on nitrate concentration. The data fit Michaelis-Menten kinetics up to a nitrate concentration of 60 *μM* (Tischner and Lorenzen, 1979). The apparent K_M was 4.3 *μM*. At higher nitrate concentrations the rate decreases (Fig. 1). No multiphasic nitrate uptake as described by Breteler and Nissen (1978) in roots and earlier by Swader et al. (1975) in *Wolffia* was found. The inhibition by substrate concentrations higher than 60 *μM* may be due to the formation of ineffective complexes of two or more substrate (NO_3^-) molecules combined with the carrier (Bisswanger, 1979). The dissociation constant of this complex (ES_2 K's ES+S) is estimated from a Dixon plot as 15 *μM* (Fig. 1), The smaller the K'_s is the closer in the binding of an additional NO_3^- to the uptake carrier and the more effective is the inhibition of the transport.

Lower rates of nitrate uptake were found in nitrogen-starved algae (one light-dark change without nitrogen source). The rate is also dependent on concentration, but an inhibition soon occurs with 2.4 *μM* nitrate (Fig. 2). The range, in which Michaelis-Menten kinetics is followed, is quite small. From a Lineweaver-Burk plot, a K_M of 1.2 *μM* is calculated. Thus the affinity of the uptake mechanism for nitrate is higher in nitrogen-starved cells than in nitrate-grown algae. From a Dixon plot of the substrate inhibition, a dissociation constant of K'_s =7.5 *μM* may be calculated. This points to a closer binding of excess nitrate to the transport mechanism than in nitrate-grown cells.

The rate of nitrate uptake is also dependent on temperature of the medium. From an Arrhenius plot a Q_{10}=3.3 and an activation energy of 8.3×10^4 J/mol are calculated for nitrate-grown algae (Fig. 3). The same experiment with nitrogen-starved algae gives

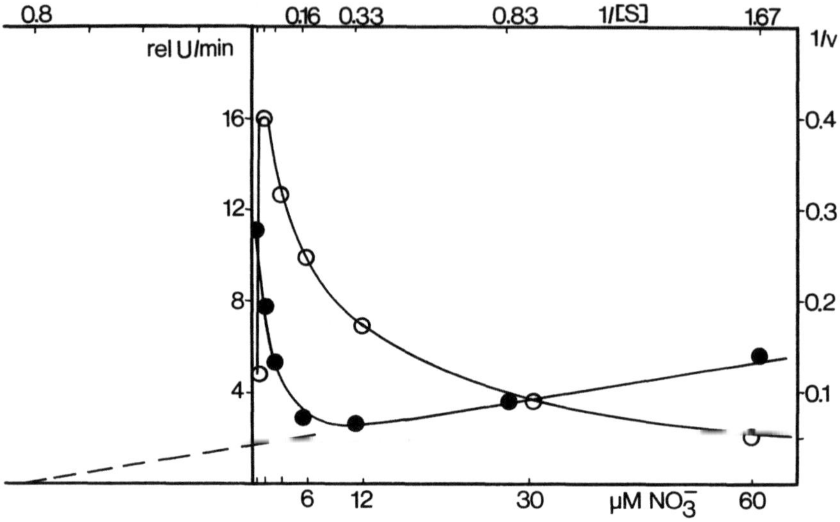

Fig. 2. The rate of nitrate uptake into nitrogen-starved algae dependent on nitrate concentration —○— *(left ordinate axis, lower abscissa)*. The data were additionally presented as Linewaver-Burk plot—●—*(right ordinate axis, upper abscissa)*

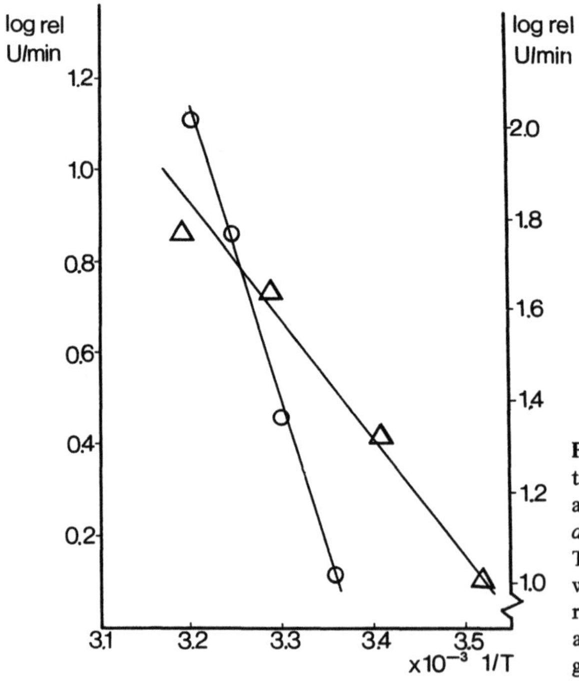

Fig. 3. The rate of nitrate uptake into nitrogen-starved —○— *(left ordinate axis)* and nitrate-grown algae —△— *(right ordinate axis)* dependent on temperature. The Q_{10}-value and the activation energy were calculated as 5.3 and 12.9x10^4 J/mol, respectively, for nitrogen-starved cultures and 3.3 and 8.3x10^4 J/mol for nitrate-grown algae

a Q_{10}=5.3 and an activation energy of 12.9x10^4 J/mol (Fig. 3). This result points to a higher sensitivity to temperature of the nitrate carrier in nitrogen-starved algae. With repect to the higher value of the activation energy one may conclude a change in the transport mechanism. It may also be that the number of carriers for nitrate is reduced in nitrogen-starved algae. These problems can be evaluated, if one succeeds in fixing a measurable competitive inhibitor for nitrate to the carrier. We found 2-nitrobenzaldehyde to be such an inhibitor. This is also true for nitrogen-starved cells, where we determined K_I to be 14–16 *μM* (Fig. 4 a,b).

3.2 The Effect of Ammonium on Nitrate Uptake into Nitrate-Grown and Nitrogen-Starved Algae

A further effect on nitrate uptake can be shown with ammonium ions. Generally one can conclude that the addition of ammonium ions to nitrate-consuming algae reduces the rate of nitrate uptake. For example 12 *μM* NH_4^+ added to 60 *μM* NO_3^- reduces the

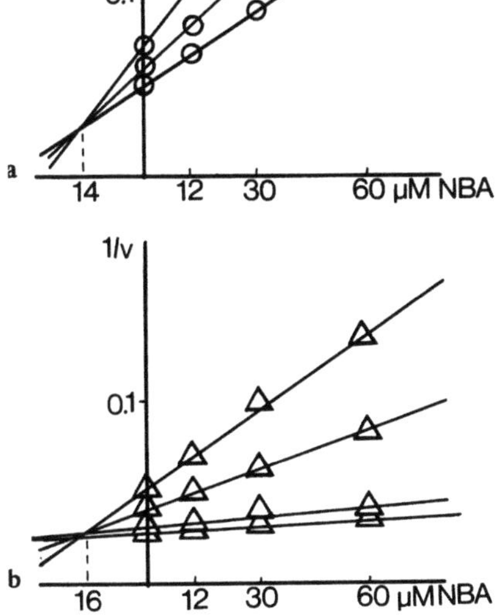

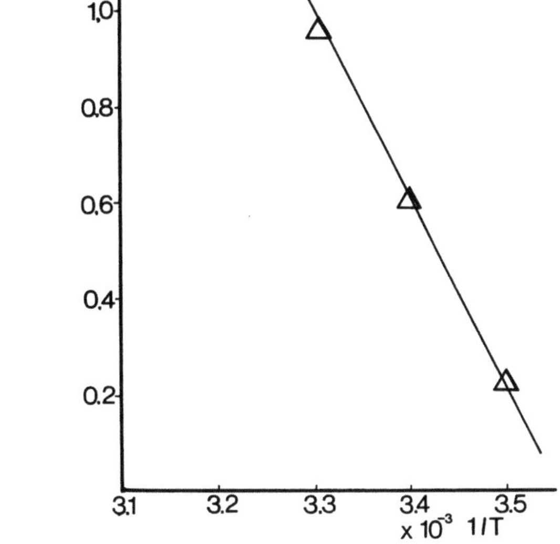

Fig. 4 a, b. The effect of 2-nitrobenzaldehyde (=NBA) on nitrate uptake into nitrogen-starved **a** and nitrate-grown algae **b** K_I were calculated

Fig. 5. The rate of ammonium uptake into nitrate-grown algae dependent on temperature. O_{10}=1.9 and an activation energy=5.1x10^4 J/mol were calculated from the data of this figure

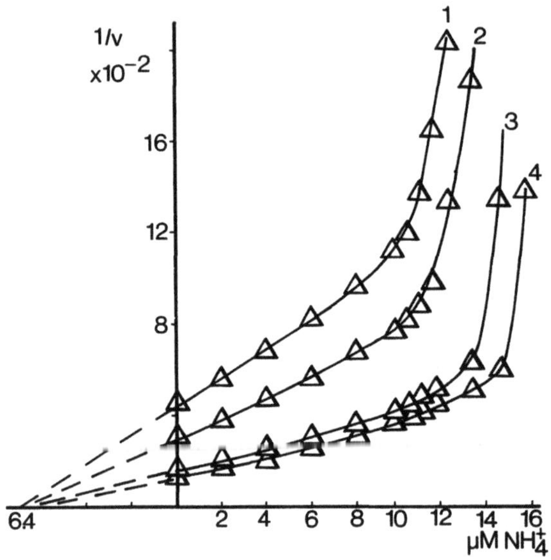

Fig. 6. The titration of the rate of nitrate uptake in nitrate-grown algae with ammonium at different nitrate concentrations. *1:* 6 μM NO_3^-; *2:* 12 μM NO_3^-; *3:* 30 μM NO_3^-; *4:* 60 μM NO_3^-; K_I was calculated to be 6.4 μM

rate to zero within 40 s. A similar result has been reported by Ohmori et al. (1977) in *Anabaena* and by Crosswell and Syrett (1979) in the diatom *Phaedactylum tricornutum*. With such a treatment no effect on nitrate reductase activity can be observed such as is reported for *Skeletonema* by Serra et al. (1978 a). Since the rate of nitrate uptake is restored immediately after the ammonium is consumed (Syrett and Morris, 1963; Tischner and Lorenzen, 1979) we can also calculate the rate of ammonium uptake. This rate is dependent on temperature (Fig. 5). From an Arrhenius plot we calculated Q_{10}=1.9 and an activation energy of 5.1×10^4 J/mol (Fig. 6). These data may point to a passive ammonium transport into the cells. As Figs. 6 and 7 demonstrate, the inhibition of nitrate uptake by ammonium is noncompetitive in both nitrate-grown and nitrogen-starved algae.

The sensitivity of nitrate uptake to ammonium is reduced in nitrogen-starved algae. Addition of 18 μM NH_4^+ together with 6 μM NO_3^- has no effect on the rate of nitrate uptake. But the sensitivity of the nitrate carrier is restored within 10–15 min after the addition of nitrate to the cultures (Fig. 8). Parallel with the rate of nitrate uptake the activity of the nitrate reductase increases (Fig. 9), showing a close relation between these processes. The nitrate reductase activity is regulated by nitrate uptake (Tischner and Lorenzen, 1979; Ullrich, 1979). This was also shown by Shaner and Boyer (1976) using maize leaves.

4 Discussion

As one can conclude from the data presented, the characteristics of the nitrate uptake system in nitrate-grown algae and in nitrogen-starved cells are different in some manner. The system in cells grown without nitrate has a higher affinity to nitrate and is more sensitive to temperature. It seems that during nitrogen starvation the nitrate

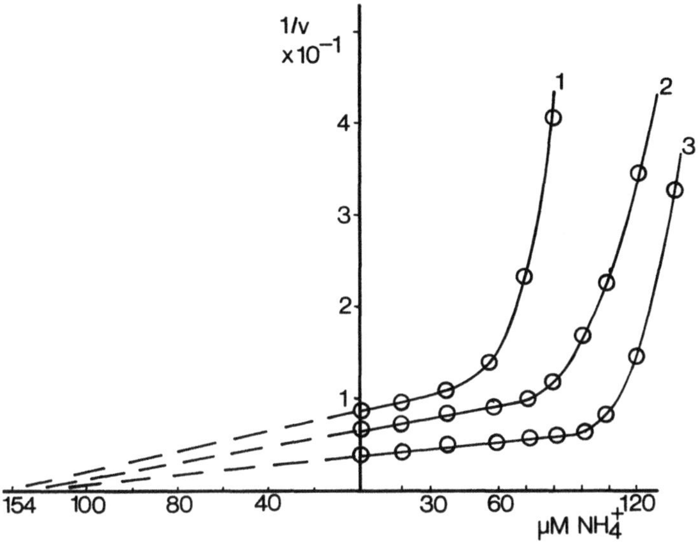

Fig. 7. Titration of the rate of nitrate uptake in nitrogen-starved cells with ammonium at different nitrate concentrations. *1:* 6 μM NO_3^-; *2:* 3 μM NO_3^-; *3:* 1.5 μM NO_3^-. K_I was calculated to be 154 μM

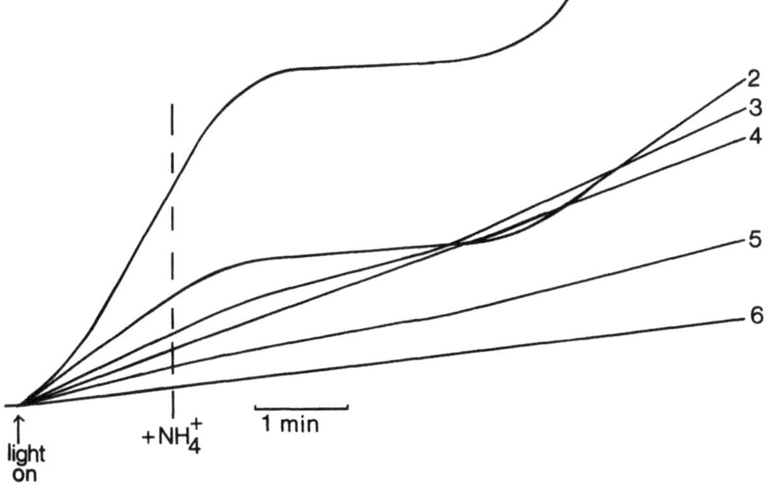

Fig. 8. The effect of NH_4^+ on nitrate uptake into nitrogen-starved cells during recovery from nitrogen starvation by nitrate. Nitrogen-starved cells were illuminated for 1 h and fed with nitrate for different times. *1:* nitrate-grown algae; *2:* 1 h presence of NO_3^-; *3:* 30 min presence of NO_3^-; *4:* 20 min presence of NO_3^-; *5:* 10 min presence of NO_3^-; *6:* nitrogen-starved cells

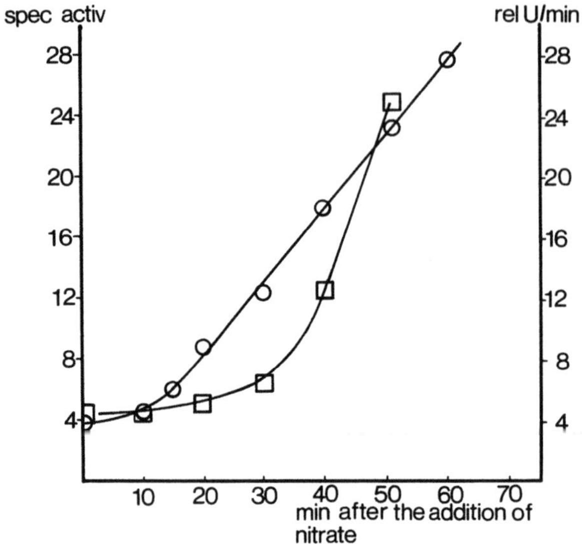

Fig. 9. Increase of nitrate reductase activity and the rate of nitrate uptake dependent on recovery from nitrogen starvation. The cells were grown in nitrogen-deficient medium for 1 LD. At zero time nitrate was added and samples were taken as indicated. — ◻ — nitrate reductase activity; — ○ — rate of nitrate uptake

carrier is changed to higher effectivity. This is also indicated by the strongly reduced sensitivity to ammonium. It seems reasonable that cells thus treated are able to use nitrate and ammonium simultaneously. Reduction equivalents, which are necessary for nitrate reduction, are available from the high level of carbohydrates stored in the cells (Thomas et al., 1976) and from photosynthesis. Thus nitrogen starvation may be overcome quite quickly. This becomes evident during recovery from nitrogen starvation by nitrate. The sensitivity to ammonium is restored parallel to nitrate uptake and nitrate reductase activity. A similar result is reported for nitrogen-depleted dwarf bean by Breteler et al. (1979). Also in nitrogen-starved *Skeletonema,* ammonium severely inhibits the major part of nitrate uptake. Serra et al. (1978b) differentiate between a diffusion process and a carrier-dependent part of nitrate uptake.

The kinetic data for the uptake of ammonium point to a passive transport of this cation over the investigated concentration range. Q_{10} for nitrate and ammonium uptake are quite different in contrast to the results reported by Clarkson and Warner (1979) for roots of *Lolium* sp. Of course, nitrate uptake can be completely eliminated by addition of particularly high concentration of ammonium. But evaluation of such experiments (Figs. 6 and 7) demonstrates that ammonium inhibits nitrate uptake in a noncompetitive way. This observation supports our results and implies that the uptake of the both ions occurs in different sections of the plasmalemma. But the inhibitory effect of NH_4^+ occurs directly on the nitrate carrier (Ullrich, 1979).

In both algae high concentrations of nitrate inhibit nitrate uptake. This substrate inhibition can be interpreted by a binding of a second nitrate ion to the nitrate carrier, forming an ineffective complex. This step can be described by $ES{+}S{\rightarrow}ES_2$. The dissociation constant $K'_S = \frac{ES x S}{ES_2}$ is small if ES_2 is large and vice versa. From the data presented one can conclude that the nitrate transport mechanism in nitrogen-starved cells binds excess nitrate more strongly than in nitrate-grown cells. This is probably due to the lower K_M and the general shift to higher effectiveness (low sensitivity to ammonium, for example).

In view of further experiments it may be helpful to quantify the nitrate carrier. One way to do this is to use a competitive inhibitor. We found 2-nitrobenzaldehyde to be such an inhibitor. The tests with other substances like nitromethane, nitroethane, nitrobenzene, and nitroglycerine were not successful. Most of them were ineffective, others were taken up simultaneously with nitrate. It is not quite clear if only 2-nitrobenzaldehyde is a competitor or if other configurations may show better effects. Both reactive groups, aldehyde- and nitrite respectively, must be present. Benzaldehyde and nitrobenzene are ineffective.

Acknowledgment. We want to thank Mrs. G. Rohde for excellent technical assistance.

References

Bisswanger H (1979) Theorie und Methoden der Enzymkinetik. Verlag Chemie, Weinheim, Deerfield Beech, Florida Basel
Breteler H, Nissen P (1978) Poster 40 A FESPP meeting Edinburgh
Breteler H, Hänisch Ten Cate ChH, Nissen P (1979) Physiol Plant 47:49–55
Clarkson DT, Warner AJ (1979) Plant Physiol 64:557–561
Crosswell RC, Syrett PJ (1979) Plant Sci Lett 14:321–325
Lorenzen H (1961) Ber Dtsch Bot Ges 73:58–59
Ohmori M, Ohmori K, Strotmann H (1977) Arch Microbiol 114:225–229
Serra JL, Llama MJ, Cardenas EC (1978a) Plant Physiol 62:987–990
Serra JL, Llama MJ, Cardenas MJ (1978b) Plant Physiol 62:991–994
Shaner DL, Boyer JS (1976) Plant Physiol 58:505–509
Ssymank V (1972) Arch Mikrobiol 82:311–324
Swader JA, Stocking CR, Chin Ho Liu (1975) Physiol Plant 34:335–341
Syrett PJ, Morris J (1963) BBA 67:566–575
Thomas RJ, Hipkin CR, Syrett PJ (1976) Planta 133:9–13
Tischner R, Lorenzen H (1979) Planta 146:287–292
Ullrich W (1979) Ber Dtsch Bot Ges 92:273–284
Ullrich W, Eisele R (1977) In: Thellier M, Demarty M, Dainty J (ed) Coll Int CNRS, Rouen, pp 307–313

Nitrogen Metabolism in Photosynthetically Inhibited Plants

C. FEDTKE[1]

The inhibition of photosynthesis by a great number of herbicides presents a threat not only to weed but also to crop plants. We have studied the consequences of a partial inhibition of photosynthesis by methabenzthiazuron [MBT, N-(benzothiazol-2-yl)-N,N'-dimethylurea, trade name Tribunil] in the respective crop plant, which is wheat (*Triticum aestivum* L. "Kolibri"). Wheat plants respond to a treatment with MBT via the roots by a rapid decrease in the concentration of soluble reducing sugars. With a delay of one or two days a decrease of the malate and an increase of the nitrate pools has been found (Fig. 1) (Fedtke, 1974).

The increased nitrate concentration in the treated tissue may be based in part on a temporary reduction of the in vivo nitrate reductase activity (Fedtke, 1973; Aslam and Huffaker, 1973), but an increased uptake of nitrate has also been found in herbicidally photosynthesis-inhibited plants (Pulver and Ries, 1973; Mohandas et al., 1978). An increased nitrogen supply to and also an increased nitrate reduction in these herbicide-treated plants may also be concluded from the observation of increased nitrate reductase activity and increased levels of soluble amino acids and soluble protein (Ries et al., 1967; Fedtke, 1974). Increased percentage protein content has also been observed in the seeds produced from treated plants (De Datta et al., 1972; Patil and Kale, 1975). It therefore appears from these results that, under conditions of carbohydrate stress, the plant increases the capacity for the uptake and reduction of nitrate and for the synthesis of protein.

Of course, carbohydrate stress may not only be caused by photosynthesis-inhibiting herbicides, but also by low light intensities. Figure 2 shows basically similar responses as compared to herbicide treatment (Fig. 1) when soybean plants are treated with low light intensities. This low light similarity also extends to the chloroplast structure and pigment composition which represents the shade type in plants treated with photosynthesis-inhibiting herbicides (Fedtke, 1979b). Nitrate accumulation in plants growing in low light intensities is well known (George et al., 1971).

A third possibility to reduce the carbohydrate availability in plants is to lower the CO_2 concentration. When corn plants which have been grown on ammonia are placed in a nitrate nutrient solution and are allowed to take up nitrate at different ambient CO_2 concentrations (Purvis et al., 1974), again an inverse relationship between nitrate and malate accumulation is observed (Fig. 3). Similar results have been obtained repeatedly (Pace et al., 1979; Aslam et al., 1979). The increased nitrate concentrations may be

1 Bayer AG, PF-A, Biologische Forschung, 5090 Leverkusen/FRG

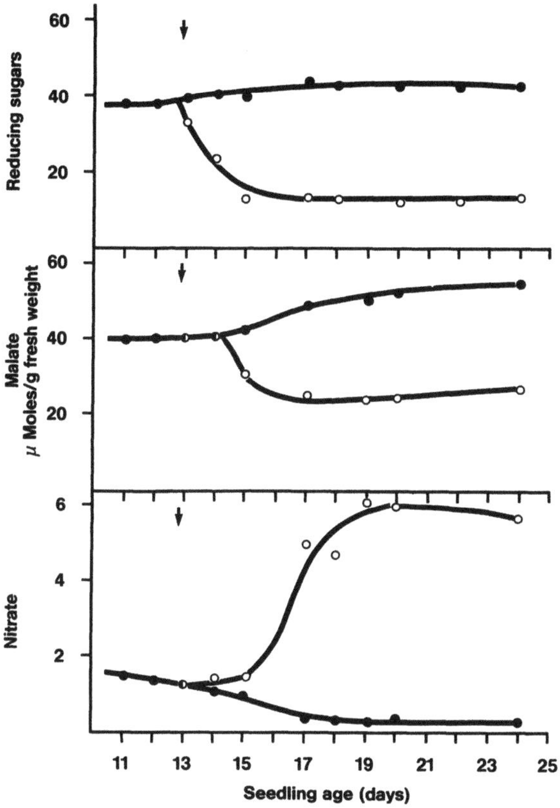

Fig. 1. Influence of methabenzthiazuron on soluble reducing sugars, malate and nitrate content in spring wheat. –•– control, –○– methabenzthiazuron 3 mg l^{-1} at ↓

based on increased nitrate transport into the leaf (Neyra and Hageman, 1976; Pace et al., 1979).

Special attention has been paid to the inverse relationship between malate and nitrate levels. An early interpretation was the possible function of the malate anion in maintaining ionic and pH balance during active nitrate reduction (Palmer and Bacon, 1967). Ben-Zioni et al. (1970, 1971) have extended this model to include the countertransport of KNO_3 and potassium malate which could serve two purposes: circulating K^+ as the accompanying cation and supplying malate to the root which, after decarboxylation, could be exchanged for NO_3^- from the soil without a simultaneous cation uptake (Fig. 4). However, though a role of K^+ in nitrate uptake and/or translocation does appear to exist (Frost et al., 1978), the amount of K^+ recirculating in the vascular tissue appears to be too low to support a sizable potassium malate/KNO_3 countertransport (Kirkby and Knight, 1977; Breteler and Hänisch Ten Cate, 1978). Furthermore, Neyra and Hageman (1976) came to the conclusion that there exists neither a direct correlation between nitrate reduction and malate synthesis nor that an oxidation of extra malate in the root would stimulate nitrate uptake. The validity of the Ben-Zioni model has, however, been confirmed for castor oil plants (Kirkby and Armstrong, 1980).

In herbicide-treated plants, an increased malate transport into the root might be expected if the countertransport model was correct. Table 1 shows results with MBT-

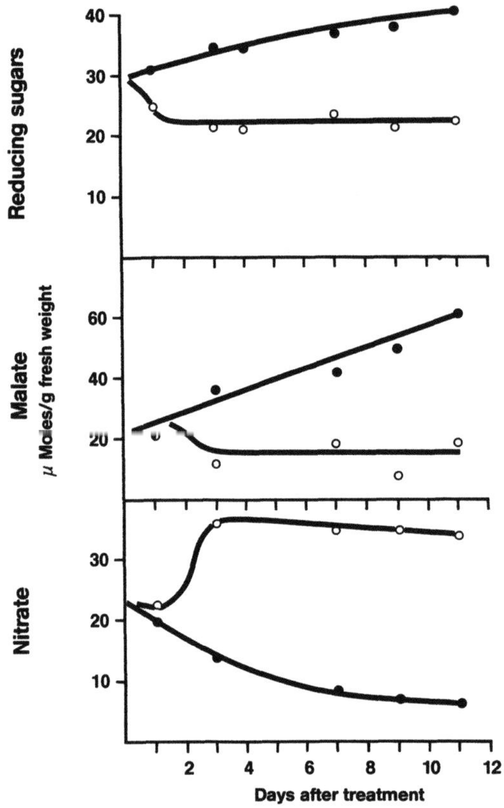

Fig. 2. Influence of low light intensity on soybeans (cf. Fig. 1; the seedlings have been deprived of their cotyletons when seven days old and were placed in low light seven days later) (Fedtke, 1979a). –●– 14 klx throughout, –○– 1 klx from day 0

treated wheat plants. The chlorophyll a/b ratio and the contents of reducing sugars, amino acids, nitrate and malate show that the physiology of the treated plants is changed as usual. However, the specific activity of malate in treated roots rose much more slowly than in control roots, indicating that an increased transport of malate into the roots did not exist in MBT-treated plants.

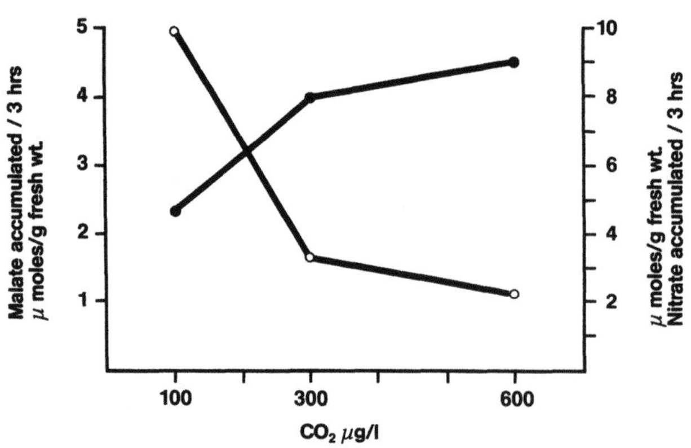

Fig. 3. Inverse relationship of malate and nitrate accumulation in corn plants in different ambient CO_2 concentrations. (Redrawn from Purvis et al., 1974). –●– malate, –○– nitrate

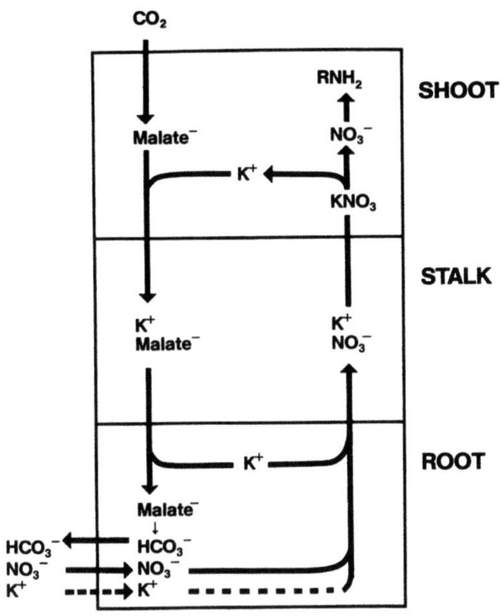

Fig. 4. A possible mechanism for the regulation of nitrate uptake by the root by the nitrate reduction in the shoot. (After Ben-Zioni et al., 1971; and Kirkby and Knight, 1977)

There remains, then, a possible role for malate in maintaining pH and electroneutrality, and a second possible role in supplying NADH for nitrate reduction. Very likely malate and other organic acids play a key role in the ionic balance of plant cells reducing nitrate. Neyra and Hageman (1976), when plotting the nitrate content of 25 corn varieties against their content of malate plus aconitate, found a good negative correlation ($r = -0.75$). However, a stoichiometric synthesis of malate during nitrate reduction appears unlikely. One reason is that in effect malate apparently supplies much of the NADH required for in vivo nitrate reduction (Neyra and Hageman, 1976), although other substrates might be relevant under specific conditions (Sawhney et al., 1978,

Table 1. Physiological status and malate labeling in methabenzthiazuron-treated spring wheat

	Control	Methabenzthiazuron 2 mg l^{-1}
Chlorophyll a/b	3.51	2.89
Reducing sugars μmol/g fw	41.2	16.4
Amino acids μg/g fw	1,640	2,008
Nitrate mg/g fw	1.02	3.34
Malate μmol/g fw		
root	7.5	3.3
shoot	30.4	21.5
Malate labeling cpm/μmol		
root	1,125	71
shoot	2,540	675

$^{14}CO_2$ in the air was supplied for 60 min in the light, then the plants were kept for an additional 6.5 h in ambient air. fw = fresh weight

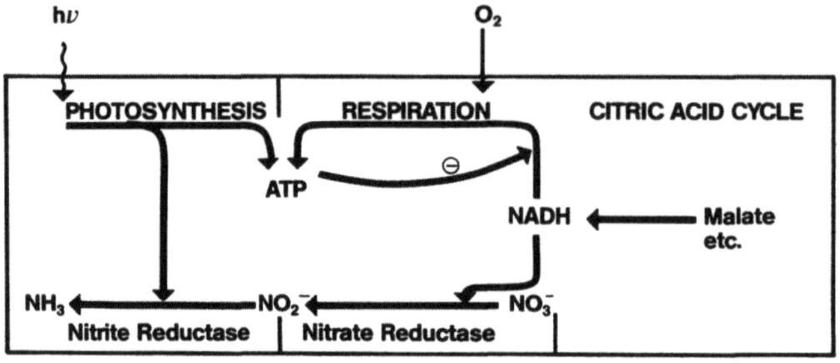

Fig. 5. A possible mechanism for the regulation of nitrate reductase in vivo. (After Sawhney et al., 1978)

Aslam et al., 1978; Woo et al., 1980). Malate also substitutes for the illuminated leaves in the induction and maintainance of root nitrate reductase and leads to an increased export of amino acids out of the roots (Deane-Drummond et al., 1979).

Sawhney et al. (1978) have suggested a model for the regulation of in vivo nitrate reduction (Fig. 5) which is based on the presently available evidence. From their observations they suggest a regulation of the flow of NADH by the cytoplasmic energy charge. In the dark a low energy charge would favor the oxidation of NADH by the respiratory chain, which would avoid the accumulation of nitrite. In the light, however, with a higher energy charge, an increased use of cytoplasmic NADH by nitrate reductase would favor nitrate reduction. In the presence of photosynthesis-inhibiting herbicides, which generally have little effect on cyclic photophosphorylation, nitrite accumulation might consequently occur (Klepper, 1979), greatly enhanced by anaerobiosis which would further increase the availability of NADH for nitrate reductase (Fedtke, 1977). The model cannot easily explain, however, the nitrite accumulation found in leaves treated with respiratory uncouplers.

References

Aslam M, Huffaker RC (1973) Physiol Plant 28:400–404
Aslam M, Huffaker RC, Rains DW, Rao KP (1979) Plant Physiol 63:1205–1209
Ben-Zioni A, Vaadia Y, Lips SH (1970) Physiol Plant 23:1039–1047
Ben-Zioni A, Vaadia Y, Lips SH (1971) Physiol Plant 24:288–290
Breteler H, Hänisch Ten Cate CH (1978) Physiol Plant 42:53–56
Deane-Drummond CE, Clarkson DT (1979) Plant Physiol 64:660–662
De Datta SK, Obcemea WN, Jana RK (1972) Agron J 64:785–788
Fedtke C (1973) Pestic Sci 4:653–664
Fedtke C (1974) Ber Dtsch Bot Ges 87:155–160
Fedtke C (1977) Pestic Sci 8:152–156
Fedtke C (1979a) Weed Sci 27:192–195
Fedtke C (1979b) Z Naturforsch 34c:932–935
Frost WB, Blevins DG, Barnett NM (1978) Plant Physiol 61:323–326
George JR, Rhykerd CL, Noller CH (1971) Agron J 63:413–415
Kirkby EA, Armstrong MJ (1980) Plant Physiol 65:286–290
Kirkby EA, Knight AH (1977) Plant Physiol 60:349–353
Klepper LA (1979) Plant Physiol 64:273–275
Mohandas S, Wallace W, Nicholas DJD (1978) Phytochemistry 17:1021–1028
Neyra CA, Hageman RH (1976) Plant Physiol 58:726–730

Pace GM, Volk RC, Jackson WA (1979) Plant Physiol Suppl 63:25
Palmer MJ, Bacon JSD (1967) Biochem J 102:304–312
Patil VY, Kale VR (1975) Qual Plant 2:187–192
Pulver EL, Ries SK (1973) Weed Sci 21:232–237
Purvis AC, Peters DB, Hageman RH (1974) Plant Physiol 53:934–941
Ries SK, Chmiel H, Dilley DR, Filner P (1967) Proc Natl Acad Sci USA 58:526–532
Sawhney SK, Naik MS, Nicholas DJD (1978) Biochem Biophys Res Commun 81:1209–1216
Woo KC, Jokinen M, Canvin DT (1980) Plant Physiol 65:433–436

Investigations on the Reduction of Aliphatic and Aromatic Nitro Compounds by Clostridium Species and Enzyme Systems

L. ANGERMAIER, F. HEIN, and H. SIMON[1]

1 Introduction

During our studies on the use of *Clostridium* species as hydrogenation catalysts on a preparative scale (Simon et al., 1979), we observed that many of them such as *C. pasteurianum, C. butyricum, C. propionicum, C. barkeri, C. kluyveri, C.* species *La 1,* and other anaerobes *(Proteus mirabilis, Peptostreptococcus anaerobius, Micrococcus lactilyticus)* consume hydrogen in the presence of pNB or NE.

Nitro compounds are used in large quantities as intermediates for the production of bulk chemicals and many nitro compounds are used as chemotherapeutic agents. An extensive literature exists about the biological reduction of nitro compounds in animal tissues, by yeasts, prokaryotic microorganisms and plants (Nason, 1956; Peterson et al., 1979). For many years now nitro compounds have been used as indicators for different respiration and fermentation processes. The clinical use of chloramphenicol raised the interest in activities of nitroaryl reductases in bacteria. *E. coli* and other bacteria are able to inactivate chloramphenicol by reduction of the nitro group to an amino group (Smith and Worrel, 1949). Nitroaryl reductase activities in mammals, especially in liver, have been described (Peterson et al., 1979). Such studies were also conducted relating to the mutagenic and cancerogenic activities of nitro compounds such as 4-nitrochinolin-N-oxide (Poirier and Weisburger, 1974).

Some nitroaryl reductases have been partially purified and characterized. While our work was in progress Peterson et al. (1979) published mechanistic studies with partially purified nitroaryl reductases from *E. coli* showing that nitrofurazone undergoes an initial 1-electron (oxygen-sensitive) or 2-or more electron (oxygen-insensitive) reduction. Lindmark and Müller (1976) reported on the reduction of metronidazole by reduced Fd and Chen and Blanchard (1979) on the reduction by Fd and flavodoxin. Besides this hardly any report on activities about nitro group reductases in strict anaerobes and especially on the reduction of aliphatic nitro compounds has appeared. Therefore we were interested in the enzyme systems of *Clostridium* species capable of reducing alphatic and aromatic nitro compounds.

Nonstandard abbreviations: NE = 2-nitroethanol; pNB = para-nitrobenzoate; Fd = ferredoxin; H_2/H_2ase/Fd = a system of H_2 (1 bar), hydrogenase and ferredoxin

1 Lehrstuhl für Organische Chemie und Biochemie, Technische Universität München, Lichtenbergstraße 4, 8046 Garching/FRG

2 Materials and Methods

C. species *La 1* (DSM 1460) and *C. kluyveri* (DSM 555) were grown as previously described (Bader et al., 1980; Bader and Simon, 1980), *C. pasteurianum* (DSM 525) containing Fd or flavodoxin according to Schönheit et al. (1979). Most manipulations and all assays and hydrogenation experiments were conducted under strict oxygen exclusion using methods described by Tischer et al. (1979). NE, 2-(N-hydroxyimino)ethanol and 2-(N-hydroxyamino)ethanol were determined by gas liquid chromatography on a 5 m PEG 400 (10% on Chromosorb G AW-DMCS) column at 120 °C, glycol and 1.4-butandiol on a 1.25 m PEG 4,000 (10% on the same carrier) column at 140° and 180 °C. 2-Aminoethanol was determined with fluorescamine (Fluram, Hoffmann-La Roche) (Weigele et al., 1972). 2-(N-hydroxyimino)- and 2-(N-hydroxyamino)ethanol were synthesized according to Kahr and Berther (1960) and Borch et al. (1971), respectively.

Clostridium Fd from *species La 1* was isolated according to Schönheit et al. (1978) and the hydrogenase from *C. kluyveri* by a modification of the procedure of Chen and Mortenson (1974). Nitrite was determined as described by Shinn (1941). The ESR-spectra were run on a Varian E 3 EPR spectrometer.

Standard assay of hydrogen uptake: in two-neck Warburg vessels in a total volume of 2.1 ml containing 50 m*M* phosphate buffer pH 8.5, 50 m*M* potassium chloride, 0.02% sodium azide, 7 m*M* 2-mercaptoethanol, crude cell extracts or hydrogenase, Fd and further additives as indicated were incubated at 35 °C under an atmosphere of hydrogen with shaking (120 strokes/min). After equilibration the substrate was added from a side arm. One neck of the vessel was closed with a rubber stopper. Aliquots could be withdrawn and solutions added by gas tight syringes.

3 Results and Discussion

3.1 Common and Different Behavior of p-Nitrobenzoate and 2-Nitroethanol

Intact cells as well as crude extracts of *C. kluyveri* and *C. pasteurianum* reduce pNB and NE in the presence of hydrogen gas. Six electrons are necessary for the reduction of the nitro group to the amino group (Scheme 1). In the presence of crude extracts of *C. kluyveri* pNB takes up 2 mol hydrogen (4 electrons) relatively rapidly. The further hydrogenation proceeds very slowly (Fig. 1). Three mol of hydrogen are consumed by NE in the presence of whole cells or crude extracts of *C. kluyveri* (Fig. 2). Table 1 shows the rate of hydrogen uptake as well as the NADH consumption catalyzed by crude extracts of *C. kluyveri* or *C. pasteurianum* in the presence of pNB or NE, respectively. pNB is reduced about 5-12 times faster than NE. The latter cannot be reduced by NADH. *C. kluyveri* contains much higher NADH-dependent nitroaryl reductase activities than *C. pasteurianum*. NADPH is no cosubstrate.

Table 1 clearly reveals that the enzyme systems which transfer electrons from NADH to aromatic nitro groups are not able to do this with aliphatic nitro groups. This difference could be related to the fact that nitroaryl compounds such as pNB or m-dinitrobenzene show at neutral pH polarographic half wave reduction potentials of ±0.0 and

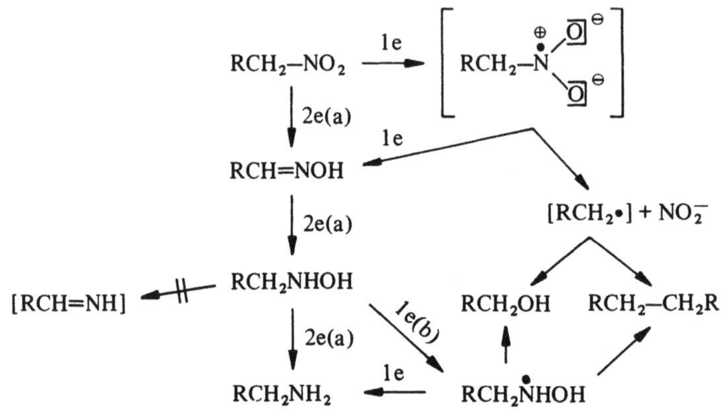

Scheme 1. Possible mechanistic routes of the reduction of nitroalkanes by *C. kluyveri* or the system H_2/H_2ase/Fd under conditions given in Fig. 2. The products not in parenthesis have been observed. *a* The simultaneous transfer of two electrons is not proven. *b* A nitrogen radical is formed

-0.1 V (Meites and Zuman, 1974) and aliphatic nitro compounds such as nitroethane or 3-nitropropionate potentials of about -0.5 V (Miller et al., 1949).

The initial rate of the NADH-dependent reduction of pNB by the crude extract of *C. kluyveri* shows a distinct maximum in 50 m*M* glycine buffer at pH 10.5. At pH 12 the activity is still 40% of that at pH 10.5 and twofold higher than that at pH 6.0–8.2 (not shown).

Whole cells of *C. kluyveri* reduce different aliphatic compounds such as NE, D,L-2-nitro-1-propanol, 3-nitro-2-butanol and 2-methyl-2-nitro-1.3-propandiol with relative rates of 10:4:5.6:6.6. These differences are rather small compared with the structural

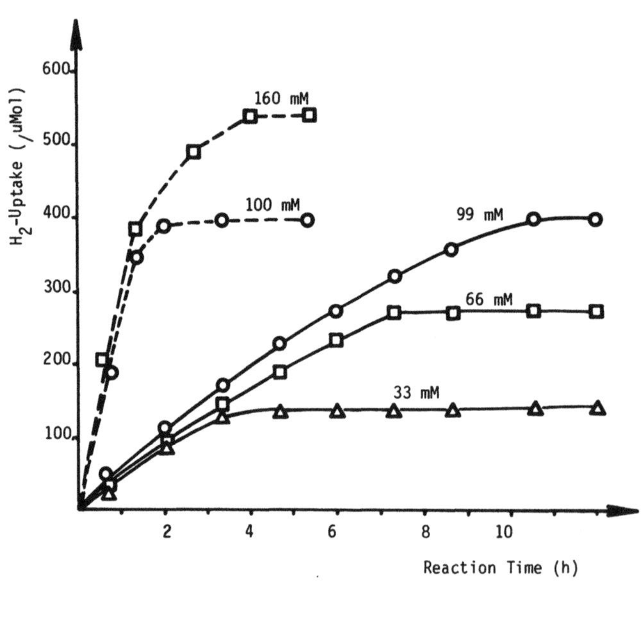

Fig. 1. Time course of hydrogen uptake catalyzed by crude extract of *C. kluyveri (broken lines)* as well as by the system hydrogenase, ferredoxin *(continuous lines)* in the presence of the indicated concentrations of p-nitrobenzoate. The crude extract of 280 mg wet packed cells was incubated in a total volume of 2.1 ml phosphate buffer (100 m*M*) pH 7.0. The hydrogenase (3.8 U), ferredoxin (7 μ*M*) system was in 2.1 ml phosphate buffer (50 m*M*) pH 8.5

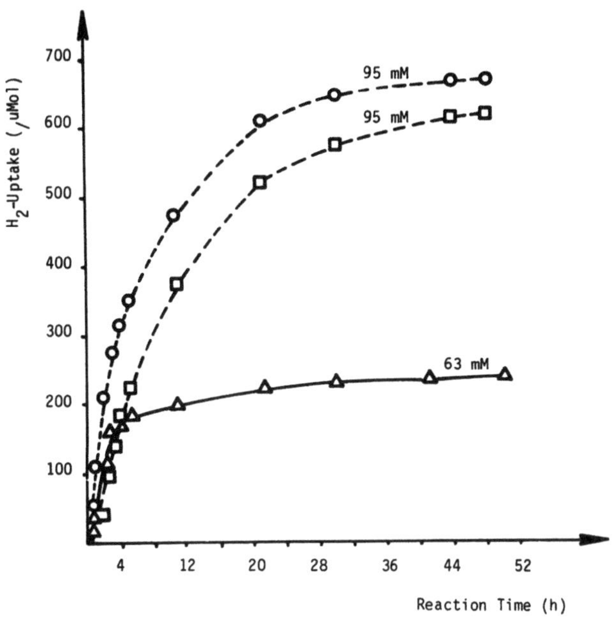

Fig. 2. Time course of hydrogen uptake catalyzed by 570 mg wet packed cells of *C. kluyveri* (□–□), the supernatant of the same amount of cells (○–○) or the stabilized system hydrogenase (17 U), ferredoxin (5 μ*M*) (△–△) in the presence of the indicated concentrations of 2-nitroethanol. The experiments were conducted in 2.1 ml phosphate buffer (50 m*M*) pH 8.5. The hydrogenase, ferredoxin system contained in addition $FeSO_4$, Na_2S (each 0.9 m*M*) and 2-mercaptoethanol (67 m*M*) for stabilization

Table 1. Hydrogen uptake as well as NADH consumption catalyzed by crude extracts (prepared from 1 g wet packed cells of *Clostridium*) in the presence of p-nitrobenzoate (35 m*M*) or 2-nitroethanol (35 m*M*) in 50 m*M* phosphate buffer pH 8.5

Organism	μmol H_2/h		μmol NADH/h	
	pNB	NE	pNB	NE
C. kluyveri	780	170	6,900	<0.1
C. pasteurianum containing ferredoxin	820	70	140	<0.1
C. pasteurianum containing flavodoxin	350	60	340	<0.1

differences between these substrates. No rate difference can be observed with the two enantiomers of 2-nitro-1-propanol and the different forms of 3-nitro-2-butanol.

The hydrogen uptake or NADH consumption catalyzed by crude extracts of *C. kluyveri* in the presence of sulfate, sulfite, nitrate, nitrite, and hydroxylamine is in any case less than 5% of that observed in the presence of pNB.

3.2 The System Hydrogen, Hydrogenase, Ferredoxin

The above-mentioned observations led us to the idea that in the system hydrogen/*Clostridium* reduced Fd or flavodoxin could be the electron donors for nitro compounds. The initial rates of hydrogen uptake of NE catalyzed by hydrogenase and Fd together with supplements can be seen from Table 2. Hydrogenase in the presence of hydrogen is not

Table 2. Initial rates of the reduction of 2-nitroethanol (95 m*M*) by the system hydrogenase, ferredoxin in 2.1 ml 50 m*M* phosphate buffer pH 8.5

No.	Hydrogenase (U)	Ferredoxin (μM)	Fe^{2+} (m*M*)	S^{2-} (m*M*)	$HSCH_2CH_2OH$ (m*M*)	H_2 uptake (μmol/5 min)
1	2.9	5	–	–	5	6.7
2[a]	3.9	5	0.25	0.25	22	5.8
3[a]	2.9	5	0.5	0.5	39	5.5
4[a]	3.9	5	2.0	2.0	141	4.6
5	3.9	–	2.0	2.0	141	<0.2
6	–	5	0.5	0.5	5	<0.2
7[b]	2.9	5	0.5	0.5	39	<0.2

[a] The total amounts of hydrogen consumed in 7 h are 170, 190, and 210 μmol for experiments 2, 3, and 4, respectively

[b] Without 2-nitroethanol

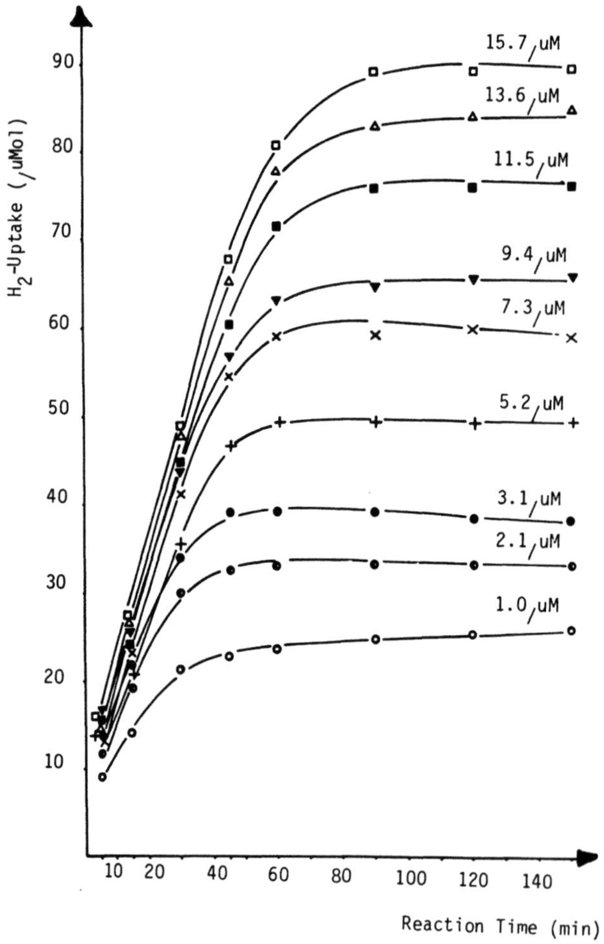

Fig. 3. Dependence of the time course of hydrogen uptake from the ferredoxin concentration in a nonstabilized system of H_2ase, ferredoxin. A volume of 2.1 ml buffer (as in Fig. 2) contained hydrogenase (5.5 U), nitroethanol (95 mM) and the indicated concentrations of ferredoxin

able to reduce nitro compounds. However, the system H_2/H_2ase/Fd reduces NE as well as pNB (Figs. 1 and 2). Again pronounced differences can be observed with NE and pNB. NE inactivates the H_2/H_2ase/Fd system relatively rapidly. Aliquots taken from such incubations show a nearly unchanged activity of hydrogenase. Therefore one has to assume that Fd is inactivated. The more Fd is present, the higher the reduction rate is and the longer the reduction of NE proceeds (Fig. 3). In the presence of 1 μM Fd the reduction rate of NE decreases drastically already after 25–30 min. Under otherwise identical conditions in the presence of 15 μM Fd the system is stable for about 70 min. In the experiment with 15 μM Fd the hydrogen uptake was about 4 times higher than in that with 1 μM Fd. However, with 7 μM Fd, solutions with concentrations of 33, 66, and 99 mM pNB, respectively, are reduced at a rather constant rate until 4 electrons are taken up by pNB. In the case of 99 mM pNB this takes about 10 h (Fig. 1). As already shown (Fig. 2) in the presence of NE crude extracts of *C. kluyveri* are a good deal more stable than the system H_2/H_2ase/Fd. A crude extract with a Fd concentration of about 5 μM is still active after 20 h. The system H_2/H_2ase/Fd can be stabilized by the addition of Fe^{2+}/S^{2-} and 2-mercaptoethanol. This effect of stabilization can be seen in Fig. 2 (lower curve) in comparison with Fig. 3. After 5 h the hydrogen uptake is still rather fast in a 63 mM solution of NE if besides hydrogenase and 5 μM Fd the aforementioned factors are added. Reconstitution of apoferredoxin to Fd with ferrous and sulfide ions has been described by Hong and Rabinowitz (1970).

The question arises whether NE or reduction intermediates are responsible for the inactivation of Fd. The incubation of 62 mM NE under nitrogen for various times followed by the substitution of the nitrogen by hydrogen shows no hydrogen uptake. However, the addition of Fe^{2+} and S^{2-} activated the system. If the supplementation is conducted 15 h after the switch from nitrogen to hydrogen the activation is not possible any more (Fig. 4).

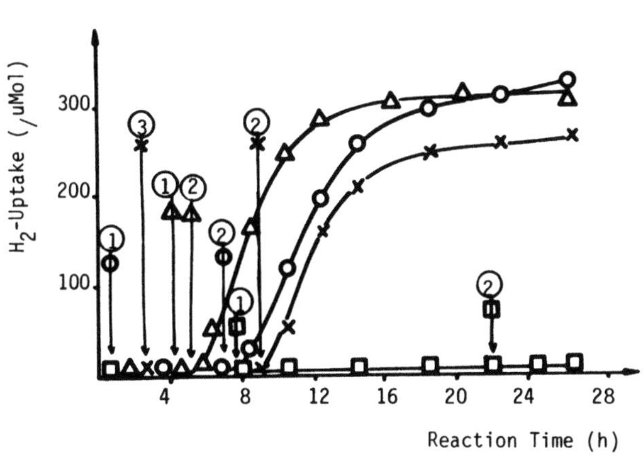

Fig. 4. Time course of hydrogen uptake catalyzed by hydrogenase (4.4 U) and ferredoxin (6 μM) in the presence of 2-nitroethanol (62 mM), 2-mercaptoethanol (7 mM) in 2.1 ml buffer (as in Fig. 2) after different kinds of preincubation. The supplementation with $FeSO_4$ and Na_2S was 1.9 mM in each case. Additions or other changes of the systems are indicated by *circles with arrows* and a *symbol* corresponding to one of the four curves. ① = substitution of nitrogen by hydrogen, ② = supplementation, ③ = addition of hydrogenase and nitroethanol to ferredoxin under hydrogen. *Circles in the same height* belong to the same experiment

3.3 Mechanism and Intermediates

The time course of the reduction of NE and the transient concentrations of the intermediates 2-(N-hydroxyimino)ethanol and 2-(N-hydroxyamino)ethanol as well as the formation of 2-aminoethanol catalyzed by the system H_2/H_2ase/Fd can be seen in Fig. 5. NE disappears in 4–5 h accompanied by a sequential increase of the concentrations of first 2-(N-hydroxyimino)ethanol, maximum at about 5 h, and then 2-(N-hydroxyamino)ethanol, maximum at about 10 h. No significant amounts of 2-aminoethanol are formed before 5 h. After 80 h 2-(N-hydroxyamino)ethanol is still present but the main product is 2-aminoethanol. The glc-analysis of these substances is rather difficult. The accuracy is not better than ±10%. In experiments identical with those shown in Fig. 5 the following by-products in the indicated concentrations have been observed: Nitrite (2–2.5 m*M*), ethylene glycol (0.5 m*M*) and 1.4-butanediol (0.08 m*M*). pNB leads not to the formation of detectable amounts of nitrite, i.e., less than 0.002 m*M* nitrite are formed from pNB. Another difference between NE and pNB is the fact that in the presence of crude extract the former takes up six electrons rather readily and the latter only four. Therefore one can assume that p-(N-hydroxyamino)benzoate is more stable than 2-(N-hydroxyamino)ethanol or it is catalytically or enzymatically transformed to another product. One reason for the different behavior could be that the former cannot be dehydrated to the imine (Scheme 1) which could be reduced to the amine. However, hydrogenation experiments with crude extracts of *C. kluyveri* and 2-(N-hydroxyamino)ethanol conducted in tritium-labeled water resulted in a tritium incorporation of less than 1.5% of that expected in the absence of an isotope effect. Since the intramolecular isotope

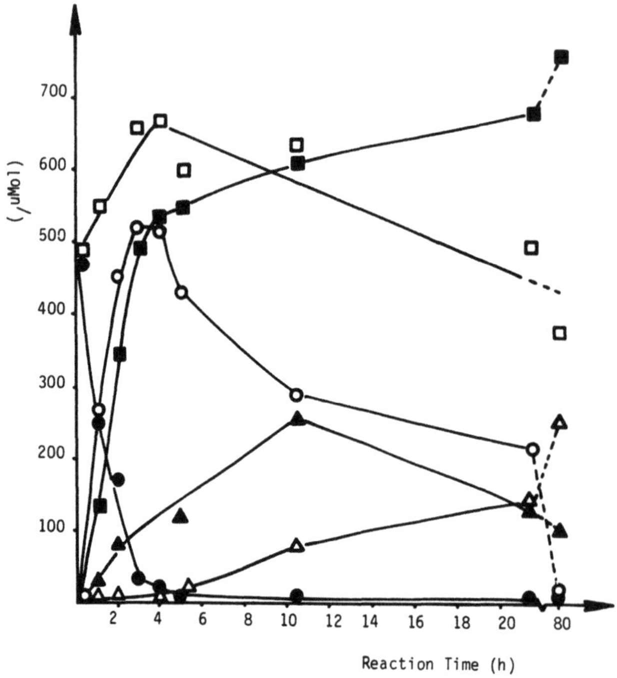

Fig. 5. Time course of hydrogen uptake (■–■), the disappearance of 2-nitroethanol (●–●), intermediate concentrations of 2-(N-hydroxyimino)ethanol (○–○), 2-(N-hydroxyamino)ethanol (▲–▲), the formation of 2-aminoethanol (△–△) as well as the sum of substrate, intermediates, and product (□–□). A total volume of 6.4 ml phosphate buffer (50 m*M*) pH 8.5 contained 2-nitroethanol (63 m*M*), hydrogenase (52 U), ferredoxin (5 μ*M*), $FeSO_4$, Na_2S (each 0.9 m*M*) and 2-mercaptoethanol (67 m*M*)

effect of the splitting of H-O-T is usually <7 we exclude the imine as an intermediate. Therefore the above-mentioned difference between NE and pNB is not due to the conversion of the 2-(N-hydroxyamino)ethanol to the imine.

The by-products nitrite, ethylene glycol, and 1.4-butanediol of the reduction of NE are indicative for radical reactions (Scheme 1). Under conditions under which pNB in the system H_2/H_2ase/Fd shows the same kind of ESR spectrum of a nitrogen radical (not shown) as observed by Mason and Holtzman (1975) during the reduction of pNB by rat hepatic microsomes no radical could be observed with NE. This could be explained by a too low concentration of the radical of NE in the steady state due to a decay fast compared with its formation. As mentioned above, the polarographic half-wave reduction potentials of aliphatic nitro groups are about 0.4 V more negative than those of nitroaryl compounds. That means the radical of the latter can be much better stabilized than that of the former and therefore its half-life may be longer. Whereas the nitroaryl radical seems to react only in one way, the NE radical, if it is formed, may encounter two fates, as shown in Scheme 1. Another possibility would be that NE is already a branching point. In a certain ratio NE may react in a two-electron step leading to 2-(N-hydroxyimino)ethanol and in a one-electron step leading to the radical anion. According to Mortenson and Nakos (1973) clostridial ferredoxins are able to transfer 1 or 2 electrons simultaneously. As already mentioned it was shown recently that, depending on the enzymes which were partially purified from *E. coli*, nitrofurazone undergoes an initial 1-electron or 2-or more electron reduction (Peterson et al., 1979). The N-hydroxyimino intermediate is further reduced to the N-hydroxyamino derivative. A radical could not be observed under our conditions. However, this is the case when 2-(N-hydroxyamino)ethanol is incubated with the H_2/H_2ase/Fd system. Therefore some of the observed ethylene glycol and 1.4-butanediol may result from the decay of this radical. The question arises whether nitrite and/or intermediate radicals are responsible for the fast inactivation of the unsupplemented H_2/H_2ase/Fd system. Nitrite (4.0 m*M*) inactivated Fd, however, did not inactivate hydrogenase. The inactivation of Fd by nitrite can be prevented or nitrite-inactivated Fd can be reactivated by concentrations of ferrons sulphate and sodium sulfide substoichiometric to those of nitrite (not shown).

3.4 NADH-Dependent Nitroaryl Reductases

Crude extracts of *C. kluyveri* chromatographed on DEAE-Sepharose reveal three protein areas which are able to reduce pNB in a NADH-dependent reaction. Figure 6 shows a typical elution pattern. Although these fractions are far from pure it seems possible to exclude a few known enzymes which seem not to be responsible for the nitroaryl reductase activities. Besides three nitroaryl reductases Fig. 6 shows the activities of hydrogenase, butyryl-CoA dehydrogenase and, indicated by arrows, the maxima of three further dehydrogenases. The maxima of nitroaryl reductases I, II, and III seem to be different from those of three forms of aldehyde dehydrogenase, two forms of alcohol dehydrogenase and one form of 3-hydroxybutyryl-CoA dehydrogenase, respectively. The maximum of the NADH-dependent nitroaryl reductase III coincides almost with that of butyryl-CoA dehydrogenase. This latter dehydrogenase requires for its reduction by NADH a flavin containing electron transfer factor. Since crotonyl-CoA cannot be reduc-

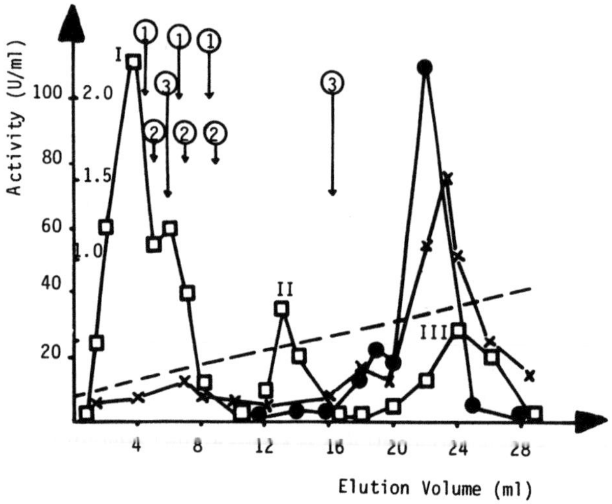

Fig. 6. Elution pattern of enzyme activities of a lysate of *C. kluyveri* on a DEAE-Sepharose column (1x10 cm). In 4.0 ml phosphate buffer (50 m*M*) pH 8.0 containing dithionite (2 m*M*) 2 g wet packed cells have been incubated for 40 min at 35 °C with lysozyme, DNase, RNase, and 0.8 mg protamine sulfate. After sonification (5 times 2 min) and centrifugation 3.9 ml of this extract containing 108 mg protein was put on the column and eluted with a linear gradient of KCl from 80–800 m*M* in phosphate buffer (50 m*M*) pH 8.0. The resolution of this elution pattern is rather high. Fractions of about 270 μl were collected and all of them were tested in the areas of maxima of enzyme activities. NADH-dependent nitroaryl reductases (□–□) *(right hand scale)*, butyryl-CoA-dehydrogenase (x–x) and hydrogenase (•–•) *(left hand scale)*. The position of the enzyme maxima for aldehyde dehydrogenase, alcohol dehydrogenase, and 3-hydroxybutyryl-CoA dehydrogenase are indicated by the symbols ①, ②, and ③ respectively

ed by NADH and the protein of this peak, this factor is not eluted together with the butyryl-CoA dehydrogenase.

Furthermore, hydrogenase is not able to reduce nitro compounds (Table 2) and also not enoate reductase, an iron-sulfur flavoprotein recently described by us (Tischer et al., 1979).

The nitroaryl reductase II exhibits its rate maximum at an astonishing high pH value of 11.5 (not shown). In conclusion it seems quite possible that one or more of the nitroaryl reductases may be enzymes occurring in *C. kluyveri, C.* species *La 1, C. pasteurianum* and probably other anaerobes which have been unknown.

Acknowledgments. We thank Dr. J. Bader for the growth of the different microorganisms and for the continuous supply with *C. kluyveri,* furthermore, Miss I. Mayer and Mr. R. Feicht for skillful technical assistance.

References

Bader J, Simon H (1980) Arch Microbiol 127:279–287
Bader J, Günther H, Schleicher E, Simon H, Pohl S, Mannheim W (1980) Arch Microbiol 125:159–165
Borch RF, Bernstein MD, Durst HD (1971) J Am Chem Soc 93:2897–2904

Chen JS, Blanchard DK (1979) Anal Biochem 93:216–222
Chen JS, Mortenson LE (1974) Biochim Biophys Acta 371:283–298
Hong JS, Rabinowitz JC (1970) J Biol Chem 245:6574–6581
Kahr K, Berther C (1960) Chem Ber 93:132–136
Lindmark DG, Müller M (1976) Antimicrob Agents Chemother 10:476–482
Mason RP, Holtzman JL (1975) Biochemistry 14:1626–1632
Meites L, Zuman P (1974) In: Electrochemical data, part 1, vol A. John Wiley and Sons, New York London Sydney Toronto
Miller EW, Arnold AP, Astle MJ (1949) J Am Chem Soc 70:3971–3975
Mortenson LE, Nakos G (1973) In: Lovenberg W (ed) Iron sulfur proteins, vol 1. Academic Press, London New York, p 46
Nason A (1956) In: McElroy, Glass B (eds) Inorganic nitrogen metabolism. Johns Hopkins Press, Baltimore, pp 109–136
Peterson FJ, Mason RP, Hovsepian J, Holtzman JL (1979) J Biol Chem 254:4009–4014 and literature cited therein
Poirier LA, Weisburger JH (1974) Biochem Pharmacol 23:661–669
Schönheit P, Wäscher C, Thauer RK (1978) FEBS Lett 89:219–222
Schönheit P, Brandis A, Thauer R (1979) Arch Microbiol 120:73–76
Shinn MB (1941) Ind Eng Chem Anal Edn 13:33–35
Simon H, Bader J, Rambeck B, Krezdorn E, Tischer W (1979) In: Dellweg H (Hrsg) Anaerobe Fermentationen – klassische Prozesse mit neuen Aussichten. Verlag Versuchs- und Lehranstalt für Spiritusfabrikation and Fermentationstechnologie, Berlin, pp 325–334
Smith GN, Worrel CS (1949) Arch Biochem 24:216–223
Tischer W, Bader J, Simon H (1979) Eur J Biochem 97:103–12
Weigele M, DeBernardo SL, Tengi JP, Leimgruber W (1972) J Am Chem Soc 94:5927–5928

Fine Structure Analysis of the Gene Region for N_2-Fixation (nif) of Klebsiella pneumoniae

A. PÜHLER and W. KLIPP[1]

1 Introduction

Klebsiella pneumoniae, a free-living N_2-fixing organism, carries the genes for nitrogen fixation *(nif)* in a cluster on the chromosome (Streicher et al., 1972). By genetic engineering techniques in vivo this chromosomal *nif* region was transferred to *Escherichia coli* (Dixon and Postgate, 1972), integrated into the chromosome (Cannon et al., 1974), combined with the *E. coli* fertility plasmid F (Cannon et al., 1976) and subsequently with the resistance plasmid RP4 (Dixon et al., 1976). Plasmid pRD 1 consisting of RP4 and the *his nif* region of *K. pneumoniae* is the most important *nif* plasmid constructed in vivo: Due to the broad host range of RP4 it can be transferred to a large variety of gram negative bacteria (Cannon and Postgate, 1976; Postgate and Krishnapillai, 1977).

In order to get more information about the *Klebsiella nif* region we started molecular investigations of plasmid pRD 1. In this paper we report the physical organization of plasmid pRD 1 and the in vitro cloning of the entire *nif* region with multicopy plasmid vectors in *E. coli.* In addition, we present a method which allowed us to determine the extension and the exact location of nearly all of the coding regions within the *Klebsiella nif* region.

2 The Physical and Genetic Organization of Plasmid pRD 1

Plasmid pRD 1 has a contour length of 49 μm or a molecular weight of 101 Mdals (Pühler et al., 1979a). It carries all the genetic markers of plasmid RP4: the resistance genes against ampicillin (Ap), kanamycin (Km) and tetracycline (Tc) and the transfer region (Tra^+). In addition, the *his nif* region of *K. pneumoniae* is located on this plasmid. Since plasmid pRD 1 is very unstable – the *his nif* region is lost spontaneously at high frequency (Pühler et al., 1979a) – we decided to clone the entire *nif* region onto multicopy plasmid vectors in *E. coli.* For this purpose it was necessary to localize first the *his nif* region on the restriction map of plasmid pRD 1. We therefore digested pRD 1 DNA with the endonucleases *Eco*RI and *Hin*dIII and separated the resulting DNA-fragments by agarose gel electrophoresis. 20 *Eco*RI and 16 *Hin*dIII fragments could be identified. By combining several methods we were finally able to construct the *Eco*RI and *Hin*dIII restriction map of plasmid pRD 1 (Pühler et al., 1979c; Pühler, 1980). These maps are

1 Lehrstuhl für Genetik, Fakultät für Biologie, Universität Bielefeld, 4800 Bielefeld/FRG

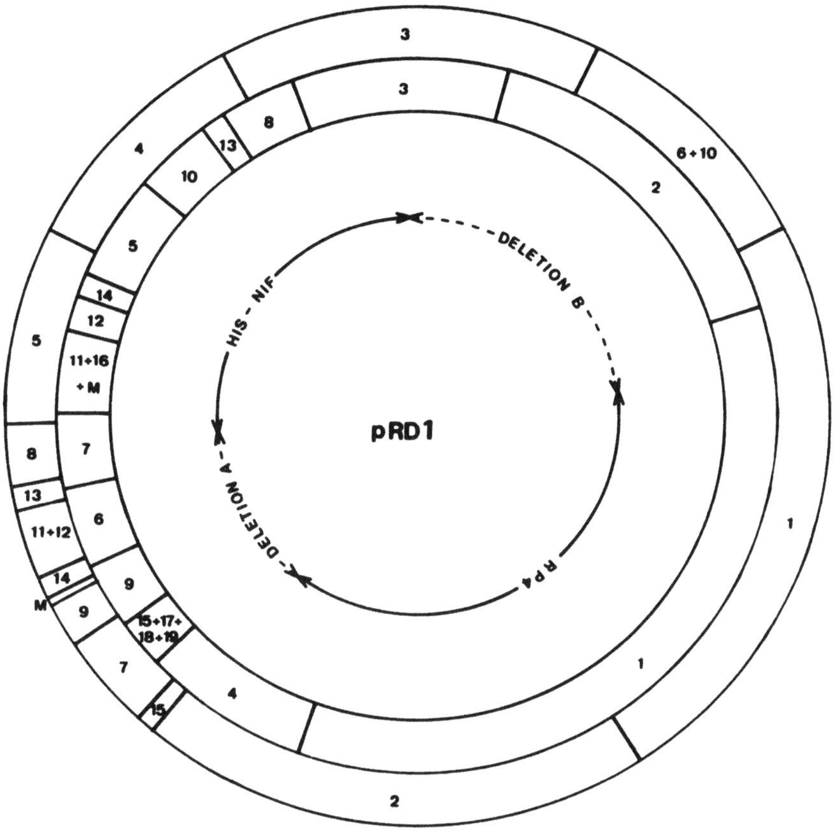

Fig. 1. The *Eco*RI and *Hin*dIII restriction maps of plasmid pRD1. The *outer circle* shows the *Hin*dIII, the *inner circle* the *Eco*RI sites. The numbers of the *Eco*RI and *Hin*dIII fragments are chosen according to the size of the fragments. The symbol *M* indicates minifragments. The correlation between the physical and the genetic map is described in the text

shown in Fig. 1. It is evident that most of the *Eco*RI and *Hin*dIII sites are clustered in one third of these maps. The localization of the *his nif* region was achieved by isolation of deletion mutants of pRD1. Two of them, one carrying the deletion A and the other the deletions A and B helped us to identify the *his nif* region on the *Hin*dIII fragments number 5, 4, and 3 (Pühler et al., 1979b). These *Hin*dIII fragments belong to that part of the pRD1 map where *Hin*dIII and *Eco*RI sites are located highly clustered. The RP 4 could be localized opposite to *his nif*.

3 The in vitro Constructed nif Plasmid pWK 120

After the identification of the *his nif* region on the physical map of plasmid pRD1 we started with cloning experiments. Our strategy was the following: Since we know that *Hin*dIII fragment 5 was able to complement *his* mutations in *E. coli* we assumed that

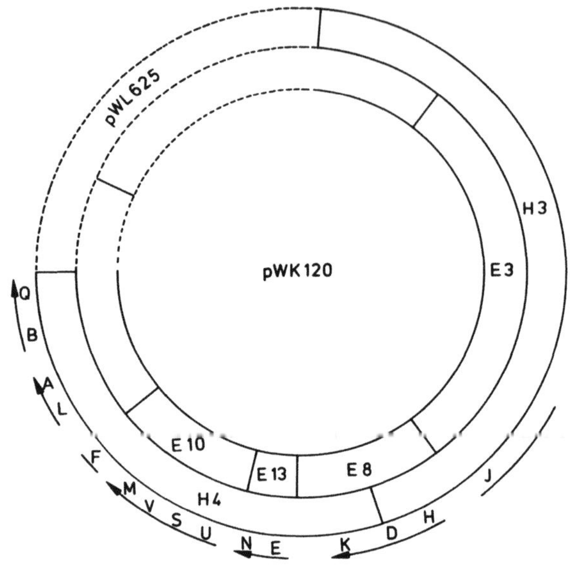

Fig. 2. The *nif* plasmid pWK120. pWK120 was constructed by in vitro cloning of the *Hin*dIII fragments H3 and H4 of plasmid pRD1 into the plasmid vector pWL625. The *Eco*RI and *Hin*dIII maps are given. Fifteen known *nif* genes and their operon structure are drawn in correlation to the physical map of pWK120

the entire *nif* region is located on the *Hin*dIII fragments 4 and 3. We therefore cloned these fragments separately in the *Hin*dIII site of plasmid pWL625 (Goebel et al., 1977). The *Hin*dIII site is located within the kanamycin resistance gene, which is inactivated by the inserts. The next step was to combine the two *Hin*dIII fragments 4 and 3 in the same alignment as in plasmid pRD1 and then to clone these combined fragments in the *Hin*dIII site of pWL625. The resulting plasmid was called pWK120. This plasmid confers nitrogen fixation capacity on *E. coli* C cells. This was tested by growing *E. coli* C (pWK120) in nitrogen-free medium. Also the acetylene reduction assay gave a positive result, indicating an active nitrogenase in the *E. coli* C (pWK120) strain. The construction of plasmid pWK120 is described in more detail by Pühler et al. (1979b,c).

Plasmid pWK120 has a contour length of 17 μm or a molecular weight of 34 Mdals. Since pWL625 is a multicopy plasmid, pWK120 exists in *E. coli* in a high copy state. We measured more than 60 copies per cell (Pühler et al., 1979b). The restriction maps of pWK120 are shown in Fig. 2. The *Hin*dIII and *Eco*RI restriction sites are indicated. In addition, 15 *nif* genes distributed on 7 *nif* operons are given in relation to the physical map of plasmid pWK120. The locations of the *nif* genes and their operon structure is taken from Cannon et al. (1980).

4 Theoretical Considerations About Mapping of nif Coding Regions by a Combination of Various Methods

The expression of *Klebsiella nif* genes is highly regulated. *K. pneumoniae* as well as *E. coli* cells allow the synthesis of nitrogen fixation enzymes only under anaerobic conditions and requirement for reduced nitrogen. So far, this regulation is not well understood. But it is assumed that an activator protein stimulates the transcription of the different *nif* operons. Genetic engineering techniques now provide methods to substitute *nif* promotors against constitutively expressed promotors of diverse origins. Thus it should be possible to express parts of the *Klebsiella nif* region in *E. coli* in the presence of oxygen and with nitrogen sources in the medium. We performed such experiments and were able to identify *nif* gene products in minicells of *E. coli*. These experiments gave us the molecular weights of nearly all of the *nif* proteins. In addition, by using a special transposon mutagenesis technique we could determine the extension and the exact location of most of the *nif* coding regions. The different steps of these experiments are outlined schematically in Fig. 3.

In Fig. 3a the map of the plasmid vector pACYC184 (Chang and Cohen, 1978) is shown. It mediates to its host resistance against chloramphenicol (Cm) and tetracycline (Tc). The coding region for Cm carries a unique *Eco*RI site. This site can be used for *Eco*RI cloning. The gene product of the Cm gene is a chloramphenicol acetyl transferase

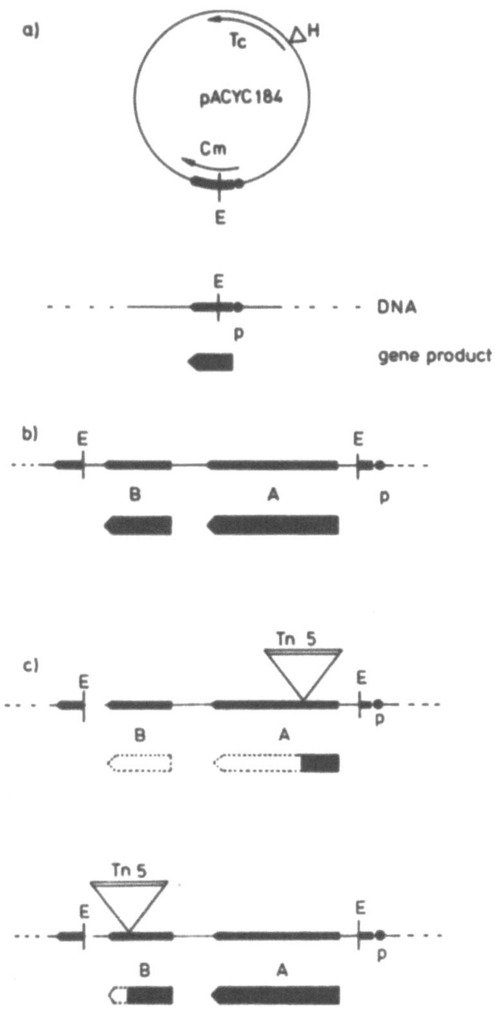

Fig. 3a–c. Scheme for determination of coding regions located on cloned *Eco*RI fragments in pACYC184. In **a** the vector plasmid pACYC184 and the expression of the Cm gene are shown. **b** demonstrates the expression of a cloned *Eco*RI fragment under the control of the Cm promotor of pACYC184. The determination of coding regions located on a cloned *Eco*RI fragment is achieved by analyzing different Tn5 insertion mutants as shown in **c**. *E Eco* RI site; *H Hin* dIII site; *Tc* tetracycline resistance; *Cm* chloramphenicol resistance; *p* promotor; *A* and *B* two assumed genes on a cloned *Eco* RI fragment

(cat) (Le Grice and Matzura, 1980). The Cm gene is transcribed under the control of a strong promotor that can be shown by the minicell technique. Minicells of *E. coli* are chromosome-free cells which are produced by a special filament-forming *E. coli* mutant. If such a minicell-producing strain harbors a multicopy plasmid, this plasmid segregates into minicells. The genes of this plasmid are expressed – and the synthesized gene products can be studied by radioactive labeling. This method was employed for pACYC 184 and the cat protein could be identified as a 22 K protein band on a polyacrylamide gel.

After cloning an *Eco* RI fragment into the Cm gene this gene is inactivated (Fig. 3b). If we assume that this cloned fragment carries two coding regions (A and B) without promotors and terminators for transcription, these genes can be transcribed by the Cm promotor and the gene products identified with the *E. coli* minicell system. This method worked well for all *Eco* RI fragments of the *Klebsiella nif* region.

Using now transposon (e.g., Tn 5) induced mutations in the cloned fragment it is possible to map exactly the extension and the location of a special coding region. If Tn 5 is inserted into gene A the gene product B is not synthesized. From gene A only a truncated polypeptide is made. On the other hand if Tn 5 is inserted into gene B the gene product A is synthesized and in this case a truncated polypeptide is made from gene B (Fig. 3c). The insertion sites of transposon Tn 5 into the cloned fragment can be determined exactly by restriction analysis. Knowing the Tn 5 insertion sites and the molecular weights of the wild type and the truncated gene products, the coding regions of gene A and B can be localized on the cloned fragment. As will be shown in the following chapters the methods presented were successfully employed to map the *Klebsiella nif* coding regions.

5 Subcloning of the Klebsiella nif Region and Expression of nif Genes in Minicells of E. coli

Before subcloning the *Klebsiella nif* region it was necessary to map further restriction sites within the *nif* region. The endonucleases *Eco* RI, *Sma* I, and *Pst* I were used for this purpose. Their restriction sites are indicated on the physical map of the *nif* region in Fig. 4. The different *Eco* RI fragments obtained after an *Eco* RI digestion of *nif* DNA were subsequently cloned in the *Eco* RI site of the Cm gene of plasmid pACYC 184 (Fig. 3). It should be noted that for each *Eco* RI fragment two hybrid plasmids were constructed which contain the cloned fragment in two possible orientations. The hybrid plasmids which allow transcription of *nif* genes under control of the Cm promotor were called pWK 23, pWK 25, pWK 26, pWK 27, and pWK 28.

All these hybrid plasmids were transformed to a minicell-producing strain. Minicells containing these plasmids were isolated and the gene products of the cloned fragments were determined by polyacrylamide gel electrophoresis. For further demonstration of our experimental results we have chosen the hybrid plasmid pWK 25 which contains the *Eco* RI fragment E 10 with the *nif* operon USVM. In the first lane of the protein gel of Fig. 6 the gene products are shown which are encoded by the fragment E 10. Peptides with the following molecular weights could be identified: 45K, 42K, 32K, 28K, 25K, and 18K. There exists now the problem that an *Eco* RI fragment with four known genes

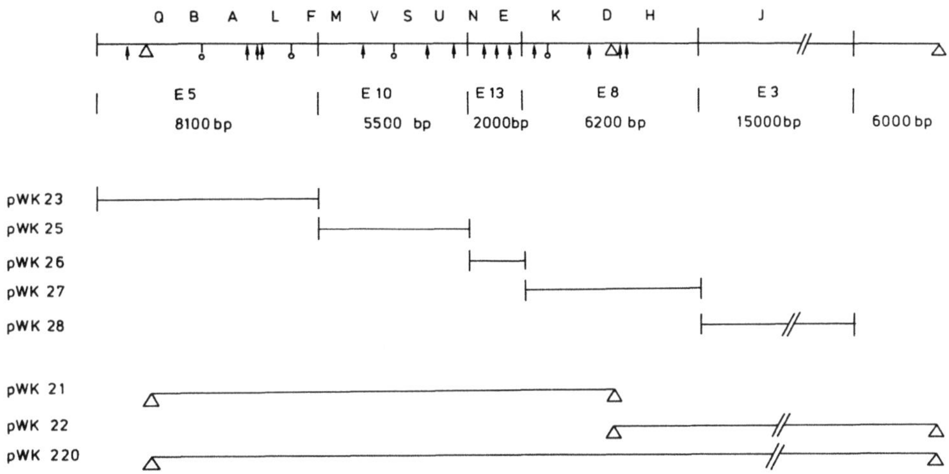

Fig. 4. Subcloning of the *Klebsiella nif* region. The *Hin*dIII (△), *Eco*RI (|), *Sma*I (♁) and *Pst*I (↑) sites of the *Klebsiella nif* region are indicated in correlation to the genetic map. The *Eco*RI fragments E5, E10, E13, E8, and E3 originally located on pRD1 were cloned into pACYC184, resulting in the hybrid plasmids pWK23, pWK25, pWK26, pWK27, and pWK28. In addition the *Hin*dIII fragments H4, H3, and H4 + H3 were cloned in the same plasmid vector giving rise to pWK21, pWK22, and pWK220, respectively.

produces six different polypeptides. The answer to this question is given in the next section.

Besides *Eco*RI cloning also *Sma*I and *Pst*I fragments were cloned in the plasmid vector pACYC177 (Chang and Cohen, 1978) and pBR322 (Bolivar et al., 1977), respectively. With both vectors *nif* genes can be transcribed under the control of constitutive plasmid promotors. In addition, the *Hin*dIII fragments 4, 3, and 4 + 3 were cloned in the *Hin*dIII site of pACYC184 resulting in the hybrid plasmid pWK21, pWK22, and pWK220 (Fig. 4). The hybrid plasmid pWK220 confers again nitrogen fixation capacity on *E. coli* C.

6 Determination of the Coding Regions on an EcoRI Fragment Carrying nif USVM

For the determination of the coding regions on fragment E10 a saturation transposon mutagenesis was necessary. We used for this purpose the kanamycin transposon Tn5 (Berg et al., 1975) and identified 39 different insertion sites on the cloned fragment E10 (Fig. 5a). The methods for isolation the Tn5 insertions and mapping the insertion sites are straight forward and shall be described elsewhere. In Fig. 5b the polypeptide analysis of different Tn5 mutagenized E10 fragments are shown. It is obvious that downstream from a Tn5 insertion site no polypeptides are produced, whereas all polypeptides upstream from the insertion site are made. If a Tn5 insert is located within a coding region a smaller polypeptide is synthesized with a molecular weight directly proportional to the truncated coding region. By this analysis it was possible to determine the coding re-

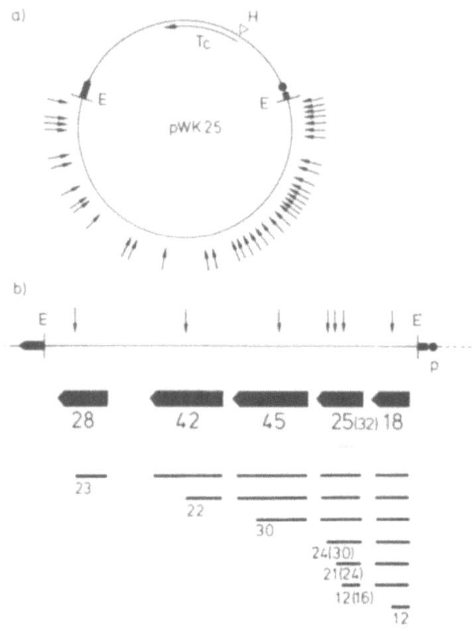

Fig. 5a,b. The coding regions of the *Eco*RI fragment E 10. In **a** the result of the saturation mutagenesis of the cloned fragment E 10 is shown. *Arrows* indicate insertion sites of transposon Tn5. The determination of coding regions located on E 10 is depicted in **b**. The gene products of some Tn5 mutagenized E 10 fragments are indicated. Molecular weights of wild-type and truncated gene products in Kdals determined by polyacrylamide gel electrophoresis can be taken from **b**. (For symbols see Fig. 4)

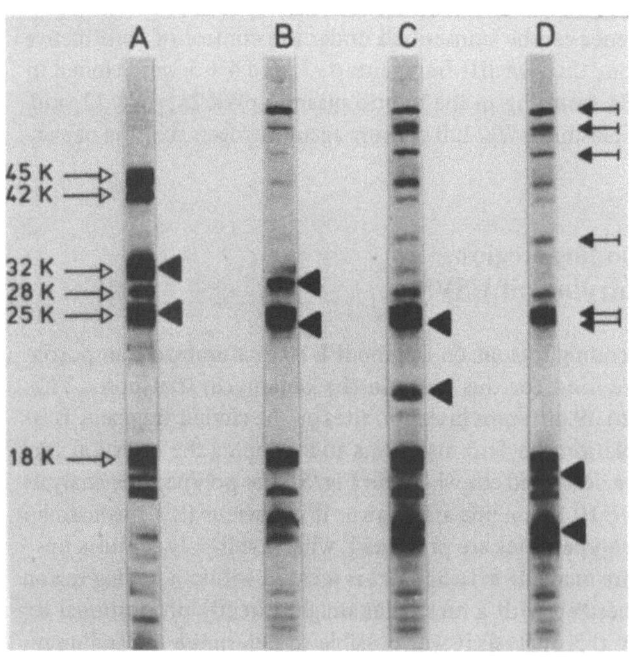

Fig. 6. Polyacrylamide gel electrophoresis of gene products encoded by wild-type and Tn5 mutagenized E 10 fragments. In lane *A* the wild-type proteins of fragment E 10 and their molecular weights are given. Lane *B, C*, and *D* show gene products of E 10 fragments with three different Tn5 inserts in the coding region of the "double" protein 25K (32K). For comparison see Fig. **5b**. The wild-tpye "double" protein in *A* and the truncated "double" proteins in *B, C*, and *D* are indicated (▲). In lane *D* Tn5 specific proteins are marked by *arrows* (←)

gions for the 45K, 42K, 28K, 25K, and 18K polypeptide. The 32K protein is a byproduct of the 25K protein. This can be shown directly by the protein gel in Fig. 6. When the 25K protein is shortened by a Tn5 insertion the 32K product is shortened, too. From this data we suggest that both proteins are synthesized by the same coding region. We speculate that the 32K product is a modified version of the 25K protein, e.g., caused by additional side chains.

7 The Coding Regions of Klebsiella nif DNA

The methods used to map the coding regions on the *Eco*RI fragment E10 were also employed to analyze the other *nif* fragments E5, E13, E8, and E3. Also *Sma*I and *Pst*I fragments of the *Klebsiella nif* region were handled in a similar manner. The complete data are summarized in Fig. 7. There, the restriction map of the *Klebsiella nif* region with *Hin*dIII, *Eco*RI, *Sma*I, and *Pst*I sites are shown in relation to the coding regions. In addition, the genetical map of the *Klebsiella nif* region according to the results of Cannon et al. (1980) is drawn in scale. A comparison of the genetical map with our coding region map delivers the following molecular weights for the different *nif* gene products:

nif Q: ?	? X: 18 K
nif B: ?	*nif* N: 50 K
nif A: 60 K	*nif* E: 40 K
nif L: 50 K	? Y: 24 K
nif F: <10 K	*nif* K: 60 K
nif M: 28 K	*nif* D: 60 K
nif V: 42 K	*nif* H: 35 K (39 K)
nif S: 45 K	*nif* J: 120 K
nif U: 25 K (32 K)	? Z: 100 K

These results lead to the following comments and conclusions:

- For the coding regions X, Y, and Z no genes determined by genetical procedures are known. We were able to show that a deletion of Z does not affect the Nif^+ phenotype. For the other two coding regions X and Y which are located with the *Klebsiella nif* region similar experiments were not performed till now.
- No polypeptides could be determined for the genes *nif* Q and *nif* B. It is assumed that this is due to a transcriptional stop in fragment E5 present after the coding region of *nif* A.
- Another stop signal for transcription is assumed in fragment E8 following the coding region of gene Y.
- Both genes *nif* U and *nif* H produce a double band on polyacrylamide gel. It is not clear to us why the gene products of *nif* U and *nif* H band in each case at two different positions.
- Our results indicate that all coding regions of the *Klebsiella nif* DNA are transcribed in the same direction from the right to the left as shown in Fig. 7. This is also true for the monocistronic transcription units *nif* F and *nif* J.

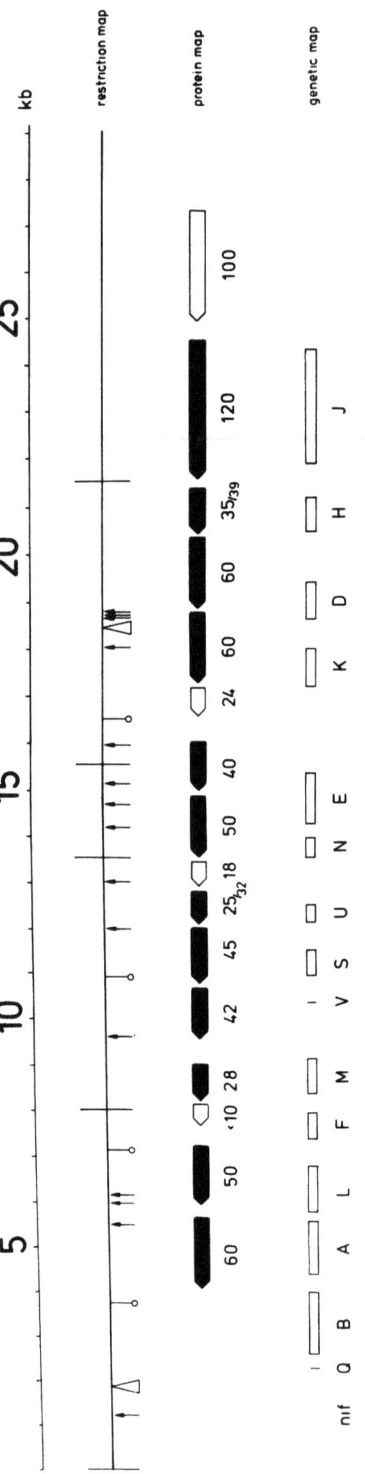

Fig. 7. The coding region of the *Klebsiella nif* region. A detailed restriction map of the *Klebsiella nif* region is given (△: *Hin* dIII, |: *Eco* RI, ♁: *Sma* I, ↑: *Pst* I). In relation to the physical map the summarized coding regions are drawn in scale. The molecular weights (in Kdals) of the corresponding gene products are indicated. For comparison the genetic map of the *Klebsiella nif* region determined by Cannon et al. (1980) is also presented

- The operon structure of the *nif* genes was first determined by transposon mutagenesis and complementation analysis (Merrick et al., 1978; MacNeil et al., 1978). Our results support this operon structure: We assume transcriptional stops after operons *nif* A L and *nif* H D K Y. On the other hand our results suggest that instead of the *nif* U S V M operon two transcription units *nif* X U S V and *nif* M exist. This suggestion was confirmed by Elmerich (1979, unpubl. results).
- Our molecular weight determinations of gene products for *nif* N, *nif* E, *nif* K, *nif* D, *nif* H, and *nif* J are in good agreement with those performed by Roberts et al. (1978) who used different *K. pneumoniae nif* mutants and two-dimensional protein gel electrophoresis. However, we have found different molecular weights for the gene products of *nif* F and *nif* S.

8 Conclusions

In this paper we report experiments with *Klebsiella nif* DNA and show that cloning of two *Hin*dIII fragments of plasmid pRD1 into multicopy plasmid vectors (pWL625, pACYC 184) resulted in a Nif$^+$ phenotype. By combination of various methods like sub cloning of the *nif* region, transposon saturation mutagenesis of the cloned fragments we were able to determine the molecular weights of *nif* gene products and to map the exact location and extension of the different *nif* coding regions.

These methods are of general use for any cloned fragment. We therefore decided to clone *Rhizobium nif* DNA which can be identified by specific hybridization techniques with radioactively labelled *Klebsiella nif* DNA (Jäckel and Pühler, unpubl. results). These *Rhizobium nif* regions will be analyzed *in E. coli* using methods described in this paper. In addition, we try to transfer special Tn5-induced mutations of cloned *Rhizobium nif* regions back to *Rhizobium* in order to determine the phenotype of such mutations.

References

Berg DE, Davies J, Allet B, Rochaix J (1975) Proc Natl Acad Sci USA 72:3628
Bolivar F, Rodriguez RL, Greene PJ, Betlach MC, Heyneker HL, Boyer HW, Crosa JH, Falkow S (1977) Gene 2:95–113
Cannon FC, Postgate JR (1976) Nature (London) 260:271–272
Cannon FC, Dixon RA, Postgate JR (1974) J Gen Microbiol 80:227–239
Cannon FC, Dixon RA, Postgate JR (1976) J Gen Microbiol 93:111–125
Cannon FC, Riedel GE, Ausubel FM (1980) to be published
Chang ACY, Cohen SN (1978) J Bacteriol 134:1141–1156
Dixon RA, Postgate JR (1972) Nature (London) 237:102–103
Dixon R, Cannon F, Kondorosi A (1976) Nature (London) 260:268–271
Goebel W, Lindenmaier W, Pfeiffer F, Schrempf H, Schelle B (1977) Mol Gen Genet 157:119–129
Le Grice SFJ, Matzura H (1980) FEBS Lett 113:42–46

MacNeil T, MacNeil D, Roberts G, Supiano M, Brill WJ (1978) J Bacteriol 136:253–266
Merrick M, Filser M, Kennedy C, Dixon R (1978) Mol Gen Genet 165:103–111
Postgate JR, Krishnapillai V (1977) J Gen Microbiol 98:379–385
Pühler A (1980) Arzneim Forsch/Drug Res 30:570–575
Pühler A, Burkardt HJ, Cannon FC, Wohlleben W (1979a) Mol Gen Genet 171:1–6
Pühler A, Burkardt JH, Klipp W (1979b) Mol Gen Genet 176:17–24
Pühler A, Burkardt HJ, Klipp W (1979c) In: Timmis KN, Pühler A (eds) Plasmids of medical environmental and commercial importance. Elsevier/North-Holland Biomedical Press, Amsterdam New York, pp 435–447
Roberts G, MacNeil T, MacNeil D, Brill W (1978) J Bacteriol 136:267–279
Streicher SL, Gurney EG, Valentine RC (1972) Nature (London) 239:495–499

In vitro Associations Between Non-legumes and Rhizobium

D. HESS[1]

There are several possibilities to improve biological nitrogen fixation:

1. The improvement of existing symbioses or associations of higher plants with nitrogen-fixing microorganisms
2. The establishment of new symbioses or associations of higher plants with nitrogen-fixing microorganisms
3. The transfer of the *nif*-operon from microorganisms into higher plants
4. The selection of free-living microorganisms with higher nitrogen-fixing abilities.

Due to energetical restrictions, the last possibility seems to be limited. The present paper deals with the second point, especially with the establishment of new associations of non-legumes with *Rhizobia.*

1 "Symbiosis" Between Non-leguminous Tissues and Rhizobium

Since the first publications in 1975 (Child, 1975; Scowcroft and Gibson, 1975) it has been well established that nonleguminous tissue cultures are able to induce nitrogenase activity in *Rhizobia* (Gibson et al., 1976; Rao, 1976; Schetter and Hess, 1977). As demonstrated by the group of Reporter (Reporter and Hermina, 1975; Reporter, 1976, 1978; Bednarski and Reporter, 1978; Storey et al., 1979), soybean cells in suspension culture and conditioned media from cultured soybean cells induced nitrogenase activity in *Rhizobia.* In the transfilter apparatus for use with suspension cultures, however, nitrogenase activity declined within a few days (Reporter and Hermina, 1975). In a simple transfilter apparatus for use with solid media (Fig. 1), *Portulaca grandiflora* tissues associated with *Rhizobium* sp. "cowpea" 32H1 maintained nitrogenase activity for up to 30 days (Fig. 2). This improved system seemed best suited for studying the transport of fixed nitrogen and its utilization by the plant cells (Lustig et al., 1980a).

In an earlier investigation, Scowcroft and Gibson (1975) had claimed an uptake of ^{15}N fixed by rhizobia into plant cells. In their system, however, plant cells and bacteria were not separated, so that convincing evidence for a possible use of the fixed nitrogen by the plant cells was lacking. Using the *Portulaca-Rhizobium* transfilter system it was possible to clarify this question: preliminary results indicate that the nitrogen fixed by the bacteria migrates in the form of ammonia to the plant cells. Within the *Portulaca* cells, first an increase in glutamic acid was observed (Fig. 3). Thereafter, the nitrogen

---

1 Lehrstuhl für Botanische Entwicklungsphysiologie, Universität Hohenheim, Emil-Wolff-Str. 25, 7000 Stuttgart 70/FRG

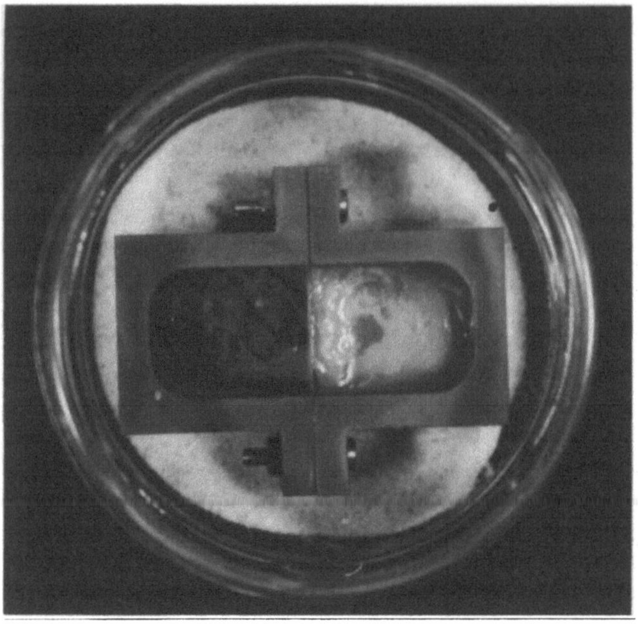

Fig. 1. Transfilter apparatus for use with solid media. The apparatus was fashioned from polyvinylchloride. It consisted of two identical chambers with a volume of about 6 ml, which were connected by screws. *Rhizobia (right)* and *Portulaca* callus (*left*) were separated by a dialysis membrane with a pore size of 15–20 Å. The photograph was taken 20 days after initiation of the transfilter culture. *Rhizobia* are concentrating along the membrane. (From Lustig et al., 1980a)

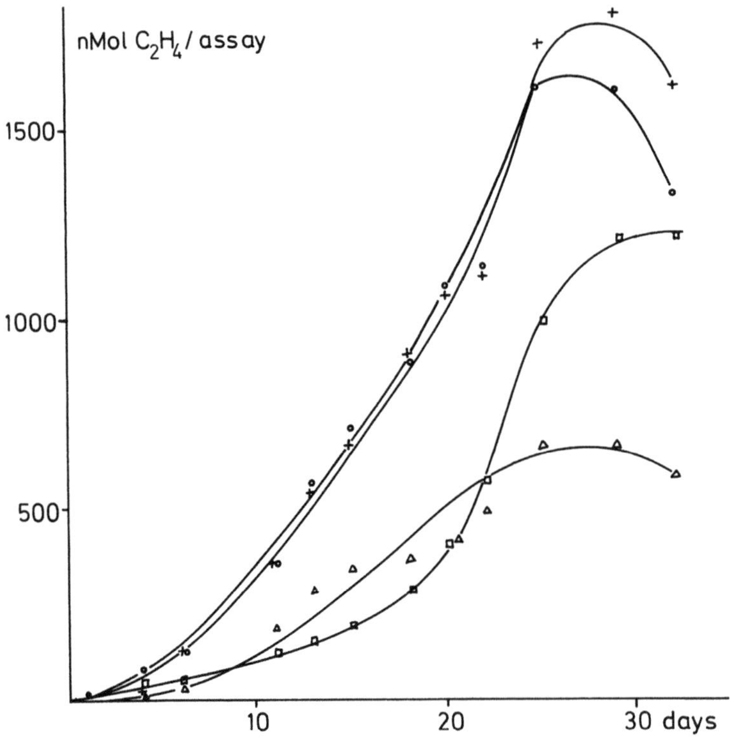

Fig. 2. Time course of nitrogenase activity in associations between *Portulaca* callus cultures and *Rhizobium* sp. "cowpea" 32H1 in a transfilter culture. The different *curves* represent 4 parallel assays. The acetylene activity was first determined 10 days after initiation of the transfilter culture. Bacteria alone showed no nitrogenase activity. (From Lustig et al., 1980a)

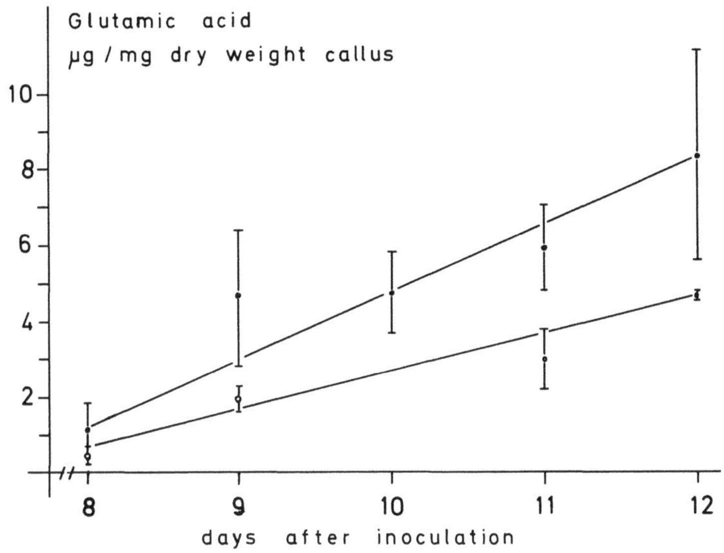

Fig. 3. Time course of glutamic acid accumulation in *Portulaca* callus tissues associated in a transfilter culture with *Rhizobium* sp. "cowpea" 32H1. Controls without bacteria. •—• experiment, ○—○ control. (From Lustig and Hess, in prep.)

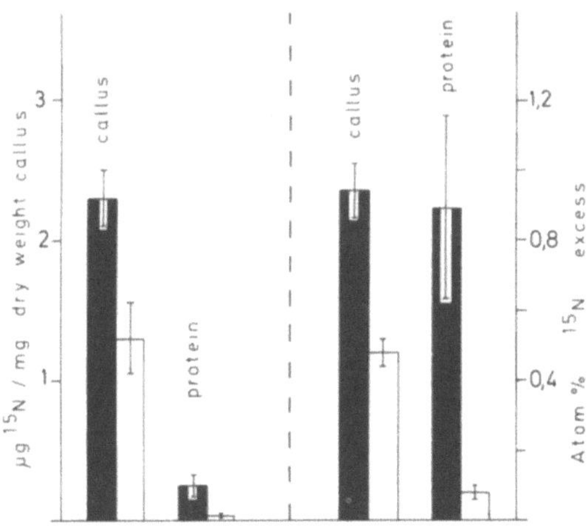

Fig. 4. Use of ^{15}N fixed by *Rhizobium* sp. "cowpea" 32H1 by *Portulaca* tissues in transfilter cultures. *Left* μg^{15}N within the callus and within callus proteins, respectively, per mg dry weight callus. *Right* Atom% ^{15}N accumulation within the callus and within callus proteins, respectively. *Black* experiment *(Portulaca* callus with *Rhizobia). White* control *(Portulaca* callus without *Rhizobia).* Experiments and controls were incubated 96 h in an atmosphere of about 10% O_2 and 90% N_2, containing 54 atom% ^{15}N. (From Lustig and Hess, in prep.). The ^{15}N detected in the protein controls was within the limits of the method. The ^{15}N found in the callus controls was probably due to a small amount of water, containing ^{15}N and remaining in the sample despite prolonged lyophilization (the callus cultures were lyophilized until we observed no further loss of weight). To compare the experimental data only with controls incubated in air (for instance Scowcroft and Gibson, 1975) seems insufficient. (From Lustig et al., 1980a)

was used for the synthesis of other substances, for instance betalaines. Using $^{15}N_2$ it could be demonstrated that 10% of the nitrogen imported into the plant cells was incorporated into proteins (Fig. 4, Lustig et al., 1980b; Lustig and Hess, in prep.).

These results demonstrate that the plant cells benefit from the nitrogen fixed by the bacteria, a fact interesting especially from the viewpoint of a practical use of such associations.

2 In vitro Associations of Non-leguminous Plants with Rhizobium

2.1 Petunia hybrida, a Model System

For a possible practical use, first of all one has to replace the tissue cultures by plants. To obtain a first insight, we started our investigations with *Petunia hybrida*, a model system with which we have been well acquainted for 20 years.

In a first series of experiments we detected that 13-day-old *Petunia* plants were able to induce nitrogenase activity in *Rhizobium* 32H1 (Hess and Götz, 1977). Further stud-

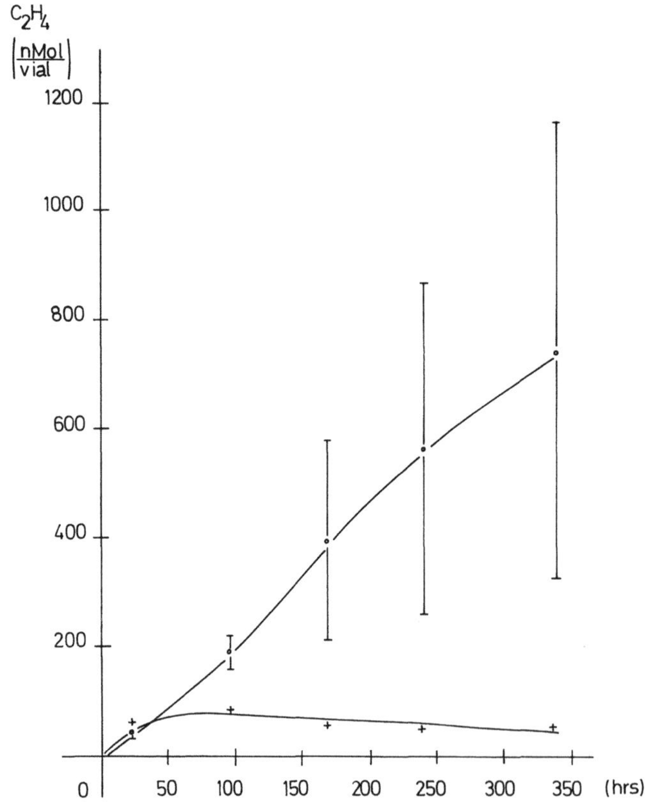

Fig. 5. Time course of nitrogenase activity in associations of 7-week-old *Petunia hybrida* plants and *Rhizobium* sp. "cowpea" 32H1. o: test vials +: control bacteria. (From Götz, 1980)

ies revealed that much older petunias had the same ability. The inducing activity continued over a period of at least 15 days (Fig. 5). Following removal of the plants, nitrogenase activity rapidly declined (Götz, 1980).

Light microscopical observations showed root and root hairs covered with bacteria. In most cases, *Rhizobia* hat attached to the root hair surface in a polar end-on manner (Fig. 6). To detect a possible uptake of *Rhizobia* into the cells, EM studies are initiated.

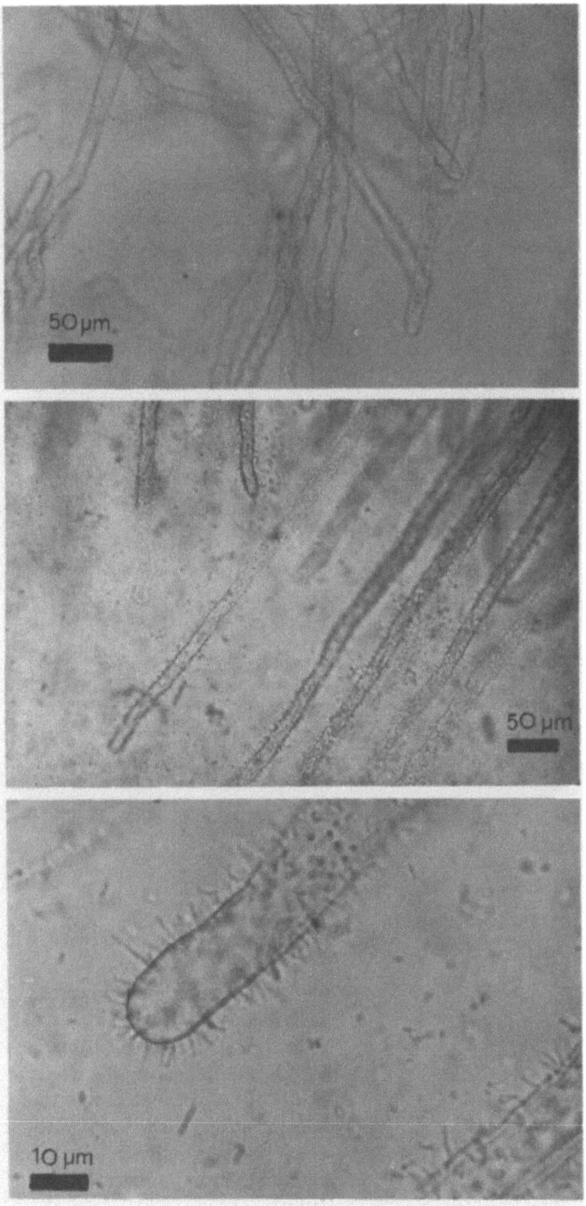

a

b

c

Fig. 6a–c. Roots hairs of *Petunia* plants not inoculated with *Rhizobium* 32H1 **(a)**, and inoculated with *Rhizobium* 32H1 **(b)** and **(c)**. (From Götz, 1980)

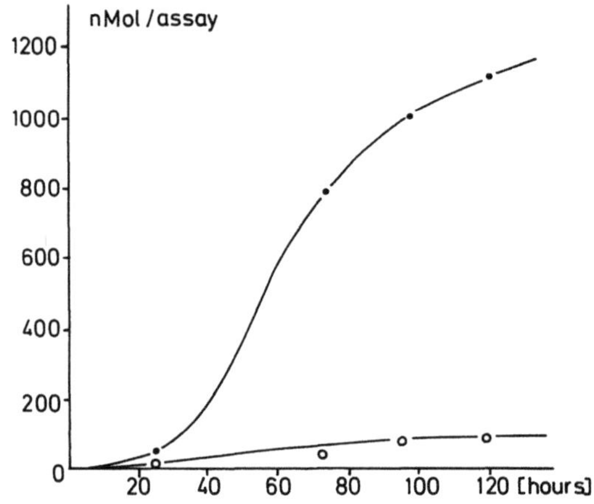

Fig. 7. Time course of nitrogenase activity in associations of 13-day-old *Petunia* plantlets with *Rhizobium* sp. "cowpea" 32H1. •: medium with, ○: medium without 1.0 mg/l 6-benzyl-adenine. (From Hess and Feuereissen, 1980)

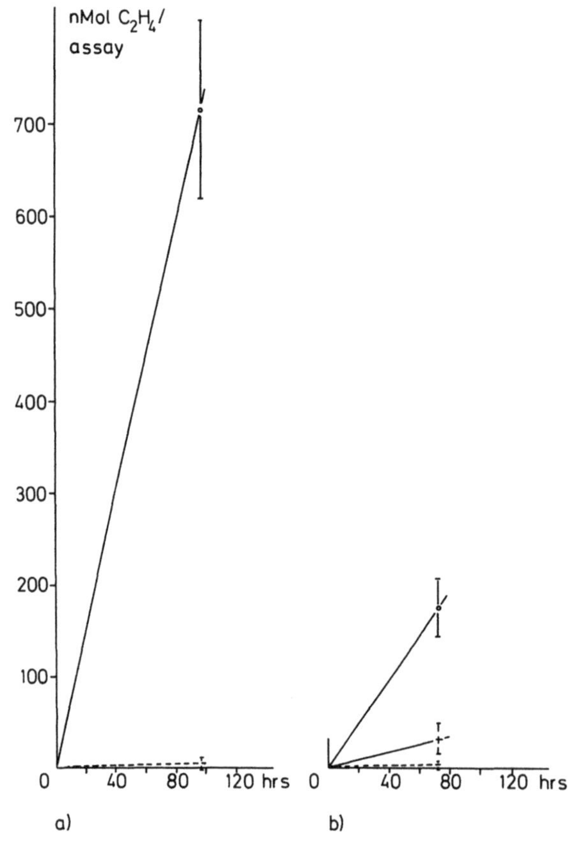

Fig. 8a, b. Nitrogenase activity. **a** in associations of 23-day-old wheat plants 96 h after addition of the acetylene. ○: wheat-*Rhizobia* associations, –: control bacteria. **b** 76 h after removing the wheat plants out of the association vials. ○: association vials without wheat, +: excised wheat plants, –: control bacteria. (From Götz and Hess, 1980)

There are difficulties in comparing the efficiency of in vitro associations with natural associations or symbioses. We will come back to this point somewhat later in discussing our wheat experiments. In any case, it seemed useful to stimulate nitrogenase activity in the *Petunia-Rhizobium* association. We studied the influence of several plant growth regulators. IAA and gibberellic acid showed only a 2- to 3-fold increase of nitrogenase activity compared with controls without the hormones. 6-Benzyl-adenine, however, increased nitrogenase acitivity more than 10-fold (Fig. 7). Apparently all the growth substances tested influenced the plant and not the bacterial partner (Hess and Feuereissen, 1980).

2.2 Triticum aestivum var. "Selpek", a Plant of Nutritional Value

Even in times of severe famine, the nutritional value of petunias seems rather limited. Therefore, all our experiments with the model plant *Petunia* were followed by appropriate experiments with wheat.

Young wheat plants were associated with *Rhizobium* 32H1. Under the conditions used, wheat alone showed no endogenous ethylene production and no nitrogenase activity. *Rhizobia* alone had a very low or no nitrogenase activity. In associations of wheat with *Rhizobia,* however, a comparatively high nitrogenase activity developed. Following removal of the wheat plants nitrogenase activity greatly declined, indicating the necessity of inducing wheat (Fig. 8, see also Götz and Hess, 1980).

a b

Fig. 9a,b. Wheat roots 14 days after initiation of the association. **a** controls without bacteria; **b** experimental assay with *Rhizobia*. (From Götz and Hess, 1980)

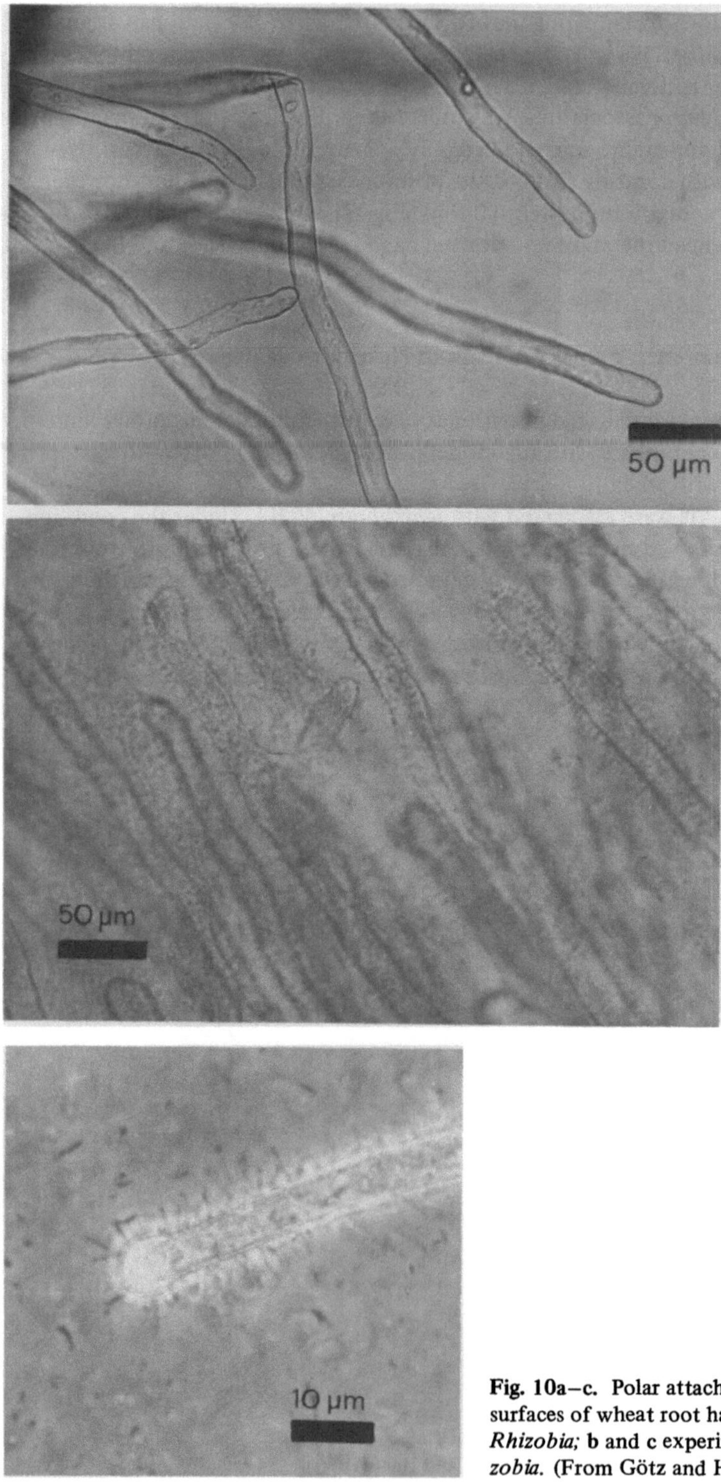

Fig. 10a–c. Polar attachment of *Rhizobia* to the surfaces of wheat root hairs. **a** control without *Rhizobia;* **b** and **c** experimental assays with *Rhizobia.* (From Götz and Hess, 1980)

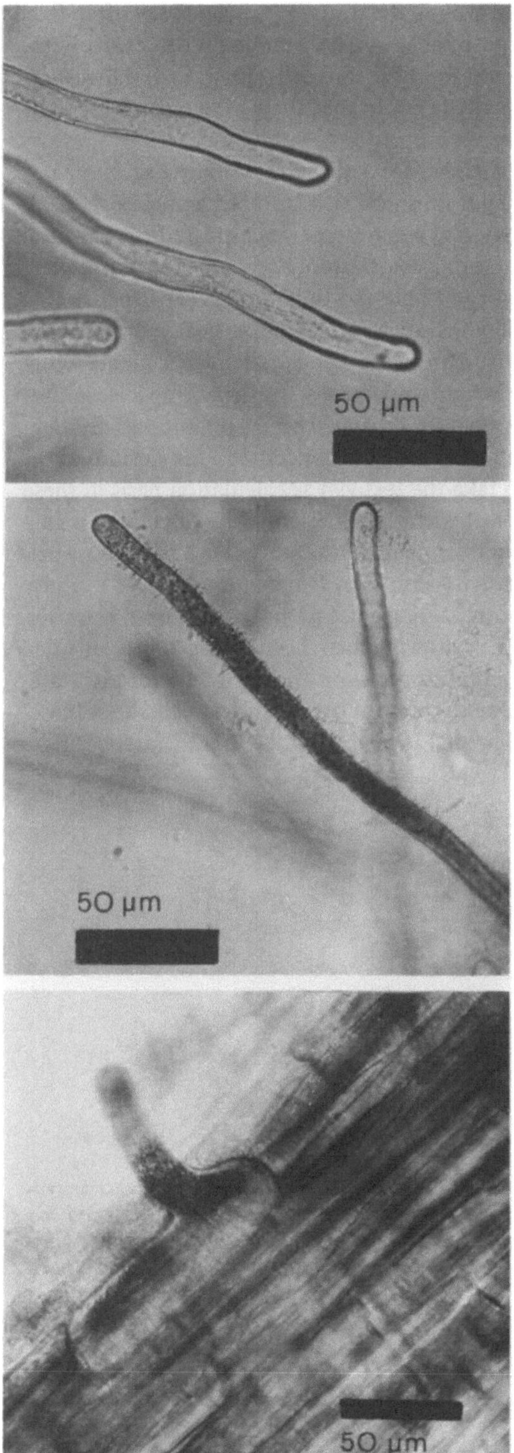

a

b

c

Fig. 11a–c. Attachment and uptake of *Rhizobia* into wheat root hairs. The root hairs were stained using tetrazoliumchloride. **a** control without *Rhizobia;* **b** root hair covered with bacteria; **c** *Rhizobia* within a developing root hair. (Hess and Götz, in prep.)

As with *Petunia*, *Rhizobia* concentrated around the roots (Fig. 9) and very often fixed to the surface of the root hairs in a polar manner (Fig. 10). Staining with tetrazolium-chloride revealed the presence of rhizobia within wheat root cells (Fig. 11). EM investigations still have to be done, but there remains little doubt that they will confirm the light microscopical observations (Hess and Götz, in prep.).

Studies concerning the stimulation of nitrogenase activity in wheat-*Rhizobium* associations using growth regulators are not yet finished (Hess and Mertens, in prep.).

To judge the efficiency of nitrogen fixation in our system, we compared the artificial association wheat-*Rhizobium* with the natural association *Sorghum-Azospirillum* (Döbereiner, 1977). At present we are not yet able to study the wheat-*Rhizobium* system under field conditions. Therefore, comparative studies had to be performed under in vitro conditions. Wheat and *Sorghum nigricans* were associated on agar media with *Rhizobium* 32H1 and *Azospirillum lipophorum*, respectively. Both plant species induced nitrogenase activity in both bacteria, wheat in *Azospirillum* even more rapidly than *Sorghum* (Fig. 12). As far as quantitative differences are concerned, one should take into consideration that it was necessary to take the test vial as calculation basis in these kinetic studies. Only the last nitrogenase test within a series could be calculated on a dry weight basis (plant dry weight). Doing this, there was no difference between wheat and *Sorghum* in inducing nitrogenase activity either in *Rhizobium* (Fig. 13) or in *Azospirillum*. The activity obtained by both plant species with *Azospirillum*, however, was 100 times higher than with *Rhizobium*. Following this, the inducing ability of wheat and *Sorghum* would be nearly the same, and *Azospirillum* would be a better associate for both plant species than *Rhizobium*. One should not overlook, however, that these studies were done on fully supplemented media. It might well be that differences be-

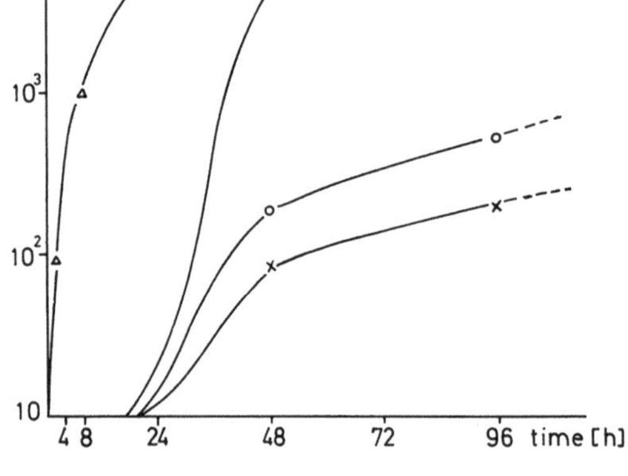

Fig. 12. Time course of nitrogenase activity in associations of *Triticum aestivum* var. "Selpek" and *Sorghum nigricans* with *Azospirillum lipoferum* 108 st and *Rhizobium* 32H1, respectively. □: *Sorghum* + *Azospirillum*, △: *Triticum* + *Azospirillum*, o: *Sorghum* + *Rhizobium*, x: *Triticum* + *Rhizobium*. Control values (bacteria alone): *Azospirillum* 0.0 nmol ethylene per assay, *Rhizobium* 10 nmol ethylene per assay. (From Hess and Kiefer, in prep.)

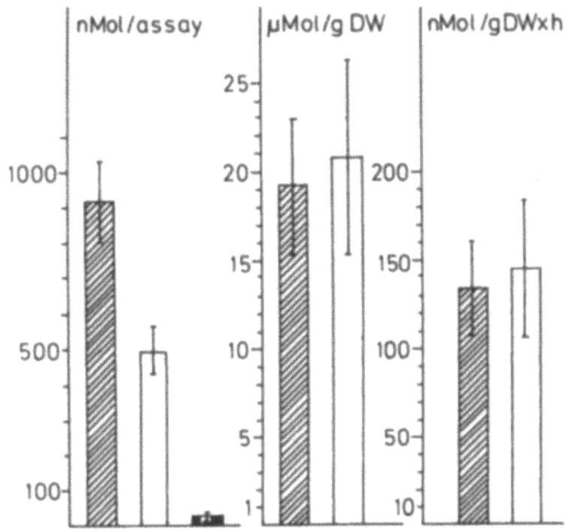

Fig. 13. Nitrogenase activity in associations of *Rhizobium* 32H1 with *Triticum aestivum* and *Sorghum nigricans,* respectively. *From left to right* ethylene production in nmol/assay, in μmol/g dry wt. (DW), and in nmol/g dry wt. x h. Tests 6 days (h=144) after addition of acetylene. At this time there was no further increase in nitrogenase activity. Taking the plant dry weight as calculation basis there were no differences between *Triticum* and *Sorghum.* (From Hess and Kiefer, in prep.) ▨ *Triticum + Rhizobium,* □ *Sorghum + Rhizobium,* ■ *Rhizobium*

tween the C4-plant *Sorghum* and the C3-plant *Triticum* would become apparent on reduced media (Hess and Kiefer, in prep.).

Our final experiment with the wheat-*Rhizobium* system will be to test a possible transfer into field conditions. Our last steps on this rather long way were to replace the agar by more natural media. In associations kept on vermiculite plus liquid culture medium and on soil plus liquid culture medium nitrogenase activity developed as well. The next step will be to reduce the supplementation of the liquid medium added as far as possible.

In the investigations outlined here we were approaching a possible practical use step by step: from tissue cultures to plants, from the model plant to the plant of nutritional value, from agar media to soil. At present, however, it is not yet possible to predict whether associations of this kind will function satisfactorily under field conditions. It might turn out that one has to introduce additional gene material into the plant partner, for instance leghemoglobin genes. Of course, this remark leads to another field, that of speculations. Sometimes, however, and especially in biology, the speculations of today are the realities of tomorrow.

References

Bednarski MA, Reporter M (1978) Appl Environ Microbiol 36:115–120
Child JJ (1975) Nature (London) 253:350–351
Döbereiner J (1977) In: Ayanaby A, Dart PJ (eds) Biological nitrogen fixation in farming systems of the tropics. John Wiley, Chichester, pp 3–12
Gibson AH, Child JJ, Pagan JD, Scowcroft WR (1976) Planta 128:233–239
Götz E-M (1980) Z Pflanzenphysiol (in press)
Götz E-M, Hess D (1980) Z Pflanzenphysiol (in press)
Hess D, Feuereissen S (1980) Biochem Physiol Pflanz 175:689–698
Hess D, Götz E-M (1977) Z Pflanzenphysiol 85:185–188
Lustig B, Plischke W, Hess D (1980a) Z Pflanzenphysiol 98:277–281

Lustig B, Plischke W, Hess D (1980b) Experientia (in press)
Rao R (1976) Plant Sci Lett 6:77–83
Reporter M (1976) Plant Physiol 57:651–655
Reporter M (1978) Plant Physiol 61:753–756
Reporter M, Hermina N (1975) Biochem Biophys Res Commun 64:1126–1133
Schetter C, Hess D (1977) Plant Sci Lett 9:1–5
Scowcroft WR, Gibson AH (1975) Nature (London) 253:351–352
Storey R, Rainey K, Pope L, Reporter M (1979) Plant Sci Lett 14:253–258

Limitations of Symbiotic and Associative Nitrogen Fixation by Developmental Stages in the Systems Rhizobium japonicum with Glycine max and Azospirillum brasilense with grasses, e.g., Triticum aestivum

D. WERNER, J. WILCOCKSON, R. STRIPF, E. MÖRSCHEL, and H. PAPEN[1]

1 Summary

By the oxygen enhancement effect of nitrogenase activity four stages (I–IV) of nodule development in *Glycine max* are separated. This effect correlates with an increase of leghemoglobin content above 7×10^{-2} µmol per g nodule fresh weight. Nodules fall into distinct size classes. Four maxima can be separated for the nodule population in stages II to IV, indicating that age and size are rather different parameters to characterize nodule development. A drop in nitrogenase activity in stage II is correlated with the exhaustion of starch granules in the amyloplast-mitochondrion associations and with space-filling multiplication of the bacteroids in the infected cells. During stage III with a constant nitrogenase activity at a low level, the bacteroids continue to accumulate poly-3-hydroxybutanoic acid (PHBA). In pure culture of *Rhizobium japonicum* maximum nitrogenase activity of 100–150 nmol $C_2H_4 \cdot$ mg protein$^{-1} \cdot h^{-1}$ appears suddenly at a certain culture age as well as depending on the oxygen concentration. Some activities of other enzymes of N-metabolism are compared for bacteroids from effective and ineffective nodules with nitrogenase repressed and derepressed free-living cells. This indicates that the *Rhizobium* cells in pure culture can behave similarly to the bacteroids in the nodule cells. However, we have almost no information about the actual composition of the infection vacuoles surrounding the bacteroids in the nodule cells.

Our incomplete understanding of the limitations of associative nitrogen fixation is summarized with the example of *Azospirillum brasilense* and grasses, e.g., *Triticum aestivum.* Root growth and cytology under microaerophilic conditions, root exudation, competition of other chemoorganotrophic microorganisms with diazotrophs, influence of the various soil factors on nitrogenase activity during bacterial development, and uptake by the plants of nitrogen fixed by the bacteria are some of the most relevant questions.

2 The Symbiotic System Glycine max with Rhizobium japonicum

Quantitative assessment of symbiotic nitrogen fixation, e.g., in legumes is complicated by the fact that the bacteroids inside the infected nodule cells, the bacteroid zone in the organ nodule, the complete nodule, nodule number per plant, the whole root system

1 Botanisches Institut der Philipps-Universität Marburg, Abt. Zellbiologie und Entwicklungsphysiologie, Fachbereich Biologie, Lahnberge, 3550 Marburg-L/FRG

of the plant, and the size of the leaf system do not develop synchronously. Maximum nitrogenase activity in soybeans per mg protein of bacteroids is in the range of 400–500 nmol $C_2H_4 \cdot h^{-1}$ (Bergersen, 1977). Per mg nodule fresh weight activity is 20–50 nmol $C_2H_4 \cdot h^{-1}$, depending on the time of decapitation of the plants from the root system (Stripf and Werner, 1978a, 1980). Expressing activity per plant (grown at 1200 μl CO_2 per 1 air) it is 100–150 μmol $C_2H_4 \cdot h^{-1}$ (Hardy et al., 1977).

Nitrogenase activity depends on the age of the nodules (Fig. 1). This is made obvious by use of an infection method giving no increase in nodule number (25–30 per plant) once nitrogenase activity has commenced (Stripf and Werner, 1978a). After infection of ten-day-old seedlings with a defined titer of *Rhizobium japonicum* plants were culti-

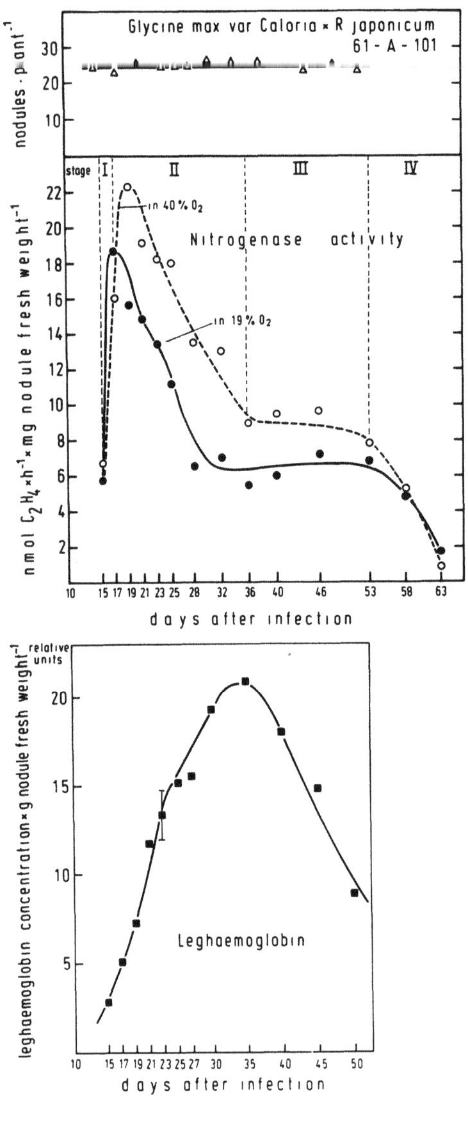

Fig. 1. Nodule number, oxygen enhancement of nitrogenase activity (incubation under 40% O_2 compared to 19% O_2) and leghemoglobin concentration (1 relative unit is equivalent to 10^{-2} μmol) in *Glycine max* var. Caloria infected with the effective strain 61-A-101 of *Rhizobium japonicum* 15–63 days after infection. (Stripf and Werner, 1980)

vated in a phytotron and watered with N-free nutrient solution. Nitrogenase activity of effective nodules certainly affects the number of nodules formed, since the number of ineffective nodules increases almost 10 times up to 140 per plant in the var. Caloria (Stripf and Werner, 1978b) (Fig. 2). Increasing the oxygen concentration from 19% to 40% enhances nitrogenase activity in the nodules only, however, at stage II and stage III of nodule development (Fig. 1). This enhancement effect is found in nodules, with a leghemoglobin concentration above 7 relative units (1 r.u. equals 10^{-2} μmol) (Fig. 1). Another complication in using nodule weight as reference value is the finding, that with soybean nodules from several varieties we have from day 20 up to day 50 at least 4 maxima and four minima in the frequency of nodules size classes No. 1-12 (Fig. 3). Nodules from these size classes have a different composition, as exemplified for the leghemoglobin content (Fig. 4) which, as Fig. 1 suggests, is likely to affect nitrogenase activity.

Oxygen concentration affects and limits nitrogenase activity of *Rhizobium japonicum* in pure culture even more than in the symbiotic state. A drop from maximum activity at 2% oxygen in the gas phase to zero at 5% O_2 (Werner and Stripf, 1978) is found as well as an enhancement of nitrogenase activity by increasing the gas flow rate (1% O_2

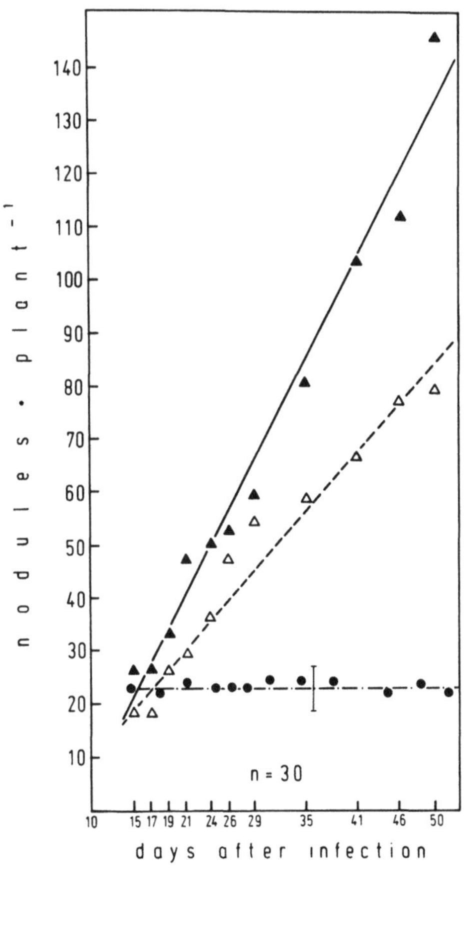

Fig. 2. Nodule number in *Glycine max* var. Caloria (–▲–) and var. Mandarin (–△–) infected with the ineffective strain 61-A-24 of *Rhizobium japonicum.* Nodule number produced by the effective *Rhizobium japonicum* strain 61-A-101 (–●–) varies with the two soybean varieties only as indicated by the *vertical line.* (Stripf and Werner, 1978a)

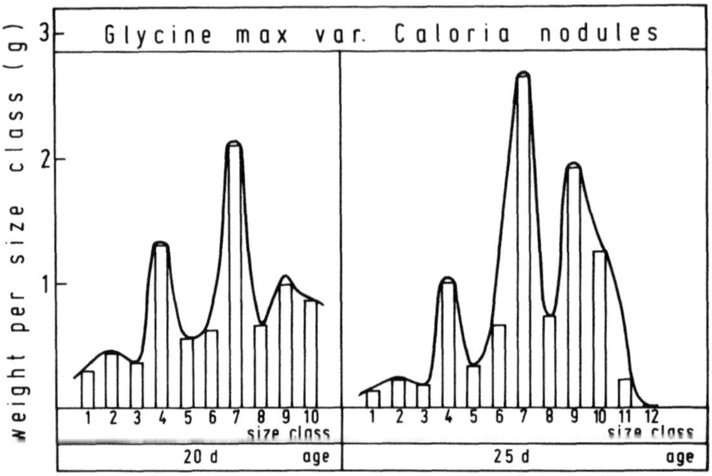

Fig. 3. Fractionation of effective nodules of 50 plants of *Glycine max* var. Caloria into 10–12 size classes by weight per class, 20 and 25 days after infection. Size class No. 1 is equivalent to a maximum diameter of 1.6 mm, size class No. 12 to a minimum diameter of 4.0 mm. (Stripf and Werner, 1980)

in 99% N_2) from 5 l/h^{-1} to 15 l/h^{-1} above 500 ml of cell suspension stirred in a 1 l culture vessel (Fig. 5). It is interesting to note that nitrogenase activity at both gas flow rates appears at the same growth phase of the culture, indicating that oxygen concentration primarily does not so much affect the time for development of nitrogenase derepression, but N_2 fixation itself.

Several carbon and nitrogen sources affect to a different extent not only the time for derepression of nitrogenase in pure culture but also maximum as well as duration of the C_2H_2-reduction activity (Wilcockson and Werner, 1978, 1979). Most of these studies

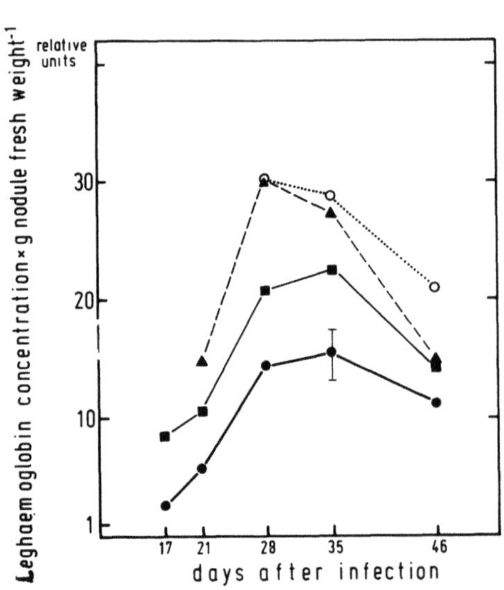

Fig. 4. Leghemoglobin concentration in four size groups of nodules of *Glycine max* var. Caloria, infected with ***Rhizobium japonicum*** 61-A-101: (–●–) <1.8 mm; (–■–) 2.2–2.6 mm; (–▲–) 3.0–3.6 mm; (–○–) >4.0 mm nodule diameter. (Stripf and Werner, 1980)

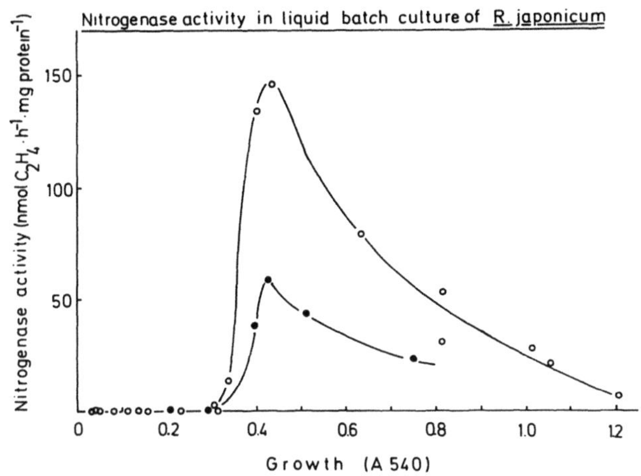

Fig. 5. Specific nitrogenase activity of *Rhizobium japonicum* 61-A-101 in liquid batch culture at two different gas flow rates (1% O_2 in 99% N_2) versus growth (O.D. at 540 mm): (–•–) gas flow rate 5 l/h; (–○–) gas flow rate 15 l/h

were performed with a growth and test system on agar plates under air, which needs a certain minimum growth of the bacterial colonies to reduce the oxygen concentration in the cell mass to tolerable and yet sufficient concentrations. One example is given in Fig. 6. Twenty m*M* concentrations of succinate as the sole carbon source (besides glutamine as a nitrogen and carbon source) allows only a very short period of nitrogenase activity. The addition of a second carbon source (50 m*M* glycerol) increases maximum activity and the period of nitrogen fixation. Replacement of glycerol by arabinose enhances both. The inclusion of glycerol as a third carbon source brought about a situation of a still-rising nitrogenase activity about ten days after the growth of the culture (in Fig. 6) has ceased. This situation is in fact found in nodules of *Glycine max,* about 20 days after infection, when the infected cells are completely packed with bacteroids

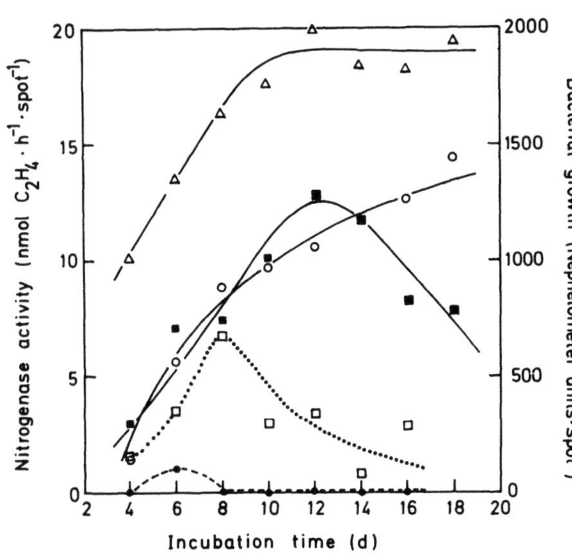

Fig. 6. Nitrogenase activity and growth of *Rhizobium japonicum* 61-A-101 on agar surfaces exposed to air with several combinations of carbon sources at days after incubation (Wilcockson and Werner, 1979): (–△–) growth with 50 m*M* arabinose, 50 m*M* glycerol, 20 m*M* succinate; nitrogenase activity with (–■–): 50 m*M* arabinose, 20 m*M* succinate; (–□–): 50 m*M* glycerol, 20 m*M* succinate; (–•–): 20 m*M* succinate; (–○–): 50 m*M* arabinose, 50 m*M* glycerol, 20 m*M* succinate

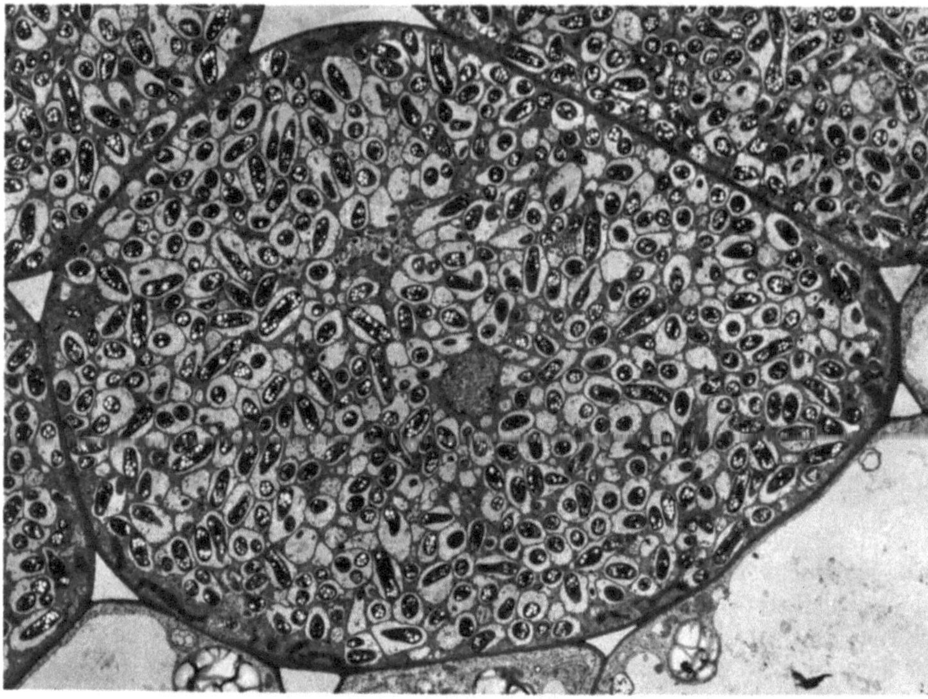

Fig. 7. Nodule cell of *Glycine max* var. Caloria, 20 days after infection with *Rhizobium japonicum* 61-A-101x4,600. (Werner and Mörschel, 1978)

inside infection vacuoles (Fig. 7) and the bacteroids are not multiplying, however nitrogenase continues, although at a somewhat reduced rate per mg nodule. The drop of nitrogenase activity at stage II (Fig. 1) correlates with the exhaustion of the starch granules in the amyloplast-mitochondrion associations inside infected cells of the nodules (Werner and Mörschel, 1978). These specific associations were not found in the ineffective nodules (Werner et al., 1980a). Effective and ineffective bacteroids are also very different in the rate of PHBA accumulation (Werner et al., 1980a) and somewhat different in the specific activity of 3-hydroxybutyrate dehydrogenase (Table 1). Bacteroids from effective nodules are more similar to *Rhizobium japonicum* cells with derepressed nitrogenase in the free-living state than to ineffective bacteroids in specific activities of some other enzymes of nitrogen metabolism (Table 1). Alanine dehydrogenase (absent in the plant cytoplasm of nodule cells) is eight to ten times more active in the rhizobia with nitrogenase activity. Glutamine synthetase, on the other hand, is three to four times more active in the nitrogenase-repressed cells. In the plant cytoplasm of the nodules, this enzyme is, however, about eight times more active in effective nodules. The very low activities of glutamate dehydrogenase in the bacteroid fraction may be merely an impurity from other nodule compartments, as indicated by zero activity in rhizobia from pure culture. Despite these similarities of bacteroids to *Rhizobium* cells from pure culture experiments we have to emphasize that we have almost no data about the in vivo

Table 1. Enzyme activity (mμmol substrate x mg protein^{-1} x min^{-1}) in effective and ineffective bacteroids from nodules of *Glycine max* var Caloria and free-living cells of *Rhizobium japonicum* with repressed and derepressed nitrogenase

Enzyme-activity (m-units)	Bacteroids (effective)	Bacteroids (ineffective)	*Rhizobium japonicum* 61-A-101 (nitrogenase derepressed free-living cells)	*Rhizobium japonicum* 61-A-101 (nitrogenase repressed free-living cells)
Nitrogenase (nmol $C_2H_4 \cdot h^{-1} \cdot$ mg protein^{-1})	400–500	0	100–180	0
Alanine DH	250	30	320	75
3-Hydroxybutyrate DH	200	100	360	240
Glutamine synthetase				
Transferase assay (+Mn^{2+})	350	1,200	480	1,600
Biosynthetic assay (+Mg^{2+})	14	48	20	64
Glutamate synthase	40	40	40	20
Glutamate DH	<20	<20	0	0

environment of the bacteroids in the infection vacuoles. This important compartment of the legume nodule has not been isolated intact in a preparative scale so far permitting an analysis of its low molecular and macromolecular components.

3 The Associative System Azospirillum brasilense with grasses, e.g., Triticum aestivum

Evaluation of associative nitrogen fixation in terms of plant physiology is much more difficult than symbiotic nitrogen fixation. Data on nitrogenase activity in pure culture of the bacteria are available (Okon et al., 1976; Burris et al., 1977). However, no universally acceptable basis for associative nitrogenase activity such as root weight, root size, root respiratory activity or plant weight is available. We also find a large variation of associative nitrogen fixation on a unit area basis ranging from insignificant (Barber et al., 1979; Tjepkema and van Berkum, 1977) to over 40 kg $N_2 \cdot ha^{-1} \cdot year^{-1}$ (Smith et al., 1976) up to 400 kg (von Bülow and Döbereiner, 1975).

Data from pure culture experiments, mostly lasting only 24–72 h, cannot be easily compared to associations with plants, establishment of which lasts several days to weeks due to the slow development (in terms of microbiology) of the root system. In Figs. 8 and 9 we give two examples of long-lasting (8–14 days) pure culture experiments with *Azospirillum brasilense,* using two different cultivation systems. We cannot discuss here in detail whether a liquid batch culture under microaerophilic conditions (Fig. 8) or a culture on (agar) surfaces exposed to air is more similar to an association of bacteria on root surfaces. In the liquid culture we find two phases of nitrogenase activity, together with two growth phases and two phases of PHBA accumulation (Papen and Werner, 1980). Use of nitrogenase activity at only two or three days of the 8-day period certainly would give misleading results in the assessment of total nitrogen fixation during this

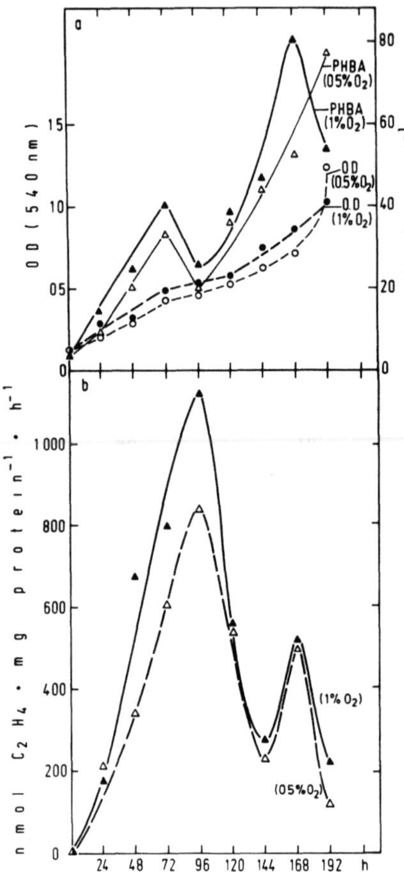

Fig. 8. Biphasic growth, PHBA (poly-3-hydroxy-butanoic acid) accumulation and specific nitrogenase activity in *Azospirillum brasilense* in long-lasting batch culture at 0.5% and 1.0% O_2 concentration. (Papen and Werner, 1980)

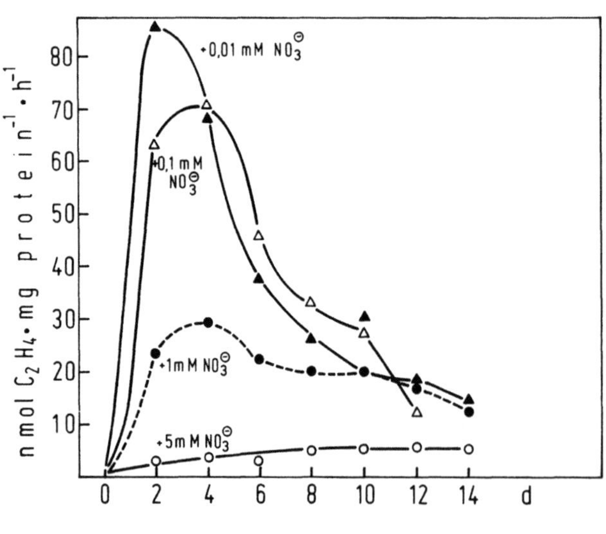

Fig. 9. Specific nitrogenase activity in *Azospirillum brasilense* on agar surfaces exposed to air, as affected by 0.01, 0.1, 1.0, and 10 m*M* concentrations of nitrate at days after incubation. The data with 0.01 m*M* nitrate are the same as those with no nitrate added

Table 2. Limitations of associative nitrogen fixation by developmental stages and ecological factors

System: *Azospirillum brasilense* and grass, e.g., *Triticum aestivum*

1. Plant growth in general
2. Soil factors and nitrogenase activity of associated bacteria
3. Root growth under microaerophilic conditions
4. Root exudation
5. Competition of chemoorganotrophic microorganisms in the rhizosphere
6. Topology of the association at the root surface or in the root cortex
7. Uptake of nitrogen by the plants in situ, fixed by the diazotrophs

period. Figure 9 illustrates that nitrate at various concentrations affects specific nitrogenase activity during a 14-day period to a rather different extent (Werner et al., 1980b).

We will discuss here only a few other limitations of the associative systems as summarized in Table 2.

All genetic and environmental factors, determining plant growth in general, of course affect the association via the plant. This is also true for all soil factors determining growth and nitrogenase activity of the diazotrophic bacteria, besides the plant influence. Since microaerophilic conditions are a prerequisite for maximum nitrogen fixation, the effect of reduced oxygen tension on root growth and physiological activity are of special importance (Werner et al., 1980b). Root exudation as the major continuous supply of carbon and energy for the bacteria (Sauerbeck et al., 1976; Barber and Martin, 1976) is under many circumstances the most important limitation. Composition of the exudates, together with the prevailing soil factors, determines also the outcome of the competition of diazotrophic bacteria with other chemoorganotrophic microorganisms in the rhizosphere (Patel and Brown, 1969). We have not much information on topology of the association at the root surface and in the intercellular spaces of the root cortex, especially as regards the selection of some bacterial species against others. Also very scarce is reliable information on the uptake by plants of nitrogen fixed by the associated bacteria, especially about the quantity transported in the form of ammonia, excreted continuously by the bacteria, comparable to bacteroids in legume nodules. Decay of the bacterial cells in the rhizosphere and the plant influences on this process are also poorly understood.

References

Barber DA, Martin JK (1976) New Phytol 76:69–80

Barber LE, Russell SA, Evans HJ (1979) Plant Soil 52:49–57

Bergersen FJ (1977) In: Newton W, Postgate JR, Rodriguez-Barrueco C (eds) Recent developments in nitrogen fixation. Academic Press, London New York, p 309–320

Bülow JFW von, Döbereiner J (1975) Proc Natl Acad Sci USA 72:2389–2393

Burris RH, Okon Y, Albrecht SL (1977) In: Hollaender A (ed) Genetic engineering for nitrogen fixation. Plenum Press, New York London, p 445–450

Hardy RWF, Criswell JG, Havelka UD (1977) In: Newton W, Postgate JR, Rodriguez-Barrueco C (eds) Recent developments in nitrogen fixation. Academic Press, London New York, p451–467

Okon Y, Albrecht SL, Burris RH (1976) J Bacteriol 127:1248–1254
Papen H, Werner D (1980) Arch Microbiol (in press)
Patel JJ, Brown MO (1969) Plant Soil 31:273–281
Sauerbeck D, Johnen B, Six R (1976) Landwirtsch Forsch Sonderh 32/I:49–58
Smith RL, Bouton JH, Schank SC, Quesenberry KH, Tyler ME, Milam JR, Gaskins MH, Littell RC (1976) Science 193:1003–1005
Stripf R, Werner D (1978a) Z Naturforsch 33c:373–381
Stripf R, Werner D (1978b) Ber Dtsch Bot Ges 91:683–692
Stripf R, Werner D (1980) Z Naturforsch 35c:776–782
Tjepkema J, Berkum P van (1977) Appl Environ Microbiol 33:626–629
Werner D, Mörschel E (1978) Planta 141:169–177
Werner D, Stripf R (1978) Z Naturforsch 33c:245–252
Werner D, Mörschel E, Stripf R, Winchenbach B (1980a) Planta 147:320–329
Werner D, Stripf R, Abramowski R, Fiedler U (1980b) Z Pflanzenphysiol (in press)
Wilcockson J, Werner D (1978) J Gen Microbiol 108:151–160
Wilcockson J, Werner D (1979) Arch Microbiol 122:153–159

Regulation of Nitrogenase Biosynthesis in Free-Living and Symbiotic N_2-Fixing Bacteria: a Comparison

H. HENNECKE[1]

1 Introduction

Nitrogenase activity in N_2 fixing cells has usually been detected by the simple acetylene reduction assay. Only very recently have studies been initiated to measure the formation of *nif* mRNA and *nif* polypeptides. Experiments of this kind yield more direct information about the regulatory constraints acting on *nif* gene expression and avoid all the complications imposed by the further series of steps necessary for the production of an active nitrogenase complex, such as subunit modification and assembly, incorporation of cofactors (Mo, Fe, S), the functioning of an electron transport system and the availability of sufficient ATP as an energy source.

This article attempts to summarize some of the recent findings on the regulation of *nif* gene expression exemplified by studies of a member of the free-living N_2 fixing bacteria, *Klebsiella pneumoniae,* and by studies of a symbiotic N_2 fixing microorganism, *Rhizobium japonicum.*

2 Regulation of Nitrogenase Biosynthesis in Klebsiella pneumoniae

The constituent polypeptides of the nitrogenase complex of *K. pneumoniae* (Zumft, this vol.) are coded for by the genes *nif*$\overrightarrow{\text{HDK}}$ which form one transcriptional unit. The expression of *nif*HDK and of an additional set of a least 12, but probably 14, *nif* genes is necessary to produce a N_2-fixing phenotype (Nif^+) (Brill; Postgate et al.; Pühler and Klipp, this vol.). How the expression of all these genes is regulated is an intriguing problem. One set of genes, *nif*LA, is of particular interest with respect to regulation: the *nif*A gene product appears to function as a positive activator for *nif* transcription (Roberts et al., 1978; Merrick et al., 1980).

2.1 Control by Fixed Nitrogen

In *K. pneumoniae* dinitrogen fixation occurs only under otherwise nitrogen-limiting conditions. Hence, "fixed" nitrogen compounds such as NH_4^+ were found to strongly inhibit the derepression of all known *nif* genes, as detected by pulse labeling of cells and subsequent analysis of the radioactive *nif*-specific proteins on gels (Eady et al., 1978;

1 Lehrstuhl für Mikrobiologie der Universität München, 8000 München/FRG

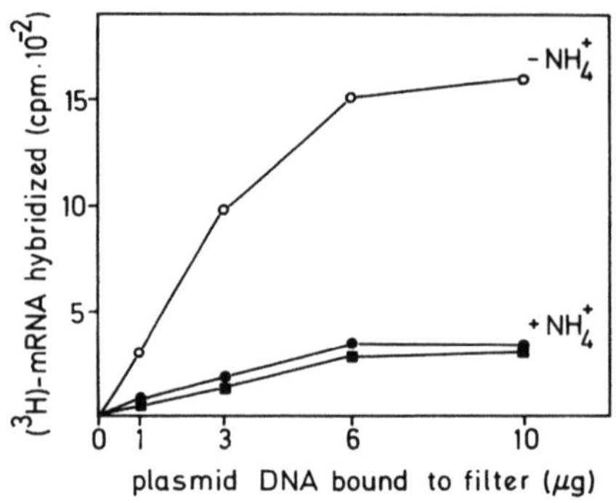

Fig. 1. Hybridization of *K. pneumoniae* mRNA to *nif*DNA. *K. pneumoniae* strain M5A1 was grown anaerobically in the presence (•, ■) and in the absence (○) of NH_4^+, pulse labeled with (^{3}H)-uridine, and the total cellular mRNA was isolated as described by Miller (1972). The mRNA was annealed to denatured pSA 30 DNA (○, •) which carries the genes *nif*HDK (Cannon et al., 1979), by nitrocellulose filter hybridization (Miller, 1972). Hybridization of mRNA from NH_4^+-grown cultures to the vector plasmic pACYC184 (Chang and Cohen, 1978) is shown as control (■)

Roberts et al., 1978) and by the expression of *lacZ*, the structural gene for β galactosidase, that had been fused to promoters of various *nif* genes (Merrick et al., 1980). Earlier evidence indirectly suggested that NH_4^+ prevents *nif* gene expression at the level of transcription (Tubb and Postgate, 1973). The validity of this assumption could be tested directly with help of a mRNA-DNA hybridization experiment (Fig. 1). Total mRNA was first pulse labeled with (^{3}H)-uridine and isolated from cultures grown in the presence and absence of NH_4^+, and then hybridized to a recombinant plasmid, pSA30 (Cannon et al., 1979), carrying almost exclusively the structural genes of the nitrogenase complex (*nif*HDK). Figure 1 shows that no *nif*-specific mRNA was detected in cultures grown in the presence of NH_4^+.

The mechanism of NH_4^+ repression is still uncertain. It appears clear that a functional glutamine synthetase, the *gln*A gene product and crucial enzyme in the NH_4^+ assimilation pathway, is required (Streicher et al., 1974; Tubb, 1974). But as yet no rigorous experiment has conclusively proven that glutamine synthetase functions as a transcriptional activator for *nif*. It has been argued that products of the NH_4^+ assimilation pathway like glutamine/glutamate or glutamine-$tRNA^{Gln}$/glutamate-$tRNA^{Glu}$ might act as (co)repressors (Shanmugam and Morandi, 1976; Shanmugam and Hennecke, 1980). Recently, new genes besides *gln*A involved in nitrogen control, specifically *gln*F, *gln*G (or *gln*R), and *nim*, were found in *Klebsiella, Salmonella,* and *E. coli* (Kustu et al., 1979; Pahel and Tyler, 1979; Close and Shanmugam, 1980). The elucidation of their functions could help to support or modify the glutamine synthetase model.

2.2 Redox Control

The inability of *K. pneumoniae* to fix N_2 under aerobic conditions must in part be due to inactivation of its O_2-sensitive nitrogenase, but O_2, like NH_4^+, has also been shown to repress nitrogenase biosynthesis (St. John et al., 1974; Eady et al., 1978). *Klebsiella pneumoniae* can substitute NO_3^- for O_2 as a terminal electron acceptor. When nitrate is added to anaerobic cells, they convert their metabolism to that of anaerobic respiration (Haddock and Jones, 1977). Under such conditions NO_3^- (and also NO_2^-) were found to markedly repress the biosynthesis of nitrogenase polypeptides in mutant strains (SK-25, SK-561) in which nitrogenase biosynthesis is insensitive to the NH_4^+ internally produced from NO_3^- or NO_2^- (Hom et al., 1980). It is therefore possible that the respiratory controls exerted by O_2 and NO_3^- share a similar, maybe even a common, mechanism which, in turn, is different from that of NH_4^+ control. With help of *nif*::*lac* fusion strains Hill, Kennedy, and coworkers (Merrick et al., 1980) recently demonstrated that the *nif*L/A genes or products play a crucial role in oxygen control. They succeeded in generating mutations in or very near *nif*L, enabling the mutant strains to escape from oxygen repression. Preliminary results showed that these mutant strains were also able to escape from NO_3^- repression (Hill, Kennedy, Hennecke, unpubl. observations).

2.3 Temperature Control

The growth temperature represents a third control system, besides NH_4^+ and O_2, governing the expression of *nif* genes in *K. pneumoniae* strain M5A1. At temperatures above 37 °C *K. pneumoniae* does not fix N_2 but is well able to grow on other nitrogen sources (Postgate, 1974; Hennecke and Shanmugam, 1979). In vitro, the nitrogenase enzyme complex is not thermosensitive (Hennecke and Shanmugam, 1979) whereas in vivo the biosynthesis of the constituent polypeptides of the nitrogenase complex clearly is thermosensitive (Fig. 2). When cells grown at 30 °C under nitrogen-limiting conditions

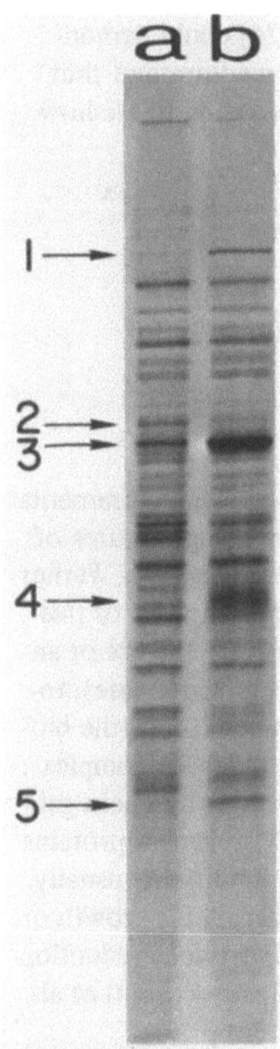

Fig. 2a, b. Repression of nitrogenase biosynthesis in *K. pneumoniae* strain M5A1 by high temperatures. The figure shows the autoradiogram of the electrophoretic analysis of the (^{14}C)-labeled total SDS-soluble cell protein. Cultures were grown in minimal medium with 100 μg/ml L-glutamine under anaerobic conditions and pulse labeled at an OD_{420} of 1.0. **a** Growth at 37° C; **b** growth at 30 °C. The *arrows* indicate the positions of the following proteins: *1* *nif*J gene product (M.W. 120,000); *2* "60K" protein; *3* nitrogenase component 1; *4* nitrogenase component 2; *5* 18,000 M.W. protein (*nif*Y gene product?). (Modified from Hennecke and Shanmugam, 1979)

(in the presence of 100 μg/ml L-glutamine) were shifted to 39 °C there was a rapid decrease of the rate of de novo biosynthesis of Kp 1 and Kp 2 and the *nif*J gene product (Hennecke and Shanmugam, 1979). In order to extend these studies to the biosynthesis of other *nif* gene products besides those that can be unequivocally identified on gels, the expression of β-galactosidase in various *nif*:: *lac* fusion strains in response to high temperature was investigated. The preliminary results imply that the expression of the *nif*LA genes is far less thermosensitive than the expression of all the other *nif* transcriptional units (Kennedy, Dixon, Hennecke, unpubl. observations). It remains to be seen whether those important regulatory gene products encoded by *nif*LA are naturally thermosensitive. Another open question is why *K. pneumoniae*, having an optimum growth temperature of 37 °C, "prefers" to repress the nitrogen fixation pathway at this temperature.

2.4 Molybdenum Control

This control circuit is by far the least understood. *K. pneumoniae* grows well in the absence of molybdenum in media containing excess NH_4^+, but growth on N_2 is completely dependent on the presence of Mo in the medium. Brill et al. (1974) could demonstrate that Kp 1 and Kp 2 are inactive in a molybdenum-depleted medium and that neither component can be detected immunologically. It will be interesting to see how Mo functions as an inducer of nitrogenase biosynthesis.

3 Regulation of Nitrogenase Biosynthesis in Rhizobium japonicum

3.1 Derepression of nif Genes Under Asymbiotic Conditions

Considerable progress has been made in the elucidation of the physiological requirements needed for the formation of whole-cell nitrogenase activity in free-living cultures of "slow-growing" *Rhizobium* species (Bergersen et al., 1976; Gibson et al., 1976; Werner and Stripf, 1978). *Rhizobium japonicum* is the most prominent representative of this group. The main requirements for maximal rates of nitrogen fixation are the use of an appropriate carbon source (e.g., gluconate) and a nitrogen source (e.g., glutamate), together with a very low dissolved oxygen concentration. Under such conditions the biosynthesis of the constituent polypeptides of the *R. japonicum* nitrogenase complex (αRj 1, βRj 1, Rj 2) can be measured after pulse labeling of the cells and subsequent gel electrophoresis of total cell protein and autoradiography (Fig. 3). Nitrogenase proteins are among the major proteins synthesized under microaerophilic conditions, usually constituting between 6% and 8% of the total protein. During microaerophilic growth of *R. japonicum* these peptides were detected even before the onset of acetylene reduction activity (during day 2) reaching maximum synthesis between day 3 and 4 (Scott et al., 1979). All three *nif* proteins seem to be coordinately synthesized.

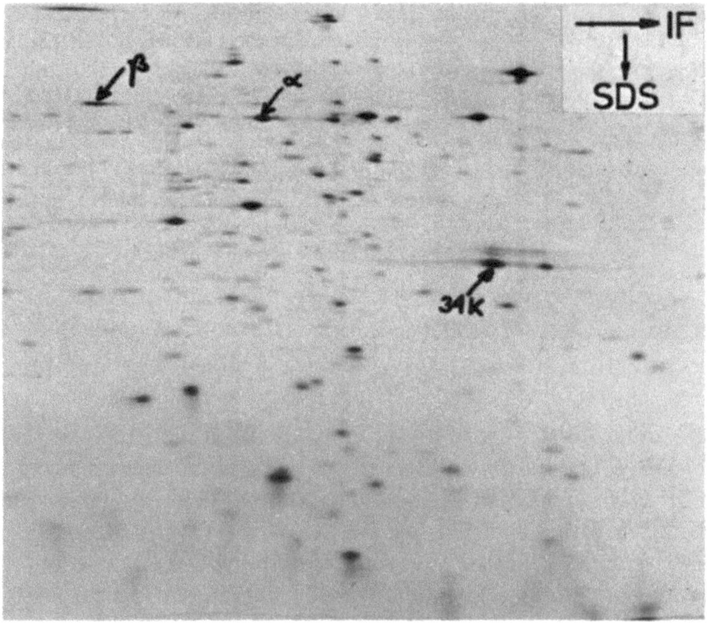

Fig. 3. Derepression of nitrogenase polypeptides during growth of *R. japonicum* strain 110 under microaerophilic conditions. The picture shows the autoradiogram of the two-dimensional gel electrophoretic analysis of in vivo (^{35}S)-pulse-labeled *R. japonicum* proteins. The *arrows* indicate the positions of the α and β polypeptides of nitrogenase component 1 (Rj1) and of a 34,000 M.W. protein assigned as component 2 (Rj2). These proteins were identified with the help of purified reference proteins (Rj1) and by specific immunoprecipitation. (Modified from Scott et al., 1979)

3.2 Oxygen Control and the Possible Involvement of Cyclic Mononucleotides

Oxygen appears to repress nitrogenase biosynthesis in *R. japonicum* (Scott et al., 1979) as it does in *K. pneumoniae.* Using the pulse-labeling technique described above, it was shown that the rates of *nif* derepression decreased rapidly when a microaerophilic culture induced for nitrogenase was exposed to O_2. No *nif*-specific cross-reacting material could be detected immunologically in aerobically grown *R. japonicum* cultures (Fig. 4d). In contrast to *K. pneumoniae* the use of NO_3^- as the terminal electron acceptor did not result in an inhibition of *nif* polypeptide synthesis (Scott et al., 1979). However, no whole cell acetylene reduction activity was observed under such growth conditions, most probably as a consequence of the accumulation of toxic amounts of NO_2^-.

A finding of considerable interest was made when the pool sizes of cyclic mononucleotides were measured during microaerophilic-aerobic shift experiments. The intracellular levels of cyclic GMP increased from 0.25 pmol/mg cell protein under microaerophilic conditions to 2.6 pmol/mg cell protein under aerobic conditions, suggesting that the cellular pool size of cyclic GMP may also be under redox control (Lim et al., 1979). A dramatic decrease in the rate of nitrogenase polypeptide synthesis was observed when cyclic GMP at a concentration as low as 0.1 m*M* was added to an early exponential phase

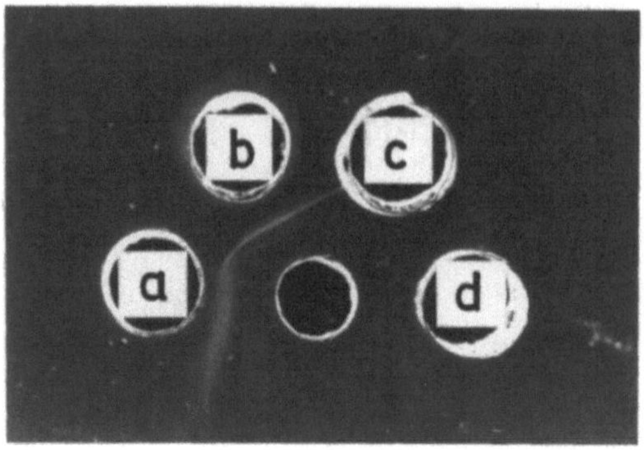

Fig. 4. Immunological detection of *R. japonicum* nitrogenase component 1 (Rj 1) in crude cell extracts by Ouchterlony double diffusion. The wells contained: (*center*) antibodies against purified Rj 1; *a* crude cell extract from microaerophilic culture induced for nitrogenase; *b* purified Rj 1; *c* crude cell extract from microaerophilic culture grown in the presence of 0.1 m*M* cGMP; *d* crude cell extract from aerobic culture. (From Lim et al., 1979)

culture of *R. japonicum*, and, consistent with this finding, no *nif*-specific cross-reacting material could be detected in *R. japonicum* cultures grown under microaerophilic conditions in the presence of 0.1 m*M* cyclic GMP (Fig. 4c). Thus, a correlation between oxygen control and the cyclic GMP-mediated control is possible. However, the mechanism of control is obviously very complex as the rates of synthesis of numerous other proteins, besides the *nif* polypeptides, are dramatically altered when a microaerophilic culture is exposed to either cyclic GMP or oxygen (Lim et al., 1979; Scott et al., 1979).

3.3 Is There Regulation by Fixed Nitrogen?

With help of (^{15}N) analysis, O'Gara and Shanmugam (1976) could demonstrate that free-living cultures of *R. japonicum* export most of the fixed N_2 into the medium as NH_4^+, a situation analogous to that found for isolated bacteroids.

Hence, unlike other nitrogen-fixing bacteria, *Rhizobium*, when grown under free-living conditions, appears to be unique in that nitrogenase synthesis is derepressed in the presence of NH_4^+ (Keister, 1975; Gibson et al., 1976). However, reports conflict. Bergersen and Turner (1976) suggest that the discrepancies may be a consequence of varied oxygen concentrations in different experimental systems. Using the pulse-labeling technique, we found that under our culture conditions, i.e., where the oxygen concentration in the gas phase was maintained at 0.1% (Scott et al., 1979), *nif* polypeptide synthesis was derepressed in the presence of NH_4^+ (7.5 m*M*). Higher concentrations of NH_4^+ drastically affected total protein synthesis, possibly due to inhibition of oxidative phosphorylation (D.B. Scott, unpubl. observations).

Nevertheless it is well established that inorganic nitrogen sources such as NO_3^- and NH_4^+ inhibit the development of the N_2-fixing symbiosis in legume root nodules (Fred et al., 1932) but this type of repression certainly involves the host plant metabolism and/or the host-symbiont recognition process (Dazzo and Brill, 1978).

3.4 Limitation of Symbiotic N_2 Fixation by Energy Supply

Hardy and Havelka (1975) realized that the rate-limiting step for symbiotic N_2 fixation is the amount of available photosynthate provided by the host plant. We performed the

Table 1. The effect of different carbon sources on nitrogenase synthesis in *R. japonicum* (strain 110). The rate of nitrogenase synthesis (Rj 1) was determined from the amount of radioactivity incorporated into the α and β polypeptides (see Fig. 3) after pulse labelling of cells grown under microaerophilic conditions in the presence of glutamate (0.1%) and the appropriate carbon source (0.4%). (From Scott et al., 1979)

Carbon Source Used for Growth	Nitrogenase Synthesis (% of total protein)	Nitrogenase Activity (nmol C_2H_4 formed $\cdot h^{-1} \cdot mg^{-1}$ cell protein)
Mannitol	<0.5	1.5
Glycerol	<0.5	2.5
Gluconate	6.1	65.5
Glutamate	5.7	< 0.1
Glutamate + 2% H_2	n.d.[a]	18.3

[a] n.d. – not determined

experiment in Table 1 in order to test whether the type of carbon source used for growth of *R. japonicum* under asymbiotic conditions can have a direct influence on *nif* gene expression. The results show that mannitol and glycerol do not support the expression of *nif* genes, whereas gluconate and glutamate allow *nif* polypeptide synthesis. However, glutamate may not provide enough reductant to support whole cell nitrogenase activity. This process can partly be restored by the addition of H_2 into the gas phase. These few experiments demonstrate that *R. japonicum* must contain control circuits which regulate *nif* gene expression in response to the quality of the available carbon source. This would enable the cells to avoid wasting energy synthesizing nitrogenase components under a condition – energy limitation – unfavorable to the reduction of N_2.

4 Conclusion and Future Prospects

The comparison of the regulation of *nif* gene expression in *K. pneumoniae* and in free-living cultures of *R. japonicum* has revealed similarities such as the oxygen control and dissimilarities such as the control, or lack of it, by fixed nitrogen, as well as peculiarities characteristic for each organism. For *Klebsiella pneumoniae* one can expect a good understanding of *nif* gene regulation within a few years, with help of *nif*::*lac* fusion strains, regulatory mutants, and recombinant plasmids carrying part or all of the *nif* genes.

For *Rhizobium*, studies similar to those described in this article are promising and could lead to a better description of the regulatory phenomena acting on *nif* gene expression during the developmental stages of symbiosis. It should also be helpful to isolate *nif* regulatory mutants which, in the case of *R. japonicum*, can be screened for and analyzed in the absence of the host plant. Very recently, we were able to clone nitrogen fixation genes from *Rhizobium japonicum* on small amplifiable plasmids (Hennecke and Mielenz, to be published). Such plasmids could be a useful tool for future studies of *nif* gene expression in vitro and could serve to directly construct new nitrogen-fixing *Rhizobium japonicum* strains with altered regulatory properties.

Acknowledgments. Most of the own work presented in this article was done at the Plant Growth Laboratory of the University of California at Davis. I wish to thank R. Valentine, K.T. Shanmugam, S.T. Lim, B. Scott, and S.M. Hom for their collaboration.

References

Bergersen FJ, Turner GL (1976) Biochem Biophys Res Commun 73:523–531
Bergersen F, Turner GL, Gibson AH, Dudman WF (1976) Biochim Biophys Acta 444:164–174
Brill WJ, Steiner AL, Shah VK (1974) J Bacteriol 118:986–989
Cannon FC, Riedel GE, Ausubel FM (1979) Mol Gen Genet 174:59–66
Chang ACY, Cohen SN (1978) J Bacteriol 134:1141–1156
Close TJ, Shanmugam KT (1980) J Gen Microbiol 116:501–510
Dazzo FB, Brill WJ (1978) Plant Physiol 62:18–21
Eady RR, Issack R, Kennedy C, Postgate JR, Ratcliffe HD (1978) J Gen Microbiol 104:277–285
Fred EB, Baldwin IL, McCoy E (1932) Root nodule bacteria and leguminous plants. University of Wisconsin Press, Madison
Gibson AH, Scowcroft WR, Pagan JD (1976) In: Newton W, Postgate JR, Rodriguez-Barrueco C (eds) Recent developments in nitrogen fixation. Academic Press, London New York, p 387
Haddock BA, Jones CW (1977) Bacteriol Rev 41:47–99
Hardy RWF, Havelka UD (1975) Science 188:633–643
Hennecke H, Shanmugam KT (1979) Arch Microbiol 123:259–265
Hom SSM, Hennecke H, Shanmugam KT (1980) J Gen Microbiol 117:169–179
Keister DL (1975) J Bacteriol 123:1265–1268
Kustu S, Burton D, Garcia E, McCarter L, McFarland N (1979) Proc Natl Acad Sci USA 76:4576–4580
Lim ST, Hennecke H, Scott DB (1979) J Bacteriol 139:256–263
Merrick M, Dixon R, Eady R, Hill S, Kennedy C, Iaccarino M (1980) In: Stewart WDP, Gallon R (eds) Nitrogen fixation; Proc Phytochem Soc Meet. Academic Press, London New York (in press)
Miller JH (1972) Experiments in molecular genetics. Cold Spring Harbor Laboratory, New York
O'Gara F, Shanmugam KT (1976) Biochim Biophys Acta 437:313–321
Pahel G, Tyler BM (1979) Proc Natl Acad Sci USA 76:4544–4548
Postgate J (1974) In: Quispel A (ed) The biology of nitrogen fixation. North-Holland Publ Comp, Amsterdam Oxford, p 663
Roberts GP, MacNeil T, MacNeil D, Brill WJ (1978) J Bacteriol 136:267–279
Scott DB, Hennecke H, Lim ST (1979) Biochim Biophys Acta 565:365–378
Shanmugam KT, Hennecke H (1980) In: Subba-Rao N (ed) Recent advances in biological nitrogen fixation. Oxford Int Bookhouse, Oxford New Delhi (in press)
Shanmugam KT, Morandi C (1976) Biochim Biophys Acta 437:322–332
St John RT, Shah VK, Brill WJ (1974) J Bacteriol 119:266–269
Streicher SL, Shanmugam KT, Ausubel F, Morandi C, Goldberg RB (1974) J Bacteriol 120:815–821
Tubb RS (1974) Nature (London) 251:481–485
Tubb RS, Postgate JR (1973) J Gen Microbiol 79:103–117
Werner D, Stripf R (1978) Z Naturforsch 33:245–252

Characterization of the Nitrite-Oxidizing System in Nitrobacter

H. SUNDERMEYER and E. BOCK[1]

1 Properties of Nitrobacter X_{14}

Nitrobacter X_{14} was isolated from soil of the old Botanical Garden in Hamburg. It is a facultative lithotrophic nitrite-oxidizing bacterium. As well as *Nitrobacter agilis* (Smith and Hoare, 1968; Bock, 1976) *Nitrobacter* X_{14} was able to grow mixotrophically and heterotrophically. In contrast to other strains of *Nitrobacter* the doubling time of autotrophically grown cells was not 13 h but 40 h. After nitrite consumption the cell yield with 3 mg protein/l was lower than in *Nitrobacter agilis* (5.5 mg protein/l). Under mixotrophic conditions with nitrite, pyruvate, and yeast extract-peptone (py–ni cells) growth was much better. The doubling time was 10–11 h and the cell yield amounted to 17 mg protein/l when the nitrite was consumed. When *Nitrobacter* X_{14} was grown heterotrophically with pyruvate as carbon and energy source and yeast extract-peptone to meet nitrogen requirements (y–py cells) cells doubled within 15–20 h, when grown with pyruvate and casamino acids (cas–py cells) the doubling time increased to more than 20 h. Compared to *Thiobacillus intermedius* (London and Rittenberg, 1966) *Nitrobacter* X_{14} should be regarded as an organism highly specialized to mixotrophic growth.

In addition *Nitrobacter* X_{14} cells were characterized by cytochrome difference spectra, light and electron microscopy and GC content. The cytochrome difference spectra of py–ni cells and y–py cells showed the characteristic absorption bands of *Nitrobacter.* The difference spectra of cas–py cells were different. The α-band of cyt a_1 was missing and a shoulder near 562 nm indicated the presence of a cytochrome of the b-type (cf. Fig. 5).

As shown by light microscopy the cells were pleomorphic, mostly coccoid to pear-shaped, budding cells were sometimes seen. Ultrathin sections were examined by means of electron microscopy. Py–ni cells (Fig. 1) and y–py cells (Fig. 2) are characterized by intracytoplasmic membranes which form a polar cap. In heterotrophic cells the double membranes of the ICM were less well arranged than in mixotrophic cells. PHB

Abbreviations: ATPase = adenosine triphosphatase; cyt.=cytochrome; CM=cytoplasmic membrane; ICM = intracytoplasmic membrane; PHB = poly-β-hydroxybutyric acid; RuBPCase = ribulose-1,5-bisphosphate carboxylase; SDS = sodium dodecylsulfate; py–ni cells = cells grown mixotrophically with nitrite, pyruvate and yeast extract-peptone; cas–py cells = cells grown heterotrophically with pyruvate and casamino acids; y–py cells = cells grown heterotrophically with pyruvate and yeast extract-peptone

1 Institut für Allgemeine Botanik, Mikrobiologie, Universität Hamburg, Jungiusstraße 6–8, 2000 Hamburg/FRG

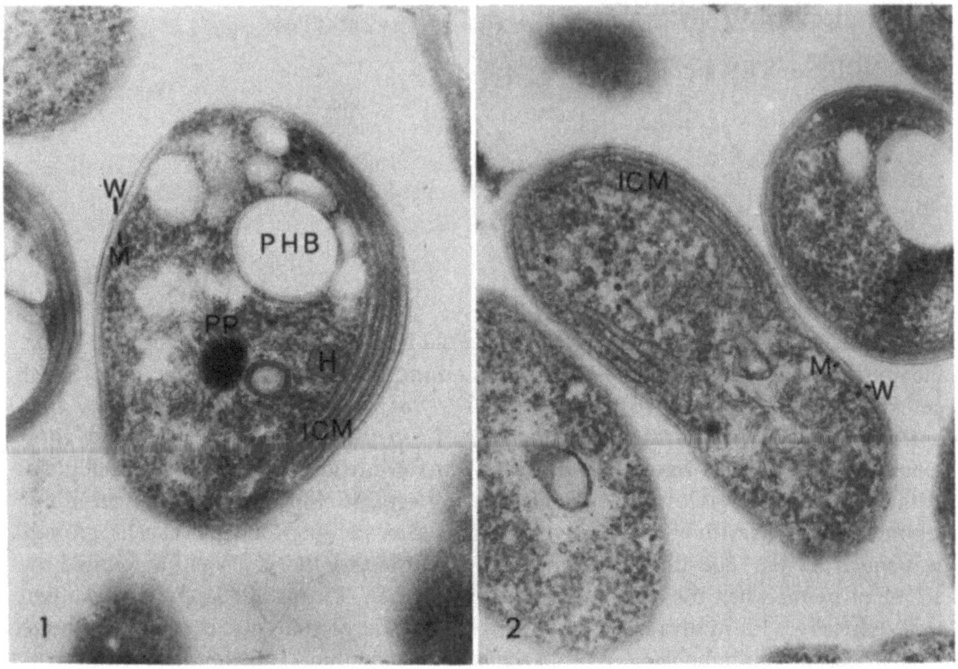

Figs. 1 and 2. Ultrathin sections of mixotrophically grown py–ni cells (1) and heterotrophically grown y-py cells (2) of *Nitrobacter* X_{14}. *W* cell wall; *M* cytoplasmic membrane; *ICM* intracytoplasmic membranes; *PHB* poly-β-hydroxybutyric acid; *PP* polyphosphate; *H* hexagonal particle (carboxysome). x 50,000

and polyphosphate granules as well as hexagonal particles with RuBPCase activity (carboxysomes, Shively et al., 1977) were present. Cas–py cells had the same ultrastructure as y–py cells (not demonstrated).

According to Watson and Mandel (1971) the GC content of *Nitrobacter* amounts to 60.2%–61.2%. The GC content of *Nitrobacter* X_{14} was 61.2% independent of growth conditions. Mixotrophic and heterotrophic cultures were routinely checked for purity by light microscopy and by a dilution technique. Every single cell was able to start nitrification when inoculated in nitrite media.

For investigation of the nitrite oxidizing system *Nitrobacter* X_{14} was chosen because doubling times were shortest and the cell yield was highest compared to 10 different *Nitrobacter* strains tested.

2 Isolation of Membranes

After sonication of mixotrophic and heterotrophic cells the membranes (ICM and CM) were isolated by differential centrifugation and by sedimentation in 10%–50% or 10% to 60% (wt/wt) linear sucrose gradients for 19 h. Figure 3 shows the procedure for membrane isolation. Six fractions of the gradients were tested for nitrite oxidase,

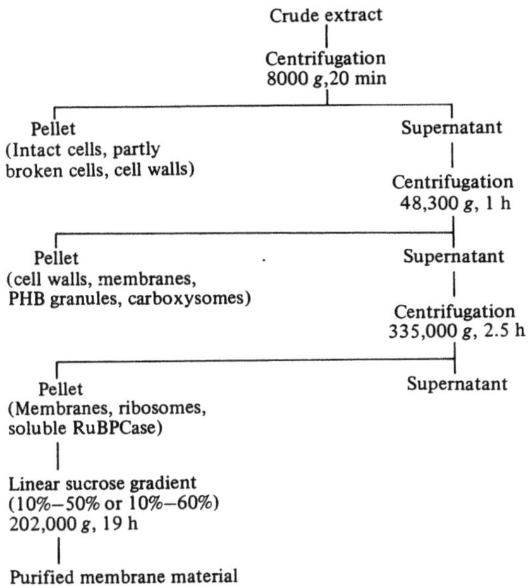

Fig. 3. Diagram of the purification procedure of *Nitrobacter* membranes

ATPase and RuBPCase activity. The nitrite oxidase activity was assayed in the presence of ferricyanide [$K_3(Fe(CN)_6)$] as an artificial electron acceptor under aerobic conditions. The determination of ATPase was done according to the method of Abrams et al. (1960) and Davies and Bragg (1972); RuBPCase activity was measured as described by Shively et al. (1977). The presence of ribosomes was indicated by SDS gel electrophoresis. In addition all fractions were examined by electron microscopy using the negative staining technique.

The membrane fraction was identified by means of assays for nitrite oxidase activity, cytochrome difference spectra and electron microscopy. The membrane fraction was always brownish in color because of its high cytochrome content.

3 Characterization of Membranes

According to Aleem (1968) and O'Kelley et al. (1970) the nitrite-oxidizing system of *Nitrobacter* is membrane-bound. For biochemical characterization of the nitrite oxidase the membranes of mixotrophically grown cells of *Nitrobacter* X_{14} were isolated as described. Optimum conditions for the enzyme are a temperature of 30 °C and a pH value of 8.0. The apparent Michaelis constant amounted to 1.5 m*M* (I. Kraft, Diplomarbeit, Hamburg Univ., 1979). This result is in accordance with that of O'Kelley et al. (1970), who estimated a K_m of 1.2 m*M* at pH 8.0. As demonstrated by storage experiments the nitrite oxidase is a labile enzyme. Nitrite was found to be essential during purification of the enzyme. In buffer without nitrite the activity was lost within a few hours, with nitrite the enzyme was stable for two days at 4 °C. During the procedure of purification, taking 30 h, nitrite oxidase activity was assayed in each preparation step. The specific activity of the nitrite oxidase increased during purification. It was highest in the membrane fraction of the sucrose gradient. With 0.6 μmol NO_2^- oxidized/mg protein min^{-1}, the specific activity was higher than that obtained by O'Kelley et al. (1970), amounting to 0.028 μmol NO_2^-/mg protein min^{-1}. According to Faull et al. (1969) the nitrite oxidase activity was 0.24 μmol NO_2^-/mg protein min^{-1}, achieved by ammonium sulphate fractionation, respectively, 0.035 μmol NO_2^-/mg protein min^{-1}, using differential centrifugation.

In addition to nitrite oxidase activity, the membrane fraction involved ATPase, NADHoxidase and succinate dehydrogenase activity (H. Uhlig, Diplomarbeit, Hamburg Univ., 1980). As indicated by electron microscopy, cell wall components were not present in the membrane fraction. Because of its higher density the wall material sedimentated through the gradient and formed a colorless pellet. By the technique used it was not possible to separate the CM from the ICM.

4 Comparative Studies of Membranes of Cells with and Without Nitrite Oxidase Activity

In order to obtain more information on the nitrite-oxidizing system, the membranes of heterotrophically grown cells with low (y–py cells) or without nitrite oxidase activity (cas–py cells) were isolated as described. Figure 4 shows typical UV absorption profiles

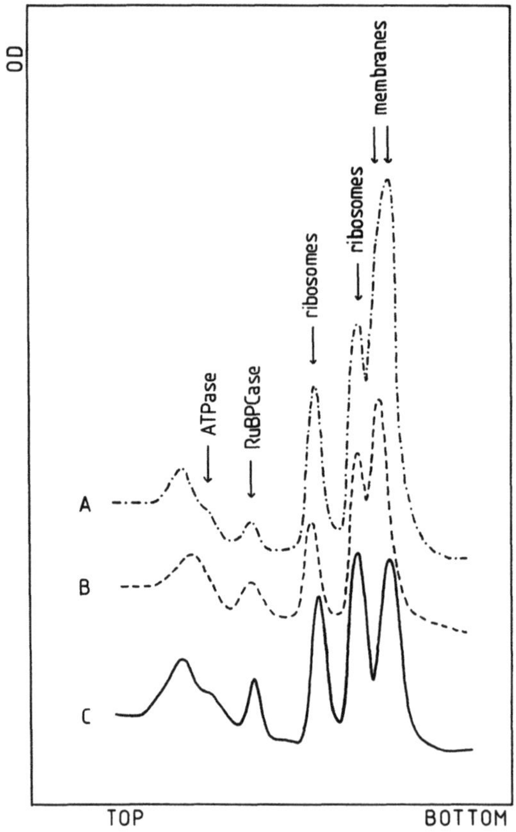

Fig. 4. UV absorption profiles of the 335,000 g pellet of cellfree homogenates of differently grown *Nitrobacter* cells after a 19 h run on a linear 10%–50% sucrose gradient. *A* mixotrophically grown py–ni cells; *B* heterotrophically grown cas–py cells; *C* heterotrophically grown y–py cells

of sucrose gradients obtained from the 335,000 *g* pellet of mixotrophic py-ni cells and heterotrophic y-py and cas-py cells. The UV profiles were similar in all three preparations. RuBPCase activity was concentrated at about 21% sucrose in the second peak. The specific activity amounted to 0.49 μmol CO_2 fixed/mg protein min^{-1} in particulate fractions of py-ni cells, to 0.21 μmol CO_2/mg protein min^{-1} in those of y-py cells and it was 0.029 μmol CO_2/mg protein min^{-1} in those of cas-py cells. Thus, RuBPCase was mostly enriched in fractions of mixotrophic cells and less in those of heterotrophic cells. ATPase activity was found in the shoulder of the first fraction and in the membrane fraction, which could be easily detected by its brownish color. Nitrite oxidase activity was exclusively found in the membrane fraction of the gradients. The specific activity of nitrite oxidase was 0.6 μmol NO_2^- oxidized/mg protein min^{-1} in py-ni membranes, 0.08 μmol NO_2^-/mg protein min^{-1} in y-py membranes - in the following called nitrite membranes -; nitrite oxidase activity could not be determined in cas-py membranes. The membrane peak also revealed traces of RuBPCase activity. As demonstrated in Fig. 4 the membranes were more or less contaminated by ribosomes. Membranes of py-ni and y-py cells were always located deaper in the gradient than membranes of cas-py cells. After 19 h centrifugation the membranes are in equilibrium with 40% sucrose as demonstrated by Uhlig (pers. comm.). Thus, the membranes of cas-py cells are less dense than membranes of py-ni and y-py cells.

The membranes with and without nitrite oxidase activity were characterized further by cytochrome difference spectra (Fig. 5). The cytochrome difference spectra of mem-

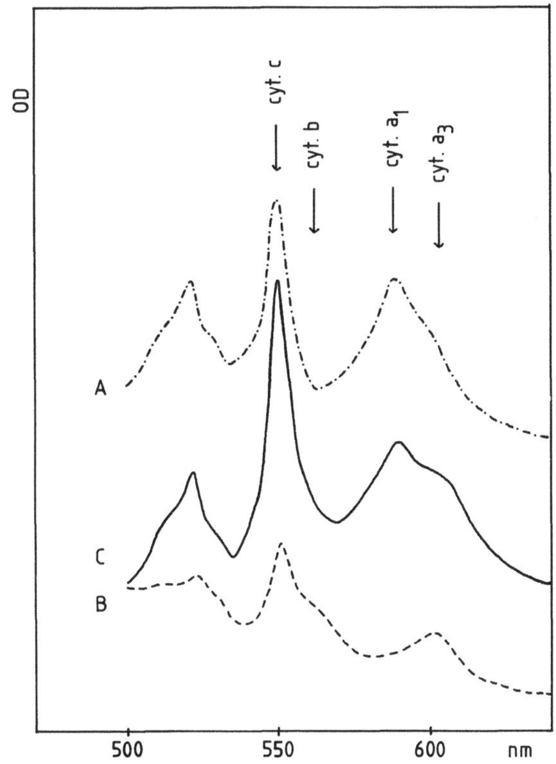

Fig. 5. Cytochrome difference spectra of *Nitrobacter* membranes. *A* mixotrophically grown py–ni cells; *B* heterotrophically grown cas–py cells; *C* heterotrophically grown y–py cells

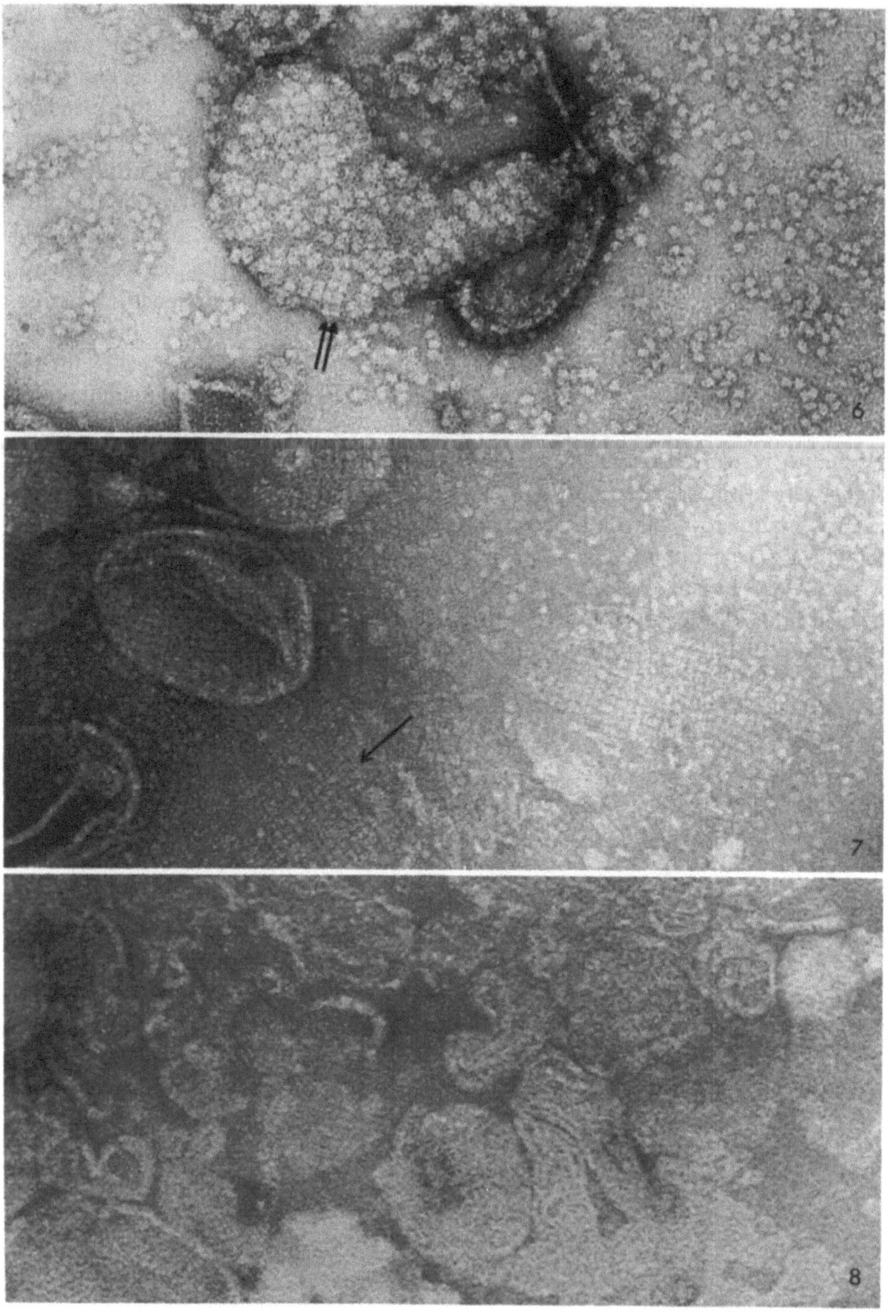

Figs. 6–8. Isolated membrane fragments of *Nitrobacter* cells with high (6), low (7), and no (8) nitrite oxidase activity. Particles of 90 Å in diameter form rows (↑↑) on py-ni membranes and are detached (6), particles of 70–80 Å in diameter form a regular mesh with and without (↑) membrane basis and are detached (7). Membranes without particles (8) x 200,000

branes of py–ni, y–py, and cas–py cells were in accordance with those of intact cells (not demonstrated). In addition there were no differences between the cytochrome spectra of membranes of py–ni and y–py cells. The α-bands of cyt c and cyt a_1 were located at 550 nm and 590 nm, the β-bands at 522 nm and there was a peak at 605 nm, indicating the presence of cyt a_3. The cytochrome difference spectra of cas–py cells were different. Cyt c was present, but there was an additional shoulder near 562 nm, which could be a cytochrome of the b-type. The α-band of cyt a_1 was missing.

In addition the membrane fractions were examined by electron microscopy using a special negative staining technique. Membranes of *Nitrobacter* X_{14} did not form vesicles but flat spread structures. As demonstrated by Figs. 6 and 7 the nitrite membranes are characterized by particles forming a layer on a membrane basis. The particles of 70–90 Å in diameter are arranged in rows (Fig. 6) or in a complex structure of a molecular mesh (Fig. 7). The molecular mesh is also found without the membrane basis (Fig. 7). The distance from center to center of the repeating subunits amounts to approximately 86 Å. A lot of particles are detached from the nitrite membranes. The particles of py–ni membranes were larger with 90 Å compared to those of y–py membranes with 70–80 Å. Tsien and Laudelout (1968) also described so-called submicroscopal particles, 90–100 Å in diameter, which were found on the membrane surface and which seemed not to be arranged in a regular pattern.

We assume that the particles have a function in the nitrite-oxidizing system, because cas–py membranes have no adequate membrane-bound particles. The membranes without nitrite oxidase activity (cas–py cells) look totally different (Fig. 8). The membranes are low in contrast and without any regular structure.

5 Conclusions

As demonstrated by the previous experiments the nitrite oxidase is a very complex enzyme. To our knowledge nitrite is oxidized according to the following equations:

$$NO_2^- + H_2O + 2\ \text{cyt}\ a_1^{3+} \rightleftharpoons NO_3^- + 2\ H^+ + 2\ \text{cyt}\ a_1^{2+} \quad (1)$$

$$2\ H^+ + 2\ \text{cyt}\ a_1^{2+} + 0.5\ O_2 \rightleftharpoons H_2O + 2\ \text{cyt}\ a_1^{3+} \quad (2)$$

$$NO_2^- + 0.5\ O_2 \rightarrow NO_3^- .$$

In the first reaction oxygen is not involved, nitrite is oxidized to nitrate by water, protons, and electrons are released, and the electrons are transferred to cyt a_1, an essential component of the nitrite-oxidizing system (Aleem et al., 1965). Then they are transported to oxygen. This second reaction is the energy-yielding step, which is coupled to the electrontransport phosphorylation.

As we could demonstrate, nitrite oxidase activity was only found in intact nitrite membranes, also containing other enzyme activities, but never in membrane-free fractions. The cytochrome difference spectra of nitrite membranes showed a typical cytochrome with cyt c, and cyt a_1. Cytochromes of the c- and a-type are essential parts of the nitrite-oxidizing system. The cas–py membranes without nitrite oxidase activity revealed a lack of cyt a_1 and a cyt of the b-type seemed to be present. Thus, the cytochrome composition of the electrontransport chain of membranes without nitrite oxidase activity differs markedly from that of membranes with nitrite oxidase activity.

In contrast to membranes without nitrite oxidase activity, nitrite membranes showed a higher density in sucrose gradients. As revealed by electron microscopy, the inner surface of nitrite membranes is covered with a dense layer of regularly arranged particles. This might explain the asymmetry of nitrite membranes which can be seen in ultrathin sections. On the other hand the protein layer on the nitrite membranes might be responsible for the higher density of nitrite membranes compared to cas–py membranes. In intact nitrite membranes the particles seemed to be arranged in a complex structure of a molecular mesh. At high ionic strength e.g., during drying in the negative staining technique, the complex structure is more or less destroyed and the particles are detached from the membranes. The different size of the particles of py–ni cells compared to those of y–py cells may be explained by either swelling of the particles of py–ni during the preparation technique (artifact) or a difference in size on account of different proteins. In contrast to nitrite membranes, membranes without oxidase activity never possessed the particulate structure. In ultrathin sections these membranes did not appear asymmetric.

The presence of cytochromes is not detectable in the EM preparations of membranes, because they are embedded in the lipid bilayer tightly bound by hydrophobic bindings (integrated proteins), whereas the particles are loosely bound (peripheral proteins). We assume that the particles have a function in nitrite oxidation according to Eq. (1). For nitrite oxidation the presence of cytochromes is essential according to Eq. (2), because the electrons have to be removed immediately. Thus the nitrite-oxidizing system is only intact if both parts are present. The investigations, especially on the protein composition, are in progress.

References

Abrams A, McNamara T, Johnson BF (1960) J Biol Chem 235:3659–3662
Aleem MJH (1968) Biochim Biophys Acta 162:338–347
Aleem MJH, Hoch GE, Varner JE (1965) Proc Natl Acad Sci USA 54:869–873
Bock E (1976) Arch Microbiol 108:305–312
Davies PL, Bragg PD (1972) Biochim Biophys Acta 266:273–284
Faull KF, Wallace W, Nicholas DJD (1969) Biochim J 113:449–455
London J, Rittenberg SC (1966) J Bacteriol 91:1062–1069
O'Kelley JC, Becker GE, Nason A (1970) Biochim Biophys Acta 205:409–425
Shively JM, Bock E, Westphal K, Cannon GC (1977) J Bacteriol 132:673–675
Smith AJ, Hoare DS (1968) J Bacteriol 95:844–855
Tsien HC, Laudelout H (1968) Arch Microbiol 61:280–291
Watson SW, Mandel M (1971) J Bacteriol 107:563–569

Contributed Papers: Sulfur

The Role of Thioredoxins for Enzyme Regulation in Cyanobacteria

A. SCHMIDT[1]

Assimilatory sulfate reduction in plants and green algae needs the sulfonucleotide adenosine-5′-phosphosulfate (APS) as sulfate donor for further reduction (Schiff and Hodson, 1973; Schmidt, 1979b). The enzyme specific for APS was named APS-sulfotransferase, since it transfers the activated sulfate from APS onto suitable thiols (Schmidt, 1972). APS-sulfotransferases from higher plants and green algae are characterized by pH-optima between 8.5 and 9 and by their ability to react with mono- and dithiols (Schmidt, 1976). In bacteria the sulfonucleotide 3′-phosphoadenosine-5′-phosphosulfate (PAPS) is the sulfate donor for further reduction and thioredoxin is needed as cofactor in bacterial PAPS-sulfotransferase reactions (Tsang and Schiff, 1976; Schmidt, 1979b). This demonstrates that the biochemical pathway for assimilatory sulfate reduction in bacteria and higher plants is different. This finding initiated studies on the distribution of APS- and PAPS-dependent sulfotransferases, and the suggestion was made that oxygenic photosynthetic organisms should use APS for further reduction (Tsang and Schiff, 1975). It will be shown here that thioredoxin-dependent APS- and PAPS-sulfotransferases are found in cyanobacteria.

APS- and PAPS-dependent sulfotransferases could be detected in phototrophic bacteria (Schmidt and Trüper, 1977) and in cyanobacteria (Schmidt, 1977). Some examples selected from about 20 strains are summarized in Table 1. In cyanobacteria APS-dependent systems are found in *Plectonema,* which are partly active with monothiols as well. On the other side APS- and PAPS-sulfotransferases are found, which are active only with the dithiol DTE and in the presence of a protein-cofactor identified as thioredoxin (Schmidt and Christen, 1978). Examples for thioredoxin-dependent PAPS-sulfotransferases are *Synechococcus* 6301 *(Anacystis nidulans)* and *Synechocystis* 6714 and for thioredoxin-dependent APS-sulfotransferases *Synechococcus* 6312 and *Chroococcidiopsis* 7203 (Fig. 1). These systems seemed to be similar to the PAPS-sulfotransferase from *Escherichia coli,* which needed a heat-stable protein fraction as cofactor identified as a thioredoxin by Tsang and Schiff (1976); therefore the PAPS-dependent system of *Synechococcus* 6301 was analyzed in detail. In these cases, where in Table 1 the thioredoxin requirement is marked with "yes" the data were obtained with thioredoxin B prej ared from *Synechococcus* 6301. Note that the PAPS-sulfotransferase from *Synechc coccus* 6301 is not inhibited by 5′-AMP, however, it is inhibited by 3′-5′-ADP.

The following procedure has been worked out for a large-scale purification of thioredoxins from *Synechococcus* 6301: Cells are broken by French-press treatment, the cell extract is heated for 5 min and the supernatant is acidified to pH 2. The supernatant

1 Botanisches Institut, Universität München, Menzinger Str. 67, 8000 München 19/FRG

Table 1. Properties of sulfotransferases from different organisms

Organism	*Spinacia oleracea*	*Chlorella pyrenoidosa*	*Plectonema* 73110	*Rhodospirillum rubrum*	*Porphyridium cruentum*	*Cyanophora paradoxa*	*Chroococcidiopsis* 7203	*Synechococcus* 6312	*Synechococcus* 6301	*Synechocystis* 6714
Sulfonucleotide needed	APS	APS	APS	APS	APS	PAPS	APS	APS	PAPS	PAPS
Without thiol	1	1	4	0.3	3	4	1.4	0	0	0.7
+DTE	100	100	100	100	100	100	100	100	100	100
+BAL	64	1	81	48	34	142	50	31	5	15
+GSH	98	99	22	136	80	22	4	2	1	6
+Cysteine	22	2	14	4	15	33	4	5	1	4
+Mercaptoethanols	13	2	31	3	18	4	5	2	1	7
Thioredoxin dependence	no	no	no	no	no	no	yes	yes	yes	yes
Inhibitory nucleotide	5′-AMP	5′-AMP	5′-AMP	5′-AMP	?	5′-AMP 5′-ADP APS	5′-AMP	5′-AMP	3′-5′-ADP	?

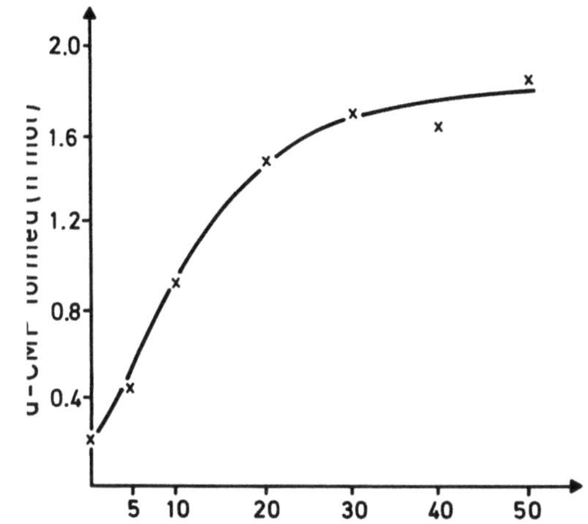

Fig. 1

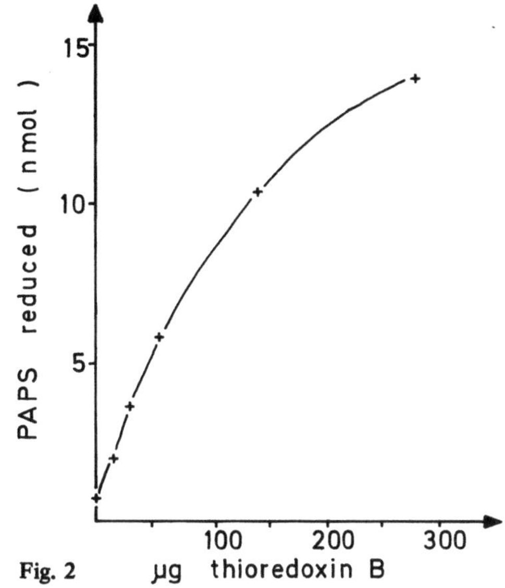

Fig. 2

Fig. 1. Stimulation of the ribonucleotide reductase of *Anabaena* 7119 by thioredoxin B from *Synechococcus* 6301

Fig. 2. Stimulation of the PAPS-sulfotransferase reaction from *Synechococcus* 6301 by thioredoxin B from the same organism

is concentrated by ammoniumsulfate precipitation. Chromatographic separation on DEAE-cellulose, CM-cellulose, and gel-filtration followed by isoelectric focusing yielded a pure protein designated as thioredoxin B. Separation on CM-cellulose yielded a second thioredoxin fraction designated as thioredoxin A. This thioredoxin was characterized using the same techniques. However the thioredoxin A fraction still contains several proteins, therefore K_m-data could not be determined for thioredoxin A. Since thioredoxins are defined as electron donors for the reduction of ribonucleotides to deoxyribonucleotides, the activity of these thioredoxins in an assay for ribonucleotide reduction is necessary for an identification as a thioredoxin. This was shown for thioredoxin B in cooperation with Prof. Follmann (Wagner et al., 1978) and for thioredoxin A (unpubl. data). The thioredoxin-dependent reduction of CMP to d-CMP with a ribonucleotide reductase obtained from *Anabaena* 7119 and thioredoxin B from *Synechococcus* 6301 is shown in Fig. 1 (Gleason and Follmann, unpubl. results). This demonstrates that thioredoxin B is active in a cyanobacterial ribonucleotide reductase with an apparent K_m of 20 μM.

Thioredoxins from *Synechococcus* were detected as a factor needed for activity of the PAPS-sulfotransferase from this organism; such data are shown in Fig. 2. Without addition of thioredoxin practically no activity is found, and for the activation the apparent K_m for thioredoxin B was 10.7 μM. In this PAPS-sulfotransferase system thioredoxin is needed for activity, even at a concentration of 100 m*M* DTE. Thus the thioredoxin requirement cannot be circumvented by a high thiol content. This absolute requirement for thioredoxin of this PAPS-sulfotransferase reaction can be used as a tool for thioredoxin determinations. This PAPS-sulfotransferase will accept thioredoxins from *Escherichia coli, Synechococcus, Scenedesmus,* and spinach (Wagner et al., 1978), which shows that this enzyme is nonspecific with respect to thioredoxins from different organisms. With this assay system thioredoxin was detected in a variety of cyanobacteria (Schmidt and Christen, 1979) and seven different fractions with thioredoxin activity were obtained from spinach with this assay system. Three of these were obtained from chloroplasts (Köst-Reyes and Schmidt, unpubl. results). It is not clear whether, thioredoxin B is the acceptor thiol for the activated sulfate (carrier). It is suggested that thioredoxin might be a reductant for a low molecular weight fraction active in the PAPS-sulfotransferase assay which can be detected by isoelectric focusing of crude *Synechococcus* and spinach extracts with an isoelectric point of about 2. Methods to isolate and characterize this low molecular weight material have to be worked out; however its properties seem to be related to the low molecular weight fraction described by Wilson and Bierer (1976) from yeast. Thus it could be possible that thioredoxins function as an electron carrier in the PAPS-sulfotransferase system; however, detailed investigations are necessary to elucidate this possibility.

Table 1 demonstrates that APS-dependent sulfotransferases can be detected in cyanobacteria which are similar to the PAPS-sulfotransferase system described in their requirement for DTE, suggesting the involvement of thioredoxin. Such APS-sulfotransferases were partly purified from *Synechococcus* 6312 and *Chroococcidiopsis* 7203. The enzyme from *Chroococcidiopsis* is specific for the sulfonucleotide APS and it needs the presence of thioredoxin for activity. The data for the *Chroococcidiopsis* APS-sulfotransferase were obtained with the thioredoxin B fraction of *Synechococcus* (Fig. 3); the apparent K_m for that thioredoxin was calculated as 5.4 μM. The thioredoxin-dependent APS-sulfotransferase demonstrates a new type of sulfotransferases detected here for the first time. It might be a link between thioredoxin-dependent PAPS-systems in bacteria and thioredoxin-independent APS-system in plants.

The last thioredoxin-activated reaction to be discussed here is related to the regulation of carbon metabolism. The group of Buchanan discovered in spinach that the fructose-1-6-bisphosphatase is activated by a low molecular weight protein identified as a thioredoxin (Wolosiuk and Buchanan, 1977; Buchanan et al., 1979). Therefore the thioredoxin B fraction from *Synechococcus* was analyzed also using spinach fructose-1-6-bisphosphatase. A comparison of the *Synechococcus* thioredoxin B fraction and a thioredoxin isolated from spinach using the PAPS-sulfotransferase assay showed that the spinach thioredoxin had good activity whereas the *Synechococcus* thioredoxin showed only little activation (Wagner et al., 1978). This suggested some specificity of the homologous thioredoxin fructose-bisphosphatase system. Therefore a fructose-bisphosphatase was isolated from *Synechococcus* 6301 and the activity of this enzyme was tested with the thioredoxin fraction obtained from the same organism. This cyanobacterial fructose-bisphosphatase is activated by thioredoxin A and B; the data for thioredoxin B are shown in Fig. 4. It can be deduced from these data that thioredoxin activates this enzyme with an apparent K_m for thioredoxin B of 1.7 μM. The fructose-bisphosphatase from *Synechococcus* can be activated also by thiols in the absence of thioredoxin; thus it is an enzyme needing thiols for activity. The low apparent K_m for thioredoxin B (1.7 μM) suggests that it could also be active in vivo, since the apparent K_m-data obtained for GSH (14 mM) or cysteine (50 mM) are about 10,000 fold higher (Schmidt, 1979a).

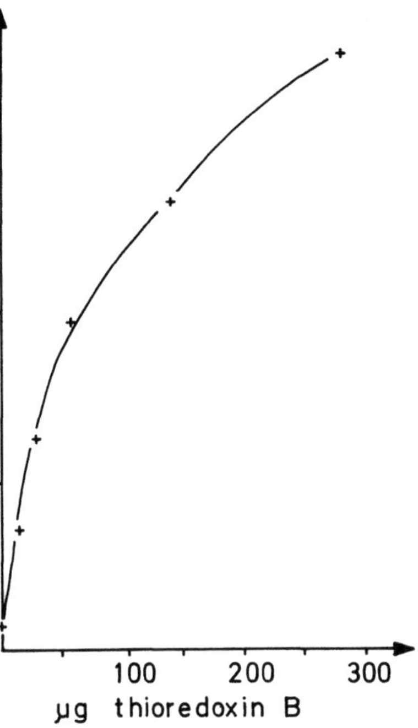

Fig. 3

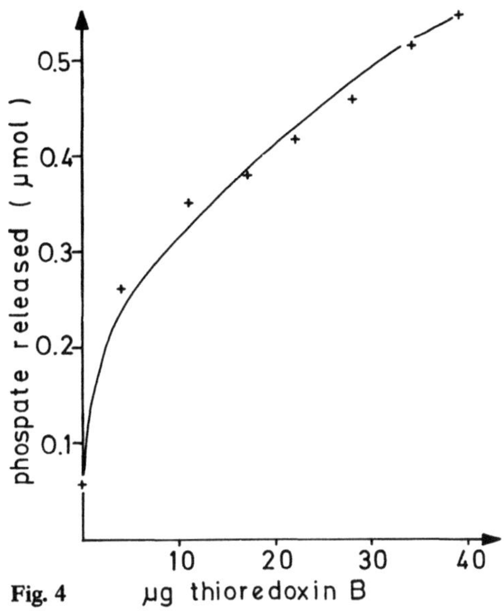

Fig. 4

Fig. 3. Stimulation of the APS-sulfotransferase reaction from *Chroococcidiopsis* 7203 by thioredoxin B from *Synechococcus* 6301

Fig. 4. Stimulation of the fructose-1-6-biosphosphatase from *Synechococcus* 6301 by thioredoxin B from the same organism

During the separation of thioredoxin on CM-cellulose two different thioredoxin fractions were obtained designated as thioredoxin A and thioredoxin B. The examples shown in the figures were obtained with thioredoxin B. These reactions were analyzed also with thioredoxin A, however, no K_m-data can be stated for thioredoxin A since this fraction still contained several proteins. All data obtained are summarized in Table 2.

In the first three lines of Table 2 the different properties of these two thioredoxin fractions are given. They are distinguished by their behavior during chromatography on CM-cellulose. Under the conditions used thioredoxin B is bound, whereas thioredoxin A is not bound. The isoelectric points obtained differ by 0.5 pH units. The molecular weight of thioredoxin B was determined to be 11,800 by gel filtration. As can be seen from the data of Table 2 thioredoxin B is active in all enzyme systems tested. Thus it has properties of the bacterial type of thioredoxins (Follmann and Buchanan, 1980). Thioredoxin A has been found to be active in the reactions tested except the APS-sulfotransferase of *Chroococcidiopsis* 7203. This APS-sulfotransferase was found to be inactive with thioredoxin A.

In crude extracts of *Synechococcus* cells grown on sulfate thioredoxin A has about 10% of all thioredoxin present. These data are preliminary, since we have recently found by the isoelectric focusing technique a small molecular weight substance active in the PAPS-sulfotransferase assay system, which might interfere in these assays. So far it re-

Table 2. Properties of thioredoxin A and B from *Synechococcus* 6301

	Thioredoxin A	Thioredoxin B
Chromatography on CM-cellulose	Not bound	Bound
Isoelectric point	4.1	4.6
Molecular weight	>12,000	11,800
K_m for the PAPS-sulfotransferase from *Synechococcus* 6301	950 µg/ml	10.7 µM
K_m for the APS-sulfotransferase from *Chroococcidiopsis* 7203	Not active	5.4 µM
K_m for the fructose-1-6-bisphosphatase from *Synechococcus* 6301	241 µg/ml	1.7 µM
K_m for the ribonucleotide reductase from *Anabaena* 7119	Not determined	20 µM
K_m for the ribonucleotide reductase from *Escherichia coli*		5 µM

mains to be established why different thioredoxins are present in the same cell and whether these thioredoxins have different functions related to specific enzymes.

It is tempting to speculate that these two different thioredoxins are reduced by different thioredoxin reductases specific for either $NADP^+$ or ferredoxin. This would facilitate a switch from the bacterial type of thioredoxin reductases coupled to $NADP^+$ to the plant type thioredoxin reductase of chloroplasts, which uses ferredoxin as an electron donor (Wolosiuk and Buchanan, 1977; De La Torre et al., 1979; Suske et al., 1979). In crude *Synechococcus* extracts the PAPS-sulfotransferase activity can be stimulated by the addition of either ferredoxin or NADPH. Attempts to purify one of such postulated thioredoxin reductases has failed so far.

The sulfate reduction sequence is regulated in *Synechococcus* 6301, which can be shown by the growth of *Synechococcus* on cysteine. The same was found previously for the higher plant *Lemna minor* (Brunold and Schmidt, 1978). This regulation of assimilatory sulfate reduction in *Synechococcus* could be achieved either by the regulation of the PAPS-sulfotransferase activity, or by a regulation of the sulfate activation sequence, or by a regulation of the thioredoxin content of the cell. These possibilities have to be analyzed in a detailed study. It is evident, however, from our studies of sulfur nutrition in *Synechococcus* that other physiological activities of the cell are affected by the sulfur source as well, which is shown in the control of C-phycocyanin formation by homocysteine (Schmidt, unpubl.). This indicates some interaction of sulfur metabolism and physiology in areas normally not connected to a common nutrient.

Acknowledgement. I am indebted to Prof. Dr. H. Follmann for the ribonucleotide reductase measurements and to Prof. Dr. R.Y. Stanier for a generous gift of *Chroococcidiopsis* 7203.

References

Buchanan BB, Wolosiuk RA, Schürmann P (1979) TIBS 4:93–96
Brunold Ch, Schmidt A (1978) Plant Physiol 61:342–347
DeLaTorre A, Lara C, Wolosiuk RW, Buchanan BB (1979) FEBS Lett 107:141–145

Follmann H, Buchanan BB 1980 (in press)
Schiff JA, Hodson RC (1973) Annu Rev Plant Physiol 24:381–414
Schmidt A (1972) Arch Mikrobiol 84:77–86
Schmidt A (1976) Planta 130:257–263
Schmidt A (1977) FEMS Microbiol Lett 1:137–140
Schmidt A (1979a) Plant Physiol 63:S–11
Schmidt A (1979b) In: Cribbs M, Latzko E (eds) Encyclopedia of plant physiology, new series, vol VI. Photosynthesis II. Springer, Berlin Heidelberg New York, pp 481–496
Schmidt A, Christen U (1978) Planta 140:239–244
Schmidt A, Christen U (1979) Z Naturforsch 34c:1272–1274
Schmidt A, Trüper HG (1977) Experientia 33:1008–1009
Suske G, Wagner W, Follmann H (1979) Z Naturforsch 34c:214–221
Tsang ML-S, Schiff JA (1975) Plant Sci Lett 4:301–307
Tsang ML-S, Schiff JA (1976) J Bacteriol 125:923–933
Wagner W, Follmann H, Schmidt A (1978) Z Naturforsch 33c:517–520
Wilson LG, Bierer D (1976) Biochem J 158:255–270
Wolosiuk RA, Buchanan BB (1977) Nature (London) 266:565–567

Recent Results on the Assimilatory Sulfate Reduction: APS-Kinase and the Reduction of APS to Cysteine in Higher Plants

J.D. SCHWENN and H. URLAUB[1]

Biosynthesis of PAPS by algal or higher plant material has occasionally been reported (references[a] in Table 1) and, so far, three different types of enzymatic assays for the APS-kinase have been introduced (references[b] in Table 1). The presence of APS-kinase (EC 2.7.1.25, ATP: adenylylsulfate-3′-phosphotransferase) in plants seems to be ascertained. But, as compared to other enzymes involved in the assimilatory reduction of sulfate, detailed investigations of the APS-kinase are not available.

In plants and algae accumulation of PAPS has been observed with cell-free extracts from *Chlorella* (Hodson et al., 1968; Hodson and Schiff, 1969; Tsang et al., 1976) and with higher plant fragmented chloroplasts (Burnell and Anderson, 1973; Mercer and Thomas, 1969). Although intact chloroplasts were found to form comparably small amounts of PAPS (Schmidt and Schwenn, 1971) it appears that the enzyme is located inside the chloroplasts as all other enzymes participating in assimilatory sulfate reduction (Schmidt, 1978; Schwenn and Trebst, 1976). Burnell and Anderson (Burnell and Anderson, 1973) proposed that the enzyme is bound to the chloroplast lamellae system which is supported by our data on the distribution of enzymes between stroma and lamellae (Schwenn and Hennies, 1975; Schwenn et al., 1976).

Very few data have been published on purified APS-kinase. Previous investigations were focused on the enzyme from liver, yeast or *Nitrobacter* (Robbins and Lipman, 1958a,b, Varma and Nicholas, 1971). A partially purified APS-kinase from the alga *Chlamydomonas reinhardii* has been obtained recently (Schwenn and Jender, 1980b,c). The enzyme had an approximate molecular weight of 120 kdaltons. It was saturated at extremely low concentrations of APS (K_m app.: <6 μM) and of ATP (K_m app.: <50 μM). These kinetic constants are in the same range as found for the enzyme from yeast (Robbins and Lipman, 1958b). The high affinity for APS of the APS-kinase is most efficiently suited to complete the sulfate activation because only very low concentrations of APS can be expected at equilibrium of the ATP-sulfurylase (EC 2.7.7.4) (Siegel, 1975). 3′5′PAP was found to increase V_{max} of APS-kinase rather than to protect PAPS from degradation.

The kinetic investigation of APS-kinase activity in *C. reinhardii* demonstrated that ATP is required for the biosynthesis of PAPS (Fig. 1A). The onset of PAPS biosynthesis was nonlinear. The time course of PAPS formation at the initial phase of enzymatic action indicated the existence of an APS-enzyme complex. The assumption of a binding by the enzyme of APS is supported by data obtained from gelchromatography. The

1 Ruhr-Universität Bochum, Lehrstuhl Biochemie der Pflanzen, Postfach 102148, 4630 Bochum-Querenburg/FRG

Table 1. APS-Kinase activity in algae and higher plants

Ref.	Source	Fraction	Substrates (m*M*)				Thiol (m*M*)	Miscellaneous properties
			ATP	SO_4	APS	Mg^{2+}		
1[a]	*Chlorella pyrenoid.*	Crude extr.	~30.0	~15.0	–	24.3	50 MSH	3′5′PAP inhibits further metabolism of PAPS
2[a]	*Phasaeolus*	Chloroplast	10.0	(a)	–	5.0	None	Chloroplasts sonicated
	Z. mays	fragments	10.0	–	(a)	5.0	None	
3[a]	*Spinacea*	Intact chloroplasts	–	0.6	–	2.0	None	Photosynthetic conditions
4[a]	*Chlorella*	Crude extr. desalted	25.0	12.5	–	15.0	1.25 GSH or Cys	
5[b]	*Spinacea*	Crude extr. isol.	13.3	26.6	–	20.0	None	Coupled to PAPS: Phenolsulfotransferase, required 10 m*M* 3′AMP, membrane bound
		chloroplasts	10.0	5.0	–	10.0	(b)	
6[a]	*Ochromonas*	Crude extr.	25.0	(a)	–	12.5	12.5 MSH	
7[b]	*Spinacea*	Crude extr.	2.5	–	0.02	10.0	None	Photometric assays of APS-dependent ADP-formation coupled to NADH oxidation; APS recycled
8[b]	*Triticum, Sorghum, Helianthus*	Crude extracts	25.0	40.0	–	20.0	0.5 MSH (d)	Required 25 m*M* 3′ or 5′AMP "Renilla" bioluminescence assay
9[a]	*Pisum*	Chloroplasts	5.0	5.0	–	5.5	None	Fluorimetr. method coupled to PAPS: Phenolsulfotransferase
10[a]	*Spinacea*	Chloroplasts osmot. shocked	7.5	4.0	–	15.0	None	Associated with the chloroplast thylakoids
11[a]	*Rhodella*	40%–60% AS precip. desalt.	8.0 (e)	(a)	–	12.0	None	Minute activity
12[b]	*Anabaena*	Crude extr. desalted	10.0	–	2.1	31.0	None	Coupled assay as in Ref. 7
13[a]	*Spinacea*	Washed chloroplasts	5.0	5.0	–	10.0	None	5′AMP K_i = 1.3 m*M*, ADP K_i = 0.5 m*M*

Table 1 (continued)

Ref.	Source	Fraction	Substrates (m*M*)				Thiol (m*M*)	Miscellaneous properties
			ATP	SO_4	APS	Mg^{2+}		
14[b]	*Chlamydomonas r.*	30%–60% AS, several chrom. steps, part. purified	1.0	–	0.01	5.0	None (d)	Measurement of initial velocity, K_mATP: 50 μ*M* K_mAPS: 6 μ*M*, 3′5′PAP stimulates, app. mol. weight 120,000, APS-sulfotransferase activity

[a] Detection of reaction product PAPS
[b] Assay for APS-kinase activity

(a) Carrier-free ^{35}S-sulfate; (b) chloroplast prep. containing max. 2 m*M* β-MSH; (c) 3′nucleotidase regenerates APS from PAPS; (d) β-MSH has been used for prep. of extracts; (e) from assay system for ATP-sulfurylase

(1) Hodson et al., 1968; Hodson and Schiff, 1969; Tsang et al., 1976; (2) Mercer and Thomas, 1969; (3) Schmidt and Schwenn, 1971; (4) Schmidt, 1972; (5) Burnell and Anderson, 1973; (6) Mercer et al., 1974; (7) Burnell and Whatley, 1975; (8) Stanley et al., 1975; (9) Hoppe, 1976 (unpubl.); (10) Schwenn et al., 1976; (11) Møller and Evans, 1974; (12) Sawhney and Nicholas, 1976b; (13) Schwenn and Depka, 1977; (14) Schwenn and Jender, 1980a,b

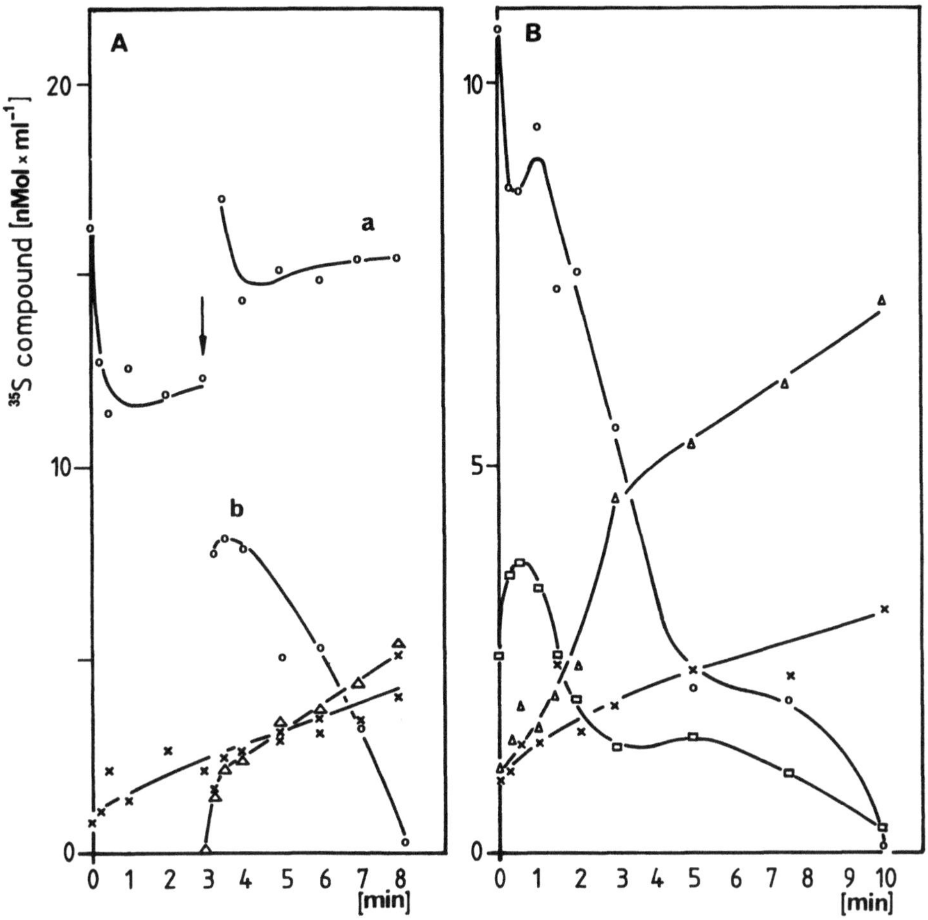

Fig. 1A, B. Kinetic of APS-kinase from *Chlamydomonas reinhardii.* **A** ATP-dependent biosynthesis of PAPS from APS in a cell-free system. The *arrow* indicates the time when ATP has been added to start the reaction. *a* control without ATP; *b* in the presence of ATP. **B** Occurrence of an enzyme-substrate intermediate during the action of isolated APS-kinase. (Data from Schwenn and Jender, 1980b). x–x: sulfate; o–o: APS; □–□: intermediate; △–△: PAPS

enzyme-substrate complex sustained rechromatography on Sephacel and Sephacryl columns (Schwenn and Jender, 1980c). The kinetics of its formation during the action of APS-kinase is only conceivable as the intermediate between APS and PAPS (Fig. 1B). The concentration of APS decreased rapidly when this intermediate was formed. When PAPS became detectable, the concentration of this compound declined concomittantly, whereas the concentration of the intermediate remained unaltered under conditions which did not allow PAPS biosynthesis.

The APS-protein complex was found to release free sulfite upon the addition of reducing thiols (glutathione, dithioerythritol and dithiothreitol). In this respect it is not to be distinguished from the APS-binding protein of *Chlorella* (Abrams and Schiff, 1973), e.g., APS-sulfotransferase as designated by Schmidt (Schmidt, 1976). This sulfotransferase activity copurified with APS-kinase throughout several chromatographic steps. Since the ratio of specific activity of APS-kinase versus APS-sulfotransferase activity was unchanged during the purification, it seems possible that both activities are associated with an identical protein. APS-sulfotransferase, therefore, may represent a side activity of the APS-kinase which can be described as the reductive cleavage of APS by a thiol when no ATP is available. If, indeed, this reaction accounts for the APS-sulfotransferase its direct participation as the sulfate group-transferring enzyme in the assimilatory sulfate reduction as proposed in the literature becomes questionable. Furthermore, the exchangeable sulfite formed by an APS-sulfotransferase reaction would not be identical with the intermediary "bound sulfite" ($XSSO_3H$) of the assimilatory sulfate reduction by higher plants.

Available evidence indicates that the chloroplasts of higher plants contain the enzymatic apparatus for the reduction of sulfate to cysteine. Ultrasonic treatment (Burnell and Anderson, 1973) and immunological methods (Schwenn and Hennies, 1975) revealed that APS-kinase and the thiosulfonate reductase are membrane-bound enzymes. However, the aspect of a complete sulfate-reducing enzyme complex attached to the plastid membrane system remained unexplored. With emphasis upon the intracellular distribution of enzymes catalyzing the reduction of APS to cysteine, cell suspension cultures of *Catharanthus roseus* grown under S-autotrophic conditions have been investigated (Urlaub and Jankowski, 1980). The reduction of APS to divalent sulfide sulfur was assayed in a reaction which was coupled to the biosynthesis of cysteine. A purified O-acetyl-L-serine sulfhydrylase (EC 4.2.99.8) has been used to trap the reduced sulfide sulfur and thereby "pull" the reductive step.

Extracts of *C. roseus* were found to contain a particulate enzyme (termed "APS-reducing activity", app. MW $\simeq$700 kdaltons) which in the presence of O-acetyl-L-serine, OAS sulfhydrolase, and a reduced ferredoxin-generating system can catalyze the incor-

Table 2. Distribution of enzymes related to assimilatory sulfate reduction from *C. roseus* cell suspension cultures (Urlaub and Jankowski, 1980)

"Soluble" (S 28)	"Particulate" (P 100)
1. ATP-sulfurylase (EC 2.7.7.4)	n.d.
2. APS-kinase (EC 2.7.1.25)	APS-sulfotransferase activity [a]
3. OAS-sulfhydrylase (EC 4.2.99.8)	n.d.
4. Glutathione-reductase (EC 1.6.4.2)	Traces
Dithiol-reductase, unspecif.	Traces
5. n.d.	APS-reducing activity
Sulfite reductase (EC 1.8.99.1)	n.d. [b]
and (EC 1.8. 7.1)	
6. S:Sulfoglutathione-reductase (b)	n.d.
7. Fd:$NADP^+$ reductase (EC 1.18.1.2)	10%–25%

S 28: supernatant of cell homogenate centrifuged at 28,000 *g*, individual activity recovered after gel chromatography. P 100: pellet of S 28 after differential centrifugation with final step at 100,000 g

[a] APS-kinase assay in "soluble" and APS-sulfotransferase assay in "particulate"

[b] Released from APS-reducing particles by sonication or exposure to low ionic strenght. n.d.: not determined

poration of the sulfur atom of APS into cysteine. The particulate enzyme system was enriched by differential centrifugation and investigated for individual enzyme activities (Table 2). ATP-sulfurylase and OAS-sulfhydrylase were retained predominantly by the soluble fraction. Untreated particles exhibited, in addition to the APS-reducing activity, an APS-sulfotransferase activity, residual amounts of glutathione-reductase (EC 1.6.4.2), unspecific $NADPH_2$-dependent disulfide reductase, and ferredoxin: $NADP^+$-oxidoreductase (EC 1.18.1.2). Sonicated particles immediately lost their APS-reducing activity and partially their glutathione- and disulfide reductase. Exposure to low ionic strength or to chelating reagents resulted in a more gradual loss of APS-reducing activity. In both cases, this loss was accompanied by the release of a ferredoxin-dependent sulfite reductase (EC 1.8.7.1) which was virtually absent from the untreated fragments. In view of the fact that untreated particles formed $AP^{35}S$, bound sulfite ("$XS^{35}SO_3H$") and bound sulfide ("$XS^{35}SH$") from ATP and ^{35}S-sulfate when provided with a $NADPH_2$-regenerating system and ATP-sulfurylase, it is suggested that the "complex" found in homogenates from *Catharanthus* cells contains at least the binding site(s) for the sulfite- and sulfide moiety and the "bound sulfite" reductase, e.g., thiol sulfonate reductase. Moreover, as judged from the apparent APS-sulfotransferase activity of this enzyme aggregate, APS-kinase might also be located in these particles.

Thus, the structural organization of the sulfate-reducing apparatus in higher plants has also to be considered since in vitro concentrations of intermediates obtained with an enzyme complex do not reflect the true activity of the individual enzyme in situ.

References

Abrams WR, Schiff JA (1973) Arch Mikrobiol 94:1–10
Burnell JN, Anderson JW (1973) Biochem J 134:565–579
Burnell JN, Whatley FR (1975) Anal Biochem 68:281–288
Hodson RC, Schiff JA (1969) Arch Biochem Biophys 132:151–156
Hodson RC, Schiff JA, Scarsella AJ, Levinthal M (1968) Plant Physiol 43:563–569
Mercer E, Thomas G (1969) Phytochemistry 8:2281–2285
Mercer E, Thomas G, Harrison D (1974) Phytochemistry 13:1297–1302
Møller ME, Evans LV (1976) Phytochemistry 15:1623–1626
Robbins PW, Lipman F (1958a) J Biol Chem 233:681–685
Robbins PW, Lipman F (1958b) J Biol Chem 233:686–690
Sawhney SK, Nicholas DJD (1976a) Plant Science Lett 6:103–110
Sawhney SK, Nicholas DJD (1976b) Planta 132:189–195
Schmidt A (1972) Z Naturforsch 27b:183–192
Schmidt A (1976) Planta 130:257–263
Schmidt A (1978) In: Pirson A, Zimmermann MH (eds) Encyclopedia of plant physiology, new series, vol VI. Photosynthesis II. Springer, Berlin Heidelberg New York, pp 481–496
Schmidt A, Schwenn JD (1971) In: Forti G, Avron M, Melandri A (eds) Proc 2nd Int Congr Photosynth, Stresa. W Junk Publ, The Hague, pp 507–514
Schwenn JD, Depka B (1977) Z Naturforsch 32c:792–797
Schwenn JD, Hennies HH (1975) In: Avron M, Proc 3rd Int Congr Photosynth. Elsevier Scientific Publ, Amsterdam Oxford New York, pp 629–635
Schwenn JD, Jender HG (1980a) J Chromatogr 193:285–290
Schwenn JD, Jender HG (1980b) Phytochemistry (in press)
Schwenn JD, Jender HG (1980c) Planta (in press)

Schwenn JD, Trebst A (1976) In: Barber J (ed) The intact chloroplast. Elsevier/North Holland Biomedical Press, Amsterdam, pp 315–334
Schwenn JD, Depka B, Hennies HH (1976) Plant Cell Physiol 17:165–176
Siegel LM (1975) In: Greenberg DM (ed) Metabolic pathways, vol VII: Metabolism of sulfur compounds. Academic Press, London New York, pp 223
Stanley PE, Kelley BC, Touvinen OH, Nicholas DJD (1975) Anal Biochem 67:540–551
Touvinen OH, Kelley BC, Nicholas DJD (1976) Z Allg Mikrobiol 16:551–555
Tsang ML-S, Lemieux J, Schiff JA, Bojarski T (1976) Anal Biochem 74:623–626
Urlaub H, Jankowski G (1980) Planta (in press)
Varma AK, Nicholas DJD (1971) Arch Mikrobiol 78:99–117

Aspects of S- and N-Metabolism in Tissue Cultures

L. BERGMANN[1]

1 Introduction

Cell suspension cultures of higher plants are grown in most cases as batch cultures in media that contain mineral salts and carbohydrates in concentrations which allow the cells to grow and multiply for a certain time until the media become exhausted and growth ceases. Therefore, to maintain the cultures, they have to be transferred to fresh media at regular intervals. To keep the number of such transfers within reasonable limits the level of minerals and carbohydrates in the culture media is made high enough at the start to meet the demands of the cells for a longer period. Depending on the amount of nutrients and the size of the inoculum 30- to 250-fold increases in the total cell dry weight can thus be attained, which correspond to 5-8 cell generations.

It is obvious that cells in such cultures have to maintain growth under quite different conditions. At the start of a culture they are exposed to an excessive initial supply of macronutrients and at the end of the growth period they are subjected to growth on very low levels of nutrients as, for example, nitrogen or sulfur. Under these conditions the cells remain not unchanged in their composition. Normally, at the beginning of a culture period they exhibit a relatively high nitrogen content. This declines when the cultures get older and the dry weight increases faster than the protein content (cf. King and Street, 1973). Thus in suspension cultures of *Ipomea* grown with NH_4NO_3 60% of the final total nitrogen has been accumulated by the time when only 25% of the final dry weight has been attained, as shown by Rose and Martin (1975).

This situation is very similar to that which we meet in crop plants where high amounts of fertilizer are applied at the beginning of the growth season and are often used up at the time of seed filling so that the plants have to draw on their nitrogen reserves. An analysis of the changes in the composition and metabolism of batch cultures of plant cells could therefore be useful to give some insight into the regulation employed by higher plants to adapt their metabolism to variations in the environment. In this paper I shall discuss experiments on the regulation of sulfur assimilation and on the effect of sulfur starvation on the nitrogen metabolism in tobacco cell suspension cultures.

1 Botanisches Institut der Universität Köln, II. Lehrstuhl, Gyrhofstr. 15, 5000 Köln 41/FRG

2 Regulation of Sulfate Assimilation by ATP-Sulfurylase

The regulation of the assimilation of sulfate in higher plants is far from being solved. Since the level of free cysteine and methionine in higher plant cells is normally low and relatively constant even when the external sulfate supply varies over a wide concentration range (cf. Datko et al., 1978) there must be a tight control over this biosynthetic pathway.

Reuveny and Filner (1977) have recently made the interesting suggestion that the sulfate assimilation in tobacco cells is regulated by the amount of ATP-sulfurylase formed in the cells. They have based this suggestion on the observation that the level of ATP-sulfurylase, the first enzyme of sulfate assimilation, is low and remained essentially constant with time in tobacco cells grown on sulfate, cysteine, or methionine as sole sulfur source. If grown in medium free of sulfur or grown with a slowly assimilated sulfur source, like djenkolic acid, the level of the enzyme in the tobacco cells increased 5- to 10-fold, and decreased again after addition of sulfate.

Reuveny and Filner have shown that the changes in the ATP-sulfurylase activity measured in extracts of the tobacco cells could not be attributed to activation or inactivation mechanisms. They have further demonstrated that the increase in ATP-sulfurylase activity in cells grown on sulfur-free medium and the decay of ATP-sulfurylase activity following the addition of a repressing sulfur source are inhibited by cycloheximide at concentrations which inhibit protein synthesis.

The simplest interpretation of these data is that in the medium with sulfate as a sulfur source the cells make sufficient of the endproducts, presumably cysteine and methionine, to fulfill their needs for protein synthesis and to repress the ATP-sulfurylase. If the level of cysteine and/or methionine falls, the ATP-sulfurylase becomes derepressed and the enzyme level rises to meet the demand.

The experiments of Reuveny and Filner were done with heterotrophic, chloroplast-free cell cultures. Since in green, chloroplast-containing cells adenosine-5-phosphosulfate, the product of the reaction catalyzed by ATP-sulfurylase, is needed not only as an intermediate in sulfate reduction but also for the synthesis of chloroplast sulfolipids, it is difficult to conceive a regulation of sulfur assimilation in chloroplast-containing cells alone via ATP-sulfurylase. Also experiments in which changes in the rate of sulfate assimilation were correlated with enzyme changes are missing in the paper of Reuveny and Filner, although such a correlation would certainly strengthen their hypothesis.

We have, therefore, repeated the experiments of Reuveny and Filner using a cell strain of *N. tabacum* var. Samsun that develops photosynthetically active chloroplasts when grown in the light. This strain also synthesizes different amounts of glutathione and protein depending on the culture conditions employed, and for this assimilates quite different amounts of sulfate. Figure 1 shows the changes of ATP-sulfurylase activity in photoheterotrophically grown cell suspensions of this strain during normal growth and when sulfate in the medium becomes limiting. During exponential growth of the cultures the specific activity of the enzyme remained at a constant level; after depletion of the sulfate in the medium the ATP-sulfurylase activity began to rise and increased for 48–72 h, the increase resulting in a 4-fold augmentation of the specific activity. Transfer of the cells to fresh media resulted in a sudden decline in enzyme activity, the ATP-sulfurylase activity reaching the low level of cells before sulfur starvation. As the cells

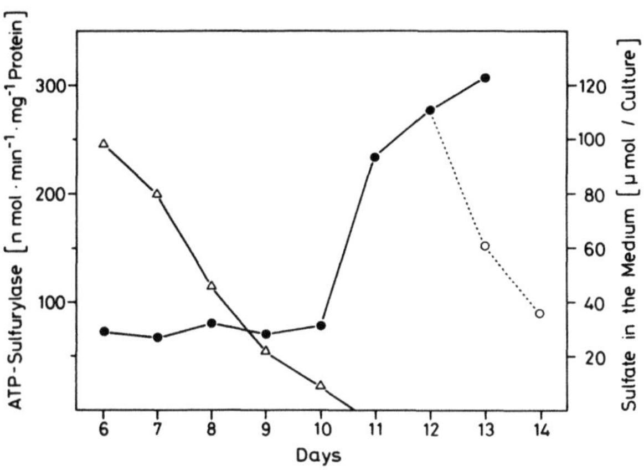

Fig. 1. Development of ATP-sulfurylase activity in photoheterotrophic cell suspension cultures of *N. tabacum* grown with limiting amounts of sulfate. Fluorescent light 3000 lx, 23 °C. ATP-sulfurylase activity estimated according to Schwenn et al. (1979). ●–●–●–● ATP-sulfurylase activity; ○–○–○–○ ATP-sulfurylase activity after transfer of cultures to fresh sulfate-containing media; △–△–△–△ Sulfate in the medium

in this experiment grew with a generation time of 48 h, the decline in enzyme activity cannot be caused by dilution only, but indicates a degradation of the ATP-sulfurylase.

This experiment demonstrates that green, chloroplast-containing cells react in the same way as do chloroplast-free tobacco cells and confirms the results of Reuveny and Filner on the regulation of the ATP-sulfurylase level in those cells. However, the activity of the ATP-sulfurylase in our tobacco cells did not parallel the rate of sulfate assimilation in the cells, which points to a different mechanism for the fine control of sulfate assimilation.

As we have recently reported (Bergmann and Rennenberg, 1978), photoheterotropic cell suspension cultures of *N. tabacum* release substantial amounts of glutathione into the culture media. If the cells are grown with ammonium plus nitrate or with glutamine as nitrogen source and are adequately supplied with sulfate, more than 50% of the sulfate assimilated is transported into the medium and concentrations as high as 1 m*M* can accumulate outside the cells. In contrast, cell suspensions grown either with nitrate as nitrogen source or grown heterotrophically with ammonium plus nitrate release very small amounts of glutathione, the glutathione level in the media rising only to 10 µ*M* or 40 µ*M* (Rennenberg and Bergmann, 1979).

Since green and nongreen cells in the cultured tobacco cell system show no differences in the intracellular glutathione level and exhibit similar growth rates when cultivated on ammonium plus nitrate, the differences in the rate of glutathione efflux indicate great differences in the rate of sulfate assimilation between the cells. This difference is especially large in the case of photoheterotrophic cells grown on nitrate and cells supplied with ammonium plus nitrate, since nitrate-grown cells, besides the minute glutathione production, exhibit a smaller rate of protein formation than cells grown with ammonium plus nitrate.

Therefore, in the cultured tobacco cell system it is possible to compare the ATP-sulfurylase activities between cells which are engaged in sulfate assimilation to different extents. If the suggestion of Reuveny and Filner holds true, that the sulfate assimilation of the tobacco cells is regulated by the amount of ATP-sulfurylase synthesized by the cells, these differences in sulfate assimilation should be reflected in the level of ATP-sulfurylase in the cells.

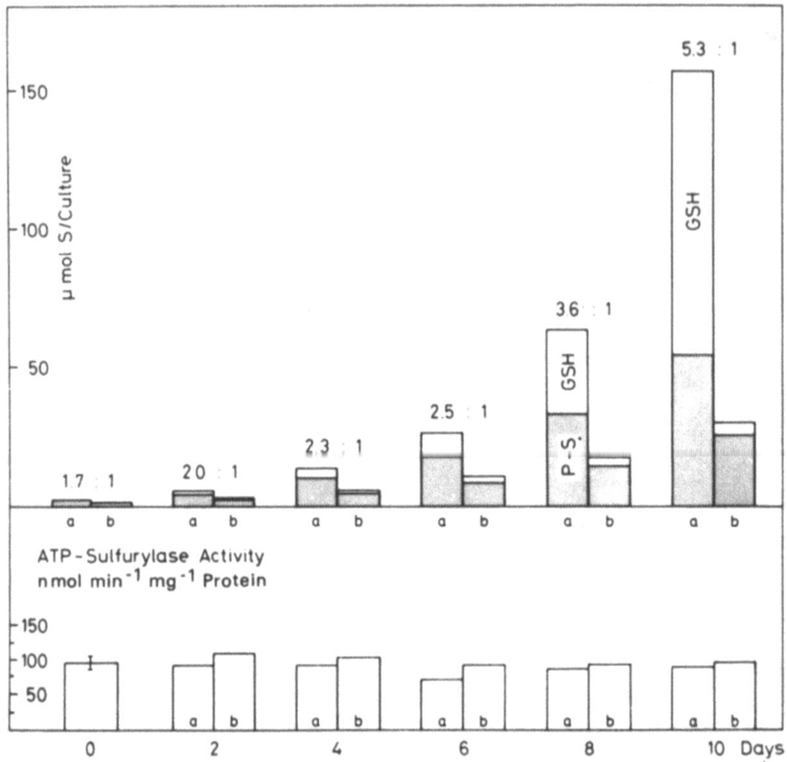

Fig. 2. ATP-sulfurylase activity and assimilated sulfur in photoheterotrophic suspension cultures of *N. tabacum* grown on nitrate and ammonium plus nitrate suspension cultures of *N. tabacum* were grown photoheterotrophically (3000 lx, 23 °C) with 1.73 m*M* sulfate and 20 m*M* NH_4^+ plus 40 m*M* NO_3^- *a* or with 60 m*M* NO_3^- alone *b*. At times indicated glutathione in the medium and cell protein were determinated and the protein-S calculated assuming a S:N relation of 1:25. ATP-sulfurylase activity has been estimated according to Schwenn et al. (1979)

As can be seen from the data in Fig. 2, such a correlation between the enzyme activity and sulfate assimilation has not been observed. The results in Fig. 2 show the activity of extractable ATP-sulfurylase of cells grown photoheterotrophically with nitrate or ammonium and nitrate and the amount of sulfur incorporated into cell protein and in glutathione at various times of culture. Although the amount of sulfur assimilated by the cells grown with ammonium is two to five times higher than in cells cultivated with nitrate, cells from both cultures show the same specific activity of ATP-sulfurylase which equals the enzyme activity of cells in which the ATP-sulfurylase is "repressed" by sulfate.

If the levels of ATP-sulfurylase measured in the cell-free extracts reflect the amount of enzyme in intact cells, then the amount for the enzyme in the repressed state appears to be sufficient for the assimilation of quite different amounts of sulfate, as shown by the current results. The discrepancies between the enzyme activities measured in vitro and the sulfate assimilated in vivo are also clearly visible in a comparison of the ATP-sulfurylase per culture and the amount of reduced sulfate at various times of culture. The results of such a comparison in which heterotrophic cells have also been included

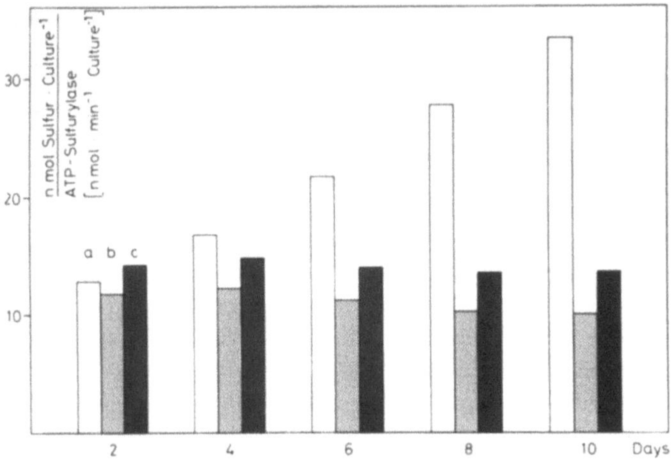

Fig. 3. Sulfate assimilation and ATP-sulfurylase activity in photoheterotrophic and heterotrophic suspension cultures of *N. tabacum*. For details see Fig. 2; *a* NH_4NO_3 light; *b* NO_3^- light; *c* NH_4NO_3 dark

demonstrate that the enzyme activities assayed in vitro in different cultures do not parallel the amount of sulfate assimilated in those cultures (Fig. 3).

The failure to demonstrate a correlation between the enzyme activities measured in cell-free extracts and the sulfate assimilation observed in vivo indicates that the rate of sulfate assimilation in the tobacco cells cannot be controlled only by changes in the level of ATP-sulfurylase in the cells but must be regulated by other systems.

This situation is similar to the situation observed in the case of nitrate reduction. The amount of nitrate reductase is under strong regulatory control, as has been convincingly demonstrated by many authors (Beevers and Hageman, 1969) and in a general way levels of nitrate reductase measured in cell-free extracts parallel changes in the in vivo reduction of nitrate. But in cases where more subtle changes are involved such a correlation is not observed and measurements of the nitrate reductase level in vitro are not an accurate reflection of the functioning of the enzyme in the intact cells (Oaks et al., 1979, Huffaker and Rains, 1978). No doubt, controls on both enzyme synthesis or degradation and on enzyme activity are used by plant cells to coordinate and integrate the various metabolic pathways. We should, therefore, be careful not to limit our view by restricting our interest to only one of those possibilities, but try to uncover the whole machinery used by the cells. This point has to be stressed here, because Schwenn et al. (1979) have reported an increase in the ATP-sulfurylase activity in suspension cultures of *Catharanthus roseus* during the first part of the exponentional growth phase and a decline during the following phase, a result that does not agree with the situation observed in tobacco cells.

3 Glutathione Production in Tobacco Cells

Having demonstrated that the tobacco cells are able to assimilate and reduce more sulfate than needed for protein synthesis and to use the reduced sulfur in large quantities for the biosynthesis of glutathione, I will now turn to the biological significance of this production. The surplus production of glutathione is restricted to chloroplast-containing cells and induced by high ammonium and sulfate concentrations. When the sulfate supply in the medium is exhausted, the glutathione level in the medium decreases. Since the total protein-N in the cultures during this period increases, the decrease of glutathione in the medium is supposedly caused by an uptake of GSH by the cells and use of the cysteine incorporated in it for protein synthesis. This suggests that glutathione may serve as a storage and transport form for reduced sulfur.

Two observations are consistent with this suggestion. Firstly we have shown (Rennenberg et al., 1979) that the amount of sulfur reduced in leaves of mature tobacco plants can exceed their own requirement and enables an export of reduced sulfur. The export of reduced sulfur occurs in the phloem with glutathione as the main transport form. Secondly, the observation that glutathione is often the main organic sulfur compound found in plant cells the concentration of which changes with variations in the sulfur supply fits into this picture. According to Smith (1975) 60%-67% of the soluble organic sulfur in tobacco cell suspension cultures are glutathione, cysteine amounts to 21%-26% and methionine only to 11%-14%. Smith's data clearly show that the glutathione content of the cells declines from roughly 150 nmol/g fr.wt. to 47 nmol/g fr.wt. after depletion of the available sulfur. The same can be observed in cells of our tobacco strain where the glutathione level declines from 100 nmol/g fr.wt. to 30 nmol/g fr.wt. and less in cells that are subjected to sulfur deficiency (Bergmann and Rennenberg, 1978).

Since cells normally strive to keep their intracellular glutathione concentration constant (Kosower and Kosower, 1976) the changes in the glutathione level observed in tobacco cells imply that part of the glutathione must be in a pool, the fluctuations of which do not affect the metabolic state of the cells. It is natural in plant cells to suspect the vacuole of the cells to be a compartment where glutathione could be stored. Dr. Ulbrich in our laboratory has therefore prepared vacuoles from protoplasts of tobacco leafs and compared the contents of protoplasts and vacuoles. As the figures in Table 1 show, the vacuoles contain 17% of the glutathione found in protoplasts and various amounts of free amino acids which indicate that the vacuoles have retained their contents during isolation procedures. The amount of glutathione detected in the vacuoles is small but significant. Since it has been found in leaves of normally grown plants its occurence must have a physiological meaning – we believe it to be a store of cysteine.

The suggested use of glutathione as a reservoir for cysteine, that can be released and incorporated into protein, poses the interesting question how the breakdown of glutathione in the cells is regulated and glutathione degraded without disturbance of the vital intracellular glutathione level. To answer this question it is important to know whether glutathione is taken up and channeled into the metabolism as an intact molecule, or is degraded at the cell surface and only the amino acids taken up as reported for kidney cells (Hahn and Oberrauch, 1978).

By incubating cell suspension cultures with labeled glutathione, the uptake of glutathione and the incorporation of sulfur from glutathione into cell protein has recently been demonstrated by Rennenberg et al. (1980). From his data in Table 2 it can be seen that 87% (94.5%) of the radioactivity lost from the medium after 2 h of incubation are found inside the cells. 54.7% (61.1%) are incorporated into protein and 4.6% (4.0%) are

Table 1. Protoplasts were isolated by enzymatic treatment of leaf tissue of mature tobacco plants. For the release of the vacuoles the protoplasts were osmotically shocked and the vacuoles isolated and collected by centrifugation on a discontinuous Ficoll gradient. The GSH content of the protoplasts and vacuoles has been analyzed according to Bergmann and Rennenberg (1978), the amino acids were estimated by an automated acid analyzer. The data shown are the means of four analyses

	Distribution of Glutathione and Amino Acids in Tobacco Leaf Cells		
	Protoplasts nmol/10^6	Vacuoles nmol/10^6	Percentage in vacuoles
GSH	5.8	1.0	17.2
Asp	22.7	5.8	25.3
Thr	10.1	2.6	25.2
Ser	26.7	3.7	13.7
Asn	12.2	3.3	27.0
Glu	10.6	2.5	23.6
Gln	8.7	4.7	54.6
Gly	14.7	9.6	65.3
Ala	21.4	8.0	37.4
Ile	7.5	3.2	42.6
Leu	15.0	5.6	37.3
Tyr	10.6	4.0	37.8
Phe	12.2	3.2	26.2
Gaba	104.0	32.4	31.2
Orn	6.3	1.4	22.2
Lys	17.4	5.4	31.0
His	9.0	2.6	28.9
Arg	15.6	5.9	37.8

bound to protein and can be separated by treatment with dithiothreitol (DTT). The remaining 28% of the labeled sulfur compounds are contained in the water-soluble fraction.

After 24 h of incubation the amount of labeled glutathione taken up from the medium and incorporated into protein has risen to 8.55 nMol which corresponds to 87.2% (77.6%) of the label taken up from the medium. With 2.9% (3.3%) the percentage of radioactivity attached to the protein and detached by DTT is small, whereas 21% of the label have accumulated in the water-soluble fraction. These results clearly confirm the conclusion drawn from the growth experiments, that tobacco cells are able to use cysteine incorporated into glutathione for protein synthesis, and they are consistent with the idea that glutathione can serve as a store of cysteine in tobacco cells.

They do not answer the question whether glutathione is taken up by the cells as an intact molecule, or split on the surface of the cells. To answer this question Rennenberg has analyzed the water soluble fraction which contains 20%–28% of the label taken up by the cells. The results of his analysis are given in the second part of table 2. 50%–60% of the total water-soluble radioactivity were confined to glutathione, whereas the sulfur-containing amino acids cysteine and methionine represented only 4.0% (12%), or 6.7% (1.5%) of the label.

The possibility that the labeled glutathione observed in the cells has been synthesized from cysteine taken up by the cells after degradation of the labeled glutathione on the

Table 2. Photoheterotrophically grown tobacco cells (55 mg fr.wt./ml) were incubated in sulfur-free medium for 20 h and than exposed to 25 μM ^{35}S-GSH (0.5 uCi/ml) ±5x10^{-8} M methionine sulfoximine (MSO) for 2–24 h. Incubation was stopped by removing the medium by filtration and washing the cells with GSH-containing medium. Cells were homogenized and cell contents fractionated. The water-soluble fraction was subjected to TLC on precoated plates and analyzed by a TLC scanner. The protein fraction was extracted with Dithiothreitol to remove protein-bound sulfur compounds, separated by centrifugation and solubilized with 1.5% NaOH. Radioactivity in the H_2O-, DTT- and NaOH-extracts was measured by liquid scintillation counting

I. Incorporation of ^{35}S into tobacco cells incubated with ^{35}S-GSH

Incubation time	Decrease in ^{35}S-GSH in the medium (nmol/ml suspension)	^{35}S-GSH recovered inside the cells (nmol/55 mg fr.wt.)	Distribution of ^{35}S inside the cell (nmol/55 mg fr.wt.)		
			Total water soluble	Incorporated into protein	Bound to protein
2 h	3.8 (100%)	3.30 (86.8%)	1.06 (27.9%)	2.08 (54.7%)	0.16 (4.6%)
2 h + MSO	3.5 (100%)	3.32 (94.8%)	1.04 (29.7%)	2.14 (61.1%)	0.14 (4.0%)
24 h	9.8 (100%)	10.9 (111%)	2.07 (21.1%)	8.55 (87.24%)	0.29 (2.9%)
24 h + MSO	9.9 (100%)	9.9 (100%)	1.97 (19.9%)	7.68 (77.6%)	0.33 (3.3%)

II. Distribution of ^{35}S in the water-soluble fraction

	Total ^{35}S nmol/55 mg fr.wt.	Unknown	Glutathione	Cysteine	Methionine
2 h	1.06	38.5%	50.8%	4.1%	6.7%
2 h + MSO	1.04	40.3%	49.1%	5.6%	4.8%
24 h	2.07	17.1%	68.5%	12.8%	1.5%
24 h + MSO	1.97	16.8%	69.1%	12.4%	1.7%

surface of the cells can be excluded. As the figures show, the amount of labeled glutathione found in the cells treated with methionine sulfoximine equals the amount in the untreated cells. Since the biosynthesis of glutathione in the tobacco cells is blocked by methionine sulfoximine the correspondence of the glutathione values clearly indicates that glutathione is taken up as an intact molecule and must be degraded inside the cell. This degradation most likely takes place in the same way as in animal cells with 5-oxoproline being a characteristic intermediate (Meister and Tate, 1976). Incubation of photoheterotrophically grown tobacco cells and of crude homogenates of those cells with (^{14}C-glu-U)-GSH revealed that substantial amounts of labeled 5-oxo-proline are formed during the degradation of glutathione, as recently found by Rennenberg et al. (1980).

4 Effect of Sulfur Starvation on Nitrogen Assimilation

Both nitrogen and sulfur are assimilated in plant cells largely for protein synthesis. If sulfur becomes limiting, therefore, net-protein synthesis will decline and come to a complete stop when all available sulfur sources have been consumed. It would be interesting to know how nitrogen assimilation reacts under this condition. Does it come to a halt, too, or are the cells able to assimilate and store part of the nitrogen offered to them? If so, what are the consequences of an amino acid synthesis that cannot be used for protein synthesis; are the amino acids only stored in the cells or are some of them drained away into synthesis of secondary products, like alkaloids?

Sulfur-starved tobacco cells are characterized both by a high concentration of TCA-soluble nitrogen that can amount to 60% of the total nitrogen and by a low protein content (Bergmann and Grosze, unpubl.). This change in the distribution of nitrogen has been described as a typical symptom of sulfur deficiency in several crop plants (cf. Thomas, 1958) and indicates some marked shifts in the metabolism of the cells. It is to be expected that experiments with cell cultures will give a more intelligible insight into these metabolic changes than experiments with whole plants that are hampered by our inadaequate knowledge regarding transport from and to different organs and the mobilization of endogenous storage pools. We have, therefore, started to analyze the effect of S-starvation on the nitrogen metabolism of our tobacco cells. The first results of those

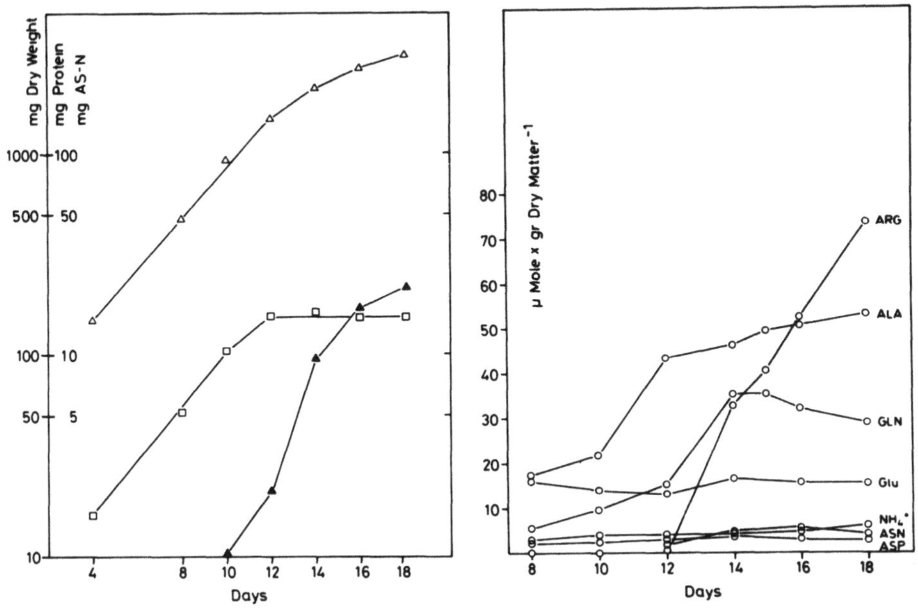

Fig. 4. Effect of sulfur starvation on growth and amino acid accumulation in cells of tobacco suspension cultures grown photoheterotrophically with nitrate at 3,000 lx and 25 °C. Free amino acids were extracted from lyophilized cells in a piric acid-lithium citrate buffer mixture, cleared by centrifugation and applied to an automated amino acid analyzer. △–△–△–△ Dry weight; □–□–□–□ Cell protein; ▲–▲–▲–▲ Amino acid-nitrogen

experiments show that in absence of net-protein synthesis nitrogen assimilation still takes place, leading to an increase in free amino acids.

Protein increase in nitrate-grown tobacco cells that were supplied with limiting amounts of sulfate comes to a complete standstill after the sulfate in the medium has been used up (Fig. 4). At the same time a sharp increase in the soluble amino acid fraction of the cells can be observed. The N-content of the free amino acids increases from 3.25 mg per culture on the 12th day to 22.4 mg on the 18th day, reaching at this time nearly the amount of protein-N in the cultures. Since the protein content of the cultures during this period remains constant, the increase in the amino acids must be due to nitrate assimilation and de novo synthesis of amino acids.

Analyzing the amino acid fraction W. Grosze in our laboratory has shown that the increase in amino acid concentration on a dry weight basis is restricted to only three amino acids: arginine, alanine, and glutamine. The levels of all other amino acids and also the level of NH_3 in the cells remain constant with exception of asparagine that shows a small rise from 2.0 μmol to 4–5 μmol per g dry wt. In contrast arginine exhibits a 100-fold increase from 0.7 μmol per g dry wt. to 73.6 μmol on the 18th day, whereas the concentrations of glutamine and alanine during this period rise only threefold. Since arginine contains four nitrogen atoms per molecule, roughly 60% of the nitrogen found in the free amino acid fraction accounts for the increase in arginine.

Feeding the cells with ammonium and nitrate during sulfur starvation drastically increases the amount of free amino acids among which the concentration of arginine rises to 180 μmol per g dry wt. and glutamine increases to 800 μmol per g dry wt. Under these conditions other amino acids also increase and the intermediates of arginine synthesis, ornithine and citrulline appear in measurable amounts in the cells, probably indicating a change in the pool sizes of this pathway.

Since arginine synthesis in plant cells is supposed to be regulated by feedback inhibition (Morris and Thompson, 1977; Miflin, 1973) and is curtailed by 98% when protein synthesis is inhibited in suspension cultures of Paul's Scarlet rose by cycloheximid (Fletcher and Beevers, 1971; Fletcher, 1975) the accumulation of arginine induced by sulfur starvation presents an interesting challenge to the plant physiologist, and probably a new approach to the study of the regulation of nitrogen and sulfur metabolism in plant cells.

References

Beevers L, Hageman RH (1969) Annu Rev Plant Physiol 20:495–522
Bergmann L, Bälz A (1966) Planta 70:285–303
Bergmann L, Rennenberg H (1978) Pflanzenphysiol 88:175–185
Datko AH, Mudd SH, Giovanelli J, Macnicol RK (1978) Plant Physiol 62:629–635
Fletcher JS (1975) Plant Physiol 56:450–457
Fletcher JS, Beevers H (1971) Plant Physiol 48:261–264
Hahn R, Oberrauch W (1978) In: Sies H, Wendel A (eds) Functions of glutathione in liver and kidney. Springer, Berlin Heidelberg New York, pp 32–40
Huffaker RC, Rains DW (1978) In: MacDonald JG, Nielsen DR (eds) Nitrogen in the environment, vol II. Academic Press, London New York, pp 1–43
King PJ, Street HE (1973) In: Street HE (ed) Plant tissue and cell culture. Blackwell Scientific Publications, Oxford, pp 191–337

Kosower EM, Kosower NS (1976) In: Arias IM, Jakoby WB (eds) Glutathione: metabolism and function. Raven Press, New York, pp 139–158
Meister A, Tate SS (1976) Annu Rev Biochem 45:559–604
Miflin BJ (1973) In: Milborrow BV (ed) Biosynthesis and its control in plants. Academic Press, London, New York, pp 49–68
Morris CJ, Thompson JF (1977) Plant Physiol 59:684–687
Oaks A, Stulen I, Boesel IL (1979) Can J Bot 57:1824–1829
Rennenberg H, Bergmann L (1979) Z Pflanzenphysiol 92:133–142
Rennenberg H, Schmitz K, Bergmann L (1979) Planta 147:57–62
Rennenberg H, Steinkamp R, Polle A (1980) Z Naturforsch (in press)
Reuveny Z, Filner, P (1977) J Biol Chem 252:1858–1864
Rose D, Martin SM (1975) Can J Bot 53:1942–1949
Schwenn JD, El-Shagi H, Kemena A, Petrak E (1979) Planta 144:419–425
Smith IK (1975) Plant Physiol 55:303–307
Thomas MD (1958) In: Ruhland W (ed) Encyclopedia of plant physiology, vol IX. Springer, Berlin Heidelberg New York, pp 37–63

Regulation of Adenosine 5′-Phosphosulfate Sulfotransferase in Higher Plants

C. BRUNOLD[1]

1 Introduction

APSSTase catalyzes the transfer of the sulfonylgroup of APS to a carrier (car-SH) to form car-S-SO_3^-. The carrier seems to be reduced glutathione in *Chlorella* (Tsang and Schiff, 1978). Using mutants blocked for sulfate reduction, APSSTase has been shown to be an enzyme of assimilatory sulfate reduction in *Chlorella* (Schmidt et al., 1974). The regulatory properties of the enzyme presented in this paper, together with enzymatic analysis (Schmidt, 1976), indicate that it is also involved in assimilatory sulfate reduction of higher plants. APSSTase has been detected in more than 50 families of higher plants (Schmidt, 1975a) and has been partially purified from spinach (Schmidt, 1976).

2 Regulation by Light

In spinach, APSSTase is localized predominantly or even exclusively in the chloroplast (Schmidt, 1976; Fankhauser and Brunold, 1978). Thus, at least the following two possibilities of regulation of APSSTase by light may be present: (1) regulation of enzyme activity by light-induced changes in the stroma and (2) increase in extractable activity during light-induced chloroplast development. Both of these possibilities are realized. The APSSTases from spinach, corn, and *Chlorella* are inhibited by 5′-AMP (Schmidt, 1975c) and it was shown for spinach APSSTase that this inhibition is competitive (Schmidt, 1976). Thus, the light-induced decrease in the chloroplast 5′-AMP-pool (Latzko and Gibbs, 1969) can increase APSSTase activity. When chloroplasts are illuminated the pH of the chloroplast stroma shifts from 7.1 to 8.1 (Heldt et al., 1973). Since the pH optimum for APSSTase from spinach is about 8.5 this light-induced increase clearly favors the enzyme activity and may be of regulatory significance in vivo (Schmidt, 1975b).

The changes in extractable APSSTase activity during light-induced chloroplast development in primary leaves of *Phaseolus vulgaris* (Wyss and Brunold, 1979) are presented in Fig. 1. Contrary to ribulosebisphosphate carboxylase activity, which was used as a control parameter in this experiment, the extractable APSSTase activity decreases dur-

Abbreviations: APS = adenosine 5′-phosphosulfate; APSSTase = adenosine 5′-phosphosulfate sulfotransferase; OASSase = O-acetyl-L-serine sulfhydrylase (E.C. 4.2.99.8); PAPS = adenosine 3′-phosphate 5′-phosphosulfate

1 Pflanzenphysiologisches Institut der Universität Bern, Altenbergrain 21, 3013 Bern/Switzerland

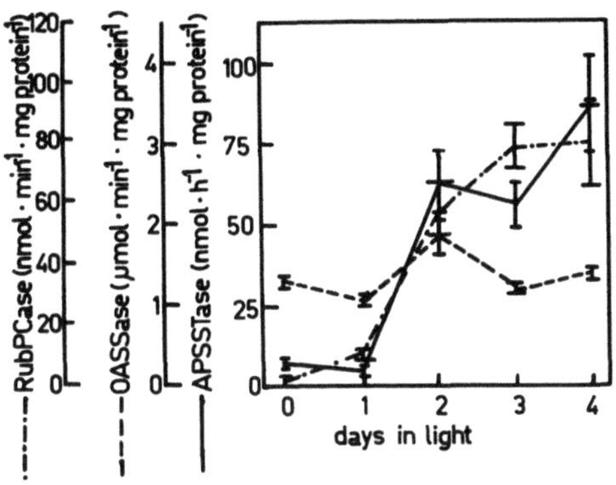

Fig. 1. Extractable activities of adenosine 5'-phosphosulfate sulfotransferase (APSSTase), O-acetyl-L -serine sulfhydrylase (OASSase) and ribulosebisphosphate carboxylase (RubPCase) of greening primary leaves of *Phaseolus vulgaris*. The mean values of six measurements are presented, *vertical lines* indicate ± standard error. Before the experiment the seedlings had been kept in the dark for 3 days. At the start of the experiment they were illuminated at 4,000 lx. (After Wyss and Brunold, 1979)

ing the first day in the light and only afterwards shows a pronounced increase. A comparable increase in APSSTase activity was observed in greening cotyledons of spinach (Fankhauser and Brunold, 1978).

3 Regulation by Sulfur Compounds

In bacteria end products of the pathway of assimilatory sulfate reduction decrease the activity of the enzymes involved (Wheldrake and Pasternak, 1965; Kredich, 1971). Cystine rather increases than decreases ATP sulfurylase (EC 2.7.7.4) activity in tomato, turnip, lettuce, and *Lemna gibba* (Ellis, 1969), indicating that different patterns of regulation of assimilatory sulfate reduction might have evolved in higher plants as compared to bacteria. In the case of ATP sulfurylase the difference in regulation may be due to the fact that in plants APS, the product of the reaction catalyzed by ATP sulfurylase is not only an intermediate in the reduction pathway leading to cysteine, but also takes part in the pathway of sulfolipid synthesis (Schiff and Hodson, 1973). APSSTase, catalyzing a reaction at the branching point of these two pathways, seems a more likely candidate for being regulated by reduced sulfur compounds. Indeed, H_2S causes a rapid loss of extractable APSSTase activity in *Lemna minor* (Brunold and Schmidt, 1976). This loss is so rapid that it cannot be explained by complete inhibition of APSSTase synthesis and dilution by growth of the existing APSSTase molecules alone. Inactivation and/or degradation of APSSTase are also involved. After removal of H_2S the extractable activity of the enzyme increases to the level of the controls within about 24 h. This restoration of activity is completely inhibited by cycloheximide, but not by chloramphenicol, indicating that protein synthesis on 80 S cytoplasmic ribosomes is involved. The activities of both ATP sulfurylase and OASSase were not affected significantly by H_2S. It had been shown before that treatment with subtoxic concentrations of H_2S inhibits the uptake and the assimilation of SO_4^{2-} (Brunold and Erismann, 1974, 1975). Isotope dilution experiments suggest that H_2S is directly incorporated into cysteine without prior oxidation to sulfate and subsequent reduction (Brunold and Erismann, 1975). In *Lemna minor* the regulation of APSSTase by H_2S may be of ecological significance, since in the natural biotope S^{2-} was detected (Kuchar, 1954). It could provide a means for almost completely stopping the reduction of sulfate when sufficient sulfur

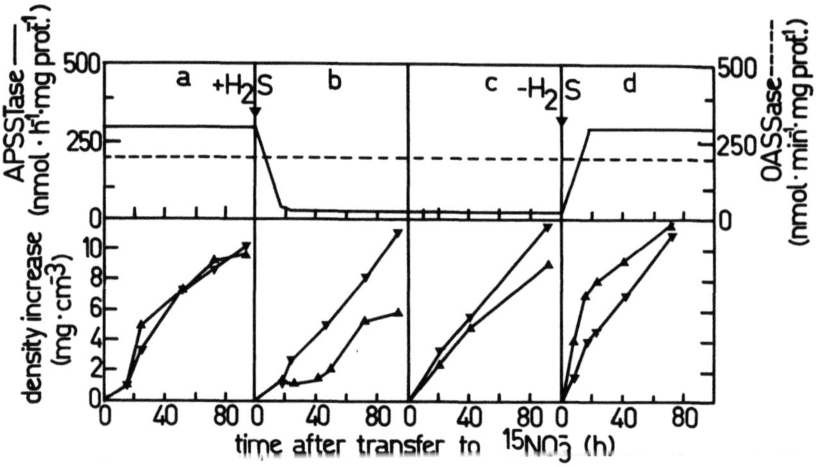

Fig. 2a–d. Activity of adenosine 5′-phosphosulfate sulfotransferase (APSSTase) and O-acetyl-L-serine sulfhydrylase (OASSase) and time course of density increase of the two enzymes (APSSTase, –▲–▲–; OASSase –▼–▼–) extracted from *Lemna minor* after transfer to $^{15}NO_3^-$ from $^{14}NO_3^-$. H_2S was present at 6.4 $\mu l \times l^{-1}$ where indicated. (After von Arb and Brunold, 1980)

at the thiol level is present in the environment to satisfy the needs of the plant. Furthermore the formation of cysteine from S^{2-} taken up from the environment could prevent, up to certain exogenous concentrations, the accumulation of S^{2-} which is a well-known cell poison.

The role of de novo synthesis in the regulation of APSSTase by H_2S was investigated, using density labeling with ^{15}N applied as $^{15}NO_3^-$ in the culture medium (von Arb and Brunold, 1980). OASSase was used as an internal marker enzyme for density labeling, since it does not show changes in extractable activity in response to H_2S. Figure 2 shows the effect of H_2S on APSSTase and OASSase activity and on the time course of density labeling with ^{15}N of the two enzymes. In *Lemna minor* cultivated in air half maximal labeling of both APSSTase and OASSase was obtained after 40–50 h (Fig. 2a). When the organisms were treated with H_2S, OASSase activity remained unchanged and the time course of density increase showed only a slight inhibitory effect of H_2S, half maximal density being reached after about 55 h (Fig. 2b). APSSTase activity, however, decreased rapidly and only a small density shift was observed during the first 50 h (about 15% of the maximal value), indicating that APSSTase synthesis was severely inhibited. In the organisms cultured permanently with H_2S, APSSTase activity was 5% of the control and a half maximal density shift was reached after about 50 h (Fig. 2c). If the rates of APSSTase synthesis, degradation and/or inactivation were the same as in air, the turnover rate of the enzyme would be increased about twentyfold, due to the twenty times smaller pool of active enzyme. This would result in faster density labeling of APSSTase as compared to OASSase. It is clear, however, from Fig. 2c that the time courses of density labeling of both enzymes are comparable, at least up to 44 h, indicating that the rate of APSSTase synthesis is lower than in air. The rapid density labeling of APSSTase which is observed after removal of H_2S indicates that the increase in extractable enzyme activity is not due to the activation of previously inactivated APSSTase molecules, but rather to a pronounced increase in the rate of APSSTase synthesis.

Cysteine has a similar effect on APSSTase of *Lemna minor* as H_2S, causing an 80% loss of extractable activity within 24 h (Brunold and Schmidt, 1978). This loss is accompanied by an inhibition of sulfate uptake, it was shown, however, that lack of sulfate is not responsible for the decrease in APSSTase activity. After the decrease, a constant low level of APSSTase activity is maintained. The initial level of enzyme activity is restored after removal of cysteine. Initially, this restoration is not inhibited by 6-methyl purine and actinomycin D, indicating that mRNA for APSSTase synthesis is present during growth of *Lemna minor* with cysteine. This is consistent with the idea that synthesis of APSSTase is necessary for maintaining the observed low level of APSSTase activity in cultures with cysteine.

Addition of S^{2-} up to 5 m*M* (Brunold and Schmidt, 1976) or of cysteine up to 10 m*M* (Brunold and Schmidt, 1978) to the assay mixture for APSSTase do not inhibit the enzyme.

The results obtained with *Lemna minor* are in contrast to those of Schwenn et al. (1978, 1979), who examined the regulation of sulfotransferases in heterotrophic cell suspension cultures of *Catharanthus roseus*. Cultivation with 0.5 m*M* cysteine as sole sulfur source gave a peak activity of APSSTase which was about five times higher than the one of control cultures with 1.8 m*M* sulfate. On the other hand, a PAPS-dependent sulfotransferase activity was low in extracts from cells cultivated with sulfur amino acids as sulfur source. These regulatory phenomena, taken together with the fact that PAPS-dependent sulfotransferases have been demonstrated in bacteria (Tsang and Schiff, 1976) and yeast (Wilson and Bierer, 1976), raises the question whether certain heterotrophic plant cells differ from autotrophic ones with respect to the physiological sulfonyl donor. It should be pointed out, however, that the assay system for APSSTase used (Schwenn et al., 1979) may not have been ideal. It involves the formation of $AP^{35}S$ from $^{35}SO_4^{2-}$ and ATP catalyzed by ATP sulfurylase which is present in the crude extracts. This enzyme, however, shows great fluctuations in extractable activity (Schwenn et al., 1979) over the time course of the experiment, so that it is not clear whether the observed changes in APSSTase activity are due to varying amounts of extractable APSSTase or to varying amounts of $AP^{35}S$ available for the APSSTase reaction. Thus, before any conclusions can be drawn from the *Catharanthus* system, it seems necessary to measure APSSTase by adding $AP^{35}S$ at optimal concentrations to the assay system. It is interesting in this connection that in heterotrophic cell suspension cultures of *Nicotiana sylvestris* addition of 0.1 m*M* cysteine to the culture medium decreases the level of APSSTase activity (measured with $AP^{35}S$) to 10% of the initial value within 6–12 h without significant inhibition of growth (Jenni et al., 1980). The added cysteine is absorbed rapidly, and after 24 h it can no longer be detected in the medium. Before cysteine is completely depleted, the activity of APSSTase starts to increase, reaching ultimately a level comparable to the initial value. This result indicates that the regulatory effect of reduced sulfur compounds on APSSTase is not restricted to the autotrophic *Lemna minor*. This is further substantiated by results from experiments with *Phaseolus vulgaris* (Wyss and Brunold, 1979). H_2S completely inhibits the light-induced increase in APSSTase activity in the primary leaves. This effect does not seem to be based on the toxicity of H_2S, but is rather a specific regulatory phenomenon, since neither the increase in ribulosebisphosphate carboxylase activity and chlorophyll, nor the growth of the leaves is affected. In green leaves containing appreciable APSSTase activity, H_2S causes a rapid decrease in extractable activity to about 10% of the initial value after about 24 h. After removal of H_2S, APSSTase activity increases to the level of the controls. When bean seedlings, after excision of the roots, were placed on nutrient solution containing cysteine, there was an inhibition of the light-induced increase in APSSTase activity at a concentration of 0.75 m*M* and a decrease in APSSTase activity in green leaves at a concentration of 0.25 m*M*, which is comparable to the 0.5 m*M* used in the experiments with *Lemna minor*.

Treatment of *Phaseolus vulgaris* with subtoxic concentrations of SO_2 revealed regulatory effects similar to H_2S and cysteine. Since SO_2 caused a prominent increase in SO_4^{2-}-concentration in the primary leaves, the question was raised of whether high sulfate concentrations affected APSSTase. When leaves are made high in sulfate by cultivating seedlings, after excision of the roots, on nutrient solution containing at least 5 m*M* Na_2SO_4 there is a decrease in APSSTase activity. This does not seem to be an unspecific effect of high salt concentrations, since treatment with various other sodium salts does not affect significantly extractable APSSTase activity (Wyss and Brunold, 1980). The mechanisms of the SO_4^{2-}-regulation of APSSTase is not clear, but an effect via reduced sulfur compounds seems possible. It was shown that cultivation of excised leaves of various species on nutrient solutions containing high sulfate concentrations induced the emission of H_2S (Wilson et al., 1978). A similar indirect effect may be present in the case of SO_2, since plants exposed to SO_2 emit H_2S in a light-dependent process (De Cormis, 1968a,b).

4 Regulation by Nitrogen Nutrition

In the formation of cysteine the pathways of nitrate and of sulfate assimilation converge. Several studies (Reuveny and Filner, 1977; Friedrich and Schrader, 1978; Rennenberg and Bergmann, 1979) revealed the regulatory interactions between these two pathways. In *Lemna minor* APSSTase activity is affected by the nitrogen nutrition. Cultivation without a nitrogen source causes a prominent decrease in APSSTase activity (Fig. 3).

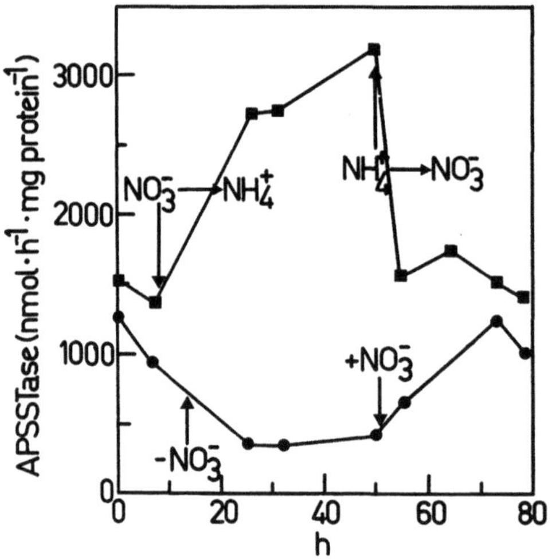

Fig. 3. Effect of nitrogen nutrition on the extractable activity of adenosine 5'-phosphosulfate sulfotransferase (APSSTase) of *Lemna minor*. Replacement of NO_3^- by NH_4^+ and vice-versa, –■–■–; omission of nitrogen source and readdition of NO_3^-, –●–●–. (Brunold and Suter, unpubl. results)

When nitrogen is again available in the form of NO_3^-, APSSTase increases to the initial level. When NO_3^- is replaced by NH_4^+ as a nitrogen source, APSSTase increases (Fig. 3). These plants contain higher levels of free amino acids, especially glutamine and asparagine, indicating a surplus of reduced nitrogen. The increase in APSSTase activity may be one of several means for coordinating the pathways of sulfate and nitrogen assimilation if reduced nitrogen is available in the environment. There is no change in the growth rate of the organisms in the presence of NH_4^+, but the plants contain significantly higher protein concentrations (Dicht et al., 1976). After replacement of NH_4^+ by NO_3^-, APSS-Tase decreases rapidly (Fig. 3).

5 Conclusions

From the results presented here, a pattern of regulation of APSSTase emerges which is reminiscent of the model proposed by Reuveny and Filner (1977) for ATP sulfurylase. On the one hand there is a regulation of extractable APSSTase activity by end products of the pathway of assimilatory sulfate reduction, on the other hand a regulatory effect on APSSTase from the nitrate reduction pathway has been revealed. These regulatory effects on APSSTase may contribute to maintain a rate of sulfate assimilation necessary to meet the needs of protein synthesis. The results from the ^{15}N density labeling experiments show that changes in APSSTase synthesis are involved in the regulation. The inhibitor studies indicate that this control of APSSTase synthesis occurs at a posttranscriptional step, but experiments using the techniques of in vitro protein synthesis are necessary to make this point clear. It would also be interesting to know whether the rate of APSSTase degradation and inactivation is increased by treatment of plants with reduced sulfur compounds.

Acknowledgments. I thank Mrs. B. Brunold and Miss M. Suter for technical assistance in obtaining the results presented from this laboratory and Dr. M. J. Emes and D. Schmutz for critical comments on the manuscript. Part of the presented work was supported by grants from the Swiss National Science Foundation to C.B. and of the Deutsche Forschungsgemeinschaft to Dr. A. Schmidt.

References

Brunold C, Erismann KH (1974) Experientia 30:465–467
Brunold C, Erismann KH (1975) Experientia 31:508–509
Brunold C, Schmidt A (1976) Planta 133:85–88
Brunold C, Schmidt A (1978) Plant Physiol 61:342–347
De Cormis LCP (1968a) Acad Sci Ser D 266:683–685
De Cormis L (1968b) Ann Physiol Veg 10:99–112
Dicht M, Kopp A, Feller U, Erismann KH (1976) Biochem Physiol Pflanz 170:531–534
Ellis RJ (1969) Planta 88:34–42
Fankhauser H, Brunold C (1978) Planta 143:285–289
Friedrich JW, Schrader LE (1978) Plant Physiol 61:900–903
Heldt HW, Werdan K, Milovancev M, Geller G (1973) Biochim Biophys Acta 314:224–241
Jenni B, Brunold C, Zryd J-P (1980) Experientia 36:724
Kredich NM (1971) J Biol Chem 246:3474–3484
Kuchar KW (1954) Arch Hydrobiol 49:329–334

Latzko E, Gibbs M (1969) Plant Physiol 44:396–402
Rennenberg H, Bergmann L (1979) Z Pflanzenphysiol 92:133–142
Reuveny Z, Filner P (1977) J Biol Chem 252:1858–1864
Schiff JA, Hodson RC (1973) Annu Rev Plant Physiol 24:381–414
Schmidt A (1975a) Plant Sci Lett 5:407–415
Schmidt A (1975b) Planta 124:267–275
Schmidt A (1975c) Planta 127:93–95
Schmidt A (1976) Planta 130:257–263
Schmidt A, Abrams WR, Schiff JA (1974) Eur J Biochem 47:423–434
Schwenn JD, El-Shagi H, Kemena A, Petrak E (1978) Zusammenfassungen der Vorträge der Botanikertagung in Marburg, p 247
Schwenn JD, El-Shagi H, Kemena A, Petrak E (1979) Planta 144:419–425
Tsang ML-S, Schiff JA (1976) J Bacteriol 125:923–933
Tsang ML-S, Schiff JA (1978) Plant Sci Lett 11:177–183
Von Arb C, Brunold C (1980) Planta 149:355–360
Wheldrake JF, Pasternak CA (1965) Biochem J 92:167–172
Wilson LG, Bierer D (1976) Biochem J 158:255–270
Wilson LG, Bressan RA, Filner P (1978) Plant Physiol 61:184–189
Wyss H-R, Brunold C (1979) Planta 147:37–42
Wyss H-R, Brunold C (1980) Physiol Plant 50:161–165

On the Enzymatic System Thiosulfate-Cytochrome c-Oxidoreductase

K. KNOBLOCH, W. SCHMITT, G. SCHLEIFER, N. APPELT, and H. MÜLLER[1]

1 Introduction

Inorganic sulfur compounds in various reduced stages serve as essential electron donors for a variety of microorganisms. The photosynthetic sulfur bacteria depend on light as their energy source in an anaerobic environment (photolithotrophs). The chemolithotrophs lack chlorophyll and use the inorganic reduced sulfur as the source of energy and electrons under aerobic conditions (Fig. 1).

The photolithotrophic sulfur bacteria (Chromatiaceae and Chlorobiaceae) represent one of the oldest forms of life that evolved on earth within its anaerobic atmosphere. The cyanobacteria (blue-green algae) likewise are considered to have covered the earth for at least a similarly long period of time. As the only group of prokaryotes, they succeeded in developing a second photosystem resulting in oxygen liberation.

Recently, cyanobacteria of different origin have been shown to perform (facultative) anoxygenic photosynthesis with sulfide as the electron donor while the second photosystem remains inoperative (Cohen et al., 1975). Among the photoheterotrophic Rhodospirillaceae, *Rhodopseudomonas palustris* could be adjusted to photolithotrophic growth on thiosulfate (Knobloch et al., 1971).

The view may become acceptable that members among the various groups of photosynthetic prokaryotes are enabled to perform anoxygenic photosynthesis with inorganic reduced sulfur compounds as the electron sources.

Of particular interest are initial steps in the process of sulfurication involving the enzymatic transfer of electrons from the inorganic sulfur towards the endogenous acceptor compound.

The first investigations dealt with cell-free extracts from autotrophs after growth on $S_2O_3^{2-}$. Based on studies on *Thiobacillus*, it was concluded that the oxidation of $S_2O_3^{2-}$ could be coupled with the reduction of cytochrome c, and the enzymatic system involved was described as being soluble (Trudinger, 1961; Aleem, 1965). On the other hand, membrane-bound sulfur-oxidizing enzyme systems were reported to operate in *Thiobacilli* (Moriarty and Nicholas, 1969). From *T. thioparus*, a $S_2O_3^{2-}$ oxidizing enzyme was purified and characterized on a molecular basis (Lyric and Suzuki, 1970). From photosynthetic bacteria, more detailed information was given concerning the endogenous cell compound which serves as the electron acceptor. It was reported that the oxidation of

1 Institut für Botanik und Pharmazeutische Biologie der Universität Erlangen-Nürnberg, Schloßgarten 4, 8520 Erlangen/FRG

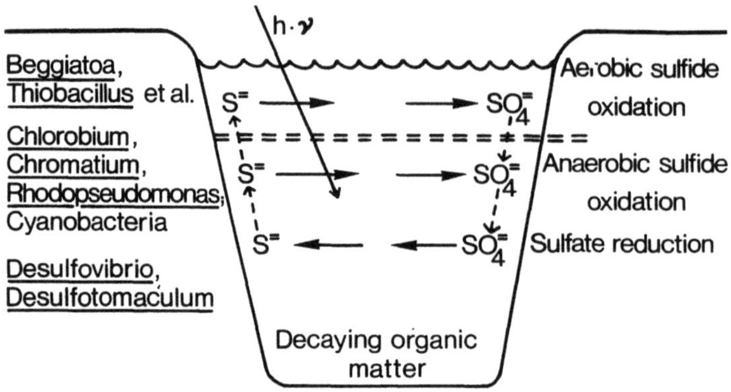

Fig. 1. An ecosystem depending on light energy and a closed inorganic sulfur cycle ("sulfuretum," modified after Postgate, 1968). (→ chemical reactions, --→ physical diffusions)

$S_2O_3^{2-}$ in *Chlorobium* is coupled with the reduction of a cytochrome c-551 involving cyt c-551 (Kusai and Yamanaka, 1973). In *Chromatium* whole cells, $S_2O_3^{2-}$ is seen to serve as an electron donor for cyt c-551 and further on to c-555 during the light reactions (van Grondelle et al., 1977).

This chapter deals with both the characterization of the $S_2O_3^{2-}$ cyt c-oxidoreductase system as well as with the description of the endogenous electron acceptor. Objects are *Chr. vinosum* and *R. palustris.*

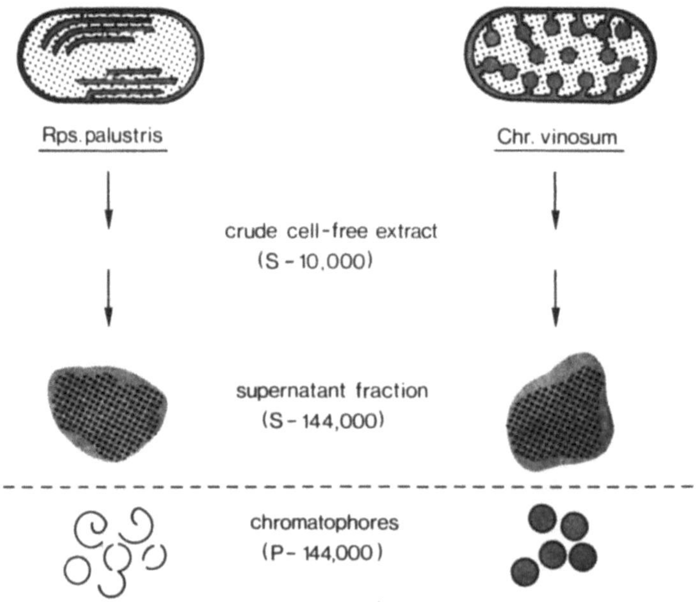

Fig. 2. Schematic intracellular membrane architecture of a *Chromatium vinosum* and a *Rhodopseudomonas palustris* cell (after Oelze and Drews, 1972).

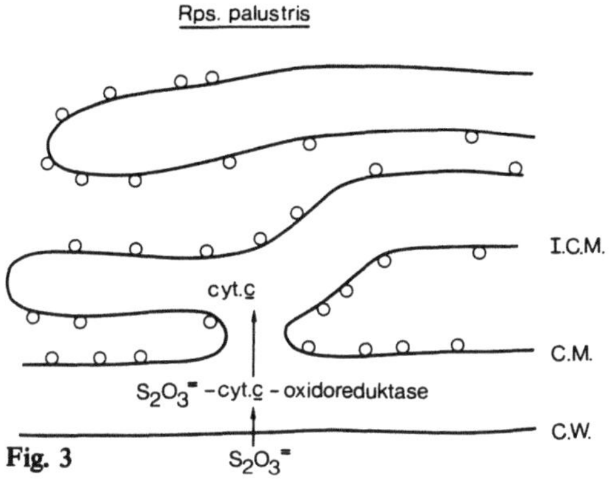

Fig. 3

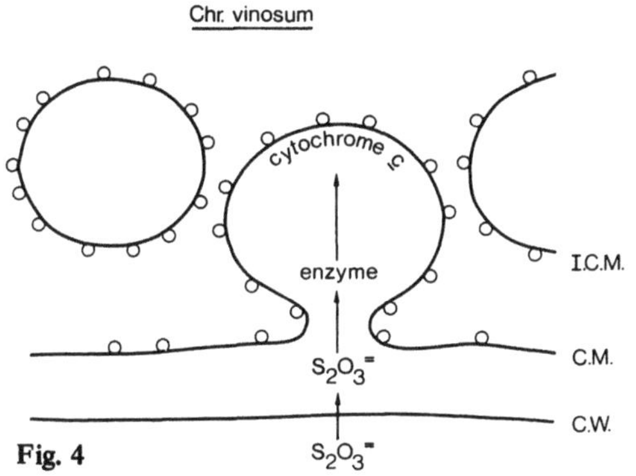

Fig. 4

Figs. 3 and 4. Intracellular distribution of the thiosulfate-cytochrome c-oxidoreductase and the acceptor cytochrome c in *Rps. palustris* and *Chr. vinosum.* (The knobs attached to the membrane represent the coupling factor or ATPase protein) (*C.W.* cell wall; *C.M.*, *I.C.M.* cytoplasmic, intracytoplasmic membrane system)

2 Intracellular Compartmentation

Cells of *Chromatiaceae* and of *Rhodospirillaceae* are supplied with a membrane system which is characteristic of the distinguished species. The cytoplasmic membrane grows to develop the intracytoplasmic membrane system (Figs. 2–4) (see Oelze and Drews, 1972). Whereas the *R. palustris* cell grows to form a continuous intracytoplasmic membrane (Figs. 2 and 3) (Tauschel and Drews, 1967), *Chr. vinosum* develops intracytoplasmic membrane vesicles which are separated from the cytoplasmic membrane (Figs. 2 and 4) (see Oelze and Drews, 1972).

Common to all photosynthetic purple bacteria remains the fact that the interior of the cell is divided into the periplasmic space and the intracytoplasmic compartment (Figs. 3 and 4).

The specific arrangement of the membrane architecture is of importance, since features of the prepared cell-free extracts and of derived fractions depend on whether the intracytoplasmic system is of vesicular or lamellar structure.

2.1 Rhodopseudomonas palustris

Due to the membrane system of lamellar type, the *R. palustris* cells yield cell-free extracts which contain membrane sheets of different size and shape. The broken membranes, in a freshly prepared state, cannot be considered to represent closed particles (Knobloch, 1974, 1975; Appelt et al., 1979). As a result, after undergoing ultracentrifugation (144,000 g), the separated chromatophores are obtained as "empty" membrane pieces (P-144), while the supernatant fraction (S-144) holds the contents of both the cell compartments (Fig. 2). It should be stressed that S-144 does not represent a "soluble" fraction, since the presence of membrane (sub)-units could be demonstrated (Knobloch, 1974) and the phospholipids present have been analyzed (Gemeinhardt and Knobloch, 1979). In *R. palustris,* the $S_2O_3^{2-}$ oxidizing system as well as the electron acceptor compound, identified as a cytochrome c, are part of S-144. The chromatophore membranes do not reveal any oxidoreductase activity (Figs. 2 and 3).

2.2 Chromatium vinosum

Cell-free extracts from *Chromatium,* when separated into S-144 and P-144 by the same procedure, exhibit a different pattern of activities. Chromatophores, resuspended from P-144, catalyze the oxidation of $S_2O_3^{2-}$ with a concomitantly occurring reduction of cytochrome c. S-144 does not contain the acceptor cyt c but, like the closed chromatophores, holds the oxidoreductase system. These findings fit the view that, similar to *R. palustris,* the oxidoreductase in *Chromatium* is located within the periplasmic compartment whereas, in contrast to *R. palustris,* the endogenous acceptor cyt c for the electrons is attached to or embedded into the chromatophore membrane (Figs. 2 and 4).

3 Enzymatic Systems

3.1 Thiosulfate-Cytochrome c-Oxidoreductase

3.1.1 Isolation and Purification

The oxidoreductase system of both the bacteria is isolated and purified from S-144 by (1) separation of the protein precipitating above 40% $(NH_4)_2SO_4$, (2) cellulose anion exchange chromatography (Fig. 5), (3) preparative isoelectric focusing (IEF) (Fig. 6), and (4) gel filtration through Sephadex G-100.

3.1.2 Characterization

When the *R. palustris* purified protein is exposed to SDS disc gel electrophoresis, one single band remains. The determination of its particle weight by elution from Sephadex G-100 yields 93,000. When the same protein has undergone polyacrylamide SDS disc gel electrophoresis, the calculation of its particle weight results in 48,000. The oxidoreductase protein from *R. palustris* is further characterized by a pI value of 5.45. Its apparent K_m value for $S_2O_3^{2-}$ corresponds to about 2 μM and its apparent K_m for the acceptor cytochrome c to ca. 2 μM.

The *Chr. vinosum* oxidreductase particle weight, if determined by gel filtration through Sephadex G-100, yields about 35,000. The pI value ranges from 4.4 to 4.7, and its apparent K_m value for $S_2O_3^{2-}$ is close to 6 μM and for the acceptor cytochrome c corresponds to ca. 0.2 μM.

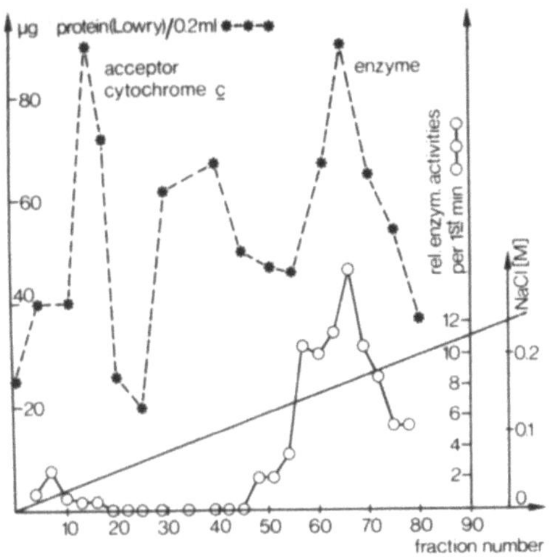

Fig. 5. Elution diagram of the protein from *R. palustris* S-144. (DEAE cellulose A-50 Sephadex column)

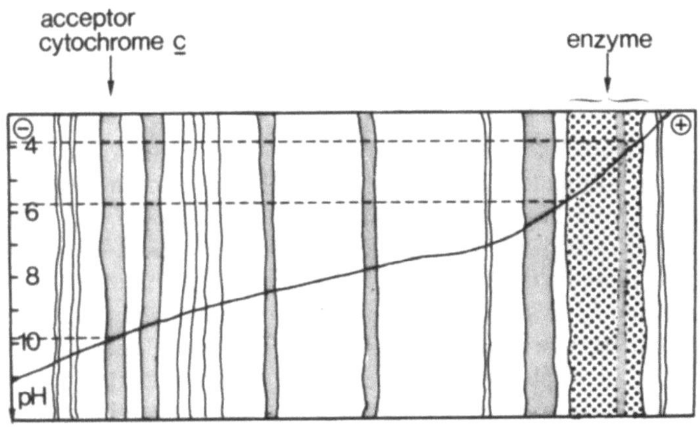

Fig. 6. Preparative (Ampholine, pH 7.0–11.0) isoelectric focusing (IEF) of *R. palustris* S-144 protein in granulated gel (Sephadex G-75 superfine). (See Appelt et al., 1979)

3.2 Acceptor Cytochrome c

3.2.1 Isolation and Purification

The acceptor cyt c of *R. palustris* is enriched from S-144. The protein precipitating above 40% $(NH_4)_2SO_4$ conc. is separated by a preparative IEF. Six fractions of cytochrome c are focused, with the one at pI 9.5–10.2 representing the acceptor compound (Fig. 6). The acceptor cytochrome also can be separated by a DEAE A-50 anion exchange column (Fig. 5). Further purification is achieved either by a second IEF using Ampholine covering a narrower pH range or also by subsequent column chromatography through Sephadex G-75 and G-100.

In *Chr. vinosum* the acceptor cyt c is connected firmly with the chromatophore membrane. It is extracted by a treatment with acetone at −18 °C (modified after Bartsch, 1971).

3.2.2 Characterization

A difference absorption spectrum of the *R. palustris* cyt c reveals absorption maxima at 550, 520, and 418 nm. The acceptor cyt c from *Chr. vinosum* is characterized by absorption peaks at 552, 522.5, and 423.0 nm.

The cytochrome from *R. palustris* exhibits a pI value close to 10, whereas the corresponding cytochrome from *Chr. vinosum* is distinguished by a pI in the range of 4–5.

Molecular weight determination of the *R. palustris* cytochrome c reveal 16,200 by use of SDS polyacrylamide disc gel electrophoresis and a value of 17,000 when column chromatography through Sephadex G-100 is applied.

The *R. palustris* acceptor cyt c is further characterized by a midpoint redox potential of E_m 7.0 = +228 mV when determined by redox titrations (Appelt et al., 1979) and of E_m 7.0 = +205 mV when calculated spectrophotometrically after Davenport and Hill (1952).

4 Conclusion

Enzymatic $S_2O_3^{2-}$ oxidation coupled to the reduction of endogenous cytochromes of type c was observed among phototrophic (Knobloch et al., 1971; Kusai and Yamanaka, 1973; Cohen et al., 1975; van Grondelle et al., 1977) and chemotrophic (Trudinger, 1961; Aleem, 1965; Moriarty and Nicholas, 1969; Lyric and Suzuki, 1970) bacteria. By investigating comparatively the oxidoreductase reactions catalyzed by the different cell-free fractions from *Chr. vinosum* and *R. palustris*, one can conclude that the oxidoreductases are located within the periplasmic space of the cells. In *Chr. vinosum*, an obligate phototroph, the oxidation of $S_2O_3^{2-}$ is coupled with an acceptor cyt c which is part of the intracytoplasmic chromatophore membrane. In *R. palustris*, a facultative anaerobic phototroph, the electrons from $S_2O_3^{2-}$ are transferred to an acceptor cyt c which can be obtained from a complete soluble supernatant cell-free fraction (S-144, 8 h). In *Chr. vinosum*, the electrons are transferred towards the chromatophore membranes to serve the generation of photosynthetic assimilatory power. In *R. palustris*, the electrons are accepted by a soluble cyt c to participate in the generation of reduced pyridine nucleotide while photophosphorylation occurs separately (Knobloch et al., 1971). So far, it has not been possible in *R. palustris* cell-free preparations to couple the $S_2O_3^{2-}$-linked reduction of cyt c with O_2 as the terminal electron acceptor in the dark; this is an attempt to simulate an electron transport pathway performed by a chemosynthetic sulfur bacterium.

References

Aleem MIH (1965) J Bacteriol 90:95–101

Appelt N, Weber H, Wieluch S, Knobloch K (1979) Ber Dtsch Bot Ges 92:365–378

Bartsch RG (1971) In: San Pietro A (ed) Meth Enzymol, vol XXIII, A. Academic Press, London New York, p 344

Cohen Y, Padan E, Shilo M (1975) J Bacteriol 123:855–861
Davenport HE, Hill R (1952) Proc R Soc London Ser B 139:327–345
Gemeinhardt F, Knobloch K (1979) Hoppe-Seyler's Z Physiol Chem 360:265–266
Grondelle R van, Duysens LNM, Wel JA van der, Wal HN van der (1977) Biochim Biophys Acta 461:188–201
Knobloch K (1974) Ber Dtsch Bot Ges 87:119–128
Knobloch K (1975) Z Naturforsch 30c:342–348
Knobloch K, Eley JH, Aleem MIH (1971) Arch Mikrobiol 80:97–114
Kusai S, Yamanaka T (1973) Biochim Biophys Acta 325:304–314
Lyric RM, Suzuki I (1970) Can J Biochem 48:355–363
Moriarty DJW, Nicholas DJD (1969) Biochim Biophys Acta 184:114–123
Oelze J, Drews G (1972) Biochim Biophys Acta 265:209–239
Postgate JK (1968) In: Nickless G (ed) Inorganic sulphur chemistry. Elsevier Publishing Co, Amsterdam, p 259
Tauschel HD, Drews G (1967) Arch Mikrobiol 59:381–404
Trudinger PA (1961) Biochem J 78:673–680, 680–686

Subject Index